T0281676

BIOLOGICAL ACTIONS
OF
SEX HORMONES

BIOLOGICAL ACTIONS
OF
SEX HORMONES

BY

HAROLD BURROWS
C.B.E., Ph.D., F.R.C.S.

CAMBRIDGE
AT THE UNIVERSITY PRESS
1945

CAMBRIDGE UNIVERSITY PRESS
Cambridge, New York, Melbourne, Madrid, Cape Town,
Singapore, São Paulo, Delhi, Mexico City

Cambridge University Press
The Edinburgh Building, Cambridge CB2 8RU, UK

Published in the United States of America by Cambridge University Press, New York

www.cambridge.org
Information on this title: www.cambridge.org/9781107625501

© Cambridge University Press 1945

First published 1945
First paperback edition 2013

A catalogue record for this publication is available from the British Library

ISBN 978-1-107-62550-1 Paperback

CONTENTS

PART III. ANDROGENS

PART IV. OESTROGENS

CONTENTS vii

PREFACE

In the last few years our comprehension of vital phenomena has been rapidly extending. The nature of the sex hormones, and the reactions of living tissues toward them, have been prominent in this advance, and it is now generally understood that compounds formed in the pituitary, gonads and adrenals radically affect the structure and functions of the body and the workings of the mind. To-day our knowledge of these matters is growing so fast that to keep abreast of it is not easy for those who are occupied with many other affairs. The author felt, therefore, that a co-ordinated summary of experimental inquiries in this field might be useful. In pursuing the idea attention has been confined almost entirely to biological work performed in the laboratory; the ultimate possibility of applying the experience so gained for the benefit of man has been the leading motive.

The essay can hardly be offered to the scientific world without an apology. Biological work is still largely confined to qualitative observation. Life is a changing process and in solving its problems we are often deprived of fixed and measurable data; moreover, the adaptability of living tissue to circumstance involves so many and such complex reactions that an exact prediction of the outcome of any extraneous influence cannot, as a rule, be stated in precise quantitative terms; nor can experimental results in this field be described adequately without specifying the conditions in which they were obtained. The presentation of the subject demanded by the latter drawback may, it is feared, be tedious to the reader, especially as the narrative contains many references to the literature. Sir James Paget complained of the difficulty of composing a readable scientific review, and the present writer is too modest to suppose that he has overcome the difficulty. It is hoped, however, that the matter contained in these pages may supply a trustworthy, though limited, foundation for further progress in both sex-hormone research and clinical practice.

The author would like to regard his book as a tribute to the pioneers of sex-hormone physiology, with special regard to John Hunter (1728–93), the first and greatest of them. More than a hundred years before the term hormone had been invented, Hunter showed that the accessory reproductive organs are largely dependent for their development and even for their existence on some influence derived from the gonads.

The writer regrets that much good work done in foreign lands has been given inadequate consideration, but it may be unnecessary or impossible to insert every detail into a picture; for general portrayal it is perhaps enough to draw the salient features as far as possible with fidelity of outline and correct emphasis.

My own experimental work on the sex hormones has been done, under the auspices of the British Empire Cancer Campaign, at the Chester Beatty Research Institute of the Royal Cancer Hospital (Free). To the authorities of these institutions, to Professor Kennaway the director, and to my other colleagues

I should like to express gratitude for their interest and encouragement. I have also to thank, most cordially, several donors for generous gifts of chemical material; among these donors are Dr Girard of Paris, Dr Laqueur of Amsterdam, Dr Macbeth of Organon, Limited, Mr Smart and Dr Miescher of the Ciba Company. To Dr Macbeth I am greatly indebted for extensive references to relevant literature. I should like also to acknowledge valuable advice from Dr F. H. A. Marshall. To Dr J. N. Goldsmith I am especially grateful; not only has he prepared an exceptionally helpful index but he has given great help with proof-reading and advice on various details. Lastly, I must mention the unremitting help which I have received from my wife, without whose aid this book could not have been written.

H.B.

London, 1944

PART I. GONADOTROPHINS

Chapter I. *The Nature and Functions of Gonadotrophins*

Introductory remarks. Inactivation, excretion, sources and distribution of gonadotrophin. Are the pituitary gonadotrophins identical with those produced by the placenta? Are FRH and LH distinct compounds? The action of gonadotrophins on the ovary and testis. The interstitial-cell-stimulating hormone. Puberty and the awakening of sexual activity.

Introductory Remarks

THE gonadotrophins are so called because they govern the development and biological activities of the gonads.* Cushing & Goetsch (1915) showed that pituitary deficiency, whether naturally or artificially produced, is accompanied by atrophy of the reproductive and adrenal glands. Noble (1938a) has described the progressive atrophy of these organs and of some of the accessory genital structures which follows hypophysectomy in male and female rats (Table 1).

TABLE 1.   Progressive atrophy of reproductive organs after hypophysectomy in male and female rats (Noble, 1938a)

Males

Intervening time (weeks)	Number	Average weights of organs in mg.			
		Testes	Prostate	Seminal vesicles	Adrenals
1	12	1,204	120	52	12
2	12	854	109	56	12
4	8	366	87	42	8
6	5	273	54	34	5

Females

Weeks	Number	Average weights of organs in mg.		
		Ovaries	Uterus	Adrenals
1	4	24	142	28
2	6	22	129	23
4	4	17	117	14
6	3	15	82	8

The changes in the gonads and adrenal cortices after removal of the pituitary include shrinkage of nuclei and cytoplasm with arrest of secretion. These results can be prevented, or if already present can be reversed, by injecting extracts made from the anterior lobe of the pituitary of other animals into the muscles or subcutaneous tissues.

The chemical nature of the gonadotrophins has not been exactly determined. They are soluble in water, give the general reactions of proteins, and are precipitated without denaturation by ethyl alcohol. According to Askew & Parkes (1933) the ovulation-producing hormone of pregnancy urine is inactivated by heating to 100° C. in water, but loses none of its activity if kept at that temperature for 1 hour when dry; the results are unaffected by the exclusion of oxygen. From

* γονή = gonad, τροφή = nourishment.

these and other observations it seems that the gonadotrophins are proteins or are so closely associated with proteins that their activity disappears when the latter are destroyed. The gonadotrophins also contain carbohydrate in the form of mannose and galactose. Among gonadotrophins from different sources Gurin (1942) has detected differences in the carbohydrate content.

The effects of proteolytic enzymes on gonadotrophin.   Evans, Simpson & Austin (1933 b) found that the gonadotrophin of pregnant mares' serum was but little affected by pepsin during 4 hours at 37° C. when the pH was between 4 and 5, though at a pH of between 1·8 and 2 its activity was destroyed. They note that other samples lost their potency when subjected to this pH in the absence of pepsin. Trypsin at a pH of 8·5 inactivated the hormone. Bates, Riddle & Lahr (1934) obtained FRH from beef pituitaries and digested it with trypsin at 37° C. and pH 8·0 for 2 hours. The material after this treatment was injected in four equal daily doses into immature ring doves whose testes were weighed 96 hours after the first dose, and compared with those of untreated birds of the same age. The results show that the hormone had been destroyed by the tryptic digest (Table 2). (See also Riddle, Bates, Lahr & Moran, 1936.)

TABLE 2.   The effect of trypsin on a gonadotrophic extract
rich in FRH (Bates, Riddle & Lahr, 1934)

Material injected	Dose (mg.)	Average weight of doves' testes 96 hours after the first injection mg.	
		Uninjected control doves	Injected doves
Untreated FRH	4	8·7	49·2
	4	6·5	34·0
FRH after digestion with trypsin	4	6·5	6·8
	4	7·8	10·7
	8	6·3	6·6
	8	6·3	8·8

Van Dyke (1936) says that the gonadotrophin of pregnancy urine is rendered inert by boiling, ultra-violet light, hydrogen peroxide or trypsin, but not by pepsin, though according to Fevold (1937) it is destroyed by pepsin. Collip (1937) states that pituitary gonadotrophin is inactivated by prolonged boiling and is sensitive to alkali, losing its potency at a little beyond pH 8. Thyrotrophic hormone, he says, shows approximately the same reactions. Using extracts of sheep's pituitary McShan & Meyer (1938, 1939) found that LH* is largely if not entirely destroyed by trypsin when exposed to it for 3½ hours at 37° C. and pH 8·0, and is relatively resistant to ptyalin, whereas FRH is resistant to trypsin and destroyed by ptyalin. Ch'en & Van Dyke (1939) found that tryptic digestion abolished most of the luteinizing action of extracts of sheep or horse pituitary, but large doses of such digested extracts still caused some luteinization in the ovaries of hypophysectomized immature rats, showing that the destruction was not complete.

* For brevity and ease of discussion follicle ripening and luteinizing hormones will be referred to as FRH and LH respectively, as though their separate identities had been established.

Chow, Greep & Van Dyke (1939) incubated extracts of fresh pig pituitaries at 37° C. with various proteolytic enzymes; some of their results are given in Table 3, and show that the gonadotrophin used was inactivated by trypsin, chymotrypsin and pepsin, but not by papain or carboxypeptidase.

TABLE 3. The effect of proteolytic enzymes on pituitary gonadotrophin (Chow, Greep & Van Dyke, 1939)

Enzyme	pH of digest	FRH	LH	Thyrotrophin
Pepsin	4·57	+	−	+
Papain	4·57	−	−	−
Trypsin	8·69	+ −	+	+
Chymotrypsin	8·69	+ −	+	+
Carboxypeptidase	8·69	−	−	−

NOTE. + = inactivated; − = not inactivated; + − = partially inactivated.

Abramowitz & Hisaw (1939) have also investigated the action of proteolytic enzymes on three different gonadotrophins, namely purified FRH and LH extracted from the pituitaries of sheep and a chorionic gonadotrophin derived from the urine of pregnant women. Their findings suggest some differences in the proteolytic reactions of the extracts which were tested (Table 4).

TABLE 4. Proteolysis of pituitary and chorionic gonadotrophins (Abramowitz & Hisaw, 1939)

Enzyme	pH of digest	Pituitary gonadotrophin		Chorionic gonadotrophin
		FRH	LH	
Papain	7·1	+ −	−	+
Trypsin	7·1	+	+	+
Chymotrypsin	7·6	+ −	+ −	+
Crude ptyalin	7·1	+	−	+

NOTE. + = inactivated; − = not inactivated; + − = partially inactivated.

Li (1940) ground and treated the pituitaries of gonadectomized rats with trypsin at pH 9·6 and incubated the material for 2 hours at 38° C., after which it was assayed on 21-day-old female rats, eight doses being given in 4 days, and the findings were compared with those obtained by pituitaries which had not been treated with trypsin (Table 5). His results show that the extract used which was rich in FRH was to a large extent inactivated by trypsin; the high degree of alkalinity of the digest will be noted.

TABLE 5. The effect of trypsin on the gonadotrophic potency of the rat's pituitary (Li, 1940)

Sex of pituitary donor	Total dose (mg.)	Pituitary treated by trypsin	Mean weight of ovaries in test rats (mg.)	Mean weight of uterus in test rats (mg.)
Female	5	−	69·76	82·10
Female	10	+	23·73	63·77
Male	5	−	66·65	82·89
Male	10	+	22·81	40·35

The results of proteolysis which have just been mentioned, though they are not all in complete agreement, suggest that the gonadotrophins, if not themselves protein, are dependent for their activity on a close association with protein. Spielman & Meyer (1937), having examined the electrophoretic properties of chorionic gonadotrophin, believe that it probably consists of a specific principle combined with a non-specific carrier. They arrived at this conclusion by observing that the hormone may be still active biologically in spite of a change of its isoelectric point.

The gonadotrophins undergo bacterial decomposition, and cannot be given very effectively by the mouth, though some degree of response may follow oral administration (Goetsch & Cushing, 1913; Goetsch, 1916).

As a rule gonadotrophins are not stored appreciably in the body; they are inactivated when introduced into the bloodstream and are excreted with the urine. An exception to the last rule will be mentioned presently (p. 5).

Inactivation, Excretion, Sources and Distribution of Gonadotrophin

Inactivation in the living body. Geist & Spielman (1934) collected blood from the two ends of the severed umbilical cord of a baby and identified gonadotrophin in blood from the placental end in a concentration of 165 r.u. per litre, while none was recognizable in the blood obtained from the foetal end. (See also Sklow, 1942.)

It was shown by Lipschütz & Vivaldi (1934) that human chorionic gonadotrophin when given intravenously to a rabbit disappears rather rapidly from the blood. Six to 8 hours after the intravenous injection of 100 r.u. only 20 per cent could be recovered from the blood, and 10 hours after injection only 10 per cent could be recovered.

Friedman & Weinstein (1937) assayed 24-hour specimens of urine from normal men for gonadotrophin and found a daily excretion of about 6 r.u. Oral ingestion of human chorionic gonadotrophin did not increase the amount excreted. Intramuscular injection of 600 or 750 r.u. was followed by an excretion in the urine of between 5 and 15 per cent of the amount given.

Stamler is quoted by Zondek (1940c) as having given intravenous injections of gonadotrophin to dogs and found 3 hours later only 38·4 per cent of the amount injected still present in the circulation. The hormone was recognized in the urine within 1 minute of its injection, and continued to be present until 20 hours later, the total amount excreted being 11·2 per cent of the original dose. In the gelding Stamler recovered from the urine only 5 per cent of a dose of gonadotrophin given intravenously. Zondek (1940c) made a detailed inquiry into the inactivation of gonadotrophin within the body. First he killed young rats weighing about 30 g. and minced them. He then added to the mince a known amount of chorionic gonadotrophin together with a phosphate buffer of pH 7·9, and placed the material in the incubator. Later he was able to extract from the mashed and incubated tissues all the gonadotrophin which had been added. If however 1,000 r.u. were injected subcutaneously into rats of the same age and the animals were killed 24 hours later, only 10 per cent of the injected gonadotrophin could be extracted from their bodies. By other experiments he showed that inactivation of the hormone did not take place in the liver, spleen or muscles.

It is conceivable that the gonads are partly concerned in the inactivation of gonadotrophin. This possibility is suggested by analogy, for Loesser (1934) injected 3,000 units of thyrotrophin into normal rabbits and was unable, by tests on guinea-pigs, to recognize any of the hormone in the rabbits' blood after the lapse of 1 hour, whereas if the same dose was given to a rabbit whose thyroid had been removed thyrotrophin could be detected in its blood 7 hours later. Loesser suggested, therefore, that the disappearance of thyrotrophin from the blood after intravenous injection might be, in part at least, effected by the thyroid gland. Pursuing this idea, Rawson, Sterne & Aub (1942) tried the effects on thyrotrophin of various tissues *in vitro*. They discovered that thyrotrophin is inactivated by thyroid gland, and to a slight extent also by thymus and lymph glands, but not by other tissues.

Excretion of gonadotrophin. In man the excretion of chorionic gonadotrophin by the kidneys was demonstrated by Aschheim & Zondek (1927, 1928) and is the foundation of their test for pregnancy. As will be shown presently, chorionic gonadotrophin may be formed apart from pregnancy; in the presence of chorionic tumours, whether in women or men, the urinary output of gonadotrophin is often very large and its recognition is a valuable diagnostic aid.

Gonadotrophin derived from the pituitary, like that from chorionic tissue, is excreted by the kidneys. Though the activity of the pituitary in producing gonadotrophin varies with changing sexual activities, in many species the hormone is formed and excreted to some extent at all ages. In women during the reproductive period of life there is a maximum excretion during each menstrual cycle at or shortly before the time of ovulation. Kurzrok, Kirkman & Creelman (1934) studied the renal output of gonadotrophin in ten young non-pregnant women and detected a suddenly enhanced output at about the middle of the menstrual cycle, apparently just preceding ovulation. The continued formation and excretion of gonadotrophin after the menopause was shown by Österreicher (1933), who assayed the urine of 149 women aged between 50 and 93 years and detected the presence of gonadotrophin in 65 per cent. He also verified its presence in the urine of five women whose ovaries had been removed (p. 33).

In the pooled urine of normal men Evans & Gorbman (1942) detected between 1 and 4·5 r.u. or between 6 and 20 m.u. of gonadotrophin per litre.

The experiments just quoted show that the fate of gonadotrophin in the body, whether derived from the pituitary or chorionic tissues of the host, or artificially introduced by injection, is of two sorts; some is inactivated within the body and some is excreted by the kidneys in an active form. Parkes & White (1933), using male and female rabbits which had been deprived of their gonads, performed the following experiment. Under anaesthesia the bladder was emptied and its outlet obstructed by ligature. Gonadotrophin was then injected into the ear vein and at various intervals afterward each animal was killed, its bladder removed and the contained urine assayed. In this way it was found that about one-third of the injected hormone could be recovered from the urine during the first 9 hours.

Pituitary and chorionic gonadotrophins may not be excreted with equal ease by the kidney. Catchpole, Cole & Pearson (1935) showed that in the pregnant mare, though much gonadotrophin is present in the blood, none is detectable in the urine. Human chorionic gonadotrophin, after injection into the pregnant mare's blood stream, appears in the urine. Gonadotrophin, from the blood of a pregnant mare, when injected into the circulation of the gelding, monkey, rabbit or rat, does not appear in the urine. For example, 3,000 r.u. of gonadotrophin prepared from the blood of a pregnant mare were given intravenously to a rabbit. After the lapse of 24 hours, assays of the blood showed that about half of the original dose was still present in the circulation. At this time none was found in the uterus, spleen, lungs, kidney or liver, and the missing hormone was not present in the urine. Gonadectomy had no influence on the rate of disappearance. Two facts seem to be indicated by these experiments. The first is that gonadotrophin is inactivated in the circulation though we do not know how or where; the other is that the gonadotrophin of the pregnant mare is different from that of the pregnant women inasmuch as the former is not excreted by the kidney, perhaps, as Zondek has suggested, because its molecules are too large. It has been thought that the production of gonadotrophin throughout gestation in the mare may perhaps take place in the pituitary and not in the placenta.

Sources of gonadotrophin. (i) *The anterior lobe of the pituitary* is the chief source of gonadotrophin, and hypophysectomy is followed by atrophy of the gonads and the accessory genital organs (Cushing & Goetsch, 1915; Smith, 1927b, and many others). After hypophysectomy the reproductive system can be maintained, or if already atrophied can be restored to a functional condition, by repeated subcutaneous or intramuscular implantations of pituitary tissue or by the injection of extracts made from pituitary glands.

(ii) *The placenta.* Collip (1930) and Collip, Thomson, McPhail & Williamson (1931) tested extracts of human placenta on immature and adult rats, and obtained positive gonadotrophic responses; confirmatory experiments have been reported by Philipp (1931), Collip, Selye & Anderson (1933) and Collip, Selye, Thomson & Williamson (1933).

Apart from direct evidence of this kind other observations have indicated that the placenta forms gonadotrophin. Evans & Simpson (1929b) noted that although the pituitary is enlarged in pregnancy and although the output of gonadotrophin is large in this condition, the gonadotrophic potency of the pituitary as tested by implantation into immature animals is not increased during gestation, as might have been supposed. Subsequently it was shown that the pituitary has little or no gonadotrophic potency during pregnancy, and it became obvious that the large amounts excreted in the urine, which are the foundation of the Aschheim-Zondek test for pregnancy, must arise elsewhere than in the pituitary. The stage of gestation at which increased amounts of gonadotrophin begin to appear in the urine is somewhere about the time when the ovum becomes attached to the uterus (Crew, 1936b; Evans, Kohls & Wonder, 1937), that is to say when a placenta is formed; and the amount falls rather abruptly after parturition and expulsion of the placenta. Crew (1936a) examined the urine by the Aschheim-Zondek test in fifty cases at periods extending from ¾ of an hour to 144 hours after delivery, and his results are given in Table 6 in which the rapid diminution of gonadotrophin in the urine after expulsion of the placenta is clearly shown.

Evans, Kohls & Wonder (1937) made repeated assays of gonadotrophin in 24-hour specimens of urine from six pregnant women. In each instance the greatest concentration, which ranged from 75,000 to 1,040,000 r.u. per litre of

TABLE 6. The reduced excretion of gonadotrophin in the
urine in 50 cases after parturition (Crew, 1936 a)

Specimen of urine	Hours after parturition	Positive to A-Z test	Negative to A-Z test
1	¾–7	45	5
2	24	26	24
3	48	15	35
4	72	2	48
5	96	0	50
6	120	0	50
7	144	0	50

urine, was observed 1 month from the beginning of the first expected but missed menstruation. This rise was followed by an abrupt fall, so that by the 65th day the hormone concentration in the urine was below 10,000 r.u. per litre and remained so till the end of pregnancy. Boycott & Rowlands (1938) by a different method found that the concentration of gonadotrophin in the urine rose rapidly from the 6th week of pregnancy, reached a maximum between the 56th and 84th days, and then declined to a fairly constant level which was maintained to the end of pregnancy. Browne & Venning (1936) found the highest concentration on the 60th day after the 1st day of the last menstrual period.

A curious observation bearing on the formation of gonadotrophin by the placenta was made by Ware, Main & Taliaferro (1938), who assayed the urine for gonadotrophin in a woman with abdominal pregnancy. The child was removed by laparotomy, the placenta being left *in situ*. The excretion of gonadotrophin in the urine continued for 47 days, during which mammary engorgement and lactation were absent. In this case the excess of gonadotrophin excreted in the urine seems attributable to the continued presence of the placenta; the foetus, at any rate, could not be regarded as the source.

Direct evidence of the formation of gonadotrophin by placental cells has been obtained by Jones, Gey & Gey (1943) who maintained *in vitro* cultures of cells from human placentae and hydatidiform moles. Assays were made when the media had been changed several times and the explants consisted entirely of new cells. Gonadotrophic responses were obtained in 20 of 29 tests made with this material on immature rats, whereas negative results followed in 18 control experiments.

Hitherto we have discussed gonadotrophin in general terms as though that which is formed by the placenta and excreted in the urine in pregnancy were identical with that normally produced by the pituitary. It is by no means sure, however, that placental and pituitary gonadotrophin are the same. Collip and his colleagues using a gonadotrophic extract of placenta found that its activities were like those of pregnancy-urine extracts, but differed from pituitary gonadotrophin inasmuch as it merely caused thecal luteinization in the ovaries of hypophy-sectomized, immature rats and guinea-pigs and did not bring about maturation of follicles or formation of corpora lutea; that is to say it consisted mainly of LH. This question will be discussed in more detail later on (p. 10).

(iii) *Chorionepithelioma*, whether in woman or man, is accompanied by a high output of gonadotrophin in the urine, as revealed by the Aschheim-Zondek pregnancy test (Zondek, 1930; Heidrich, Fels & Mathias, 1930; Ferguson, Downes, Ellis & Nicholson, 1931; Evans, Simpson, Austin & Ferguson, 1933; Montpellier & Herlant, 1933). In either sex this fact may be of use to the diagnostician.

In women, a positive test, continuing after the uterus has been apparently emptied of the products of conception, may establish the diagnosis at a time when remedial measures will be effective and from this standpoint the test may be of the greatest value. At an earlier stage, when an attempt is being made to distinguish a normal pregnancy from a hydatidiform mole, the test is not so useful, for, as Boycott & Smiles (1939) have found, the output of gonadotrophin in cases of chorionepithelioma is not necessarily greater than that in normal pregnancy, though perhaps an excessive excretion of gonadotrophin would rather suggest the presence of chorionepithelioma.

In men the test, being free from the possible complication of pregnancy, is of the greatest aid in making a differential diagnosis. Heidrich, Fels & Mathias (1930) were the first to report the use of the pregnancy test in a male. The patient was 35 and had a tumour of the testis, and among other symptoms had gynaecomastia with the secretion of colostrum. A positive Aschheim-Zondek test was obtained with both his blood and urine before death, one litre of urine containing 35,000 m.u. of gonadotrophin. After death implants of the primary tumour into immature mice also produced positive reactions, though implants of the normal testis or of the pituitary did not.

The test may be positive though the primary tumour in the gonad is so small as to be overlooked: in fact the affected testis may appear atrophic (Craver & Stewart, 1936).

Attempts to distinguish chorionepithelioma from other kinds of testicular tumour by means of the Aschheim-Zondek test have not been so successful as had been hoped.

According to Ferguson (1933 a, b; see also Ferguson, Downes, Ellis & Nicholson, 1931) an excess of gonadotrophin in the urine may accompany tumours of the testis which might be described as teratomata rather than chorionepitheliomata. It seems possible that in these cases chorionic tissue may form some small part of the teratoma.

Montpellier & Herlant (1933) reported a case of seminoma of the testicle in which large amounts of gonadotrophin were present in the urine. They believe that the nature of the gonadotrophin may to some extent reflect the nature of the testicular tumour. The urine from a patient with seminoma, they say, when injected into immature rats or mice, causes ripening of the follicles only, whereas the urine from cases of chorionepithelioma of the testis provokes the appearance of haemorrhagic follicles and corpora lutea.

Fortner & Owen (1935) think that quantitative estimations of the urinary output of gonadotrophin might help in distinguishing clinically between teratoma and chorionepithelioma and give approximate figures (Table 7).

Ferguson (1933 a) has noted the presence of gynaecomastia with secretion of colostrum in five among 117 cases of testicular tumour in which he has examined the urine for gonadotrophin. Further, he calls attention to the fact that the

TABLE 7. Urinary output of gonadotrophin in cases of
testicular tumour (Fortner & Owen, 1935)

Condition	Mouse units of gonadotrophin per litre of urine
Normal	50
Teratoma	50–10,000
Chorionepithelioma	10,000–150,000 or more

pituitary in cases of chorionepithelioma shows histological changes similar to those which accompany pregnancy. These observations have been confirmed by Entwisle & Hepp (1935) and Solcard, Le Chuiton, Pervès, Berge & Pennanéac'h (1936). According to Evans, Simpson, Austin & Ferguson (1933), the gonadotrophin present in the urine in cases of testicular tumour shows predominantly an FRH activity, although presumably it has been formed by chorion-like tissue and therefore by analogy might have been expected to show a pronounced luteinizing action.

The presence of gonadotrophin in chorionepitheliomatous tissue was confirmed by Philipp (1931), who implanted fragments of tumour from a human case into immature female mice and obtained pronounced positive responses, as Heidrich, Fels & Mathias (1930) had already reported.

(iv) *Sources outside the animal body.* Apparently the production of gonadotrophin is not confined to animals. Hisaw, Greep & Fevold (1936) extracted from brewers' yeast a water-soluble substance which had some gonadotrophic properties. In immature rats it prevented atrophy of the testes after hypophysectomy, though it did not prevent some degeneration of the interstitial glandular cells of the testis nor atrophy of the accessory generative organs. In hypophysectomized adult rats treated with the extract the testes remained in the scrotum and spermatogenesis was maintained; the accessory glands, though smaller than normal, continued to secrete.

Friedman (1938) obtained from young oat plants a substance which when given intravenously to rabbits caused ovulation. One rabbit unit of this material was extracted from between 30 and 80 g. of the dried plants.

In this connection it must be remembered that ovulation in the rabbit may be induced by the intravenous injection of material which has no gonadotrophic potency.

The distribution of gonadotrophin in the body. From the organs in which it arises, namely the pituitary, placenta and tumours of the type known as chorionepitheliomata, gonadotrophin passes into the blood and some is excreted in the urine. Because of its elimination by the kidneys together with its gradual inactivation in the body there is a falling gradient in the bloodstream. This has been demonstrated in the foetus, for blood collected from the placental end of the divided umbilical cord is rich in gonadotrophin whereas little or none is detected by ordinary means in the blood returning from the foetus (Geist & Spielman, 1934). The site of inactivation has not yet been determined. In the foetus the process appears to be rapid, for Parker & Tenney (1940) found that, in normal pregnancy, although gonadotrophin is present in about equal concentration in the maternal blood, urine and placenta, none is evident in the foetal organs, blood or amniotic fluid.* Heim (1931) reported that by means of the Aschheim-Zondek test gonadotrophin could be detected in colostrum, and that it disappeared from the milk on the 5th day after parturition. At autopsies on a man and a woman who had died with chorionepitheliomata Ewald found relatively high concentrations of gonadotrophin in the cerebrospinal fluid. Gonadotrophin has been detected in the cerebrospinal fluid during normal pregnancy, though its concentration is low. Aschheim & Zondek (1927) detected gonadotrophin in the ovaries, placenta and blood of pregnant women and in blood from the umbilical cord.

Are the Pituitary and Placental Gonadotrophins identical?

Evans & Simpson (1929b) and Engle (1929a) seem to have been the first to report a difference between pituitary and placental gonadotrophin. The former found that the weights of ovaries precociously developed in rats of 26 days under the influence of implanted pituitary gland were, within limits, nearly proportional to the amount of tissue implanted. With 4 times the minimal dose the ovaries were increased approximately 4 times in weight. With an extract of pregnancy urine if 4 times the minimal effective dose had been given the ovaries were not appreciably larger than after a minimal dose; and it was found that with 150 times the minimal dose the ovarian tissue was barely trebled. The difference lay in the number of follicles stimulated. Pituitary implants caused a much more general follicular development than that which followed the administration of extracts of pregnancy urine (see also Evans, Meyer & Simpson, 1931). Engle (1929a) noted a difference between the effects on immature female mice of pituitary implants and extracts of pregnancy urine. The former caused extensive follicular maturation and ovulation while the latter caused follicular growth and luteinization without ovulation. In an adult macaque Engle (1932a, 1933) noted that pituitary gonadotrophin induced changes in the sexual skin like those which follow the administration of oestrin, whereas no effect on the sexual skin nor stimulation of ovarian follicles was caused by placental gonadotrophin. Hamburger (1933a) has made observations like those of Evans & Simpson. He compared gonadotrophin extracted (a) from the urine of men and women who had been castrated and of women who had passed the menopause with (b) gonadotrophin obtained from the urine of pregnant women. Hormone (a) is derived from the pituitary, while hormone (b) probably is derived mainly or entirely from the placenta because the human pituitary during pregnancy has been shown to be nearly or quite free from gonadotrophin. When tested on immature female mice or rats, hormone (a) caused a large number of ovarian follicles to ripen at the same time, whereas hormone (b) affected only a few follicles, causing them to ripen and to become very large and protuberant while the remainder were not brought to maturity. Gonadotrophin prepared from patients with testicular teratomata resembled in action that obtained from the urine of pregnant women (cf. p. 9).

Reichert, Pencharz, Simpson, Meyer & Evans (1931, 1932), Evans, Meyer & Simpson (1932) and Smith & Leonard (1933) found that, in dogs and rats, prolan prepared from the urine of pregnant women failed to preserve the normal de-

* On several occasions the author, collaborating with Douglas H. MacLeod, has obtained a positive A-Z reaction in mice with human amniotic fluid.

velopment and functions of the reproductive system in males and females after hypophysectomy, a failure which contrasted with the success obtained by implants or extracts of the pituitary gland itself in similar circumstances.

Selye, Collip & Thomson (1933 b) noted that pituitary gonadotrophin can restore the ovaries in the hypophysectomized rat to a normal or nearly normal condition, while extracts prepared from pregnancy urine merely cause luteinization. Engle (1933) obtained comparable results in young rhesus monkeys.

Collip, Selye & Anderson (1933) prepared gonadotrophin from the placenta and it appeared to be identical in action on rats with that extracted from pregnancy urine but not with that from the anterior pituitary; moreover, different methods were required to extract the hormones from the pituitary and the placenta.

Leonard (1933 a, b) made tests with extracts of human pituitary gland and compared their effects with those produced from human pregnancy urine. The non-identity of these gonadotrophins was shown by three different tests. (1) The minimal dose of the pituitary extract required to cause ovulation in rabbits was $\frac{1}{4}$ r.u., while 2 r.u. of the urinary extract were required to produce the same result; (2) an amount of extract representing 1 to 10 mg. of dried human pituitary injected daily into rats for 5 days led to ovaries weighing as much as 187 mg., whereas 100 to 400 r.u. of the urinary extract gave ovaries weighing not more than 45 mg.; (3) the pituitary extract caused development of the testes of white leghorn cockerels, whereas the urinary extracts did not.

Evans, Simpson & Austin (1933 a, b) made the important observation that the inactivity of placental gonadotrophin in hypophysectomized animals can be remedied by the addition of a small amount of pituitary extract (see also Jensen, Hauschildt & Evans, 1942). Furthermore, they suggest that the pituitary component possessing this power of activation is not one of the recognized pituitary hormones. It can be boiled for 30 minutes with only a slight reduction of activity, does not become oxidized on exposure to the air and, unlike gonadotrophin, is not destroyed by formic acid. Collip, Selye, Thomson & Williamson (1933) have made similar observations, and Noble, Rowlands, Warwick & Williams (1939) have reported the same sort of phenomenon (cf. pp. 63, 74).

A curious fact bearing on this subject was elicited by Riddle (1931), who found that the pigeon's testes, though very sensitive to gonadotrophin derived from the pituitary, are insensitive to that derived from the urine of pregnant women. Schockaert (1933) obtained comparable results with chicken and Witschi & Keck (1935) reported similar results with the sparrow (*Passer domesticus*). The quiescent gonads of these birds in the winter yielded no response to the gonadotrophic hormone obtained from pregnancy urine, but when an extract from the anterior lobe of the pituitary (6 r.u. daily) was given their testes increased in volume 100 to 150 times within 18 days, and the sparrows showed the black beak and other secondary changes characteristic of the breeding season. In connection with these findings it may be mentioned that Evans & Simpson (1934) reported that the pigeon's testes failed to respond to gonadotrophin obtained from pregnant mare's blood, though this gonadotrophin is thought to be pituitary origin. Schockaert (1933) compared the actions of pituitary and placental gonadotrophins on the testes and accessory genital organs of immature rats and cockerels. Some of his results are given in Table 8 and show that in rats treated with pituitary gonadotrophin the testes are larger while the accessory genital organs are smaller than in rats treated with placental gonadotrophin. Greatly increased development of the cockerel's testes and comb were caused by pituitary gonadotrophin, whereas they were unaffected by placental gonadotrophin.

TABLE 8. The action of pituitary and placental gonadotrophins on the immature rat's testes and accessory genital organs (Schockaert, 1933)

	Treatment		
	None (mg.)	Pituitary gonadotrophin (mg.)	Placental gonadotrophin (mg.)
Weight of testes per 100 g. of bodyweight	1,686	1,348	1,104
Weight of accessory organs per 100 g. of bodyweight	667	600	1,081

Liu & Noble (1939) treated male rats after hypophysectomy with (a) an extract of pregnant mare's serum or (b) an extract of human pregnancy urine. The gonadotrophin of (a) is thought to be derived from the pituitary, whereas that of (b) is from the placenta. If treatment was begun immediately after removal of the pituitary, extract (a) maintained spermatogenesis and stimulated the interstitial cells to produce androgen enough to prevent atrophy of the accessory genital structures. In similar doses and tested in the same way extract (b) was ineffective. If treatment were delayed for 14 or 28 days after the operation and the extracts were given in much larger doses, extract (b) was more efficient than (a) as a stimulator of the interstitial cells but had less effect on spermatogenesis. Noble, Rowlands, Warwick & Williams (1939) found that the results in the female of giving placental gonadotrophin were not the same in the presence and absence of the pituitary. Given to hypophysectomized rats placental gonadotrophin caused extensive luteinization in the ovaries without stimulating follicular growth, whereas, in the presence of the pituitary, growth of the follicles was caused in addition to luteinization. Bischoff (1942) has reported a difference of chemical reactivity between pituitary and placental gonadotrophins.

Additional evidence of a difference between the gonadotrophins derived respectively from the pituitary and the placenta is afforded in the following pages and also in the section dealing with acquired resistance to gonadotrophic extracts.

Are the Follicle Ripening (FRH) and Luteinizing (LH) Hormones Distinct Compounds?

For ease of literary presentation FRH and LH are mentioned as though their separate identities had been proved. Neither of these supposedly different hormones has yet been prepared in a pure form, and until such preparation has been achieved it would be hardly justifiable to conclude that they are chemically distinct. Some unrecognized factor might account for the different reactions commonly attributed for convenience to FRH and LH respectively.

Much of what has been written in the last few pages concerning the identity or non-identity of pituitary and chorionic gonadotrophins might be used as evidence of a difference between FRH and LH; for the follicle ripening effect is a strong feature of pituitary tissue, whereas a luteinizing effect is the predominant character of chorionic extracts or substance. The matter is so important that it may be worth while, even though no precise conclusion is reached, to quote several different kinds of experiment which bear upon the present theme.

Aschheim & Zondek (Zondek, 1935) and others believed that maturation and luteinization of ovarian follicles might be caused by different agents. Though no one has yet isolated FRH or LH in pure form, extracts have been obtained showing great predominance of FRH or LH effects. Attempts to separate the two hormones, supposing them to be different substances, have followed two main lines. The first depends upon subjecting material—e.g. pituitary substance or urine—to procedures which will destroy or extract one hormone and not the other. Thus Loeb, Saxton & Hayward (1936) found that by immersing bovine pituitaries in $1\frac{1}{2}$ per cent solutions of formalin the LH content was largely inactivated, so that an extract could be prepared having a nearly pure FRH activity. Similar results have been obtained by different means (Fevold, Hisaw, Hellbaum & Hertz, 1933; Hertz & Hisaw, 1934, and others). The second method of obtaining preparations having almost pure FRH or LH effects has been by appropriate selection of the raw material. Experiments have shown that in the urine of pregnant women LH is present in a considerable preponderance, whereas in the urine of men, or of women who have undergone gonadectomy, FRH greatly predominates, and from these different urines extracts can be obtained which show almost pure LH or FRH reactions. Both of the foregoing methods have been of use in studying the biological actions of gonadotrophins.

A device which has been employed in the study of hormonal action is parabiosis. In this procedure two littermate animals are joined together side to side by a surgical operation so as to convert them, as it might be described, into Siamese twins. When two animals are thus united a slow interchange of hormones can take place between them. Martins (1930), while watching the effects of parabiosis on rats and mice, noticed that when a normal female was joined to a normal male the oestrous cycle was either unaffected or anoestrus ensued; but when a normal female was joined to a castrated male a state of constant oestrus was induced in the female, the ovaries showing an excessive follicular development and maturation without luteinization. Further, Martins found that parabiotic union of a normal female with a cryptorchid male also caused constant oestrus in the female with excessive follicular development unaccompanied by luteinization. He noticed too that cryptorchidism led to histological changes in the pituitary which, though less pronounced, resembled those following castration, including the presence of 'castration cells'. The experiment suggested that castration and to a minor degree cryptorchidism so affected the pituitary that it produced a continuous and enhanced supply of FRH and little or no LH.

Witschi, Levine & Hill (1932) sterilized male rats by X-radiation of their testes. This treatment destroyed the seminal epithelium but left the interstitial glandular tissue unaffected, so that the secondary male sexual characters were maintained. When these rats were subsequently joined parabiotically with normal females the latter exhibited constant oestrus, showing that they were receiving from the sterilized males an excessive supply of FRH and little or no LH. Further, Witschi and his colleagues found that the pituitaries of these rats which had been sterilized by X-rays showed changes, including the presence of 'castration cells', like those which follow removal of the testes. Levine & Witschi (1933) placed female rats which had been sterilized by X-rays into parabiotic union with normal females with the result that constant oestrus occurred in the latter. The same workers (Witschi & Levine, 1934) joined normal female rats with castrated males. When constant oestrus had become established the females were hypophysectomized, after which the constant oestrus persisted. The pairs were now separated with the result that anoestrus appeared in the females within 2 days and was accompanied by atresia of all the matured follicles. If the male instead of the female were submitted to hypophysectomy while the parabiotic union was maintained, the female at once returned to a condition of cyclical oestrus. These experiments confirm those of Martins and suggest that during parabiosis there was an excessive supply of FRH and that it was derived from the pituitary gland of the castrated or otherwise sterilized partner.

A more direct proof that the output of FRH is increased after castration is supplied by an experiment of Du Shane, Levine, Pfeiffer & Witschi (1935). A female rat deprived of her pituitary and joined parabiotically with a normal male or female did not obtain enough pituitary hormone from her partner to maintain her ovaries at their normal weight. If now the normal partner were castrated or spayed, the hypophysectomized twin went into constant oestrus and the ovaries contained large follicles without corpora lutea.

Hertz & Hisaw (1934; see also Fevold, Hisaw, Hellbaum & Hertz, 1933) prepared two extracts from sheep's pituitaries, the one rich in FRH and the other in LH, and tested their activities (a) on infantile rabbits of 4 weeks weighing 500 to 700 g., and (b) on juvenile rabbits of 12 to 13 weeks weighing 1,300 g. The infantile rabbits showed no ovarian response. The juveniles reacted to both the FRH and LH extracts, but in quite a different manner. FRH caused maturation of ova without luteinization, while LH caused massive luteinization.

Fiessinger & Moricard (1934), by tests carried out on infantile mice, showed that extracts of pregnancy urine caused ovarian luteinization while extracts of urine from women past the menopause caused maturation of ova. Hamburger (1933a) states that the follicle ripening and luteinizing potencies of pregnancy urine change as pregnancy advances, and this perhaps might be regarded as additional evidence that two distinct hormones are concerned in the effects. Hellbaum (1933) implanted portions of pituitary from castrated horses into sixteen immature female rats with the consequence that follicular maturation occurred in all but two; in these a slight degree of luteinization was present.

Smith, Engle & Tyndale (1934; see also Leonard & Smith, 1933, and Rogers, 1938) prepared gonadotrophic extracts (a) from the urine of women who were past the menopause or whose ovaries had been removed, and (b) from the urine of pregnant women; and they tested these preparations on hypophysectomized rats. Extract (a), in which FRH preponderated, caused in females a proliferation of granulosa cells and maturation of ova, and in males a proliferation of germinal epithelium without an increase of the interstitial tissue or a restoration of the atrophied accessory generative organs; extract (b), which consisted mainly of LH, caused hypertrophy of the interstitial tissue of the ovary with luteinization if the granulosa was well enough developed, and in males an increase of the interstitial tissue of the testis with enlargement of the accessory generative organs. Greep, Fevold & Hisaw (1936; see also Greep & Fevold, 1937, and Greep, 1937) made similar observations. In their experiments FRH given to rats from the time of hypophysectomy onward maintained spermatogenesis for 40 days, although during this time the accessory generative organs underwent atrophy. LH given in the same circumstances both maintained spermatogenesis and prevented atrophy of the accessory organs. Additional work of this kind with comparable results has been done by Fraenkel-Conrat, Simpson & Evans (1940). Leblond (1938b) prepared FRH and LH extracts from sheep's pituitaries and tested them on pigeons. Before starting the injections laparotomy was done and the left testis or the largest ovum was measured; and the weight of the testis was calculated by a method suggested by Benoit. Each pigeon received a daily dose of the extract representing 1 g. of pituitary and was killed on the 7th day. The results are shown in Tables 9 and 10.

TABLE 9.   The effect of FRH and LH respectively on the pigeon's testes (Leblond, 1938b)

Treatment	Number of birds	Average weight of left testis before treatment (calculated) (mg.)	Average weight of left testis after treatment (mg.)
None	4	762	726
FRH	3	589	958
LH	3	579	274

TABLE 10.   The effect of FRH and LH respectively on the pigeon's ovary (Leblond, 1938b)

Treatment	Number of birds	Mean diameter of largest ovum before injection (mm.)	Mean diameter of largest ovum after injection (mm.)	Mean weight of oviduct (mg.)
None	4	3	3·4	491
FRH	4	3·5	7·2	5,227
LH	3	5·8	2·4	340

As the tables show, FRH caused an increase in the average size of the testis and the largest ovum, while LH caused a decrease. The great enlargement of the oviduct is attributable to oestrin produced in consequence of the action of FRH on the ovary.

Another phenomenon which suggests that FRH and LH are not identical is described by Fevold, Hisaw & Greep (1937) as 'augmentation'. They found that when FRH is injected into 21-day-old rats in graded doses the ovaries increase in proportion with the dose only until they weigh about 45 or 50 mg.—though there is no sharply defined upper limit, and the curve of increase does not end in a horizontal line. When, on the other hand, FRH and LH are injected together the ovaries increase in weight regularly with the dosage. Ten rat units of FRH caused ovaries of 42 mg. The same amount of FRH with LH added caused ovaries of 80 mg. By increasing the doses of LH and FRH together ovaries of 600 mg. were obtained. LH alone had no effect on the weight of the ovaries of infantile rats. They conclude that LH makes the ovaries more sensitive to FRH, and explain that this augmentation of response is different from that caused by the addition of zinc salts or tannic acid. The latter act by reducing the rate of absorption of the injected hormone and so prolonging its action; to cause augmentation of effect they must be injected with the hormone, whereas LH will augment the effect of FRH whether mixed with it or introduced into another part of the body. Foster, Foster & Hisaw (1937) treated hypophysectomized rabbits with FRH and LH. If given within 14 days of hypophysectomy, FRH alone caused maturation of ovarian follicles without ovulation. If the rabbit were treated for 5 days with FRH and then given LH in addition ovulation was induced. The optimal mixture for this result was 50 parts of FRH to 1 of LH; too much LH prevented ovulation.

Saunders & Cole (1938) say that augmentation of the response to FRH can be produced as readily by egg albumin as by LH, and that the use of augmentation as a specific test for LH does not seem justified.

Leonard (1937) observed that oestrone given to immature female rats decreased the response of their ovaries to FRH but not to LH.

At this point it might appear certain that the pituitary forms at least two different gonadotrophins, one of which causes maturation of ova and the other luteinization, these two contrasting effects having been demonstrated in the rat, rabbit, guinea-pig, cat, monkey and man. Pituitary extracts exhibiting almost pure effects of one kind or the other have been prepared by Fevold, Hisaw & Leonard (1931), Wallen-Lawrence (1934), Loeb, Saxton & Hayward (1936), Dodds & Noble (1936), and others. A complete separation of FRH from LH has not, however, yet been achieved, and therefore it is still unsafe to conclude that the two are distinct entities.

To justify doubt in this matter reference may be made to experiments which reveal that a single pituitary preparation apparently may cause according to circumstances a pure FRH or a pure LH effect. Evans, Simpson & Pencharz (1935) implanted pituitaries from castrated male rats into hypophysectomized female rats, 26 days old. With a certain dose of pituitary substance a pure FRH response was obtained, but if the dose were doubled luteinization occurred. Frank, Salmon & Friedman (1935; see also Salmon & Frank, 1936) prepared gonadotrophin from the urine of women whose ovaries had been removed or who were past the menopause. Such preparations predominantly show a follicle-ripening action when tested on infantile rats or mice. Frank and his colleagues, however, found that by merely increasing the dose a luteinizing effect could be imposed.

The action of a given gonadotrophic extract may be influenced by the gonads. Oestrogens seem to assist in the luteinizing action of the pituitary, and it is possible either that oestrogens suppress the output of FRH from the pituitary while stimulating it to produce LH or that luteinization is produced by a co-operation between oestrogen and FRH.

Collip, Selye & Thomson (1933a) treated hypophysectomized immature rats with prolan (LH), and found that if the injections were begun after removal of the pituitary oestrus was not induced. If, however, the same prolan was given for 5 or more days before removal of the pituitary it caused continued oestrus; that is to say, the response of the ovaries to LH after hypophysectomy is conditioned by their state at the time of the operation.

According to Noble, Rowlands, Warwick & Williams (1939) the response of the immature rat's ovary to chorionic gonadotrophin is largely influenced by the rat's own pituitary. In normal immature rats, they say, the response is enlargement of the follicles and the formation of normal or atretic corpora lutea, whereas in hypophysectomized immature rats the only response is a diffuse luteinization of stromal and thecal tissue. These results suggest that chorionic gonadotrophin consists primarily of LH alone. Further, these workers find that preparations of pituitary gland when given to hypophysectomized rats cause only growth of the follicles without luteinization, whereas in normal rats they cause growth of the follicles and luteinization.

Witschi (1940) has recorded a special reaction to LH which does not occur in response to FRH, namely a pigmentation in the breast feathers of the weaver finch. This pigmentation represents the male breeding plumage, and Witschi suggests that its artificial induction by LH might be used for assaying this hormone.

The various experimental results just quoted neither prove nor disprove the identity of FRH and LH, nor do they indicate precisely their respective actions. The problem is complicated by the mutual influence of FRH and LH upon each other, and probably also by the effect of gonadal hormones on their actions. A similar complication was faced when the actions of androgen, oestrogen and progestin were being investigated before the individual hormones had been isolated, and before it had been learned that the biological activities of a particular gonadal hormone may depend largely on the proportion in which other gonadal hormones are or have been present.

The Action of Gonadotrophins on the Ovary with special reference to Hormonal Balance

The dominating influence of the pituitary on the ovary is shown by the results of hypophysectomy, for the operation is followed in young animals by an arrest of further ovarian development and in adults by a reversion of the ovary to the atrophic infantile condition, while every follicle already present becomes atretic. These effects of removing the pituitary can be largely prevented by the repeated implantation of pituitaries from other animals or by the injection of gonadotrophin.

Although the pituitary governs the ovary its influence is controlled. Between the two organs a balance exists by which the activity of each is kept within bounds, so that overaction of the pituitary would be checked by the ovary and vice versa.

Maturation of follicles and puberty. One consequence of this balance is a regulation of the number of follicles which mature during each oestrous cycle. The ripening of follicles does not occur until puberty has been attained, after which with every oestrous cycle one or more follicles will ripen. The number reaching maturity at each cycle in different species is limited to the requirements of the individual and is related to the size of the litters, allowance being made for wastage of ova.

When pituitary substance or FRH is supplied in excess, the regulating mechanism may be overborne, so that with immature females a precocious maturation of follicles takes place (Zondek & Aschheim, 1927 a), and in both immature and adult subjects an excessive number of follicles may come to maturity at one time.

Smith & Engle (1927) transplanted pituitaries or portions of pituitary into mice. The material used was derived from several species, namely mouse, rat, rabbit, guinea-pig, cat and pigeon, and was active from each of these sources. When given to immature mice the implants induced premature maturation of ovarian follicles with the secondary effects of precocious keratinization and opening of the vagina and enlargement of the uterus. In adult mice the pituitary implants caused a great increase in the size of the ovaries owing to the presence of an excessive number of maturing follicles. Hamburger (1933 a, b) made a similar observation. By the same method of pituitary implants Cardoso (1934) produced comparable changes in the ovary of a fish.

The effects of gonadotrophin on the ovary are not the same at all periods of life. In the very early stages of immaturity the ovarian follicles are irresponsive to gonadotrophin. This feature of the pituitary-ovarian relationship will be discussed later (p. 70).

Ovulation and luteinization. Ovulation, that is to say the discharge of ova from ripe follicles, is not an inevitable sequel to maturation. In the rabbit, ferret, sheep, pig and some other animals the mature follicles do not usually discharge their contained ova in the absence of mating or some other adequate stimulus. Heape (1905), who made these observations, noted that in the rabbit ovulation occurred about 10 hours after copulation but could be prevented by previous occlusion of the ovarian blood vessels. In the absence of mating, he noted, the ova are retained in the ovary, where they degenerate, becoming reddened and discoloured in the process. This condition is described as follicular atresia. It seems that during sexual rest the rabbit's pituitary provides effective amounts of FRH only, and that in response to the sensory impulses of mating the pituitary produces an immediate supply of LH which is the effective stimulus for ovulation (Friedman, 1929b; Casida, 1934). The rapidity with which LH becomes available in these circumstances is remarkable; to cut off its supply and so to prevent ovulation and luteinization the pituitary must be removed within about the first hour after copulation. That LH is the cause of ovulation is shown by the fact that one intravenous injection of this hormone in the oestrous rabbit will, apart from mating, cause ovulation followed by the development of corpora lutea and pseudopregnancy; moreover the release of ova from ripe follicles can be induced by LH in the absence of the pituitary (Rowlands & Williams, 1943).

The ferret's breeding season extends from April to August and during the rest of the year the animal is anoestrous. Hill & Parkes (1931) found that gonadotrophic extracts rich in LH, given between September and the end of January, caused in some ferrets ovulation followed by luteinization and pseudopregnancy. The same workers (1932) discovered that ovulation could be induced by LH in the pregnant rabbit.

Burns & Buyse (1934) by alkaline extracts of sheep's pituitary induced oviposition in immature salamanders, and the same result was obtained by Rugh (1937) in frogs, and by Hogben (1934) and Bellerby (1933) in the clawed toad (*Xenopus*).

Mankind does not appear to be excepted from this subservience of the ovary to pituitary influence. Davis & Koff (1938) gave intravenous injections of gonadotrophin prepared from the serum of pregnant mares to thirty-six women before laparotomy. Examination of the ovaries during the subsequent operation showed that ovulation had occurred in half the total number, as shown by the presence of corpora lutea in an early stage of evolution. Usually in women, as in many other mammals, ovulation occurs regularly as an immediate sequence of follicular maturation, and no special added impulse is required to cause the discharge of an ovum from a ripe follicle.

It seems probable that the action of gonadotrophin on the ovary is direct, needing no intervening mechanism. Friedman (1932) injected chorionic gonadotrophin into a follicle of a rabbit's ovary, gauze packing being used to absorb any overflow; the formation of a corpus luteum ensued at the site of injection, whereas the other ovary failed to react. When considering this experiment, however, it is to be remembered that Bouin & Ancel (1909) caused the formation of corpora lutea in the rabbit's ovary merely by puncturing ripe follicles. Zondek (1940c) states that the direct injection of gonadotrophin into the ovary causes the same effect as twice the same dose given intravenously. This experiment also might be used to support the opinion that gonadotrophin acts directly on the ovary, but it will be prudent to await further confirmation before accepting such an opinion without reserve; for there are many factors which have to be considered when ovulation and luteinization are under experimental investigation.

Marshall & Verney (1936) found that ovulation could be induced in the rabbit by faradism applied while the animal was anaesthetized by ether. With one electrode in the mouth and the other—a needle—inserted into the nape near the foramen magnum, or with one electrode in the rectum and the other in the lumbar region, faradism caused ovulation between 17 and 24 hours later. The physical conditions were A.C., 30 V., 50 cycles, and the applications in each case lasted 3 seconds, with a repetition of the dose after an interval of 7 seconds.

In a later communication, Marshall, Verney & Vogt (1939) state that intravenous doses of picrotoxin (0·9–1·1 mg. per kg.) brought about ovulation in rabbits, and Emmens (1940) found that salts of copper or cadmium given in the same way also had this effect. In these experiments, as in those of Marshall & Verney with faradism, ovulation occurred after a longer interval than when induced by mating. These observations suggest that the mechanisms in the rabbit which control the supply of gonadotrophin by the pituitary to the ovary are easily disturbed.

Ovogenesis and the formation of follicles. The pituitary has little direct control of the production of ova, the formation of which from the germinal epithelium is a response to oestrogen (p. 281). The growth of follicles depends on gonadotrophin. At the approach of puberty, before the onset of complete cycles, successive crops of follicles may undergo partial ripening and then degenerate before maturation is complete. At puberty when the full cycles have appeared one or more follicles ripen while many others achieve partial maturation and then degenerate. Whether these satellite follicles serve any special function or not their degeneration causes a considerable diminution of the number of ova contained in the ovaries. The reduction of primordial ova in the human ovary by this recurrent process was studied by von Hansemann (1913), who made complete serial sections of ovaries and counted the number of ova in every 5th section, the resulting total being then multiplied by 5. The figures he obtained are given below.

Age of individual	Total number of ova in ovary
6·5 months	30,339
1 year and 2 ,,	48,808
2 years	46,174
8 ,,	25,665
10 ,,	20,862
14 ,,	16,390
17–18 years	5,000–7,000

From these figures it seems that a persistent excess of gonadotrophin by causing superovulation would cause an abnormally rapid and irreversible disappearance of ova from the ovaries.

Superovulation and excessive fecundity. As mentioned above, an abnormal number of follicles may be caused to mature at one time by giving pituitary gonadotrophin. When such a condition has been induced, an excessive number of ova may be discharged and find their way into the oviducts. Smith & Engle (1927), after subjecting mice to implants of pituitary removed from other animals, discovered a high degree of superovulation; they counted as many as 48 ova in a single oviduct of one mouse, while in another instance there were 35 ova in the right and 28 in the left oviduct, or 63 ova in all.

This effect may be more pronounced in the young. Cole (1937) gave daily injections of 12 r.u. of pituitary gonadotrophin to adult and immature female rats. This caused in the adults a high percentage of large litters but none of the litters was larger than some of those produced by untreated rats of the same age. In the immature rats the results were more striking. Those which became pregnant were killed between the 10th and 12th day of gestation and 38 per cent contained more than twenty embryos.

Pincus (1940) performed the same sort of experiment on rabbits, and found that superovulation was caused by an extract of sheep's pituitaries. In one rabbit as many as eighty ova were discharged during a single period of ovulation. Pincus found not only that an excessive number of ova were discharged under this treatment but that an abnormally large number were fertilizable. In spite of the superfecundation induced in this way by gonadotrophin, the litters are little if at all larger than normal, because many of the surplus embryos die *in utero* (Cole, 1937; Evans & Simpson, 1940; Parkes, 1943).

Parkes & Hammond (1940) say that ovulation can be induced in sheep during oestrus by injections of an extract of equine pituitary. If these injections are given during the last 3 or 4 days of a normal oestrous cycle, superovulation occurs at the next cycle. Given at any other time in the cycle the same injections cause enlargement of the follicles without ovulation.

Allen (1941) has pointed out that the maturation of one ovarian follicle is normally accompanied by the rapid growth of numerous other follicles which do not attain full maturity. These do not rupture but become atretic and their ova degenerate. Apparently the maturation of one follicle requires the partial maturation of several others, which perhaps assist by the production of oestrogen. Possibly the superovulation which occurs under the influence of excessive supplies of gonadotrophin may be due to a ripening of these ancillary follicles.

Although observations quoted above show that an excessive number of fertilizable ova may be produced in response to pituitary extracts, it seems that the primary effect of these extracts is upon the follicle, and that the ova are affected secondarily. Like spermatozoa (pp. 200, 281), the ova seem influenced only indirectly by gonadotrophin. Smith & Engle (1927) say that ova reach their full size in rats after hypophysectomy.

The control of the ovaries by a gland situated in the head is not confined to the vertebrata. Like cross-fertilization it is a principle of wide extent in biology. Wigglesworth (1935), experimenting with the bug (*Rhodnius prolixus*), found that if an adult female has been deprived of both brain and corpus allatum she does not produce eggs, whereas if the brain is removed without the corpus allatum eggs will be produced normally. He showed also that blood from a female with a corpus allatum will induce egg formation in a female from which this organ has been removed. Weed (1936) obtained the same kind of result with grasshoppers (*Melanopus differentialis*), in which removal of the corpora allata prevents development of ova and secretion by the oviducts. So far as concerns their interactions with the gonads, a further likeness between the corpus allatum and the pituitary has been demonstrated by Thomsen (1940), using *Calliphora* and *Lucilia*. If the corpora allata were removed from these flies soon after hatching the ovaries failed to become functional. If the ovaries were removed the corpus allatum became enlarged, like the pituitary in vertebrata after spaying.

Observations by Heyl (1939) on the 'royal jelly' of the honey bee may be mentioned here. This substance is secreted, he says, in the worker bee by the supramaxillary gland which opens in the buccal region—as does Rathke's pouch in the embryo vertebrate. Ordinary worker bee grubs are given this food only on the first 2 or 3 days after hatching, being fed subsequently on honey and pollen: those destined to have fertile ovaries and to become queens receive royal jelly throughout larval life.

The control of pituitary activity by the ovary. From what has been said already the part played by the pituitary in the maturation of ova and ovulation will be more or less clear, and later the influence of the ovarian hormones, oestrogen and progestin, in checking the gonadotrophic activity of the pituitary will be made manifest (p. 51). However, it is not easy to explain completely by what mechanisms all the functional relationships between the pituitary and ovary are controlled. The problem in a general form presented itself to John Hunter (1728–93), who wondered what effect the removal of one ovary might have on the size of litters born subsequently. Having removed one ovary from a sow, he kept a record of her litters, comparing them with those of a normal sow which was used as a control. Each sow farrowed eight times. The mother with one ovary bore seventy-six piglets as compared with eighty-seven borne by the normal sow. This experiment proved that the production of young is regulated by some mechanism outside the ovary itself.

The matter was examined experimentally once more by Bond (1906), who worked with rabbits. He found that excision of one ovary was followed by enlargement of its fellow, which produced approximately the same number of corpora lutea as were formed by two normal ovaries. As in Hunter's experiment, removal of one ovary, though it slightly diminished, did not greatly affect the number of young which were born afterwards. Carmichael & Marshall (1908) repeated these experiments on rabbits and confirmed the results obtained by Hunter and Bond. They noted that if one ovary were removed, the other might attain the weight of two normal ovaries. If, in addition to removing one ovary a part of the other was excised also, the remaining fragment became greatly enlarged, the compensatory increase in size being relatively greater the larger the amount of ovary removed. The experimental work of Carmichael & Marshall on rabbits was repeated and confirmed by Lipschütz, Wagner & Tamm (1922). Hatai (1913 a) carried out similar observations on rats. In these animals also removal of one ovary was followed by an enlargement of the remaining ovary to about double the normal size. Hartman (1925) found that with the opossum removal of one ovary did not reduce the total number of ova which ripened. Arai (1920 b) removed the right ovaries from a number of rats when they were 20 days old, and killed them when 41, 55, 62 or 69 days old, intact littermates being used as controls. The ovaries having been weighed, the number of ova in each was counted by means of serial sections. It appeared that the surviving ovary might attain more than double the weight of a normal ovary because it contained more corpora lutea than each one of a pair of normal ovaries. The so-called compensatory hypertrophy of the remaining single ovary was accounted for by an increased number of maturing and degenerate follicles and corpora lutea. The stroma was not obviously increased. The average number of ova in the right ovaries removed at 20 days of age was 5,405, compared with 5,207 in the surviving left ovaries of the same rats at 41 to 69 days of age, and with 5,927 in the left ovary of the control rats at the time of death. These figures indicate that compensatory hypertrophy of the ovary does not include the development of new and additional ova. The figures also suggest that the solitary ovary becomes depleted of ova more rapidly than is the case with one of a pair of ovaries in the intact rat. Arai (1920 a) further noted that in non-pregnant rats the percentage of larger-sized ova to the total number remains nearly constant. The new formation of ova from germinal epithelium diminishes, he says, at puberty, though it may continue until the rat is 1 year old. The maximum number of ova (35,100) in the two ovaries was present at birth; 23 days later this had declined to a little over 10,000. After ovulation began there was a slow decrease in the number of ova to 2,000 at the age of 31 months.

Asdell (1924) removed one ovary from a number of rabbits and thereafter allowed them to breed freely. They were killed on the 20th day of their 4th pregnancy, and their single ovary was examined histologically. Asdell confirmed the fact that the enlargement of the remaining solitary ovary is caused by the maturation of more follicles than ripen in one of a pair of ovaries in the intact rabbit. Like his predecessors in this kind of experiment he noticed that the average number of young in subsequent litters of the rabbits with a single ovary was slightly less than normal. The total number of ova shed as revealed by counting the corpora lutea was about the same whether a rabbit had one ovary or two; the slightly diminished number of young borne by the mothers with a single ovary was attributable, he found, to a higher mortality among the embryos (Table 11).

TABLE 11.   Fecundity of rabbits after removal of one ovary (Asdell, 1924)

(The rabbits were killed on the 20th day of their 4th pregnancy.)

	Number of rabbits	Average number of ova shed	Average number of live foetuses	Average number of dead foetuses
Unilaterally spayed	8	12·1	8	1·6
Normal controls	10	11·2	9·5	0·3

Lipschütz (1927a), approaching the same problem in a different way, has found that the successful grafting of a third ovary into a rabbit does not lead to any increase in the total number of ovarian follicles which reach maturity.

The effect on the oestral cycle of removing one ovary was investigated by Papanicolaou (1920), using guinea-pigs. He found that after unilateral spaying the length of the cycle was increased only by about 1 day. Clinical experience has proved that in women removal of one ovary causes no obvious change in the duration of the menstrual cycle. Van Wagenen & Morse (1942) say that $12\frac{1}{2}$ per cent of the total amount of ovarian tissue is enough to sustain normal menstrual cycles in the Rhesus monkey.

Studying the factors which control egg-laying, Pearl (1912) counted the oocytes visible to the naked eye in different fowls. As he explained, this method only gives minimum values, but there cannot be fewer oocytes present than those counted. The total visible oocytes present in different fowls varied from 914 to 3,605; and the total number of eggs which had been laid by the different birds varied from 2 to 198. The numbers of oocytes present in the birds at these two extremes of egg-laying were 2,145 and 2,452 respectively. His observations prove that in the fowl, as in the pig, rabbit and rat, fecundity is not in normal circumstances regulated merely by the ovary.

An experiment performed by Pearl & Schoppe (1921) is of special interest in connection with the present subject. They removed between one-half and five-sixths of the ovary in eight fowls, fixed the excised fragments and counted the

TABLE 12. The formation of visible oocytes in the remains of the partially resected fowl's ovary compared with that in the ovary of a normal fowl (Pearl & Schoppe, 1921)

Condition	Number of hens	Total visible oocytes
Ovary partly resected	8	2,381 ± 143
Normal controls	20	1,793 ± 79

visible oocytes contained in them. At death after periods of between 6 and 37 months the oocytes in the regenerated remains of the ovary were counted. The number of visible oocytes produced by normal fowls of the same breed were also ascertained, and it was found that partial resection of the ovary had resulted in an increased formation of oocytes attaining visible dimensions, instead of a decrease as might have been expected (Table 12).

A supplementary experiment was carried out by Steggerda (1928) on pullets 4 months old. From some of these, portions of the ovary varying from a quarter to two-thirds of the whole were removed. In others the ovary was mutilated by a sharp hook or steel brush. In control pullets the ovary was exposed but not injured. The number of eggs laid by each fowl in the ensuing year was recorded and it was found that the ovarian mutilation had hardly affected the result, the egg-laying average being 167·4 among the hens with injured ovaries and 179·5 among those whose ovaries were intact.

One might be tempted to explain some of the facts just quoted by supposing that gonadotrophin is used up in the process of stimulating the ovary so that if only half the proper amount of ovarian tissue is present twice the normal amount of available gonadotrophin will be available for it. Selye (1940a) has obtained experimental results which he believes controvert this explanation. He removed the pituitaries from twenty female rats when they were 4 weeks old, and from ten of them he removed one ovary at the same time. Immediately afterwards treatment with gonadotrophin was begun, daily doses of 1 r.u. of FRH and 1 r.u. of LH being given to one pair, daily doses of 2, 3, 4 and so on up to 10 r.u. being given to the other pairs. When the rats were killed Selye found that no such hypertrophy of the solitary ovaries had occurred, as might be expected if the available gonadotrophin were consumed in the process of ovarian stimulation. With increased concentrations of available gonadotrophin the weights of the ovaries were increased uniformly (Table 13).

TABLE 13.   The effects of gonadotrophin on ovaries of non-spayed and half-spayed hypophysectomized rats (Selye, 1940a)

Rat units of FRH and LH injected	Half of the weight of the two ovaries of the non-spayed rats mg.	Weight of the single ovary in the half-spayed rats mg.
1	7·5	7
2	8·5	7
3	9	7·5
4	14	12
5	8·5	11·5
6	12·5	12
7	18	13
8	17·5	17
9	14·5	18
10	22·5	22

The effects of partial gonadectomy on the mutual relationship between the pituitary and ovarian functions will be referred to once again on a later page from a somewhat different aspect. The effects of the ovarian hormones, oestrogen and progestin, on the gonadotrophic potency and histological conditions of the pituitary also will be discussed later (pp. 51, 57, 404).

The control by the pituitary of the output of oestrogen from the ovary. The formation of oestrogen by the ovary is mainly the consequence of follicular maturation induced by gonadotrophin, and so the pituitary controls the production of ovarian oestrogen, the supply of which represents a speedy reaction. Fels (1930) injected rats with gonadotrophin and removed their ovaries at varying intervals afterwards. In this way he learned that if the spaying were done within 28 hours of the injection no cornification of the vagina occurred; if the ovaries were removed later than $30\frac{1}{2}$ hours after the injection, cornification of the vagina ensued, showing that oestrogen had been produced in effective amount. Zondek (1940c) repeated these experiments on rats and obtained approximately the same results.

Ovarian cysts. Apart from its influence on fecundity, a long-continued imbalance between the hormonal activities of the intact ovary and pituitary may be productive of pathological lesions. An excess of oestrogen on the one hand leads to the development of pituitary tumours, and on the other a relative excess of gonadotrophin may cause ovarian cysts. Lipschütz (1925a) removed one ovary and a large fraction of the other ovary from guinea-pigs and rabbits and found that cysts occasionally formed in the enlarged remnant of the ovary. The primary follicles in such cases, he says, may be completely used up prematurely as regards the duration of the animal's life. Lipschütz (1937 a, b) observed ovarian follicular cysts in the cat also after partial spaying. He believes that the gonadal and generative rhythm is governed by a law of 'constance folliculaire' which is fixed for each species by the concurrence of ovarian and extraovarian factors, and he accounts for the cysts by a prolongation of the follicular phase, the luteal phase being deficient or absent. Wang & Guttmacher (1927) caused similar follicular cysts in the rat's ovary by severe ovarian trauma. Smith & Engle (1927) found that the continued implantation of pituitary tissue into rats also led to the formation of follicular cysts, and Fluhmann (1933) obtained the same result in rats both with a pituitary extract and with a gonadotrophic preparation of human pregnancy serum. Hill & Parkes (1931) have reported the development of ovarian cysts in ferrets under continued treatment with gonadotrophin. Smith & Engle (1927), like Lipschütz, attribute the formation of these ovarian cysts to a disproportion between the available amounts of gonadotrophin and of reactive ovarian tissue. Witschi & Levine (1934) noted that, in female rats living in parabiosis with castrated males and thus supplied with an excess of FRH, luteinization is deficient and many of the follicles become cystic.

The experiments just quoted show that follicular cysts may be produced in the ovary either by a deficiency of ovarian substance or by an excess or irregular supply of gonadotrophin.

Ovarian cysts have been recorded by Cushing & Davidoff (1927) in cases of acromegaly.

The Action of Gonadotrophins on the Testis

The gonadotrophins, though possibly not quite identical in the two sexes, are not sex-specific; they regulate the functions of the testis as well as those of the ovary.

Hypophysectomy is followed by an arrest of testicular function and a reversion of the testis to the infantile state; and these effects can be prevented or repaired to some extent by the injection of pituitary gland or extracts made therefrom. This domination of the testicle by the pituitary is limited, and an increase of the supply of gonadotrophin much above the normal does not produce a corresponding increase of testicular growth or activity. Removal of one testis may be followed by enlargement of its fellow. Lipschütz (1922) excised one testis from each of a number of guinea-pigs with the result that the remaining testis became almost double the normal size. This enlargement is a response to gonadotrophin and represents the swelling of increased activity rather than the acquisition of additional tissue.

The male gonad in one respect differs from the ovary in its response to partial gonadectomy; for the excision of part of a testicle is not followed by great enlargement of the surviving fragment. No equivalent to ripening of follicles occurs in the testis to cause in the organ such large variations in size as take place in the ovary. Evidence has been adduced (Smelser, 1933) that unilateral castration is followed by an increased functional activity of the remaining testis so that spermatogenesis is little if at all impaired, a result which might be regarded as the analogy of that which follows unilateral spaying. Edwards (1940) has tested the effects of unilateral castration in rabbits and concludes that no increase of spermatogenesis occurs in a testis after excision of its fellow. To the author the experiments which Edwards carried out seem to have proved rather that when artificially taxed to their full capacity two testes will produce more spermatozoa than one, which is another matter.

The stimulating effect of gonadotrophin on the size and functional activity of the testes has been shown by numerous experiments. Smith (1926a) made daily transplants of rat pituitaries into male rats, beginning on the 14th day of life. This treatment caused considerable increase in the size of the testes as compared with those of littermate controls (Table 14).

TABLE 14.   The effect of repeated transplants of pituitary on the testes of immature rats (Smith, 1926a)

	Bodyweight (g.)	Weight of testes (mg.)	Weight of accessory genital organs (mg.)	Age at death (days)
Treated	44	254	480	24
Control	43	240	249	24
Treated	74	603	916	31
Control	69	339	409	31

Engle (1932b) treated immature rats with extracts (a) of sheep's pituitary and (b) of pregnancy urine. Both these preparations caused an increase in the weight of the testes as compared with controls, the pituitary being the more effective, producing testes nearly double the weight of those in the control rats. The tubules were larger in diameter than those of the controls. Neither spermatids nor spermatozoa appeared earlier than in the controls, the most striking change in the testes being a large increase in interstitial glandular tissue with advanced development of the accessory genitalia as a secondary consequence. Similar results were obtained in monkeys (Macacus).

Experiments by Smith & Leonard (1933, 1934) have shown that some at least of the effects of hypophysectomy may be counteracted by gonadotrophin. In mature hypophysectomized rats spermatogenesis could be maintained by gonadotrophin if the injections were started at the time of the operation; even fertile mating occurred in these circumstances. If the pituitaries had been removed from immature rats when they were between 31 and 34 days old, tubular development and spermatid formation were induced by gonadotrophin, but spermatozoa were not formed.

By giving gonadotrophin to adults of species which breed only at certain seasons the testes may be aroused into activity at a time when in normal circumstances they are at rest. Moore, Simmons, Wells, Zalesky & Nelson (1934) found that in the ground squirrel (*Citellus tridecimlineatus*) as soon as the mating season is over the testes recede into the abdomen and spermatogenesis ceases for the remainder of the year. If, however, one of these squirrels be given injections of gonadotrophin the testes will descend into the scrotum and spermatogenesis will be caused at any season. Moore and his colleagues removed the pituitary glands from a number of these squirrels at different times of the year, and by implanting the pituitaries into test animals they found that their gonadotrophic potency varied with the seasons, being high at the approach of the breeding period and not detectable during the months of sexual quietude. Baker & Johnson (1936) made complementary observations in the ground squirrel using gonadotrophin prepared from human pregnancy urine. Enlargement and partial descent of the testes occurred in ten and complete descent in two of the twelve squirrels, all of which were sexually inactive when the experiment began; the seminal tubules became enlarged and acquired lumina, and spermatogenesis was present in four; the testes were increased twofold or threefold in weight and the accessory genital organs, including the prostate, seminal vesicle and Cowper's gland, were enlarged and secreting. Courrier & Gros (1934) made the same sort of observations on the alpine marmot (*Marmota marmota*). When this animal is hibernating the testes are small and situated in the abdomen. At this period, after excision of one testis and epididymis, daily injections of chorionic gonadotrophin were given. Under such treatment the testes enlarged and the interstitial cells became swollen and contained granules which were stained by fuchsin and osmic acid. The accessory genital structures were hypertrophied.

In some species precocious sexual maturity can be induced in the male as in the female by gonadotrophin. Burns & Buyse (1934) gave extracts of sheep's pituitary to immature salamanders with the result that their testes increased fivefold in size and began to produce spermatozoa. Wells & Moore (1936) used the ground squirrel. The young of this animal, they say, are born in May and June, and first produce spermatozoa in January of the following year. By administering to immature squirrels fresh pituitary substance or gonadotrophic extracts prepared from the serum of pregnant mares or from the urine of pregnant women they induced spermatogenesis in August, September, October and November. Korenchevsky, Dennison & Simpson (1935) caused precocious sexual development in young male rats by gonadotrophic extracts prepared from pregnancy urine. They suggest that young male rats 22 to 25 days old might be used for the assay of gonadotrophin, the indices being enlargement of the prostate and seminal vesicles.

Price (1936) implanted pituitaries into immature rats with the result that the testes doubled in weight, the tubules became enlarged, and, although no obvious increase of interstitial tissue was observed, the endocrine functions of the testis were stimulated as shown by increased development and secretory activity of the prostate and seminal vesicles.

Descent of the testes. In some of the experiments quoted above the administration of gonadotrophin caused descent of the testes into the scrotum at a season of the year when they would in normal circumstances have remained in the abdomen. Clinically, gonadotrophin is much in use to induce testicular descent in boys whose testes have imperfectly descended. Such treatment is sometimes successful, especially in those cases in which the failure of the testes to descend is associated with other definite signs of hypogonadism. Controlled experimentation is difficult in this field because only in some of the primates, e.g. chimpanzee and man, have the testes descended permanently into the scrotum at birth. In the macaque the testes are in the scrotum at birth, soon after which they return to the abdomen or upper end of the inguinal canal and remain there until puberty, which occurs between the 3rd and 5th years (Wislocki, 1933), when they descend permanently. In this monkey they can be induced to descend prematurely by the administration of gonadotrophin (Engle, 1932 b).

Differential actions of FRH and LH on the testis. A difference has been recorded between the actions on the testes of extracts made from the pituitary and those made from the urine of pregnant women. The former, which are rich in FRH, appear to act mainly on the spermatogenic epithelium, the latter, rich in LH, act principally on the interstitial tissue and through it upon the accessory genital organs.

Bourg (1930b) gave ten daily injections of pregnancy urine to rats, with the result that the testes were enlarged together with the accessory organs, including the prostate, seminal vesicles, preputial glands, epididymis and vas deferens. The interstitial cells were swollen with abundant cytoplasm loaded with lipoid granules. The seminal tubules were unaffected.

Leonard (1933c) noticed that injections of LH into non-castrated rats led to an increase in the size of the accessory sexual glands, including the seminal vesicle and prostate, a fact which points to an enhanced output of androgens by the interstitial glandular cells of the testis. Smith, Engle & Tyndale (1934) prepared extracts (1) rich in FRH from the urine of women who were past the menopause or had undergone gonadectomy, and (2) rich in LH from the urine of pregnant women; and they compared the effects of these two preparations when injected into male rats whose pituitaries had been removed. The results were definite. FRH caused proliferation and maturation of the seminal epithelium without an accompanying increase of the interstitial glandular tissue or any stimulation of the prostate and other accessory genital organs. The LH preparation caused an increase of the interstitial glandular cells of the testis and a consequent development of the accessory genital organs. A combination of FRH and LH caused more spermatogenesis than FRH alone, and there was some response of the interstitial cells with enlargement of the accessory genitalia though this response was less than that which followed the administration of LH alone. Kuschinsky & Tang-sü (1935) gave from 25 to 50 r.u. of LH to male rats 6 to 8 days old, and continued the injections daily for 7 to 10 days. This treatment caused an increase of the interstitial glandular tissue but not of the seminiferous tubules; and there was a great enlargement of the seminal vesicles which showed that the increased interstitial tissue was functionally active. Van Os (1936) and Greep, Fevold & Hisaw (1936) obtained results which confirm the foregoing. Breneman (1936) also, administering FRH or LH to chicks 5 to 15 days old, observed similar effects; FRH caused hypertrophy of the seminal tubules and LH led to an increase of interstitial tissue.

Collip (1935) used hypophysectomized rats and found that extracts rich in FRH from the urine of castrates or women after the menopause caused proliferation of the seminal epithelium and sperm-cell production but had no obvious effect on the interstitial cells, whereas an extract of pregnancy urine rich in LH caused an increase in the size and number of the interstitial cells without affecting the seminal epithelium.

To elicit by experiment the different reactions of the testis to LH and FRH it is advisable to use either very young animals or animals which have been deprived of their pituitaries. McCahey, Soloway & Hansen (1936) reported that when adult male rats, one of which had been castrated, were placed in symbiosis, the excess of FRH derived from the castrated rat induced not only an increased spermatogenesis but caused also an increase in the size and number of the interstitial glandular cells with enlargement of the accessory genital organs.

The observations quoted above suggest that the pituitary and chorionic gonadotrophins are not identical compounds. They also seem to show that gonadotrophin prepared from the placenta or the urine requires the co-operation of the pituitary gland to acquire full biological potency. While the pituitary product may at one time consist chiefly of FRH and at another of LH, the placental product apparently is composed almost, if not entirely, of LH.

The Interstitial Cell-Stimulating Hormone (ICSH)

Evans, Simpson & Pencharz (1937; see also Evans, Simpson, Tolksdorf & Jensen, 1939, and Simpson, Li & Evans, 1942) have described a procedure for separating the gonadotrophic factors of the sheep's pituitary, and by this method they have extracted a hormone which they state is a specific interstitial cell-stimulating hormone (ICSH) which acts upon both the ovary and testis and is distinct from LH. Given to normal immature female rats ICSH does not cause precocious sexual maturity, nor ripening of ovarian follicles, nor luteinization. Given to hypophysectomized adult female rats it causes repair of the interstitial ovarian tissue. In hypophysectomized male rats, ICSH causes repair of interstitial glandular cells with secondary development of the seminal vesicles, prostate and other accessory organs. This action apparently is enhanced by the addition of FRH.

Greep, Van Dyke & Chow (1941) also have isolated from the pituitary an interstitial cell stimulating hormone (ICSH). They say it is a homogeneous protein which stimulates the interstitial tissue of the testis and ovary.

A warning may not be out of place while the effects of various agents on the testis are under consideration. Safe conclusions can hardly be made on such matters unless the location of the testis be observed throughout the experiment; otherwise effects which are the consequence of cryptorchidism or the descent of the testis into the scrotum may be wrongly attributed to a direct action on the gonad of whatever substance is being tested. Crew (1922) suggested that the increased temperature within the abdomen might account for the aspermia of the undescended testis, and his hypothesis was confirmed by the well-known experiments of Moore (1924). Obviously it is essential when considering the action of a hormone on the seminal epithelium to recognize the location of the testis throughout the experiment.

Puberty and the Awakening of Sexual Activity

When an animal has reached puberty, that is to say has become well enough developed to produce offspring, it does not always exercise this function at once. To arouse the reproductive faculty external stimuli are required, the nature of which varies widely with different species. Much thought has been given to the factors that govern seasonal breeding in domestic animals, for the problem is of economical as well as academic interest and has practical implications both for the farmer and the public. In animals which have a well-marked and limited annual breeding season, as John Hunter saw, the variations in size and appearance of the gonads and accessory genital organs at different times of the year are remarkable. Rowlands (1936) states that in the bank vole (*Evotmys glareolus*) the testes, which in the summer have a mean weight of 682 mg., shrink in the winter to less than 40 mg. Other striking contrasts could be mentioned.

Both in seasonal breeders and in those animals which retain the reproductive capacity throughout the year, the gonads after puberty can be aroused into action at any time by the administration of gonadotrophin. A temporary or seasonal abeyance of reproductive activity after puberty may therefore be attributed to a diminished output of this hormone from the pituitary. Moore, Simmons, Wells, Zalesky & Nelson (1934) used the ground squirrel (*Citellus tridecimlineatus*) for a study of the problem. They found that both the male and female have little if any detectable amount of gonadotrophin in their pituitaries during the long part of the year during which their reproductive capacities are at rest. On the approach of the breeding season the pituitary increases in weight and produces gonadotrophin abundantly. At any time during the quiet period the dormant reproductive organs can be stirred into activity by giving gonadotrophin (Wells, 1935). We may conclude that the breeding and non-breeding seasons are largely dependent on the quantity of gonadotrophin produced by the pars anterior of the pituitary gland. The subject is complex and the various species show wide differences in their responses to the same stimuli; nevertheless the vertebrates have this in common that whatever the natural agencies may be which evoke sexual behaviour at the appropriate time, they act through the medium of the anterior lobe of the pituitary gland by stimulating it to produce gonadotrophin.

Relationships between puberty and ovulation. The deciding event which determines puberty in the female may be regarded as maturation of ova. It does not follow that when this stage of development has been reached ovulation, that is to say rupture of the follicles with release of ova, will occur at once. In girls, for example, the onset of puberty is a gradual affair which begins with anovulatory cycles. During adult life in some species, including the dog, monkey and man, ovulation normally follows maturation without the necessity of any extraneous sexual stimulus acting upon the pituitary. In other species, for example the rabbit and ferret, even in adult life, maturation of ova is followed by ovulation only under the influence of some additional sexual excitation, the nature of which varies with circumstances. Mating and suckling are two common ones, either of which is followed by ovulation and the formation of corpora lutea. In the rabbit, if mature follicles are present, ovulation followed by pseudopregnancy can be induced by faradic stimulation of the cervix uteri or even by placing a healthy adult male in an adjoining cage. The experimental basis of these remarks will be mentioned later in connection with the development of corpora lutea and need no further comment here except to say that whatever the excitant of ovulation may be it acts by stimulating the pituitary to increase its output of gonadotrophin (LH). These principles cover a wide field and are applicable not only to primates and other mammals, but also to birds, reptiles, amphibia and fishes.

A plentiful supply of gonadotrophin is not, however, the only factor necessary to arouse the reproductive faculties; the gonads must be able to respond to the gonadotrophic stimulation, which they cannot do until a certain stage of development has been reached. In vertebrates the activity of the reproductive organs will depend on (1) the ability of the gonads to respond to gonadotrophin and (2) the output of sufficient gonadotrophin by the pituitary: the predominant factor after puberty being the available quantity of gonadotrophin.

Precocious puberty. The train of events leading to reproductive maturity is not yet fully understood. In mankind the onset of puberty happens occasionally during early childhood, so that menstruation, enlargement of the mammae and growth of pubic hair may occur in an infant aged only two or three years, the phenomena being accompanied by enlargement and functional activity of the ovaries (Novak, 1944). Rarely the condition is associated with a tumour of the ovary, but more commonly it seems due to a prematurely abundant supply of gonadotrophin, the cause of which remains unknown.

Circumstances affecting the responsiveness of the gonads to gonadotrophin will be discussed later, and we shall now consider some of the various influences by which the output of gonadotrophin from the pituitary is controlled.

Chapter II. *Factors which Influence the Gonadotrophic Activity of the Pituitary*

Age. Afferent nervous stimuli. Changes of external temperature. Seasons. Oestrous cycle. Pregnancy. Gonadectomy. Partial gonadectomy, cryptorchidism and sterilization by X-rays. Gonadal hormones. Gonadotrophins. Sex. Nutrition.

THE pituitary by its power to supply trophins to the gonads is the mainspring of reproductive activity, and it may be of interest to consider some of the circumstances which affect this gonadotrophic function.

Age

The pituitary may produce some gonadotrophin at an early stage of existence. Smith & Dortzbach (1929) implanted pituitaries from foetal pigs into immature female mice. With the pituitaries of foetal pigs having a crown-rump length of 14 to 15 cm., no response was obtained; those from embryos measuring 18 cm. or more gave a positive result, namely follicular maturation, distension of the uterus and opening of the vagina. The gonadotrophic content of the foetal pig's pituitary, as indicated by these experiments, increased rapidly after the foetus had attained a length of 20 cm.

Hellbaum (1935) made comparative tests of the gonadotrophic potencies of pituitaries taken from foetal, immature and adult horses, and from mares and geldings. Positive tests were given by the pituitaries of foetal and immature horses, though the effect was less than that produced by the pituitaries of adults when given in equal quantity. Smith & Engle (1927) showed that infantile rats between 5 and 30 days after birth produced some gonadotrophin, for their pituitaries, when implanted into 17-day-old mice at the rate of two to six pituitaries daily, caused opening of the vagina and other signs of precocious maturity; and Clark (1935) by similar methods found that a sharp rise in the gonadotrophic potency of the rat's pituitary occurred between the 13th and 20th day of postnatal life. Wolfe & Cleveland (1931) tested the capacity of anterior pituitary extracts prepared from immature and mature rabbits to induce ovulation in sexually mature rabbits. The extracts were injected into the marginal ear vein of the test animal, whose ovaries were examined 24 hours later. Using amounts of extract which represented equal weights of pituitary gland, they found that the pituitaries of rabbits aged from 10 to 14 weeks were as potent, or nearly so, as those of adults. On the other hand the pituitaries of rabbits 4 or 5 weeks old, given in comparable doses, always failed to cause ovulation.

Bergman & Turner (1942) also investigated the gonadotrophic potency of the rabbit's pituitary at different ages, by implanting the pituitaries into baby chicks and noting the changes induced in their gonads. By such means they found a small potency in the pituitaries of very young rabbits. After a bodyweight of 1,000 g. had been acquired the gonadotrophic potency gradually rose, its maximum being attained when the bodyweight reached 1,500 g. in the females and between 2,000 and 2,500 g. in the males.

Kallas (1929a) placed immature female rats weighing 15 to 20 g. in parabiosis, one of each pair having previously been spayed; this led to precocious puberty in the non-spayed partner, from which he concluded that the pituitary of the infantile rat is already able to produce gonadotrophin and that the ovary can react to it by maturation of follicles. Martins (1931) carried out a similar procedure with the same result, namely precocious puberty in the non-spayed partner within 6 or 7 days of the operation.

The influence of age on the output of gonadotrophin in man has been the subject of several investigations, which show that before puberty there is some formation of gonadotrophin the amount of which is increased during the period of sexual activity and may rise still further after the end of this period. Österreicher (1933) tested the urines of a large number of women who were over 50 years old for gonadotrophin and obtained positive reactions in more than 50 per cent (p. 5). As will appear, the nature of the gonadotrophin produced by women in old age is not the same as that of younger individuals, for it consists almost exclusively of FRH.

Katzman & Doisy (1934) tested on infantile mice the urine from individuals of various ages. They state that in boys and girls before puberty the amount of gonadotrophin excreted is very small, and that it increases considerably at puberty. They found that urine from women after the menopause gives an FRH response alone, causing maturation of follicles without the appearance of corpora lutea or blood spots. On the other hand the urine from sexually active women usually contains a large but variable proportion of LH. Pedersen-Bjergaard (1936) estimated the excretion of gonadotrophin in normal sexually mature women and in children. The women's urine was taken during the intermenstrual period, and it was found to contain at least 7 times as much gonadotrophin in a litre as the urine of children of 9 years.

Henderson & Rowlands (1938) assayed 100 human pituitaries from subjects of various ages for gonadotrophic (FRH) activity. They dried the pituitaries in acetone for 48 hours and after desiccation made suspensions of the powdered product in slightly alkaline water. For comparison a standard preparation was made in a similar way from the pituitaries of nine men and five women aged 50 to 65 years. The preparations were tested on immature rats, weighing between 40 and 50 g., which were injected daily for 5 days and killed 24 hours after the last injection. The amount of each pituitary required to give a significant response in the rats, as measured by the weights of their ovaries, was then compared with the amount of the standard preparation necessary to give the same response. The results were as follows. During early childhood the gonadotrophic potency of the human pituitary is slight; it increases slowly until the end of sexual activity. In women at the menopause and in men between the ages of 60 and 70 there is a sudden rise of gonadotrophic (FRH) potency, as measured by the weight of the ovaries in the test animals.

In addition to the quantitative differences in the output of gonadotrophins which accompany the various stages of life there are, as already indicated, qualitative differences. Soeken (1932) assayed the urine of a number of children and found that the small amount of gonadotrophin excreted by them consisted entirely or almost entirely of FRH; and in men of advanced years, as with women after the menopause, there is not only an increased total output of gonadotrophin but the latter consists almost entirely of FRH (Fluhmann, 1931). Perhaps the absence of LH in all these circumstances is attributable to the lack of some substance derived from the ovary or testicle.

Afferent Stimuli acting through the Central Nervous System

Although the transmission of stimuli from the hypothalamus to the pituitary is, in part at least, humeral, the nervous connections between the hypothalamus and the sensory organs of the body certainly are essential for the complete co-operation of the pituitary with the reproductive organs. Sight, hearing, smell, cutaneous sensations, and it may be other feelings, conveyed by direct nervous pathways or through the less direct psychological circuits of the brain, may have pronounced effects on the gonadotrophic activity of the pituitary. The matrimonial plumage of male birds and its display, their songs, their performances on the wing, are love spells; so too are the antics of the male hare in spring, the sexual adornments of the newt and stickleback, and the perfume emitted from glands specialized for the purpose in some of the mammals. These and other natural contrivances arouse and focus the awakening sexual powers at adolescence or the return of the breeding season. Nor is the impulse to secure and stimulate sexual attention confined to the male sex. The female too has her attractive ways, albeit these are often of a more subtle and less ostentatious kind than those used by the male.

The efficacy of the various kinds of sexual appeal differs widely among different species. The subject cannot well be discussed comprehensively because most of our knowledge concerning it has not yet been submitted to critical investigation in the laboratory, where visual impulses seem to have been the only afferent stimuli to have received an adequate and detailed study. When we consider the great part played by the olfactory sense in the life and conduct of many animals it seems certain that odours must be regarded as important agents in stimulating the reproductive faculties. Sound also plays a part. Who can doubt that the voice of the male bird is used, not only as a challenge to rivals, but as an appeal to the sexual sensibilities of his intended or acquired mate? As to cutaneous sensations, the habit of caressing is clearly a natural part of love making with many species.

The following discussion will be directed mainly to the effect of visual impulses. These may act on the pituitary simply in the form of light, or in a more complicated manner through the psychological apparatus of the brain.

(a) *Light.* A longer exposure to daylight comes naturally to the mind when we try to explain why in seasonal breeders the dormant reproductive faculties so commonly awake in the spring. To investigate the matter Rowan (1925, 1929) trapped North American migratory sparrows (*Junco hyemalis*), and keeping some of them in out-door unheated aviaries he examined their gonads at different dates. He learned that in normal circumstances the testes of these birds vary in length from 0·5 mm. in midwinter to 6·4 mm. in May. Some of the Juncos he kept in the open-air aviaries from September onward, exposing them daily to the light of 50-watt electric lamps. The duration of this artificial lighting was increased 5 minutes each successive day to imitate the lengthening days of spring. The birds remained in good condition and individuals were killed from time to time; by the end of December their testes had become much enlarged. Any increased temperature from the electric lamps had been excluded by the conditions of the experiment, and it seemed at first that the unseasonable development of the gonads must be attributed to the increased exposure to visible light. Rowan, however, thought that the abnormal condition of the gonads might in part be the consequence of the greater bodily activity and reduced hours of sleep during the artificially prolonged period of daylight.

In London, Rowan (1937, 1938) noticed starlings roosting on public buildings, and it seemed to him that the feeble light falling on them at most of their roosts must be far below the intensity shown by various investigators to be effective in causing activity in the gonads of birds. Most of these starlings were roosting over main traffic routes and were being sufficiently disturbed, Rowan thought, by the tumult below to be kept awake through much of the night. Early in February he secured specimens of these city starlings and compared their gonads with those of starlings obtained one week later from the country; and he found that the testes and ovaries of the London birds were considerably larger than those of their relatives from the country (Table 15).

TABLE 15. Size of testes in starlings roosting in London and in the country (Rowan, 1937, 1938)

	Number of birds	Date	Average weight of testes (mg.)	Average dimensions (mm.)	Maximum dimensions (mm.)	Minimum dimensions (mm.)
London starlings	12	10 Feb.	202	7·33 × 5·0	17·8 × 9·8	4·8 × 3·8
Country starlings	15	17 Feb.	26	3·23 × 2·2	4·8 × 3·0	1·8 × 1·4

Bissonette (1930 a, b, 1931) observed the influence of light and exercise on spermatogenesis in the starling, and found that increased exposure to light stimulated testicular activity but that increased exercise and wakefulness did not have this effect, and he noticed further that the influence of light on spermatogenesis in these birds depended not so much upon its exact intensity as upon its relative intensity compared with that to which they had been previously exposed.

Bullough & Carrick (1939) suggest that country starlings may not be good controls to those caught in London. Between 11 and 21 March starlings collected from a residential quarter of Leeds had large fully developed testes. Starlings from a country plantation at the same time were immature with small testes. Bullough & Carrick think that the latter birds were perhaps continental migrants.

Kirschbaum (1933; Kirschbaum & Ringoen, 1936) found that increased hours of illumination caused premature enlargement of the testes and spermatogenesis in the house sparrow (*Passer domesticus*). On the other hand, reduction of the hours of illumination in the spring, though affecting the testes, did not entirely prevent spermatogenesis; from which they conclude that increased illumination is not the only factor concerned in the awakening of the sparrows' reproductive faculties in the spring.

Benoit (1935 *a*, *b*, 1938 *b*) learnt that exposure of immature ducks to artificial white light stimulated the development of their gonads, and that this effect was produced even after division of the optic nerves or removal of the eyes, but not after hypophysectomy. Benoit showed, by implanting them into immature female mice, that the gonadotrophic potency of the ducks' pituitaries had been enhanced by the additional illumination (Table 16).

TABLE 16.   The effect of illumination on the gonadotrophic potency of the duck's pituitary as tested on immature mice (Benoit, 1935 *a*)

Treatment of mice	Number of mice	Average weight of uterus and vagina (mg.)
None	5	8·5
Two pituitary implants from non-illuminated ducks	10	12
Two pituitary implants from illuminated ducks	10	48·25

By exposing immature ducks to light filtered through coloured screens Benoit & Ott (1938) found the maximum effect was obtained with orange-red screens. Blue rays had a negligible influence (Table 17).

TABLE 17.   Gonadal development in immature ducks exposed to visible light of various wave-lengths (Benoit & Ott, 1938)

Screen	Average weight of testes (g.)
Blue (Corning, 556)	4·82
Green	20·36
Yellow (Corning, 348, 440)	44·36
Orange (Corning, 246, 428)	63·86
Red	63·58
Red limit (Corning, 348, 585)	8·98
Near infra-red (Corning, 254)	2·13
Untreated controls	2·87

Benoit (1938*a*) discovered that by removing the contents of one orbit and irradiating the pituitary through the intervening thin osseous wall the same enhanced development of the gonads could be brought about. Further, by using a quartz rod to conduct light to the pituitary itself, it was shown that the blue rays were as effective as the red in stimulating the pituitary; rays of a greater length than 8,500 Å. had little or no such effect.

There is evidence, besides that adduced by Benoit, that intact visual pathways are not essential in birds for the enhancement of gonadotrophic activity under the influence of light.

Parhon & Coban (1935) in midwinter blinded five drakes. The birds were killed together with a control on the following dates: one on 22 February, another on 3 March, and the remaining three on 7 March. As shown in Table 18 the testes of the blind drakes were larger than those of the normal controls.

TABLE 18.   Weights at death of testes of blinded drakes compared with those of normal controls (Parhon & Coban, 1935)

	Weights of testes			Weights of birds	
	Lightest (g.)	Heaviest (g.)	Average (g.)	Lightest (g.)	Heaviest (g.)
Blinded drakes	13·5	35	24·5	1,289	1,500
Normal control drakes	5·5	19	12·9	1,297	1,822

Ivanova (1935) experimented with sparrows. She used three groups: (*a*) were exposed to light for 9 hours daily; (*b*) wore silk caps with opaque screens over the eyes and were exposed to light 14 hours each day; (*c*) also were hooded but there were slits in the silk caps which admitted light to the eyes. The experiments started in the beginning of February and lasted for 6 weeks, when the sparrows were killed. The results are shown in Table 19.

TABLE 19. The effects on testes of exposing sparrows to prolonged illumination (Ivanova, 1935)

Group	Conditions	Hours of lighting	Average testicular dimensions (mm.)
(*a*)	Controls	9	2·38 × 1·80
(*b*)	Hooded and blindfolded	14	5·03 × 3·56
(*c*)	Hooded, not blindfolded	14	5·66 × 4·73

Increased illumination has been found to affect the reproductive organs in mammals. Normally the female ferret remains anoestrous from August until March. Bissonette (1932) submitted ferrets to the light of a 200-watt lamp with an intensity of 2,660 lumens daily from 5 to 11 p.m. The experiment began on 12 October. Within 38 to 64 days the females showed full oestrus, and mating took place within 59 to 70 days, though this was followed only by pseudo-pregnancy. These results have been confirmed by Hill & Parkes (1933) and Marshall (1940). Marshall & Bowden (1934, 1936) found that the effect of in-creased illumination in accelerating oestrus in the ferret was almost confined to the rays of the visible spectrum. Whitaker (1936) exposed mice (*Peromyscus leucopus noveboracensis*) to increasing artificial light with a result like that obtained with ferrets. Thirty-five of these mice were exposed to a strong artificial light for daily periods gradually increasing from 13 to 18 hours. Twenty-eight control mice of the same species were kept at the same temperature but without extra light. The animals exposed to the light became sexually active 6 to 8 weeks earlier than the controls, and anticipated them in the production of litters.

Baker & Ranson (1932) studied the effect on field mice (*Microtus agrestis*) of shortening the daily period of exposure to light from 15 to 9 hours. Reproduction was interfered with, the females being most affected. In the males the testes became smaller though most were still able to produce active sperms.

The effects of light on the response to gonadotrophin of the reproductive organs of male and female rats have been tested by de Jongh & Van der Woerd (1939). During June and July they kept one batch of rats in the light and another batch in the dark for 30 days. During the last 10 days some of the rats in each batch were given 5 r.u. of placental gonadotrophin; at the end of this period all the animals were killed and the weights of their reproductive organs were ascertained (Table 20). It will be seen that illumination caused an increase in the size of the gonads and their accessories. When, however, the experiment was repeated during the winter months (November and January) the exposure to additional light had no obvious effect on the gonads.

Before quitting the subject of increased light as a stimulus to the production of gonadotrophin by the pituitary or to its efficacy on the gonads, one or two

TABLE 20.   The influence of light on the action of gonadotrophin in rats
(de Jongh & Van der Woerd, 1939)

Treatment	Limits of organ weights in mg.			
	Ovaries	Uterus	Testes	Seminal vesicles
Light only	18–30	48–118	1,350–1,750	28–40
Dark only	10–20	25–44	320–610	6–10
Light *plus* Gonadotrophin	34–69	174–274	1,525–1,900	110–364
Dark *plus* Gonadotrophin	27–43	95–233	550–1,050	19–146

miscellaneous examples may be quoted because of their interest as curiosities. Rowan (1938*b*) mentions two examples in which birds have been exposed to increased illumination to induce them to sing at a period of the year when normally they are silent. Quoting from Miyazaki Rowan says that a practice called 'Yogai' has been in vogue in Japan from early times. Pet birds (e.g. *Zosterops*) are exposed to artificial illumination for 3 to 4 hours after sunset towards the close of the year to bring them into singing condition in January instead of in the spring. The other example Rowan has taken from Van Oordt, who says that an old custom in Holland was to put birds into the dark in June, exposing them to light again in September, so that they came into full song in the autumn and were then used as decoys by birdcatchers.

Buckner, Insko & Martin (1934) have reported that white leghorn fowls raised without exposure to light had smaller testes and larger combs than normal. If exposed to light the testes became larger and the combs became more erect but smaller.

The conduction of afferent impulses to the pituitary.   Le Gros Clark, McKeown & Zuckerman (1939) have tried to define the nervous pathway by which retinal stimuli might travel in order to arouse the pituitary of the ferret into sexual activity. In November and December, during sexual rest, the visual pathways were divided at various levels in a number of female ferrets, and the animals were then exposed to the additional illumination provided by 100-watt electric bulbs daily from 4.30 to 11 p.m. The results were judged by the appearance or non-appearance of oestrus (Table 21).

TABLE 21. The effects of dividing the visual pathways on the stimulation by light of the gonadotrophic activity of the ferret's pituitary (Le Gros Clark, McKeown & Zuckerman, 1939)

Nature of operation	Acceleration of oestrus
None (controls)	+
Division of optic nerves	−
Removal of visual cortex (occipital lobes)	+
Removal of visual cortex and destruction of superior colliculi	+
Bilateral section of optic tracts	+

From these experiments Le Gros Clark and his colleagues conclude that the primary receptor for the gonadal response to light in ferrets is the retina. Any effects of an alteration of bodily activity induced by the operation could be eliminated because the ferrets with divided optic nerves and those in which the visual cortex had been removed showed equal activity. It follows that the normal response of the pituitary to retinal stimulation depends on impulses passing either to the ventral nucleus of the lateral geniculate body or to the subthalamus by way of the accessory optic tracts.

For normal sexual function it appears that both hormonal and nervous conduction from the hypothalamus to the pituitary are required. Brooks (1938, 1940) pointed out that there are nervous connections between the hypothalamus and the pituitary and that if, in a rabbit, the pituitary stalk is severed so as to divide these nerves, ovulation does not follow mating as it would if the pituitary stalk were intact. On the other hand, several experiments have shown that the transmission of stimuli from the hypothalamus to the anterior lobe of the pituitary is largely humoral. Taubenhaus & Soskin (1941) found that pseudopregnancy could be induced in the rat by applying acetylcholine with prostigmine directly to the pituitary. The application of atropine to the pituitary prevented the induction of pseudopregnancy by electrical stimulation of the uterine cervix. Dempsey (1939) reported that in guinea-pigs normal oestral cycles occurred after section of the pituitary stalk; and it has been shown that transplantation of the pituitary to another part of the body so that all its nervous connections are severed does not prevent it from forming gonadotrophin in response to suitable peripheral stimuli. Schweizer, Charipper & Haterius (1937) removed the pituitaries from adult female guinea-pigs and at the same time grafted half of the anterior lobe of the pituitary taken from another female into the anterior chamber of each eye. The grafts became established. The ovaries underwent follicular development, and though maturation was incomplete and ovulation was absent, a constant oestral state ensued which suggested that the grafted pituitaries were forming gonadotrophin. Schweizer, Charipper & Kleinberg (1940) hypophysectomized male and female guinea-pigs and transplanted the animals' own pituitaries into their eyes. In the females the results were the same as in the earlier experiment—the transplanted pituitaries produced FRH only. In the males the reproductive organs were well maintained, the seminal vesicles were large and distended, and spermatogenesis continued. Apparently the sex of the animal from which the grafted pituitary had been obtained made no difference to the results.

(b) *Psychological stimuli.* Probably the role of psychological stimuli in arousing the sexual activity of the pituitary has an increasing importance according to the scale of cerebral development in the different classes of animal, and will be greatest in man of whom Shakespere said:

'Love looks not with the eyes but with the mind.'

However, there is evidence that, even in creatures of a relatively low grade of intelligence, what may be termed psychological impulses are significant.

An experiment by Matthews (1939) on pigeons illustrates this kind of reaction. Harper (1904) had learned that in ordinary circumstances an isolated female pigeon does not ovulate, whereas when two females are kept together in the same cage both may begin to lay eggs. To determine what stimulus leads to ovulation Matthews arranged virgin pigeons at springtime in the following conditions: (1) a male and female together in the same cage, (2) a male and female in the same cage but separated by a glass partition, (3) two females together in the same cage,

(4) a female alone in a cage with a mirror in which she could see her own reflection and (5) a female alone in a cage without a mirror. The results set out in Table 22 show that in this instance the primary stimulus inducing ovulation was a visual one, and it seems that the pathway from retina to pituitary must have traversed the higher levels of the brain.

Table 22.   Induction of ovulation in pigeons by visual stimuli
(Matthews, 1939)

Arrangement	Ovulation	Interval between beginning of experiment and egg-laying (days)
(1) Male and female together in cage	+	9
(2) Male and female in one cage but separated by a glass partition	+	12
(3) Two females together in one cage	+	52 and 56
(4) Female alone in cage with mirror	+	11
(5) Female alone in cage without mirror	−	

A comparable happening is seen in the rabbit, which when isolated from her fellows does not ovulate. If she is kept in a cage near another containing a buck or in a cage with other females she is apt to ovulate and pass through a period of pseudopregnancy.

Another phenomenon of the same kind concerns the number of eggs laid by birds in the nesting season. If, before her clutch is completed a bird is deprived, by the systematic removal of eggs, of the normal number on which she is accustomed to brood, she will often continue to lay. Jesse (1835) says that a long-tailed tit laid thirty eggs in succession because an observer continually depleted the nest. He says also that the lark, which normally lays five eggs at a nesting, will continue to lay if only one or two eggs are allowed to remain in the nest, but if three remain she will sit.

Violent disturbances of the nervous system are generally regarded as antagonistic to reproduction. The writer, when a medical student, was taught that the commonest cause of amenorrhoea in a healthy young unmarried woman was fright; the possible effect of emotional disturbance in causing miscarriage or amenorrhoea has long been recognized by the clinician. Specific examples are described by Loeser (1943).

Changes of External Temperature

The frequency with which in the animal world sexual awakening occurs in the spring leads one naturally to regard the possible effects of a rising external temperature. Great discrepancies between the reactions to heat and cold between different species are seen. In this country some fish breed in the winter and others in the summer, so that the close seasons for angling vary widely for the different species. The trout becomes sexually active in the colder part of the year. On the other hand Craig-Bennett (1931), studying the reproductive cycle of the stickleback (*Gasterosteus aculeatus*), found that, whereas increased illumination had no influence and diet but a minor one on the appearance of the secondary sexual characters, the temperature of the water was a predominant factor. In the stickleback the degree and duration of the secondary sexual manifestations varied with the temperature, a sufficient rise of which was followed by full testicular development and activity.

In mammals, too, contrasts may be seen between the effects of external temperature on the reproductive faculty of different species. With many kinds of mammal which breed only at one season the warmer part of the year is the one chosen for mating; with others cold appears to act as the stimulus. Moore, Simmons, Wells, Zalesky & Nelson (1934) investigated the sexual cycle of the ground squirrel, which in its natural environment hibernates from November to April. Exposure to light and increased warmth during this period caused no gonadal activity. If kept in cold and darkness for several months these squirrels could be induced to show the sexual awakening at any time of the year. Ogle (1934) found that mice kept at a temperature of about 18° C. showed more fertile matings and an earlier onset of sexual life than mice kept in a moist heat of 31–33° C.

It seems likely, since biological events are purposeful, that the reactions of the pituitary are subsidiary to the welfare of the young which must be born at that season most favourable to their survival. Great differences in gestational periods have to be allowed for to attain this end in mammals. The advantage gained by the adaptability of the pituitary to such requirements is obvious; the methods by which the adaptability has been acquired and the mechanisms by which the pituitary reactions are achieved remain obscure.

Seasons

In some species, though not in all, there is a bias toward reproduction at a certain time of the year without regard to surrounding conditions. If a bird or mammal normally resident in the southern hemisphere is transferred to the northern, it may at first show sexual awakening at a season inappropriate to its new surroundings but in accordance with the annual cycle to which it was accustomed in its southern home. The same is true of animals removed from the northern to the southern hemisphere. The established habit which this phenomenon denotes is not equally potent in all species. Some animals adapt themselves almost at once to the new conditions; others take longer, and some continue indefinitely to breed in the same calendar months as before (Baker & Ranson, 1938). The budgerigar is a familiar example of the last. Such an inherent seasonal bias must be borne in mind when interpreting experimental results. The subject has been reviewed by Marshall (1936, 1942) and Rowan (1938 b).

It has been thought by some that evidence of a seasonal variation in fertility could be detected in man. Certain races in particular, for example the Esquimaux, have been said to display a tendency of this kind, more children being born in one period of the year than in another. Careful inquiry, however, has failed to establish a sound basis for this belief, and it now appears indisputable that in man there is no seasonal variation in the reproductive faculty such as occurs among lower animals.

Oestrous Cycle

Variations of the gonadotrophic potency of the pituitary occur during the sexual cycle and no doubt are among its causes. When considering this fact it is to be remembered that both oestrone and progesterone have an inhibitory effect on the gonadotrophic activity of the pituitary. It may be that in some species the highest degree of inhibition is caused during the luteal stage, while in others the maximum inhibition coincides with maturation of the ova and the maximum output of oestrogen. Another point to bear in mind is that the actions of oestrone and progesterone on the pituitary are not quite identical in character.

Among experiments done to elucidate the gonadotrophic potency of the pituitary at different stages of the oestrous cycle are the following. Smith & Engle (1929) transplanted the anterior pituitaries taken from female guinea-pigs at different stages of the oestrous cycle into immature mice and found that the results varied according to the oestral stage of the donor at the time when the pituitary was taken. Schmidt (1937) obtained similar results. He implanted the pituitaries of adult virgin female guinea-pigs subcutaneously into immature female guinea-pigs weighing between 190 and 210 g. No effect on the ovaries or uterus was produced by implanting pituitaries taken from guinea-pigs in oestrus, and a maximal effect followed the use of pituitaries removed shortly before the onset of oestrus. Wolfe (1930) tested the capacity of the sow's pituitary to cause ovulation in the rabbit. The cyclical condition of the sow at death was determined by the ovarian condition. It appeared from these experiments that the gonadotrophic potency of the material used varied greatly in the course of the oestrous cycle (Table 23).

TABLE 23. Variations in the gonadotrophic potency of the sow's pituitary during the oestrous cycle (Wolfe, 1930)

Condition of pituitary-donor's ovary	Amount of pituitary required to induce ovulation in the rabbit (mg.)
(1) Absence of active corpora lutea. Follicles of 6–8 mm. diameter	1
(2) Absence of active corpora lutea. Follicles of 10 mm. diameter	20
(3) Active corpora lutea. Small resting follicles only	40

It will be seen from this table that the pituitary attained its highest degree of gonadotrophic potency in the absence of corpora lutea, and before the follicles were fully developed, and its lowest degree when the corpora lutea were in full activity. An intermediate degree was reached at oestrus. Variations in the output of gonadotrophin during the menstrual cycle in a woman have been recorded by Frank & Salmon (1935 a), who found that the greatest concentration of gonadotrophin in the blood occurred between the 9th and 12th days of the cycle, and that the greatest excretion by the kidney took place between the 10th and 14th days.

Another aspect of the oestrous cycle and the variations of pituitary activity which accompany it was shown by Lane & Hisaw (1934), who killed rats at various stages of the oestrous cycle and implanted their pituitaries into normal 22-day-old rats. From the results they concluded that in pituitaries during dioestrum FRH predominated in amount over LH and during oestrum LH was in excess of FRH.

The changing phenomena of the oestrous cycle might be regarded somewhat as follows. First the anterior lobe of the pituitary is aroused by appropriate seasonal or other conditions; the nature of the conditions effective for this purpose varying in the different species. The pituitary, thus awakened, produces FRH. This hormone reaching the ovary through the bloodstream causes the graafian follicles to ripen with a consequent production of oestrogen. The latter, having attained a certain concentration in the blood, reacts on the pituitary in two ways; first, it inhibits the output of FRH; secondly, it leads to the production of LH, which converts the follicles into corpora lutea and gradually stops the further supply of oestrogen from the ovary. The corpora lutea elaborate progestin, which facilitates the excretion of oestrogen by the kidneys and at the same time inhibits the production of FRH, so that as long as progestin is present in the blood in sufficient concentration, ripening of graafian follicles with the production of oestrogen is impeded. The end of the cycle may be regarded as coincident with a diminution or cessation of the output of progestin when the corpora lutea degenerate. In animals which are subject to a regular succession of oestrous cycles, the pituitary, now freed from the restraint of progestin, begins to elaborate new supplies of FRH and so a fresh cycle is set going. The process doubtless is not quite so simple as this, but enough facts are known perhaps to justify the description as a temporary guide.

In some species in which the entire oestrous cycle occupies but a few days, as happens for example in rats and mice, it may be that little or no progestin is formed. Certainly one can imagine the occurrence of short oestrous cycles, under the influences which are known, in the absence of progestin.

Cyclical variations in the gonadotrophic potency of the pituitary in birds have been observed by Riley & Fraps (1942), who assayed the pituitaries of fowls on immature mice and found that the pituitaries of laying hens were less potent than those of non-layers. The pituitaries of hens with maturing follicles had about the same potency as those of hens whose ovaries were in an early regressive phase.

Apart from the mechanisms just mentioned there appears to be a tendency to a rhythmical recurrence of oestrum (Marshall, 1936). Swezy & Evans (1930) observed that the cycle of ovogenesis in the rat is not interrupted by pregnancy, but continues with periods of 4 to 5 days, ripe follicles with smaller degenerating ones being present at the end of each period. From each batch of ripe follicles new small corpora lutea form which are seldom more than one-third the size of the functional corpora lutea of pregnancy. No oestral smears are obtained from the vagina during gestation, though at the end of each cycle the smears are prooestral in character. Nelson (1929) has recorded the continuance of oestral cycles in the rat during pregnancy and has noted that copulation will be permitted during each oestral phase. The writer (Burrows, 1941) has shown that in rats and mice copulation shortly before parturition may lead to impregnation and the subsequent birth of a litter. Heape (1901) and Hammond & Marshall (1925) noted that in several species pregnancy is not a bar to copulation.

Wade & Doisy (1935a) noted a cyclical variation in the rat's responsiveness to oestrogen. Oestriol in daily doses varying from 0·65 to 13·0γ were given to spayed rats for 180 days and during this period vaginal smears were taken. These revealed the occurrence of irregular oestrous cycles varying in length from 5 to 10 days, and interspaced with periods of dioestrus lasting up to 30 days or longer.

Zuckerman (1937b) has made the same sort of observation on a spayed rhesus monkey which was under treatment with daily doses of 100 i.u. of oestrone.

The mechanism of seasonal influences on the pituitary and sexual activity awaits elucidation. Heape (1901), who appears to have been the first to identify and name the different stages of oestrus, pointed out that some animals (monoestrous) have but one oestral season in the year, whereas others (polyoestrous) have repeated oestrous cycles.

Pregnancy

To test the gonadotrophic potency of the pituitary in various conditions Bacon (1930) used the pituitaries of freshly slaughtered cattle. The anterior lobes were separated from the posterior and cut into small pieces which were then implanted into infantile mice weighing not more than 10 g., precocious oestrus being the criterion of gonadotrophic effect. By this method the hormone content of the pituitary was found to be reduced during pregnancy. Hill (1932) arrived at a similar conclusion by a different method. He studied female rats which had been joined in parabiosis. When two normal females are united in this way their oestrous cycles are not affected; if, Hill says, one of them is pregnant her fellow goes into a state of continued anoestrus, a condition which suggests a deficiency of gonadotrophic hormone consequent on the pregnancy of the partner. Philipp (1930) tested by implantation the pituitaries of four men and twelve non-pregnant women, dead from various causes, for gonadotrophic potency. The test animals showed positive responses in every instance. He carried out similar tests with the pituitaries of ten women who had died during pregnancy or shortly after parturition. Nine of these women had reached the 7th month or a more advanced stage of pregnancy, and in all these cases the test animals failed to respond. The remaining case was that of a woman who aborted at about the 3rd month, and in this instance a doubtfully positive test was obtained. Lipschütz & Del-Pino (1936) likewise found little or no gonadotrophic potency in the pituitary of a pregnant woman.

Possibly the diminished production of gonadotrophin by the pituitary in the later stages of gestation is caused by placental activity. Heidrich, Fels & Mathias (1930) showed the absence of gonadotrophin from the pituitary of a man with chorionepithelioma of the testis by implanting portions of the pituitary into immature female mice, whereas implants of the tumour itself caused positive responses. The patient before death had been excreting large amounts of gonadotrophin in the urine.

Gonadectomy

Fichera (1905a) was the first to discover a physiological connection between the gonads and the pituitary. He showed that removal of the gonads in several different species and in both sexes was followed by enlargement of the pituitary accompanied by histological changes (Table 24).

Hatai (1913a, b) noticed that in normal adult female rats the pituitary is larger than in the male. After spaying no appreciable enlargement of the pituitary occurred, whereas castration in the male caused the pituitary to increase in weight by 73·62 per cent. Engle (1929b) removed the gonads of a number of male and female rats, some of which were between 20 and 30 days old, the others being

TABLE 24.   The effects of gonadectomy on the weight of the pituitary gland, as recorded by Fichera (1905 a)

Animal	Number examined	Castrated	Mean weight of pituitary glands (g.)
Cock	50	−	0·133
Capon	50	+	0·267
Bull	5	−	2·35
Ox	5	+	4·40
Buffalo	5	−	1·80
Buffalo	5	+	3·45
Female guinea-pig	2	−	0·142
Female guinea-pig	3	+	0·177
Female rabbit	2	−	0·170
Female rabbit	3	+	0·245

over 1 year old at the time of the operation. Eight or 9 months later the rats were killed and their pituitaries were implanted into immature rats and mice, the results being compared with those obtained by pituitary implants taken from normal rats. Using the increased weight of the ovaries in the test animals as an index of pituitary potency Engle found that the pituitaries of the gonadectomized rats showed a greater effect than those taken from normal rats. No difference of effectiveness was detected between the pituitaries of male and female donors. Evans & Simpson (1929 b) performed a somewhat similar experiment; they removed the testes or ovaries from rats at the age of 7 or 9 months, killed the animals 2 months later and implanted their pituitaries into female rats 24 days old. Two pituitaries were given each day for 2 days, the recipients being killed 2 days later. The effect of cryptorchidism in male rats on the pituitary potency was tested in the same way. The results are shown in Table 25, from which it appears that, though smaller, the pituitary of the normal male rat has a higher gonadotrophic potency than that of the female, and that gonadectomy in both sexes leads to an increase of the gonadotrophic potency of the pituitary. Cryptorchidism has a similar though less pronounced effect.

TABLE 25.   The effects of gonadectomy and cryptorchidism on the size and gonadotrophic potency of the rat's pituitary (Evans & Simpson, 1929 b)

	Nature of pituitary donor				
	Normal female	Spayed female	Normal male	Castrated male	Cryptorchid male
Average weight of the four implanted pituitaries (mg.)	47·6	49·3	34·4	48·8	43·6
Percentage of recipients showing oestrous vaginal smear within 4 days	16	100	100	100	100
Average weight of the two ovaries of the recipients at end of 4 days (mg.)	19·5	113·5	69·5	176	129

Severinghaus (1932) deprived male and female guinea-pigs of their gonads, and noted that the operation was followed by an enhanced gonadotrophic potency of the pituitary; and the same result was seen in the rabbit (Smith, Severinghaus & Leonard, 1933). Nelson (1935 c) spayed female guinea-pigs and castrated males or rendered them cryptorchid. After an interval of time the animals were killed and the gonadotrophic capacity of their pituitaries, as compared with those of controls, was tested by implanting them into immature rats and mice. The assays showed that the gonadotrophic capacity of the pituitary was considerably enhanced in the cryptorchid males as well as in the males and females whose gonads had been removed.

The effect of gonadectomy on the pituitary is not limited to the quantity of gonadotrophin which it excretes; the quality also appears to be changed so that FRH now greatly predominates over LH. Zondek (1932) observed an increased output of FRH in the urine of women within 2 weeks of the removal of both ovaries; and in men, too, an increase of FRH was found in the urine after bilateral castration. Hellbaum (1933) injected extracts prepared from the pituitaries of geldings into infantile rats and found that the characteristic effect was a great follicular development without luteinization. In a total of sixteen test rats, luteinization occurred only in two and in these it was slight. Though castration of the horse destroys the luteinizing capacity of the pituitary, the destruction is not complete, he says, until several years after the operation. Greep (1937) states that in the rat the pituitary contains some LH for many months after castration.

Martins & Rocha (1930, 1931) placed in parabiosis infantile female rats 19 to 30 days old with males or females of the same age which were subjected to gonadectomy at the same time. As a result, the non-spayed partner showed precocious oestrus and even ovulation within 6 or 7 days, a result which is attributable to an excessive supply of gonadotrophin (FRH) derived from the spayed or castrated partner. This experiment shows that the gonads can affect the gonadotrophic power of the pituitary before the rat has become sexually mature. Hill (1932, 1933) united female rats in parabiosis and removed the ovaries from one of each pair, with the consequence that the spayed partner remained anoestrous while the one with intact ovaries went into constant oestrum. As in the experiments of Martins & Rocha this event, it seems, must have been the consequence of an excessive output of FRH from the pituitary of the spayed partner together with a deficiency or absence of LH.

Experiments carried out by Lauson, Golden & Severinghaus (1939) have shown that in rats the enhancing effect of gonadectomy on the gonadotrophic potency of the pituitary does not entirely take place at once. Rats were spayed and after varying intervals the potency of their pituitaries was compared with that of normal female rats of the same age. In this way the pituitary potency was found to increase continually for about 60 days after gonadectomy; it then remained stationary at a high level until the 120th day, which was the limit of the experiment. The pituitaries were assayed by implantation into immature mice, the uterine weights being used as criteria of the potency of the implants. This method gives a measure of FRH capacity. The rate of increase of potency was almost linear up to the 60th day; at the end of this interval the potency was 52 times that of the normal rat's pituitary.

Partial Gonadectomy, Cryptorchidism and Sterilization by X-rays

From the various experiments which have been quoted above it appears that removal of the gonads in either sex, even though the animals are sexually immature, leads to a much enhanced output from the pituitary of FRH and a great reduction of the output of LH. From this it may be assumed that, by some direct or indirect influence, the normal gonads can reduce or inhibit the production of FRH by the pituitary. It is important to identify if possible the particular constituent of the gonad which exercises this restraint. There are three methods of investigating the problem, namely by studying the consequences of (a) partial gonadectomy, (b) cryptorchidism and (c) X-ray sterilization.

(a) *Partial gonadectomy.* The regulation of the pituitary by the gonads and of the gonads by the pituitary is revealed by the results of partial gonadectomy. Some consequences of this procedure have been discussed already (pp. 21, 26) in connection with the influence of the pituitary on the ovary and testis; it was then shown that removal of one ovary is followed by an increased development of follicles in the remaining ovary, so that the total number of follicles maturing at each cycle is little if at all changed by the operation. To this increased follicular development in the remaining ovary after removal of its fellow the so-called 'compensatory hypertrophy' of the gonad is attributable, the enlargement being due to increased functional activity rather than to the formation of new tissue to replace what has been lost. These remarks may be applied not only to unilateral gonadectomy but also to the partial excision of a single ovary, whether its fellow has been removed by operation, or whether, as in the fowl, only a single ovary is normally present. Such mutilations cause a permanent disequilibrium between the pituitary and the surviving gonadal tissue, and from this imbalance many abnormalities in the reproductive system may ultimately arise. Though the disequilibrium is permanent, most of the abnormalities caused by upsetting the balance of supply and demand within the hormonal system are reversible, and can be remedied temporarily by an artificial supply of the missing hormones. Theoretically the only way to abolish the permanent disequilibrium would be by implanting grafts of fresh living material to replace that which has been lost.

There may be an exception to the rule that the sequels of partial gonadectomy are reversible, for experiments have suggested that, with time, malignant tumours may arise as an outcome of the pituitary-gonadal imbalance, and once started the progress of such tumours has shown no sign of reversibility.

After these preliminary remarks it may be interesting to review some of the experimental work upon which the generalizations just stated are founded. When considering the work it may be borne in mind that (1) the gonads control the amount of available gonadotrophin derived from the pituitary, (2) if the gonadal tissue is partly or entirely excised this check on the available supplies of pituitary gonadotrophin will be reduced accordingly, (3) the results will be a larger total supply of gonadotrophin and a smaller amount of tissue subject to its action.

The effect of partial spaying on the remaining ovarian tissue. Lipschütz, Wagner & Tamm (1922) removed one ovary from each of a number of rabbits when they were between 4 and 6 weeks old. Six months later the single remaining ovaries were removed and found to be much larger (55 to 85 per cent) than each of the ovaries in intact littermate control animals. Further, these investigators removed from rabbits when they were a month old one entire ovary and three-quarters of the other. Six months later the surviving quarter of an ovary had attained the size of half a normal ovary. Microscopical examination showed that this enlargement was not attributable to the reformation of lost tissue, but was due mainly or entirely to the ripening of more follicles relatively to the bulk of ovarian tissue. In more prolonged experiments on female rabbits, Lipschütz (1925 a) found that follicular cystic degeneration was apt to occur in the fragment of ovary remaining after partial gonadectomy and that the walls of the cysts sometimes became luteinized. Smith & Engle (1927), it will be recalled, caused similar ovarian follicular cysts by the persistent administration of pituitary extracts (p. 25).

The effects of partial spaying on the uterus and mamma. Extended observations (Lipschütz, 1937b, 1938) on guinea-pigs and cats as well as on rabbits revealed consequences from partial spaying other than those occurring in the remaining ovarian tissue. In the guinea-pig, he found, the *uterus* may become greatly enlarged so that 3 years after partial spaying instead of weighing the normal 1 g. or thereabout it weighed in different cases between 5 and 27 g. This enlargement was accompanied by a remarkable proliferation of the uterine glands, the lumina of which were often dilated so as to cause glandular cavities or 'cysts'. Similar changes in the uterus were recorded by Wolfe, Burch & Campbell (1932) as a consequence of partial spaying in the rat.

In some of the guinea-pigs the *mammae* became enlarged so as to exceed in size those of pregnancy, though occasionally these enlarged mammae subsequently shrank.

The effects of partial spaying on the oestral cycle. Wolfe, Burch & Campbell (1932) removed one ovary and a large part of the other from rats, and they noted that, as a consequence, oestrus became prolonged so as to occupy about half of the total cycle. This extension of oestrus at the expense of anoestrus was most pronounced in those rats from which most ovarian substance had been removed. Thus there appeared to be, in spite of the greatly reduced bulk of ovarian substance, an increased amount of available oestrogen or a much enhanced reactivity towards oestrogen. Lipschütz states that the keratinizing stage of oestrus in the guinea-pig's vagina, which normally lasts for a few hours only, after partial ovariectomy may persist for 10 days or longer.

It will be observed that all these results of partial spaying might be attributed to an excessive output of FRH and consequently of oestrogen. The evidence in favour of the disturbances being the outcome of a disequilibrium between the pituitary and ovary seems complete. Lipschütz attributes the results to an irreversible imbalance between the ovary and the pituitary. He believes that the ovarian control of the pituitary is exercised not only by tertiary follicles which reach maturity, but by the secondary follicles which are destined to become atretic at an earlier stage and to contribute to the interstitial substance of the ovary. Because of the greater strain put on the surviving fragment of the gonad after partial spaying the primary follicles become used up prematurely, he suggests, and so there is a shortage of secondary atretic follicles and a consequent disequilibrium. Such an explanation seems to fit most of the known facts, though further experiments are required before it can be accepted without reserve (v. p. 21).

The effects of partial castration on the testis. More or less brief experiments (Lipschütz, 1922, 1925 a) have shown that after removal of one testicle the other becomes enlarged, but the sequels are not so striking as those which follow unilateral spaying. Moreover, according to Lipschütz, the excision of part of a testicle is not followed by a great enlargement of the surviving fragment such as happens when the ovary is mutilated in this way.

Apparently in some circumstances a more extensive removal of testicular substance may eventually lead to considerable pathological changes. Champy & Lavedan (1938) removed the testes of fowls almost but not quite completely. After this operation atypical regeneration occurred. If the operation in these circumstances is repeated, they say, and the birds are kept for 3 years or longer, large testicular tumours may arise. This happened in four of the fifteen fowls of their experiment. The tumours were seminomas or teratomas, and metastasis occurred. It is difficult to dissociate in one's mind the etiology of these tumours from that of the teratomas which have been induced in the testes of fowls by injecting zinc chloride into them, or to avoid thinking that there may be something common in the causation.

The gonadotrophic potency of the pituitary after partial gonadectomy. Regarded collectively the results of partial gonadectomy, both in the male and female, suggest that the operation is followed by an excessive production of FRH by the pituitary. The possibility has been investigated by Nelson (1935c). He removed the whole of one and the greater part of the other ovary from guinea-pigs. After intervals varying from 4 to 12 months the guinea-pigs were killed and their pituitaries were implanted into immature female rats and mice. The gonadotrophic potency (FRH) of their pituitaries, as determined in this manner, was considerably above the normal, being only slightly less than that of the pituitaries of guinea-pigs which had been completely spayed.

(b) *Cryptorchidism.* Experiments by Evans & Simpson (1929b) and Nelson (1935c) bearing on the gonadotrophic capacity of the pituitary in animals which had been artificially made cryptorchid have been mentioned earlier (p. 45). Reference will be made here to further work on the same subject. It is hardly necessary to remark that the chief defect in a cryptorchid testis is an atrophy of the seminal epithelium while the interstitial glandular tissue remains sufficiently active to maintain the accessory generative organs for a prolonged period in a functional state.

Martins (1930) produced an artificial cryptorchidism in rats and mice by operation, and then joined them in parabiosis with normal females, with the result that the latter went into a state of constant oestrus and their ovaries showed an excessive development of follicles and a failure of luteinization. On the other hand, when normal males whose testes were in the scrotum were united in the same way with normal females the oestrus cycles in the latter were not disturbed. By direct tests the pituitaries of cryptorchid rats were found to possess more gonadotrophic potency than those of normal rats (Martins, Rocha & Silva, 1931). Nelson (1935c) produced artificial cryptorchidism in guinea-pigs and after periods of 5 to 14 months killed them and implanted their pituitaries into immature rats and mice. His results confirm those which Evans & Simpson (1929b) had obtained with cryptorchid rats, the pituitaries of the cryptorchid guinea-pigs being about 70 per cent more potent than those of normal male guinea-pigs in causing precocious maturation of follicles in the ovaries of infantile rats and mice.

Du Shane, Levine, Pfeiffer & Witschi (1935) maintained female rats in parabiosis with normal males; if, in these circumstances, the male rat was rendered crypt-orchid or sterilized by X-rays, the female partner arrived at a state of constant oestrus.

In contrast with some of the foregoing results are those obtained by Cutuly, McCullagh & Cutuly (1937 *b*). These experimenters joined hypophysectomized male rats in parabiosis with (*a*) castrated, (*b*) cryptorchid and (*c*) normal male partners. The pituitary of the castrated rat was potent enough to maintain the testes and accessory genital organs of their hypophysectomized partners. The cryptorchid partners, even after 71 days of cryptorchidism, did not show any more influence than normal rats on the genital organs of their hypophysectomized fellows. In connection with this result an experiment by Nelson (1937 *a*) should be mentioned. He rendered a number of rats cryptorchid by operation. Six months later their seminal vesicles appeared normal, but after 8 months these accessory glands were markedly decreased in size. Early castration changes were present in the pituitary after 75 days of cryptorchidism. Twenty-five rats which had been cryptorchid for periods of 360 to 450 days were anaesthetized and one testis, one seminal vesicle (atrophic) and a piece of the prostate were removed. The rats were then given daily injections of gonadotrophin for 10, 20 or 30 days and then killed. At the end of this treatment there were no castration changes in the pituitary and the accessory organs were as large or larger than those of con-trols (Table 26). The results are difficult to explain if it is true that cryptorchidism produces castration changes in the pituitary accompanied by an increased output of gonadotrophin, for the rapid and ample response of the seminal vesicles suggests that the cryptorchid testis had not lost its power of responding to gonadotrophin.

TABLE 26. Showing the effect of prolonged cryptorchidism on the seminal vesicles of the rat and their response to gonadotrophin (Nelson, 1937 *a*)

Number of rats	Duration of treatment with gonadotrophin (days)	Weights of organs before administration of gonadotrophin		Weights of organs after administration of gonadotrophin	
		Testis (g.)	Seminal vesicle (g.)	Testis (g.)	Seminal vesicle (g.)
8	10	0·239	0·067	0·298	0·714
9	20	0·213	0·060	0·284	0·803
10	30	0·224	0·058	0·265	0·738

Whatever the nature may be of the restraining influence which the gonads exercise upon the pituitary there is evidence that the main influence is derived from the ovarian follicles or corpora lutea in the female and the seminal epi-thelium in the male. The experiments which have been quoted above, dealing with the effects of incomplete gonadectomy and cryptorchidism, nearly all point to this conclusion, which is further supported by the consequences which follow X-ray sterilization.

(*c*) *X-ray sterilization.* Witschi, Levine & Hill (1932) sterilized male rats by exposing the scrotum to X-rays. Three exposures were made, the total dose being 2,400 r. One month later these rats were joined in parabiosis with normal females. The effect on the females, after an interval of time during which oestrus was irregular, was to maintain them in a condition of constant oestrus. Examination of the sterilized males showed that the seminal epithelium had been destroyed while the interstitial testicular tissue seemed uninjured and was functionally active, as shown by the normal appearance of the prostate and seminal vesicles. The conclusion is that the excessive production of FRH in these rats can be attributed to the defective condition of the seminal epithelium resulting from the action of X-rays.

These three conditions, partial castration, cryptorchidism and X-ray sterilization, it may be noted, not only lead to an over-production of FRH but they cause also an enlargement of the pituitary accompanied by the same histological changes in its structure as those which follow complete gonadectomy.

Gonadal Hormones

There is little doubt that the restraint which a testis or ovary exerts on the production of gonadotrophin by the pituitary is largely caused by the known gonadal hormones. Indirect evidence pointing to this conclusion will be adduced in the section on the cytology of the hypophysis; meanwhile reference will be made to experiments which bear directly on the influence of gonadal hormones upon the gonadotrophic potency of the pituitary gland.

(a) *Oestrogen. Gonadotrophic action.* The first effect of oestrogen on the pituitary appears to be an increased output of LH (Lane & Hisaw, 1934); this has been shown, not only by the changes caused in the ovaries of adult rats by oestrin, but by the effect which the pituitaries of oestrin-treated rats cause when implanted into infantile rats. Hohlweg & Chamorro (1937) gave a single dose of 11γ of oestradiol benzoate to each of fifteen rats which were 8 weeks old and weighed 50 g. Five of the rats killed 2 days later had no corpora lutea; among five rats killed 5 days after the injection, corpora lutea were present in two and among the remaining five, which were killed $7\frac{1}{2}$ days after the injection, corpora lutea were present in four. When the same treatment was carried out on rats whose pituitaries had been removed 6 days previously no corpora lutea were produced.

Frank (1940) states that the gonadal hormones, when given in small doses and for short periods of time, increase the gonadotrophic activity of the pituitary, whereas when supplied in larger amount and during longer periods of time they suppress it. When considering such a generalization one has to remember that oestrogens and androgens have a direct gonadotrophic influence on the gonads in the absence of the pituitary as will be shown later (pp. 198, 281).

Antigonadotrophic action. The later effect of oestrogens on the pituitary is to diminish or arrest the output of gonadotrophin (FRH). Moore & Price (1930) found that the continued administration of oestrin to immature rats of 30 days caused atrophy of the testes and accessory genital organs.

Meyer, Leonard, Hisaw & Martin (1930, 1932) carried out the following experiments. (a) Thirty-four immature female rats between 30 and 40 days old were given 2 r.u. of oestrin, dissolved in oil, daily for periods extending from 30 to 70 days. Twenty-four littermates were used as controls, being given injections of oil only. At the end of the periods of injection the animals were killed. The ovaries of those which had received oestrin were 40 per cent less in weight than the ovaries of the littermate controls. (b) In a second experiment Meyer and his colleagues castrated fifteen adult male and nineteen adult female rats. Some of these were kept without further treatment as controls while the others were given 4 r.u. of oestrin daily for 31 days. At the end of this period all the rats were killed and their pituitaries were implanted into immature rats, one pituitary being given to each recipient. After a short interval the young rats were killed and it was found that the ovaries of those which had received pituitaries from the oestrin-treated castrated rats were only 35 per cent of the weight of the ovaries of those which had received pituitaries from the castrated rats which had not been given oestrin. The experiments show that oestrin inhibits the activity of the ovary, and that this effect is largely or entirely explained by an inhibition of the gonado-trophic (FRH) activity of the pituitary. Kraus (1930) made a comparable observa-tion on males. First he noted that injections of oestrin into normal male rats and mice caused atrophy of the testes accompanied by involution of the accessory generative organs, including the prostate and seminal vesicles. If, however, gonadotrophin were given in addition to the oestrin the testes and accessory organs were not atrophied. Moore & Price (1932) also found that the atrophic effect of oestrin on the rat's testis could be prevented by the implantation of fresh rat pituitaries, or by subcutaneous injections of a gonadotrophic extract from the urine of pregnant women, and that the atrophy of the prostate and seminal vesicles could be prevented by an androgenic preparation obtained from bulls' testes. Apparently oestrogens injure the male accessory organs by interfering with the supply of gonadotrophin from the pituitary so as to prevent the elaboration of androgen by the testis (see Spencer, D'Amour & Gustavson, 1932a).

Nelson (1935c) showed that the pituitaries of male and female guinea-pigs which had been deprived of their gonads, and afterwards given daily injections of oestrin in doses ranging from 25 to 100 r.u., when implanted into immature animals were less potent than the pituitaries of normal guinea-pigs; whereas gonadectomy alone leads to a considerable enhancement of the supply of gonado-trophin by the pituitary. He noticed too that the increased gonadotrophic potency of the pituitary which normally is associated with castration and cryptor-chidism could be prevented by means of ovarian grafts. Zondek (1936b) found that prolonged dosage with oestrin interferes with the functions of the anterior lobe of the pituitary in rats, causing atrophy of the testes and a consequent under-development of the penis and scrotum.

Hertz & Meyer (1937) put immature female rats in parabiosis with castrated male or female littermates, with the result that hypertrophy of the ovaries occurred in the normal partner because of an excessive supply of gonadotrophin received from the pituitary of the gonadectomized twin. This hypertrophy could be completely prevented by daily injections of 0.2γ of oestrone into the gonadecto-mized rat. The result illustrates the capacity of oestrone to check the supply of gonadotrophin by the pituitary. Hertz & Meyer found that these small doses of oestrone were as effective as 30γ of testosterone or 15γ of testosterone propionate given daily in limiting the output of gonadotrophin by the pituitary. Cutuly & Cutuly (1938) joined young rats weighing from 50 to 150 g. in parabiosis, and removed the pituitary from one of each pair, whose testes thereupon underwent involution. After castration of the other twin its pituitary produced enough gonadotrophin to cause descent of the testes and their restoration to a normal functional state in the hypophysectomized partner. If, however, daily injections of oestrone or oestradiol at the rate of 5 to 100γ were given to the castrated partner repair of the testicular condition in the hypophysectomized twin did not occur.

Yet another experiment illustrating the capacity of an oestrogen to inhibit the supply of gonadotrophin from the pituitary was done by Noble (1938b). He gave large doses of diethylstilboestrol to rats so as to render their ovaries inert. When this stage had been reached gonadotrophin was administered with the result that the ovaries resumed functional activity in spite of a continued dosage with diethylstilboestrol.

It is of interest to note that the recognized effects of oestrogen on the functions and histology of the adult rat's pituitary are independent of any direct nervous connection between the pituitary and the hypothalamus, for they are produced in the same degree after division of the pituitary stalk as in the intact animal (Uotila, 1940).

An inhibitory effect of oestrogen on the gonadotrophic potency of the pituitary has been demonstrated in man. Frank & Salmon (1935b), having verified the presence of gonadotrophin in the urine of fourteen women who had been spayed (8), or treated by X-rays to prevent conception (2), or who had passed the menopause (4), treated them by intramuscular injections of oestrone benzoate every other day until amounts ranging from 4,000 to 22,000 r.u. had been given. Under this treatment there was a speedy disappearance of gonadotrophin from the urine, an effect which lasted for periods of 28 to 70 days. Others have made similar observations. Jones & MacGregor (1936) gave oestradiol to ten women who had passed the menopause. Six received 1,500,000 m.u. in 60 days, two had 1,000,000 m.u. in 46 days and two had 500,000 m.u. in 20 days. In all the ten women the excretion of detectable gonadotrophin, which had previously exceeded 50 m.u. per litre of urine, was abolished.

A more direct investigation of the effect of oestrogen on the gonadotrophic power of the human pituitary was made by Rowlands & Sharpey-Schafer (1940). They gave daily intramuscular injections of 10 mg. of oestradiol to four women who were dying of different diseases. After death their pituitaries were dried, as also were the pituitaries of five other women who had not received oestrogen before death. The dried pituitary substance obtained from these individuals was tested on young female rats whose pituitaries had been removed 10 days previously. Each rat received five daily doses of dried pituitary and was killed 24 hours after the last injection; the effects upon the ovaries and uterus were then ascertained. The results proved that the pituitaries of the women who had been treated with oestradiol possessed much less gonadotrophic influence than those of the women who had received no such treatment. These observers reported also that 10 mg. of oestradiol given daily to a spayed woman caused a disappearance of gonadotrophin from the urine.

Inhibition of the production of FRH by oestrogen. Hisaw, Fevold, Foster & Hellbaum (1934), who regard FRH and LH as two distinct components of gonadotrophin, affirm that the immediate effect of oestrin on the pituitary is to stimulate the production of LH while suppressing the supply of FRH. During dioestrum, they say, the FRH content of the pituitary is at a maximum; with the rapid increase of oestrin production during proestrum and oestrum the FRH formation falls to a minimum while the output of LH reaches its maximum.

Hohlweg (1934) has made observations which accord with this view. He found that when oestrin is given to immature rats no increase occurs in the size of the ovaries and luteinization is absent; that is to say, oestrin has not stimulated the production of FRH, and luteinization is absent because there are no follicles in a condition to react. In mature rats oestrin causes enlargement of the ovaries with excessive luteinization, an indication that it has stimulated the production of LH. The influence of oestrin in the maintenance of corpora lutea may perhaps be attributed to this effect. Selye, Collip & Thomson (1935) reported experiments which favour the view that oestrogens cause an increased supply of LH. They gave 500γ of oestrone daily to eight pregnant rats. Gestation was prolonged and most of the embryos died undelivered. In one instance the corpora lutea were still fully developed and apparently functional at the end of the gestation period, whereas in normal circumstances they should have undergone involution before this stage. When given to lactating rats oestrin caused an increase in the size of the ovaries owing to enlargement of the corpora lutea.

According to Fevold, Hisaw & Greep (1936a) the increased production of LH by the pituitary caused by giving oestrin is a temporary effect which decreases with continued treatment. Lipschütz (1935) and Lipschütz, Palacios & Akel (1936) are of the same opinion. They gave from 5 to 20 r.u. of oestrone to normal rats daily for 6 to 10 weeks; the rats were then killed and the anterior lobes of their pituitaries, after trituration in Ringer's solution, were injected into infantile female rats. Infantile control rats were given pituitaries prepared in the same way from normal rats. While the ovaries of the control rats in every instance were stimulated to activity and showed the presence of corpora lutea in response to the pituitary injections, those which had received pituitaries from oestrin-treated rats showed a less striking response (Table 27).

TABLE 27.   The effect of prolonged treatment with oestrone on the luteinizing potency of the rat's pituitary (Lipschütz, Palacios & Akel, 1936)

Pituitary donor rats	Amount of anterior pituitary given (mg.)	Number of recipient infantile rats	Number of recipients showing luteinization in ovaries
53 normal males	8–21	13	13 (100%)
56 males treated with oestrone for 3½ to 10 weeks	14–28	22	10 (46%)

The experiments just mentioned suggest that the immediate effect of oestrogens on the pituitary is to stop the production of FRH and to cause an increased output of LH, and that a continued supply of oestrogens suppresses the output of both FRH and LH. The nature of the inhibitory action of oestrogens on the gonadotrophic potency of the pituitary awaits elucidation.

Hamburger (1936) performed some experiments which seem to show that the inhibitory action of oestrogen takes place within the pituitary and may be due to a neutralization of gonadotrophin already present in that organ. He gave daily subcutaneous injections of 50 m.u. of 'antex'—a gonadotrophic extract obtained from pregnant mare's serum—to normal and spayed rats during periods varying from 9 to 86 days; and to some of the spayed rats oestrin was given in addition. The sera of the rats treated in this way were then tested for gonado-trophic potency by injecting them into immature female mice, the weights of the ovaries in these mice being taken to indicate the degree of gonadotrophic potency of the different sera. It will be seen (Table 28) that the potency of the serum was increased in spayed but otherwise untreated rats and in normal rats treated with 'antex', and greatly increased in spayed rats treated with 'antex'. The figures also show that oestrin largely prevents an increase of potency in the sera of the spayed rats treated with 'antex'. Hamburger pursued the matter further by testing the effect of oestrin on the gonadotrophic potency of the serum of hypophysec-tomized rats which were receiving 'antex' in addition. His results indicate that the presence of the pituitary is essential for the inhibitory effects of oestrin on the gonadotrophic potency of the serum. Fevold & Fiske (1939) say that both oestrin and LH, given for a period of 8 days, will inhibit the action of FRH on the ovaries of normal immature rats. If the same treatment is applied to hypo-physectomized immature rats, oestrin will no longer prevent the action of FRH, though LH will do so. They believe that the inhibiting agent is either LH produced in the pituitary under the influence of oestrin or a factor closely associated with and inseparable from LH.

TABLE 28. The gonadotrophic potency of rat serum as affected by (1) repeated in-jections of gonadotrophin, (2) hypophysectomy and (3) spaying (Hamburger, 1936)

Condition of rats from which serum was taken	Number of rats	Previous hormonal treat-ment of rats	Number of test mice	Average weight of one mouse ovary (mg.)
Normal	29	None	31	1·1
Normal	50	Antex	49	1·8
Spayed	29	None	27	1·5
Spayed	32	Antex	33	5·8
Spayed	10	Antex + Oestrin	10	2·8
Hypophysectomized	9	Antex	10	5·4
Hypophysectomized and spayed	12	Antex	14	6·5
Hypophysectomized and spayed	9	Antex + Oestrin	9	5·9

In this table the ovarian weight is the index of gonadotrophic potency.

Oestrogens affect also the lactogenic potency of the pituitary (p. 338).

(b) *Androgen.* Androgens resemble oestrogens in the check which they im-pose on the supply of gonadotrophin from the pituitary. Moore & Price (1930) treated immature rats with androgen and noted that although their accessory genital organs were well developed under this treatment the testes were atrophic. Ihrke & D'Amour (1931) found that daily injections of an androgen obtained from bulls' testes suppressed oestrus in the female rat for as long as they were con-tinued. If, however, rats so treated were given gonadotrophin at the same time they showed oestrus within 2 or 3 days. This result suggests that androgen, like oestrogen, interferes with the supply of gonadotrophin from the pituitary. Moore & Price (1932) confirmed the results obtained by Ihrke & D'Amour; they found that injections of a testicular androgen arrested oestrus in the female rat and if, during this artificially maintained anoestrous period, gonadotrophin were given in addition to the androgen, oestrus ensued within 48 hours.

Martins & Rocha (1930, 1931) united in parabiosis immature female rats aged 19 to 30 days with castrated males of the same age. Every other day one or two testes from 30-day-old rats were pulped and injected subcutaneously into the castrated male partners of the parabiotic pairs. Without these injections the pituitary of the castrated male produced enough gonadotrophin to cause precocious oestrus in the female partner within a few days. The injections of testis pulp into the castrated male prevented this reaction in the female. The conclusion is that the injected testicular substance inhibited the excessive production of gonadotrophin by the castrated rat's pituitary. It is interesting to note that this inhibitory control by the testis over the pituitary was found to be already present in the testes of immature rats only 30 days old.

In other experiments Martins, Rocha & Silva (1931) united in parabiosis two male rats, one of which had been castrated. In consequence of the over-activity of the castrated rat's pituitary the prostate of the non-castrated partner became hypertrophied. This prostatic hypertrophy, they found, could be prevented by injecting testicular mush into the castrated rat.

Until the gonadal hormones had been obtained in pure form the precise identity of the agents to which many of the experimental results just quoted might be attributed was uncertain. Since this difficulty has been overcome by the chemists it is surprising how reliable were the conclusions drawn from the use of unpurified biological extracts. Later experiences with the use of chemically pure androgens have confirmed the opinions derived from earlier work done with cruder preparations. Only a few examples of these later experiments need be given.

McCullagh & Walsh (1935) joined a number of pairs of male rats in parabiosis and castrated one of each pair. The excessive production of gonadotrophin by the pituitary of the castrated partner led to hypertrophy of the prostate and seminal vesicles of the non-castrated partner. This effect, they found, could be entirely prevented by giving subcutaneous injections of androsterone to the castrated rat. Moore & Price (1937) treated rats with repeated injections of androsterone and then tested the gonadotrophic capacity of their pituitaries by implanting them into immature female rats. The results proved that the gonad-stimulating potency of the pituitaries of the androgen-treated rats was less than that of the pituitaries of the untreated control animals. They found further that daily injections of androsterone given to young male rats hindered the development of the testes so that they weighed only 12 to 50 per cent of the normal weight as shown by controls, a loss which could be attributed to a diminished output of gonadotrophin from the pituitary. Bottomley & Folley (1938b) found that 2·36 mg. of testosterone given daily to immature male guinea-pigs led to a great reduction in the weights of their testicles. This atrophy was prevented if gonadotrophin were given in addition to the testosterone. Hamilton & Wolfe (1938) performed experiments on rats which prove that testosterone propionate, like androsterone, suppresses the gonadotrophic capacity of the pituitary.

Hertz & Meyer (1937) put immature female rats in parabiosis with castrated males. The enhanced output of gonadotrophin from the pituitaries of the castrated males caused enlargement of the ovaries in the female partners. If testosterone or dehydroandrosterone were given to the castrated partners in sufficient amount the ovarian enlargement did not occur. Cutuly & Cutuly (1938) carried out an experiment which shows very well the inhibitory effect of a pure androgen on the gonadotrophic functions of the pituitary. They joined in parabiosis young male rats weighing between 50 and 150 g. and removed the pituitary from one of each pair. This was followed in the hypophysectomized rat by testicular atrophy, retention of the testes within the abdomen, and involution of the accessory generative organs. If now the normal partner were castrated, the scrotum of the hypophysectomized male became flushed, the testes descended and together with the entire genital tract gradually resumed a normal state. This showed that castration had led to a sufficient increase in the output of gonadotrophin by the pituitary of the castrated rat to supply the defect of this hormone in the partner whose pituitary had been removed. Testosterone propionate given to the castrated rat in daily doses ranging from 50 to 3,000 γ prevented the restoration of the generative organs of the hypophysectomized partner. Hellbaum & Greep (1943), as the result of an extended inquiry, concluded that testosterone propionate, given in daily doses of 0·5 mg., diminished the concentration of FRH in the blood of castrated rats.

(c) *Progestin.* Like oestrogen and androgen, the specific hormone of the corpus luteum curbs the gonadotrophic activity of the pituitary and thereby regulates oestrus and ovulation.

Indirect evidence. Loeb (1914) noticed that extirpation of the corpora lutea in the guinea-pig was followed by an earlier onset of the next ovulation and that this occurred even if the animal were in the early stage of pregnancy. Moreover, an artificially caused extrauterine pregnancy, which led to degeneration of the corpora lutea, was accompanied by ovulation in spite of the presence of a living embryo within the abdomen. These phenomena suggest that the luteal hormone, progestin, inhibits ovulation.

Papanicolaou (1920), also using the guinea-pig, found that removal of the corpora lutea within 24 hours after oestrus and ovulation accelerated the onset of the next oestrum, the interval between the successive oestra being reduced to 11 days instead of the normal 16–17. Removal of the corpora lutea on the 4th, 8th or 12th day after oestrus also reduced the interval. From these facts Papanicolaou concluded that the corpora lutea inhibit oestrus. Subsequently (1926) he proved by experiment that a lipoid extract of corpora lutea inhibits oestrus and ovulation in the guinea-pig. Parkes & Bellerby (1928) prepared an extract from cows' corpora lutea which inhibited oestrus in the mouse. Gley (1928) effected a similar inhibition of oestrus in the rat by means of an extract of sow's corpora lutea, and Smith & Engle (1932) were able to prevent menstrual bleeding in the monkey by daily injections of progestin.

Corner (1935) gave daily doses of 1 rb.u. of progestin to each of six rhesus monkeys with the result that no menstrual bleeding occurred so long as the administrations were continued, whereas after cessation of the injections uterine bleeding occurred within 5 to 8 days. Selye, Browne & Collip (1936) arrested the oestrus cycles in young adult rats by daily injections of 4 mg. of progesterone. This treatment caused atrophy of the ovary and enlargement of the pituitary gland (Table 29), and vaginal smears showed dioestrus throughout.

Table 29.   The effect of progesterone on the ovary and pituitary of the rat (Selye, Browne & Collip, 1936)

	Daily dose of progesterone (mg.)	Duration of experiment (days)	Average weight of ovaries (mg.)	Average weight of pituitary (mg.)
Untreated controls	0	12	39	10·5
Treated	4	12	30	13·5

During pregnancy, pseudopregnancy, and lactation, that is to say in the presence of active corpora lutea, the gonadotrophic potency of the pituitary is reduced.

Burrows (1939c) found that 1 mg. of progesterone given three times a week to young Wistar rats prevented the descent of the testes into the scrotum—an effect which is also produced by androgens and oestrogens, both of which types of hormone are known to check the supply of gonadotrophin by the pituitary.

Direct evidence that progestin curtails the supply of gonadotrophin from the pituitary has been obtained by Herlant (1939) who treated adult rats with progesterone for varying periods at the end of which their pituitaries were removed and implanted into immature female rats weighing between 30 and 40 g., each of which received three pituitaries on three successive days and were killed 2 days later. The results were as follows:

(A) Pituitary implants from rats which had been given 13 injections of 0·35 mg. during 15 days, and from rats which had received six injections of 0·5 mg. of progesterone during 10 days failed to cause an increase in the size of the ovaries and uterus in the infantile recipients as compared with controls which had received pituitaries from normal rats.

(B) Pituitary implants from rats which had received 1 mg. daily of progesterone during 4 days only caused a different effect, the ovaries and uterus in the recipients being larger than those of controls. The latter result is attributed by Herlant to an increased production of luteinizing hormone by the pituitary under stimulation by progestin.

Burrows (1939a) gave daily subcutaneous injections of 5 mg. of progesterone dissolved in sesame oil to six young adult male Wistar rats, six other rats of the same age and strain being given injections in the same way of sesame oil alone. On the 5th day the rats were killed and their pituitaries were implanted into infantile mice weighing between 10 and 13 g., each mouse receiving one pituitary. The mice were killed 4 days later and their reproductive organs were weighed. In every instance these organs were smaller in the mice which had received pituitaries from the progesterone-treated rats than in the controls (Table 30).

TABLE 30. The effect of progesterone on the gonadotrophic potency of the rat's pituitary as tested in infantile mice (Burrows, 1939a)

Source of pituitary implants	Mean weights of organs of recipients expressed in mg. per 10 g. bodyweight		
	Uterus	Vagina	Ovaries
Progesterone-treated rats	33·9	39·7	7·6
Untreated rats	60·5	58·3	10·5
No pituitary given	9·1		5·4

Biddulph, Meyer & Gumbreck (1940) showed the capacity of progestin to check the output of pituitary gonadotrophin in another manner. They placed in parabiosis two female rats, one of which had been spayed. The increased output of pituitary gonadotrophin by the spayed rat caused constant oestrus in the partner. If 1 mg. of progesterone was given daily to the spayed partner the hypersecretion of its pituitary was suppressed and the persistent oestrus in the non-spayed partner ended.

Gonadotrophins

The effects produced by the artificial administration of gonadotrophin will have to be considered again later in connection with pituitary cytology and acquired resistance to gonadotrophic extracts; our attention at present will be limited therefore to the early effects of artificially-given gonadotrophin on pituitary function.

Kuschinsky (1931) gave daily subcutaneous injections of placental gonadotrophin or of oestrin to normal and spayed female rats which were killed after 12 days of treatment, their pituitaries being then implanted into immature 24-day-old female rats weighing between 23 and 27 g. One pituitary having been implanted on each of four successive days, the recipients were killed on the 5th day and their ovaries were weighed. Using the weights of the ovaries as indicating gonadotrophic potency it will be seen that this was reduced by both gonadotrophin and oestrin in normal rats; but the gonadotrophin had this effect in non-spayed animals only, whereas oestrin caused a reduction of pituitary potency whether the ovary was present or not (Table 31).

TABLE 31. The effects of (a) oestrin, (b) placental gonadotrophin on the gonadotrophic potency of the pituitary of normal and spayed female rats, as shown by the ovaries of recipient rats (Kuschinsky, 1931)

State of donor of pituitary	Hormonal treatment of donor	Weight of both ovaries of recipient (mg.)
Normal	None	21·5
Normal	Oestrin	11·2
Normal	Gonadotrophin	10·5
Spayed	None	50·5
Spayed	Oestrin	12·5
Spayed	Gonadotrophin	54·5

Leonard (1933c) confirmed these observations and showed that gonadotrophin had comparable effects in the male. He treated a large number of male and female rats with daily injections of placental gonadotrophin for periods varying from 10 to 34 days and then implanted their pituitaries into immature mice, 20 to 22 days old. Using the ovarian weights in the test mice as indicating the gonadotrophic potency of the rats' pituitaries he found that this was significantly reduced by gonadotrophin both in male and female rats, but only if the gonads were present (Table 32). Leonard therefore concluded that the artificial supply of gonadotrophin had stimulated the gonads to produce excessive amounts of gonadal hormone which had checked the supply of gonadotrophin from the pituitary. Obviously in the absence of gonads these reactions are impossible. That the treatment of normal rats with gonadotrophin had caused an increased output of androgen and oestrogen was shown by hypertrophy of the accessory genital organs.

TABLE 32.   The effects of gonadotrophin on the gonadotrophic potency of the male rat's pituitary as indicated by the ovaries of recipient mice (Leonard, 1933 c)

Donors of pituitary, all of which had been treated with gonadotrophin	Period of treatment with gonadotrophin (days)	Percentage of deficiency in weight of ovaries of recipient mice as compared with normal
Normal males	9–17	42·4
Castrated males	12–14	3·7
Normal females	10–34	33·7
Spayed females	11–14	12·1

Sex

Is the gonadotrophic capacity of the pituitary the same in quantity and nature in the two sexes?

The question is not simple, because the functional activity of the pituitary varies during the oestrous cycle, being influenced by the gonadal hormones, which also vary in quantity and kind at different stages of the cycle. Apart from this complication, most of those who have investigated the problem experimentally have concluded that in ordinary circumstances the pituitary of the male after puberty is more potent than that of the female, and some have detected a difference in the nature of its activities in the two sexes. Evans & Simpson (1929 b) inquired into the matter by implanting pituitaries from adult male and female rats into infantile rats and estimating the gonadotrophic potencies of the implants by their effects on the ovaries and vagina. Some of their results are given in Table 33.

TABLE 33.   Comparison of gonadotrophic potencies of the pituitaries of male and female rats (Evans & Simpson, 1929 b)

Source of rat pituitary	Number of recipient infantile rats	Number of implants	Percentage of recipients giving oestrous vaginal smear on 4th day	Average weight of both ovaries at end of 4th day (mg.)
Normal females	34	4	50	19·7
Normal males	12	4	100	70
Infantile males (24 days old)	3	4	66	41

The results show that the pituitaries of the male rats had a greater follicle ripening potency than those of female rats. Even the pituitaries of infantile male rats were more potent by these tests than the pituitaries of adult females. The difference between the two sexes in adult life appeared not to depend upon temporary gonadal influences, for by the same tests the pituitary of the castrated male rat was still about $2\frac{1}{2}$ times as potent as that of the spayed female.

Ellison, Campbell & Wolfe (1932) injected saline suspensions of fresh rat pituitaries intravenously into female rabbits. Using ovulation as the criterion of gonadotrophic potency they found that the pituitary of the male rat was more potent than that of the female. With 2·5 mg. of pituitary substance obtained from male rats eight positive results were obtained in nine tests, whereas with 2·5 mg. of female pituitary there were no positive responses in twelve tests. On the basis of these and additional experiments they concluded that the male rat's pituitary possessed about twice the gonadotrophic potency of that of the female. Schmidt (1937), by the same sort of test, showed that the gonadotrophic potency of the male guinea-pig's pituitary exceeds that of the female. In man comparable observations have been made. Lipschütz & Del-Pino (1936) assayed the pituitaries of nine men and eleven non-pregnant women. The gland was removed about 10 hours after death, placed for 2 hours in acetone, and then ground and dried. The substance thus obtained was tested on infantile rats and showed that the male pituitaries were the more potent.

Clark (1935), by assays of rats' pituitaries, concluded that this relative superior potency of the male pituitary is only attained at puberty, before which the female pituitary is considerably richer in gonadotrophin than is that of the male. She observed a sharp rise in the gonadotrophic potency of the female rat's pituitary between the 13th and 20th postnatal day to a level which is not surpassed even at full maturity. In the male there is a slower, continual increase till puberty, after which the level is evenly maintained. In further experiments Clark (1935) removed the gonads from male and female rats on the day of birth and assayed their pituitaries for gonadotrophic potency between the 16th and the 18th day, at which time no difference in potency between the pituitaries of the two sexes appeared. These results suggest that the relatively increased potency of the female pituitary before puberty as compared with that of the male might be attributed to the inhibitory influence of the gonadal secretions on the pituitary, for in rats the hormones of the male gonad appear to be secreted at an earlier age than are those of the female. Pfeiffer (1936) removed the gonads from newborn male and female rats and transplanted them into the eye. In some instances the grafts were implanted in this way into rats of the opposite sex. Such ovarian grafts in males apparently did not alter the character of the pituitary, which continued to produce FRH only, but testes transplanted into newborn females suppressed the luteinizing function of the pituitary, so that the hosts only produced FRH, as though they were males.

From these experiments it appears that the pituitary shows no distinctive sexual character at birth, and that the difference in functional capacity which is seen later in life between the male and female pituitaries is due to the different gonadal influences to which they have been exposed in the early postnatal period. As Pfeiffer says, the pituitary in the newborn rat is bipotential and remains so until puberty, being capable meanwhile of male or female differentiation according to the sex of the gonad present.

Smith, Severinghaus & Leonard (1933) implanted material from the pituitaries of normal rabbits into infantile female mice, and they found that the pituitaries of males were more active in stimulating the infantile mouse ovary than were the pituitaries from females; and this difference between the sexes was still evident in the pituitaries of rabbits which had been deprived of their gonads. Nelson (1935c) castrated male and female guinea-pigs and at intervals varying from 4 to 12 months later implanted their pituitaries into immature female rats and mice. In this way he showed that in the guinea-pig, as in the rat and rabbit, the pituitary of the normal male is considerably more potent than that of the female; he found, however, that pituitaries from gonadectomized males and females were of equal potency, which was much greater than that of the pituitaries of normal animals.

Besides the quantitative difference in the gonadotrophic potencies of the pituitaries of the two sexes which the foregoing observations suggest, qualitative differences in their responses to gonadal influence have been reported. Lehmann (1927) found that transplantation of testes into gonadectomized male rats abolished the castration changes in the pituitary; ovaries did not have this effect. Martins & Rocha (1930) also found that the histological effects produced in the rat's pituitary by gonadectomy could be prevented or rectified by engrafting normal gonads provided that the donor and recipient were of the same sex. An ovary grafted into a spayed female restored to normal the pituitary of a spayed female, but not that of a castrated male. On the other hand, a testis graft could repair the pituitary defects in the male rat following castration, but not those caused in the female by excision of the ovary. In further experiments Martins & Rocha (1930; see also Martins, Rocha & Silva, 1931) approached the subject in another way. They joined male and female rats, 19 to 30 days old, in parabiosis with male and female littermates whose gonads were removed at the same time. In one group the castrated partners received subcutaneously on alternate days one or two pulped testes taken from infantile rats; in the other group no injections were given. In the non-spayed females of the latter group precocious puberty ensued. Injections of testis into the castrated male of a parabiotic pair prevented the appearance of precocious puberty in the non-spayed partner; in the pairs consisting of two females, one of which had been spayed, injections of testis did not prevent precocious puberty. Furthermore a histological examination of the pituitaries showed that the usual 'castration changes' were present in the castrated males which had not received testis grafts but were absent from those which had received grafts of testis. These results, Table 34, suggest that the increased gonadotrophic potency of the pituitary caused by gonadectomy can be counteracted by gonadal hormones if these are appropriate to the sex of the individual,

TABLE 34. The effects of injecting testis substance into the gonadectomized partner of a pair of infantile rats in parabiosis (Martins, Rocha & Silva, 1931)

Sex of normal partner	Sex of gonadectomized and injected partner	Precocious puberty in normal partner
Female	Male	−
Female	Female	+
Male	Male	−
Male	Female	+

but not otherwise; and that this selective reaction indicates a sex difference in the pituitaries.

Moore & Price (1932), experimenting with male guinea-pigs and rats, observed that the 'castration changes' in the pituitary which follow gonadectomy could be prevented, or if present, could be abolished by the administration of testicular hormone, but not by oestrogen.

Two conclusions seem justified by these experiments, namely (1) that grafts or extracts of testis will prevent 'castration changes' in the pituitary of the male but not in that of the female, thus showing a sexual difference in the pituitaries; and (2) that the substance having this effect on the pituitary is already being formed by the infantile testis.

Martins (1931) has called attention to yet another apparent difference between the functions of the pituitaries in the two sexes. He united infantile female rats of 21 to 27 days in parabiosis with castrated males and females from the same litter. In these experiments the castrated partners, whether male or female, induced precocious puberty in the normal female. The effects produced by the two sexes were however not identical. When the partner was a spayed female corpora lutea nearly always developed in the twin; when the partner was a castrated male the usual response in the twin was maturation of many follicles, corpora lutea being absent.

Hellbaum (1935) has made a confirmatory observation on the horse. Pituitaries of mares and stallions were assayed on immature rats, 21 to 22 days old, by giving each rat 12·5 mg. of desiccated and powdered pituitary substance. There was a pronounced difference in the effect produced by the pituitaries of mares and stallions respectively. The ovaries of infantile rats under the influence of mare's pituitary became solidly luteinized, whereas the stallion's pituitary caused maturation of a large number of follicles without luteinization.

Lipschütz (1935) states that when ovaries are grafted into gonadectomized guinea-pigs the effects produced in the mammae are greater in the castrated male than in the female. This inequality of effect is due, he says, to the behaviour of the graft which develops differently in the two sexes. In the spayed female corpora lutea appear in the graft, an event which never happens with an ovarian graft in a castrated male. In the latter the changes in the grafted ovary do not proceed beyond maturation of follicles; though in rats, as Goodman (1934) has shown, by giving gonadotrophic extracts prepared from pregnancy urine luteinization is caused in ovarian grafts even in the male. Lipschütz (1938) states that if a vagina be transplanted into a castrated male guinea-pig bearing an ovarian graft it will remain in an oestral condition for weeks, a result which can be correlated with the effects on the mammae and attributed, like them, to a persistent production of FRH by the host's pituitary and an absence or insufficient production of LH.

Lipschütz (1935) believes that FRH, while it does not itself cause the development of corpora lutea, greatly augments the power of LH to do so, an opinion which had already been advanced by Fevold, Hisaw & Greep (1934). When comparisons were made between the effects of giving to infantile female rats a certain amount of extract of pregnancy urine plus anterior pituitary substance taken from male and female guinea-pigs respectively, the amount of luteinization induced in the rats by the pregnancy urine plus male pituitary was much greater than that caused by pregnancy urine plus female pituitary (Table 35).

TABLE 35. The effect on the infantile rat's ovary of placental gonadotrophin plus pituitary gland from male or female guinea-pigs (Lipschütz, 1935)

Sex of pituitary donor	Average amount of pituitary substance given (mg.)	Average weight of ovaries as showing amount of luteinization (mg.)
Male	57·5	39·4
Female	60·6	20·0

Lipschütz & Del-Pino (1936) assayed human pituitaries on rats and found a similar difference between the sexes, the 'coefficient of luteinization', as they call it, being higher in the pituitary of the male. Pfeiffer (1936) grafted ovaries into male mice and, like Martins and Lipschütz, he noticed that corpora lutea did not develop in the grafts. If, however, luteinizing hormone were injected into the host the grafted ovaries underwent luteinization. Pfeiffer noted the absence of luteinization in the implanted ovaries in male mice whose testes had been excised and grafted ectopically, in non-castrated males, and in males castrated after puberty. On the other hand, if castration had been done at birth the grafted ovaries produced follicles which became luteinized, showing that the pituitary in these circumstances produces both FRH and LH. Implantations of testis into newborn females, he found, suppressed the luteinizing function of the pituitary. Ovarian transplants did not have a corresponding effect on the pituitary in the male. Pfeiffer concludes that the sex-type of the hypophysis is not genetic but depends upon hormones elaborated by the gonads during infancy.

The size of the pituitary in the two sexes.   In most of the adult animals, including man, in which comparisons have been made, the pituitary is larger in the female than in the male, even though the female has a lower bodyweight. According to Rasmussen (1931) this sexual difference in the size of the pituitary continues throughout life. Oestrogen may be the causal agent (see pp. 88, 277). Hatai (1913b) states that the greater size of the pituitary in the female rat over that of the male begins at about fifty days of age, and that gonadectomy is followed by an increase in the weight of the pituitary in the male but not in the female rat. The median sagittal plane of the pituitary fossa in a series of adult human skulls examined by Burrows, Cave & Parbury (1943) was found to be considerably larger in the male than in the female. Unfortunately the relationship of this measurement to the size of the pituitary is not known (see p. 92).

Nutrition, including vitamin deficiency and general physical development

The importance of nutrition in the breeding of domestic animals and in egg-laying by fowls is generally recognized. Nutritional deficiencies may affect the reproductive organs indirectly by their influence on the gonadotrophic activity of the pituitary, or directly by their influence on the gonads.

(a) *Chronic underfeeding.* Papanicolaou & Stockard (1920) kept female guinea-pigs on a daily ration of 20 g. of carrot, which was enough to support life but prevented the animals from attaining normal weights, and they found that oestrus was suppressed or delayed under this regimen. Their observations were confirmed by Loeb (1921), who noted in undernourished guinea-pigs an absence of mature ovarian follicles and a suppression of oestrus. The same result occurs in rats which are maintained on a diet which is qualitatively sufficient and defective only in amount (Evans & Bishop, 1922a). Mason (1933) placed male rats weighing between 105 and 112 g. on decreasing amounts of bread and milk until their weights were reduced to between 70 and 80 g., at which it was maintained. This treatment caused severe damage to the seminal epithelium, which could be repaired by a generous diet. If younger rats were underfed their testes failed to reach the mature condition. Mason & Wolfe (1930), having submitted rats to a prolonged period of underfeeding, tested the gonadotrophic potency of their pituitaries as compared with those of littermates or stock rats of the same weight at death which had been given unrestricted amounts of the same diet. The pituitaries of the underfed animals showed by this comparison a significant decrease of gonadotrophic potency. Moore & Price (1932) examined the effects of gonadotrophins on the injured generative organs of male rats which had been subjected to underfeeding, and found that these hormones, as well as testicular extracts, caused restoration of the atrophied accessory organs, though testicular extracts did not repair the seminal epithelium. Werner (1939) kept adult spayed and non-spayed female rats on a diet which was correct in quality but deficient in quantity. After periods of 1 to 4 months the rats had lost between 30 and 50 per cent of their original bodyweights, which had been between 180 and 260 g. when the restricted feeding began. The rats together with controls were now killed and their pituitaries were implanted into 22-day-old female mice at the rate of one pituitary to each mouse. The mice were killed 96 hours later and their ovaries and uteri weighed. The results show that the semistarvation had caused a considerable reduction in the gonadotrophic potency of the pituitaries in the underfed rats (Table 36).

TABLE 36. The effect of underfeeding on the gonadotrophic potency of the rat's pituitary as tested on immature female mice (Werner, 1939)

Donor of rat pituitary	Average weight of ovaries in test mice (mg.)	Average weight of uterus in test mice (mg.)
Adult, normal	4·2	29·5
Adult, underfed	2·6	9·8
Adult castrate, well fed	8·3	34·4
Adult castrate, underfed	3·9	18·4

From these experiments it may be concluded that the defects caused in the reproductive system by prolonged underfeeding are reversible and are attributable mainly to an inability of the pituitary of the underfed animal to produce a sufficient supply of gonadotrophin.

Mills (1939) states that the onset of sexual functions is affected by general physical development. The writer has observed in rats and mice that descent of the testes is correlated to some extent with the individual's rate of growth. In a single litter the date at which the testes descend in the different individuals has some correspondence with the weight of each animal; that is to say the heavier the individual the sooner will his testes descend into the scrotum. These casual observations, which at present are unsupported by arithmetical data, if established by further work, would help to illustrate the favouring effect which good nutrition has upon the production of gonadotrophin by the pituitary.

(b) *Deficiency of vitamin-B complex.*   Marrian & Parkes (1928) examined the testes of pigeons which had been kept on a diet deficient in vitamin-B, and described the changes as similar to those caused by retention of the testes in the abdomen, vasectomy, or exposure to sterilizing doses of X-rays; that is to say, there was a successive degeneration of the spermatozoa and spermatocytes in order of their maturity, with a relative or absolute increase of the interstitial tissue. In further experiments (1929 a) they compared the effects on female rats of a complete diet deficient in quantity but containing an excess of vitamin-B with those of an unlimited diet deficient in vitamin-B. The results of the two different diets on the ovaries and accessory genital structures were the same. Oestrus was suppressed, the ovaries contained atretic follicles without new corpora lutea, and the uterus resembled that of a rat which had been spayed. Rats treated in either of these two ways went into oestrus within 2 or 3 days if given daily subcutaneous injections of a watery suspension of normal rat's pituitary, each dose representing the pituitary of one male rat; and the ovaries now showed numerous corpora lutea, while the uterus was in the condition normal to oestrus. Moore & Samuels (1931) repeated these experiments, using male rats. They found that semistarvation and deprivation of vitamin-B produced similar effects on the gonads. At the end of 73 days of experimental feeding the testes of the rats in both groups appeared to be normal except perhaps for some reduction of the cytoplasm of the interstitial tissue; and spermatogenesis was in progress, though the prostate and seminal vesicles were atrophic and not secreting. After a biopsy some of the animals were given 1 r.u. each of androgen daily, and others were given pituitary extract. The consequences in both cases were alike; within 10 days the prostate and seminal vesicles were restored to a normal secreting condition. Moore & Price (1932) report similar results. Parkes (1928 b) had already shown that rats kept on a diet deficient in vitamin-B remained in a condition of anoestrus and that if, when this stage had been reached, daily injections of 10 m.u. of oestrin were given, the changes associated with oestrus appeared in the accessory organs unaccompanied by ovulation. He thus showed that a deficiency of vitamin-B does not cause anoestrus by rendering the accessory organs incapable of responding to oestrin (see also Moore, 1935 a).

Evans & Simpson (1930) caused anoestrus in rats by maintaining them on a diet deficient in vitamin-B. When this stage had been reached they gave the rats implants of normal pituitary gland, with the result that follicular maturation, oestrus and ovulation ensued. They also kept male rats on a vitamin-B deficient diet from the time of weaning. Forty days later the rats were killed and their pituitaries were tested for gonadotrophic potency on immature females in the usual way. The results showed that the pituitaries of the rats deprived of a full ration of vitamin-B were deficient in gonadotrophic potency when compared with those of littermate and other controls which had received a normal diet.

From these various experiments it appears that (1) semistarvation and (2) a deficiency of vitamin-B complex with an otherwise complete diet both cause a reduction of the gonadotrophic potency of the pituitary, with a suppression of ovarian and testicular function as a consequence of the pituitary defect. The mechanism of sterility in these instances differs from that of sterility caused by deficiency of vitamins A or E, as will appear.

(c) *Vitamin-A deficiency.* One of the earliest signs to be detected in rats which have been kept on a diet deficient in vitamin-A is persistent cornification of the vagina, a lesion which precedes xerophthalmia. Evans & Bishop (1922a) pointed out the fallacy of regarding the condition as a continuous oestrum. The state of the ovaries and the behaviour of the rats toward males show that the cornification is not associated with oestrus; moreover, the cornification does not disappear if the rat becomes pregnant (Evans, 1928), nor is it prevented by removing the ovaries. Keratinization of various epithelial structures in the body besides the vagina is an outstanding lesion caused by vitamin-A deficiency (Wolbach & Howe, 1925), but an earlier defect seen in young animals is underdevelopment of the gonads. Parkes & Drummond (1926), experimenting with rats, found that in the earlier stages of vitamin-A deficiency, though the seminal vesicles were very small, spermatogenesis occurred, provided that the testes were in the scrotum, and in females ovulation and the formation of corpora lutea took place. They thought that at this period sterility was due to physiological debility and disinclination to copulate rather than to histological changes in the gonads. In more advanced stages in males a gradual degeneration of the seminal epithelium occurs (Goldblatt & Benischek, 1927; Mason & Wolfe, 1930; Evans, 1931; Mason, 1933), until in some tubules only a few cells of Sertoli remain; and accompanying this testicular degeneration is a pronounced atrophy of the prostate and seminal vesicles (Mason, 1933). Mason & Wolfe (1930) placed a number of 25-day-old rats on a vitamin-A deficient diet, castrating half the number. When xerophthalmia was well advanced, their pituitaries, and pituitaries from normal control rats of the same weight and sex, were transplanted into immature littermate females. Using the weights of the ovaries of these recipients to indicate relative degrees of gonadotrophic potency in the grafted pituitaries, they found the pituitaries of the non-castrated vitamin-A deficient rats were 43 per cent and those of the castrated vitamin-A deficient rats 100 per cent more potent than those of the control animals. These results suggest that vitamin-A deficiency, apart from its degenerative effect on the testes, may enhance the gonadotrophic potency of the pituitary. Such a result would contrast with those caused by underfeeding or a lack of vitamin-B, which, as already noted, reduce the gonadotrophic potency of the pituitary. Van Os (1936), however, has reported experiments which indicate a diminished potency in the pituitaries of rats maintained on a diet deficient in vitamin-A. He kept male rats on a vitamin-A deficient diet until xerophthalmia appeared; the rats were then killed. To some of these rats intramuscular injections of gonadotrophin had been given during the last 16 to 19 days of life. The testes, seminal vesicles and preputial glands were weighed and examined microscopically. The testes from rats which had not received gonadotrophin showed degeneration of the seminal tubules, without any obvious change in the interstitial glandular tissue; the seminal vesicles were small, with low cuboidal epithelium, and showed but slight excretory activity. The testes and seminal vesicles of untreated control rats were normal, and so also were those which, although suffering generally from the effects of vitamin-A deficiency, had received injections of gonadotrophin. Some of the results are shown in Table 37.

TABLE 37.   The effect of gonadotrophin on vitamin-A
deficient rats (Van Os, 1936)

Rats aged 94 to 96 days at autopsy.

Number of rats	Treatment	Average weight of rats (g.)	Average weight of both testes (mg.)	Average weight of both seminal vesicles (mg.)
4	Vitamin-A deficiency	120	1,185	150
2	Vitamin-A deficiency and gonadotrophin	120	1,550	600
1	None	205	2,000	610
2	Gonadotrophin	174	2,340	1,030

From these discrepant results it appears that the effects of vitamin-A deficiency on the pituitary need further experimental elucidation.

(d) *Vitamin-E deficiency.* The degenerative changes which occur in the spermatic epithelium of animals kept on a diet deficient in vitamin-E have been described by Mason (1933), Evans & Burr (1927), and others. They consist of a liquefaction of the chromatin of the degenerated spermatozoa and their fusion into multinucleated masses. As the condition progresses the spermatocytes and spermatogonia undergo a similar process of destruction, and the Sertoli cells may be affected in the later stages though many remain as a rule even when the condition is advanced. Some, including the writer, have thought that this degeneration of seminal epithelium is accompanied by an increase of the interstitial glandular tissue. Others have failed to observe such an increase or have explained an apparent increase as attributable, not to glandular cells, but to macrophages which have accumulated in the intertubular spaces of the testis owing to the presence of intensive epithelial degeneration (Evans & Burr, 1927). Nelson (1933) tested the gonadotrophic potency of the pituitaries of male rats which had been kept on a diet deficient in vitamin-E and found the potency raised, though the increase was not so great as that caused by castration. The pituitaries of males deprived of vitamin-E showed also histological changes similar to those which follow castration. On the other hand Nelson did not find any enhancement of gonadotrophic potency in the pituitaries of females which had been maintained on a similar E-deficient diet.

Like that which follows castration, the enhancement of pituitary activity caused by vitamin-E deficiency is confined to the production of FRH, the formation of LH being impeded. This is exemplified in experiments by Rowlands & Singer (1936), who discovered that to induce ovulation and luteinization in the rabbit's ovary pituitary extracts from E-deficient male or female rats had only half the potency of similar extracts made from normal rats. Drummond, Noble & Wright (1939) found that after an E-deficient diet the pituitaries of male rats showed an enhanced gonadotrophic potency when tested on immature rats, whereas those of female rats which had been fed in the same way had a reduced potency. Observations by McQueen-Williams (1934a) do not accord with those of the other workers just quoted; he assayed the pituitaries of female rats which had been kept on an E-deficient diet and found that they had more than twice the gonadotrophic potency of those taken from normal females, so that in his experiments the deprivation of vitamin-E seems to have affected the pituitaries of female rats in the same way as those of males.

These experimental results suggest that a lack of vitamin-E exerts in males a specific deleterious action on the testes, and that the changes which ensue in the gonadotrophic potency of the pituitary are in the main secondary to the testicular injury. The pituitary also suffers from the dietary defect, for in female rats maintained on an E-deficient diet Barrie (1937) noticed degranulation of the eosinophile and basophile cells, these histological changes being accompanied by signs of pituitary deficiency, including stunting of bodily development and hypoplasia of the thyroid.

The foregoing remarks indicate that either (1) chronic underfeeding or (2) a deficiency of vitamin-B complex reduces the gonadotrophic potency of the pituitary by a direct action on that gland; whereas a deficiency of vitamin-E injures the testis directly and so, like castration, brings about an enhanced production of FRH and a lessened production of LH by the pituitary. In females, experimental results do not all agree, but the balance of evidence suggests that a vitamin-E deficiency directly causes some injury to the pituitary and so reduces its gonadotrophic potency.

Although there may be some difference of opinion about the exact manner in which the effects of vitamin-E deficiency on the testis are produced there seems to be no doubt that they are irreversible or nearly so; in this respect they are unlike those caused by underfeeding or by a lack of vitamins A and B.

A possible source of misunderstanding when the results of vitamin-E deficiency are being studied is the fact that this deficiency apparently interferes with the storage of vitamin-A and so reduces its reserves. Internal secretions other than those of the gonads may affect the gonadotrophic activity of the pituitary. The production of these internal secretions may be interfered with by defects in nutrition and so cause gonadal lesions which might be misinterpreted as direct consequences of the nutritional defect (see Biddulph & Meyer, 1941).

The influence of the thyroid on the gonadotrophic potency of the pituitary. Evans & Simpson (1930) gave 1 g. of fresh beef thyroid daily to female rats. Thyroidectomy was done on littermates of these rats and yet others were kept as controls. After a time the rats were killed and their pituitaries were weighed and assayed for gonadotrophic activity. The relative weights were as follows:

Fed with thyroid (mg.)	Untreated (mg.)	Thyroidectomized (mg.)
8·8	10·0	11·0

The pituitaries of the thyroid-fed rats, though smaller, had greater gonadotrophic potency than those of the animals whose thyroids had been removed.

For the influence of the adrenals on the pituitary see pp. 443, 448.

Chapter III. *Factors which influence the Reaction of the Gonads to Gonadotrophins*

Age. Season. Nature and mode of preparation of the gonadotrophin. Synergism and augmentation. Delayed absorption. Divisional dosage. Site of introduction. Acquired resistance to the action of gonadotrophin. Difference in responses by different species. Temperature.

Age

Tests for pregnancy in women, which depend on the presence of gonadotrophin in the urine, have shown incidentally that precocious puberty can be induced in mice and rats by supplying gonadotrophin in excess. It follows that the gonads of these animals can respond to gonadotrophin at a date which precedes the normal onset of puberty. From this and other experimental facts already quoted it may be assumed that puberty in normal circumstances depends mainly on the quantity or quality of gonadotrophin produced. Nevertheless the reactivity of the gonads must be considered, for there is, in some animals at least, an early stage of postnatal life during which the gonads will not respond to gonadotrophin in the same way as they would later on. When we review the experimental work on which these remarks are founded, it becomes apparent that the interstitial glandular elements of the gonads will react to gonadotrophin at a stage of development when the germinal elements still fail to do so.

(*a*) *Early responsiveness of the glandular interstitial cells of the gonads to gonadotrophin*. That the interstitial gonadal tissue may become susceptible to stimulation by gonadotrophin at a stage of life when the germinal tissues are inert was shown by Domm (1931), who implanted fowl pituitaries daily into immature brown leghorn chicks, beginning when they were between 28 and 59 days old. These implants caused enlargement of the gonads and accessory sexual structures, including hypertrophy of the oviduct and comb, while the ovarian follicles and seminal tubules remained unaffected; the interstitial glandular elements in the gonads of both males and females were not only enlarged, but were functionally active. Domm & Van Dyke (1932) amplified these experiments and obtained the same results, except that they found the ovarian follicles were somewhat larger than in controls, though there was no maturation. They used white leghorns, and gave them daily injections of a gonadotrophic extract ('hebin') prepared from sheep's pituitaries; the fowls were between 21 and 47 days old when the treatment began. Asmundson & Wolfe (1935), giving daily injections of pregnant mare's serum into leghorn fowls which were between 42 and 91 days old at the start of the experiment, obtained confirmatory results, that is to say the ovaries were increased in weight, though maturation was absent, and the accessory sexual structures were enlarged. In cockerels the testes were swollen owing to hypertrophy of the interstitial tissue and the tubules, but there was no spermatogenesis. Bantams, treated in the same way, showed precocious sexual behaviour, and premature mating, which was infertile, occurred. When gonadotrophin was injected into incubating eggs the same gonadal enlargement was caused in both male and female embryos owing to an enhanced development of the interstitial glandular elements, but in this experiment there was no evidence that the hypertrophied tissue had produced any gonadal hormone. The combs were not enlarged, although before chicks are hatched the comb will become enlarged if androgen is injected into the egg. No stimulation of the germinal cells was detected.

Byerly & W. H. Burrows (1938) found that the testes of newly hatched chicks become enlarged in response to gonadotrophic extracts or pituitary implants, a fact which they suggest might be used for purposes of assay.

Results comparable with those obtained in fowls were obtained by Aron (1933) in foetal guinea-pigs. He opened the abdomens of pregnant guinea-pigs and injected the embryos, through the uterine wall, with pituitary extract. The embryos were examined 48 hours later, and the treatment was found to have caused hypertrophy of the interstitial tissue in the foetal ovaries and testes. The primordial follicles of the ovary were unaffected. Selye & Collip (1933) gave daily to rats a gonadotrophic extract which was rich in LH, beginning on the 6th postnatal day. After ten injections the ovaries appeared normal to the naked eye but under the microscope showed enlargement of the thecal cells, which resembled those of corpora lutea; the granulosa cells were not luteinized and there were no signs of follicular maturation. The change caused in the thecal cells, they say, was like that caused by the same gonadotrophin when given to hypophysectomized female rats. In the course of their experiment these workers found that if injections of the gonadotrophin were continued from the 6th postnatal day onward they actually prevented follicular maturation, of which there was no sign on the 26th day, whereas if the injections were begun on the 21st day and continued till the 26th premature maturation of the follicles was induced and was followed by the formation of normal corpora lutea.

(b) *The influence of age on the response of germinal cells and follicles to gonadotrophin.* Although the germinal cells in both sexes become susceptible to the influence of gonadotrophin at a later stage in life than does the interstitial gonadal tissue, yet the germinal tissue will respond to artificially supplied gonadotrophin at a period before puberty. Corey (1928) injected fresh pituitary material, taken from adult rats, into foetal rats and young rats at various intervals after birth, and ascertained what, if any, effects had been produced on the gonads. Observations made on 156 rats treated in this way showed that prenatal injections did not hasten differentiation of the gonads. The earliest effect on the germinal cells in both sexes appeared between the 10th and 15th days of postnatal life. The same irresponsiveness in the ovaries of rats up to the 17th postnatal day was recorded by Smith, Engle & Tyndale (1934). They gave daily injections of gonadotrophin to young rats, beginning on the 6th day after birth. The doses, which were large enough to cause a pronounced response in the gonads of adult rats after hypophysectomy, did not bring about any enlargement of the ovaries or uterus, and the vagina remained closed. Similar treatment at the 21st day of life caused enlargement of the ovary, increased development of the follicles and opening of the vagina. Irresponsiveness of the rat's ovary to gonadotrophin between the 10th and 15th postnatal days has been noted also by Saunders & Cole (1936). In rats, Collip (1935) says, this refractory period lasts until the 18th day after birth. Cole (1936), testing gonadotrophin prepared from pregnant mares' serum, found that it caused premature ovulation in most rats on the 25th day; it failed to induce maturation of the follicles in 10-day-old rats, though at this age an increase of the interstitial tissue of the ovary followed the injections. Rabbits give much the same results as those just quoted. Clauberg (1932) found that the ovaries of immature rabbits weighing between 600 and 800 g. showed little response to gonadotrophin, even when given in very large doses, whereas the ovaries of rabbits of between 1,200 and 1,400 g. reacted readily.

Hertz & Hisaw (1934) studied the effects of gonadotrophic extracts prepared from sheep's pituitaries on young rabbits, and they found that rabbits 4 weeks old and weighing between 500 and 700 g. showed no ovarian response to doses of extract which were enough to cause a maximal response in rabbits of the same breed 12 or 13 weeks old and weighing 1,300 g. Rabbits of 10 weeks old gave a meagre response to the same doses.

Corey (1928) noticed that pituitary gonadotrophin fails to affect primordial follicles in the rat and is only able to stimulate follicles which already show antra. A similar observation has been made by Hertz & Hisaw (1934). These authors found additional evidence that the failure of the infantile ovary to respond to gonadotrophin is due to the condition of the ovary itself. They showed that the transference of an infantile rabbit's ovary to an adult does not render the grafted ovary responsive to stimulation by pituitary gonadotrophin.

Freed & Coppock (1936) state that in guinea-pigs the possession of antra is not essential for the responsiveness of ovarian follicles; in older guinea-pigs the smallest follicles respond to gonadotrophin. Zéphiroff & Dobrovolskaya (1940) have suggested that an antisexual factor is formed in the young. From the urine of a girl of 4 years they extracted a substance which, given to immature rats weighing between 14 and 20 g. at the rate of twenty injections a month, each injection representing 1 c.c. of urine, prevented descent of the testes and checked the development of the penis and seminal vesicles. In females the same doses prevented vaginal opening and maturation of ova.

Hypothetically a failure of response might be the result of inhibition, a lack of something essential for a response, or some other cause. At present a detailed discussion of these possibilities would hardly be profitable; though reference may be made to some observations by Freud (1938), who found that intraperitoneal injections of pituitary extract into infantile rats will prevent the development of corpora lutea under the influence of subcutaneous doses of chorionic gonadotrophin.

The following observations by Moore, Simmons, Wells, Zalesky & Nelson (1934) are also of interest in connection with this subject. Studying the seasonal changes in the pituitary of the ground squirrel, they found that, in both males and females, during the sexually quiet period no gonadotrophin could be detected in the pituitary, which however contained much demonstrable gonadotrophin on the approach of the breeding season. Castration, even though done in the non-breeding season, was followed by an increased production of gonadotrophin; from which it appears that the gonads of the adult during the season of sexual rest, like those of infancy, have an inhibitory action upon the output of gonadotrophin by the pituitary. Whatever the circumstances may be which allow the gonads at puberty to respond to gonadotrophin, there is evidence that the immature gonad like that of the adult restrains the gonadotrophic activity of the pituitary.

(c) *Irresponsiveness of the ovary to gonadotrophin after the menopause.* With advancing age another period of ovarian irresponsiveness ensues, and like that of infancy, appears to be due to an inability of the gonad to react rather than to a deficient supply of gonadotrophin, for the production of this in adequate amount continues. The menopause cannot be explained by supposing an absence of ovarian follicles, for these may still exist, in relatively small numbers, after the menopause has become established.

Perhaps the cessation of fertility in the female has not the same cause in all species. Zondek & Aschheim (1927 b) gave pituitary implants to mice in which follicular maturation and sexual activity had ceased, and in some of these mice a renewal of oestrus and ripening of follicles ensued. In these mice it seems that the termination of sexual life must have been due to a pituitary and not to an ovarian inadequacy.

The facts learned by experiment concerning the gonadal reactions to gonadotrophin at different periods of life seem to be somewhat as follows:

1. During foetal life there comes a time when the gonads become enlarged in response to gonadotrophin. This enlargement affects the non-germinal elements only and is not accompanied by their functional activity.

2. Later, gonadotrophin causes enlargement and functional activity of the interstitial tissue, and perhaps some increase in size of the seminal tubules, but as yet there is no commencing maturation of ova nor spermatogenesis.

3. Still later, but before puberty would normally appear, there is a complete response to gonadotrophin including maturation of ova, spermatogenesis, and full functional activity of the interstitial glandular elements of the gonads.

These results, considered together with the fact that the pituitaries of foetal and immature animals possess some gonadotrophic activity, show that gonadotrophin can induce precocious puberty in the normal rat and rabbit only after a certain stage of gonadal development has been reached. The stage of gonadal reactivity may, it appears, be attained at different ages in the two sexes, and is reached in some species much earlier than in others.

Possibly the reactions under consideration are of the same nature as those induced by embryological organizers in competent tissues. When subjected to an appropriate chemical organizer a tissue will at one stage react by a specific development, and at another earlier or later stage of its existence will fail to respond to the same organizer.

Season

A seasonal bias has been noted in the awakening of the reproductive faculties of some animals whose breeding activities are limited to one part of the year only. The pituitary appears, almost with certainty, to be the prime mover in this periodical sexual activity. Nevertheless it would be hazardous to assume that the ability or inability of the gonads to respond plays no part in the phenomena. Until the matter has been elucidated by experiment, the possibility of a seasonal variation in the capacity of the gonads to react to an influx of gonadotrophin may be considered.

The Nature and Mode of Preparation of the Gonadotrophin

Gonadotrophic extracts do not all act alike. Differences have been noted between placental and pituitary gonadotrophins and between pituitary gonadotrophins obtained in different circumstances. As these matters have been discussed already (pp. 10, 12) they will not receive further attention here.

The mode of preparation of gonadotrophic extracts requires a reference, not only because it is possible to prepare from a given source two separate fractions one of which has mainly a follicle-ripening influence while the chief effect of the other is luteinization, but also because the presence of impurities may greatly influence a subsequent assay by retarding absorption from the site of injection, and so causing a prolongation and enhancement of the result.

Synergism and Augmentation

The fact that placental gonadotrophin has little effect when given to animals deprived of their pituitaries (Reichert, Pencharz, Simpson, Meyer & Evans, 1931) has suggested that some synergic principle might be formed in the pituitary. This idea received added force from observations by Leonard (1933 a), who gave injections of placental and pituitary gonadotrophin separately or together to immature mice. In all these experiments the increased weight of the ovaries produced by the combined injections was greater than could be accounted for by adding together the separate actions of the two hormones. Evans, Simpson & Austin (1933 a) also found that by combining placental gonadotrophin with pituitary extracts an enhanced effect could be obtained. Furthermore, they found that this enhancement was produced equally well by preparations containing large amounts of pituitary growth hormone or of gonadotrophin. Eventually they obtained a pituitary extract which was free from detectable growth hormone and gonadotrophin and yet exerted with prolan a synergistic effect when given to rats in doses as small as 27γ. The amplifying effect of this extract may possibly have been caused by a trace of FRH. Fevold, Hisaw & Greep (1934) showed that the effect on the ovary of a given dose of LH is augmented by the addition of FRH (see Jensen, Hauschildt & Evans (1942) for further observations on this subject).

Delayed Absorption

Several workers have found that the effect of a dose of gonadotrophin may be much increased by delaying the rate of its dispersion from the site of injection. This may be done by the addition of any substance which causes a slight inflammatory reaction, for example tannic acid (Fevold, Hisaw, Hellbaum & Hertz, 1933), zinc salts (Maxwell, 1934; Saunders & Cole, 1936; Deanesly, 1939), copper (Emery, 1937), casein and egg albumin (Saunders & Cole, 1936), blood or plasma from another species (Casida, 1936), milk, lemon, thyroid, beef-liver (Hellbaum, 1936).

The writer has ascribed the enhanced effect of gonadotrophin under the influences just quoted to delayed dispersion because (1) it is known that delayed dispersion will increase the effects of injected hormones and (2) it is known that inflammation impedes the dispersion of substances from inflamed tissues. It is possible, however, that in particular instances some additional factor may have to be considered. For example, Fevold, Hisaw & Greep (1936*b*) say that the augmentative influence of copper salts on the responses to gonadotrophin cannot be attributed to inflammatory fixation only, because (1) these salts do not have an augmentative effect in the hypophysectomized rat, and (2) they will induce ovulation in rabbits when given alone. Others, too, have found it difficult to accept inflammatory fixation as a complete explanation covering every instance. Further experimental results are needed before the matter can be discussed adequately.

Divisional Dosage

The results caused by injecting a given weight of a gonadotrophic extract will vary considerably according to whether it is administered as a single dose or as a number of separate doses with an interval of time between each; the effects produced by the latter method being much in excess of those following a single dose.

Maxwell (1934) has demonstrated these facts by giving daily doses of gonadotrophin to female rats between 21 and 23 days old, for 4 days, killing them on the 6th day, and noting the condition of the vagina, uterus and ovaries. Some of his results are shown in Table 38.

TABLE 38.   Increased reaction of infantile rats to gonadotrophin as the result of (1) division of doses and (2) delayed rate of absorption (Maxwell, 1934)

Method	Period of injections (days)	Average weight of ovaries on 6th day (mg.)
Single injections per diem of 3 mg. sheep-pituitary extract	4	25
Six injections per diem of 0·5 mg. sheep-pituitary extract	4	55
Single injections per diem of 3 mg. sheep-pituitary extract *plus* ZnSO$_4$	4	76

By combining divided doses with the addition of a 10 per cent solution of zinc sulphate even larger ovaries were obtained. Maxwell states that the efficiency of pituitary gonadotrophic extracts may be increased as much as fiftyfold by a sufficiently distributed dosage and the addition of zinc salts. If the zinc salt is injected separately from the gonadotrophic extract and into another part of the body it has no effect on the ovarian responses.

Site of Introduction

Another factor which may influence the degree of response to injections of gonadotrophic extracts is the site of injection. Evans & Simpson (1930) gave 1 c.c. of an aqueous extract of placenta daily for 3 days to forty-four immature rats. The injections were given subcutaneously to twenty-two of these rats and intraperitoneally to the remaining twenty-two. The results (Table 39) show that the subcutaneous route was more effective than the intraperitoneal.

TABLE 39.   Comparative effectiveness of subcutaneous and intraperitoneal injections of placental gonadotrophin in rats (Evans & Simpson, 1930)

Route	Number of immature rats	Number with patent vagina and oestrous smear on the 4th day
Subcutaneous	22	20
Intraperitoneal	22	9

Collip & Williamson (1936) also have stated that a gonadotrophic extract of pregnancy urine is more effective when given subcutaneously than when injected into the peritoneal cavity, and Freud (1937) has made the same observation concerning a beef pituitary extract tested on young rats (Table 40).

TABLE 40. Comparative effectiveness of subcutaneous and intraperitoneal injections of pituitary gonadotrophin in rats (Freud, 1937)

Number of rats	Dose of gonadotrophic extract (mg.)	Site of injection	Mean bodyweight (g.)	Mean weight of ovaries of test rats (mg.)
5	160	Subcutaneous	48	25
4	160	Intraperitoneal	41	12

Acquired Resistance to the Action of Gonadotrophin

Experimental work has shown that repeated injections of a gonadotrophic extract may cause the recipient to become resistant to the hormonic action of that extract, either by the formation of antibodies or by some other process not yet understood. From this general statement a number of questions arise. To what extent, it may be asked, is such resistance attributable to the formation of antibodies against foreign proteins; will pituitary gonadotrophin produce a specific resistance to that formed by the placenta; do FRH and LH induce immune reactions against each other; can resistance to gonadotrophin be induced in an individual by material derived from his own tissues? A clear-cut and unreserved answer is not yet possible to every one of these questions, nevertheless a detailed consideration of them is required because of their importance; and this consideration will demand, the writer fears, a summarized mass of evidence which may be tiresome to read. The following experimental results are quoted as a groundwork for subsequent discussion. They prove that resistance to the gonadotrophic action of a potent extract may become established as a consequence of its repeated administration.

Bourg (1930a) noted that if human placental gonadotrophin be injected daily into mice the ovarian reaction gradually declines after reaching a maximum on the 5th day. Zondek (1935) carried out a similar experiment on mice with the same result. Collip, Selye & Thomson (1934) reported the outcome of another experiment of this kind. They daily injected pituitary gonadotrophin obtained from various animals into twenty-two female rats from the 21st day of life onward and killed the rats after 6 to 10 weeks of treatment. The ovaries of these rats at death consisted mainly of interstitial tissue and did not show mature follicles or fresh corpora lutea; large cells with eccentric nuclei, resembling the cells seen after castration or thyroidectomy, appeared in the pituitary together with other histological changes.

Bachman, Collip & Selye (1934) and Selye, Bachman, Thomson & Collip (1934) have described additional experiments, showing that when an injection of human placental gonadotrophin is given to a rat which has been rendered insensitive to its action, the gonadotrophin disappears from the blood of that animal more rapidly than from the blood of a normal rat. Female rats of 21 days were given daily subcutaneous injections of 100 r.u. of human placental gonadotrophin during 8 months. At the end of this period they and untreated control rats were given a single intravenous dose of 100 r.u. of the same preparation. All were killed 1 hour later, the blood of each animal being collected separately and injected in three doses of 5 c.c. each on three consecutive days into an immature rat of 21 days. With one exception the blood obtained from rats which had been previously treated with repeated doses of placental gonadotrophin evoked no response in the immature test animals, whereas the blood of all the rats which had undergone no previous treatment caused positive ovarian responses. Selye and his colleagues found also that by giving the blood of resistant rats to immature rats simultaneously with human placental gonadotrophin the action of the latter on the ovaries of the immature rats could be prevented. Wolfe (1935 a) gave daily injections of 25 m.u. of gonadotrophin extracted from human pregnancy urine to twenty mature female rats for 140 days. When the rats were killed at the end of this period he found that the pituitaries and ovaries in most of them were normal or below normal in weight, which is not what would be expected if the injections had been given for a short time only. Thompson & Cushing (1937) gave daily injections of a gonadotrophic extract of sheep-pituitary glands to a female collie dog. After this treatment the dog's serum was found to prevent the action on rats of gonadotrophin prepared from (1) the pituitaries of sheep and pigs, (2) the urine of women who had passed the menopause, (3) the urine of pregnant women and (4) the serum of pregnant mares. With infantile rats puberty was delayed as long as the antigonadotrophic serum was injected. Similar inhibitory results were obtained in several species including man. Rowlands & Parkes (1937) found that prolonged treatment of a goat with human placental gonadotrophin conferred on the goat's serum a power to counteract in rats and mice the activity not only of human placental gonadotrophin but of a gonadotrophic extract prepared from the human pituitary. Rowlands (1937) administered bovine pituitary gonadotrophin to rabbits daily for 8 months. Male and female adult rats of 140 g. were then injected daily for 14 days with varying doses of serum from these rabbits, and killed 24 hours after the last injection. In the females there was an inhibition of ovulation and an absence of corpora lutea; in the males there was an atrophy of the testes and accessory reproductive organs. The effects produced, according to Rowlands, were similar to those which follow hypophysectomy. The antagonistic body was present in the globulin fraction of the antiserum. Collip, Selye & Williamson (1938) say that the daily subcutaneous administration to rats of gonadotrophin from pigs' pituitaries eventually leads to an ovarian atrophy comparable with that occurring after hypophysectomy and accompanied by changes in the pituitary resembling those which follow castration. Hertz & Hisaw (1934) gave ovine and equine pituitary gonadotrophin twice daily to female rabbits; under this treatment the ovaries at first were increased in size, but after 10 days of continued treatment they began to decrease until at about the 15th day they had reverted to their original dimensions. This refractory state of the ovaries persisted in rabbits 10 weeks old but was only temporary in adults. Hisaw, Hertz & Fevold (1936) found that repeated injections of a gonadotrophic extract from horse and sheep pituitaries, given to a monkey, caused a fivefold increase in the size of the ovaries at the end of 15 days, but at the end of 20 days the granulosa cells began to degenerate and ultimately disappeared and the ovaries progressively decreased in size in spite of continued injections of the gonadotrophic extract. This refractory state was only temporary in adults but persisted in juvenile monkeys. Twombly & Ferguson (1934) made preparations of gonadotrophin from the urine of men with teratomata of the testicle and injected 100 m.u. of the extract daily for $3\frac{1}{2}$ months into female rabbits. At the end of this time the rabbits' sera were tested. Infantile mice of 6 to 8 g. were given 4 m.u. of the gonadotrophic extract and 0·5 c.c. of the rabbit serum in five doses during a period of 30 hours. In thirty-nine mice treated in this way there was no response. On the other hand, twenty-four infantile mice receiving the gonadotrophic extract alone showed large corpora lutea and patent vaginas; thirteen mice treated with the hormone together with normal rabbit serum also showed these positive responses.

(a) *To what extent can resistance which has been artificially induced against a gonadotrophic extract be ascribed to the formation of antibodies in response to foreign protein or protein-like material?* In all the experiments just mentioned the gonadotrophin used had been obtained from a species different from the one into which it was injected. In most of the work in which antigonadotrophic effects have been induced this has been the rule. A few exceptions in which animals appear to have become resistant to gonadotrophins derived from their own species will be considered later (p. 82).

We shall now refer to experiments which afford evidence that the resistance produced by repeatedly giving a particular gonadotrophin is largely due to an immunity developed against a foreign protein or protein-like body, and that it is not, as so much of the literature seems to imply, a reaction which can be described correctly as antigonadotrophic.

Selye, Collip & Thomson (1934*b*) reported that the persistent treatment of female rats with gonadotrophic extracts of human pregnancy urine led at first to enlargement of the ovaries, and that later the ovaries became subnormal in size and contained regressing corpora lutea without maturing follicles. It was apparent that resistance had developed against the preparation used. If now gonadotrophin prepared from pig's pituitary were given, the ovaries became greatly enlarged and corpora lutea developed. Fluhmann (1935) found that a gonadotrophic extract prepared from human pituitaries repeatedly injected into rats for a period of 100 days caused them to produce substances which inhibited the action of human gonadotrophic extracts whether prepared from the pituitary or the placenta but had no inhibitory action against a gonadotrophic extract prepared from sheep's pituitaries. Meyer & Gustus (1935; see also Gustus, Meyer & Dingle, 1935) state that the persistent administration of gonadotrophin obtained from the serum of pregnant mares to a rhesus monkey causes resistance to that hormone but not to human pituitary or chorionic gonadotrophin, or to gonadotrophin obtained from the pituitaries of sheep. The same sort of result was reported by K. W. Thompson (1937), who discovered that extracts made from ovine pituitaries, which produced antigonadotrophic reactions in rats, guinea-pigs, rabbits, dogs, monkeys and a horse, failed to cause any such reaction in sheep. Zondek & Sulman (1937) gave 250 r.u. of human placental gonadotrophin 4 times a week subcutaneously and twice a week intravenously to rabbits; at the end of 2 months the rabbits' ovaries were atrophic and luteinized and their serum showed antigonadotrophic activity toward human gonadotrophin when tested on infantile female rats. The antigonadotrophic factor was formed in both male and female rabbits whether their gonads were present or not. From this and other experiments Zondek & Sulman concluded that resistance to foreign proteins accounted for 99·5 per cent of the antigonadotrophic effect. They also detected some evidence of organ specificity; that is to say, placental gonadotrophin produced a resistance more readily against placental than against pituitary gonadotrophin, though the extracts used had been derived from the same species. Twombly (1936), having made a detailed inquiry into the subject, concludes that the so-called antihormones are antibodies formed in response to foreign proteins. He found that gonadotrophic extracts, which had been inactivated by heat or by ageing, still retained unabated their protective action against fresh gonadotrophic extracts derived from the same source. He gave 500 r.u. of a gonadotrophic preparation of human pregnancy urine to rabbits daily for 12 days, at the end of which time their sera afforded complete protection to infantile mice against the same gonadotrophic preparation, and on examining the rabbits' sera he obtained strong precipitin reactions. In another experiment he largely inactivated the gonadotrophic preparation by heating it in boiling water, so that the potency of a sample was reduced from 1,250 r.u. to between 2 and 4 r.u. per c.c. This given daily to a rabbit for 18 days caused its sera also to protect infantile mice against the action of fresh uninjured gonadotrophic extract. In contrast with these results, and supporting the conclusions drawn from them, were some observed in women. Three women were given 100 r.u. of placental gonadotrophin prepared from human pregnancy urine for 2 weeks, 6 weeks and over a year respectively. None of their sera caused any protective effect against the same preparation when given to mice. De Fremery & Scheygrond (1937), like Twombly, believe that the presence of active gonadotrophin may not be essential for the production of resistance to gonadotrophic extracts; for a rabbit serum could be made antigonadotrophic by repeated injections of a 'negligibly gonadotrophic' extract. They made an extract from the urine of male factory workers, 3·9 mg. of which when given to immature female rats caused no gonadotrophic response. Daily intravenous injections of 10 mg. of this extract were given to a rabbit during 30 days. Six immature female rats were then each given 1·2 c.c. of this rabbit's serum and at the same time 5 r.u. of 'pregnyl'—a preparation of human placental gonadotrophin—the serum being injected into one side of each rat and the gonadotrophin into the other side. The effect of the latter was completely inhibited by the rabbit serum, as was the case also when the hormone and the protective rabbit serum were given simultaneously in the same way to male rats.

Sulman (1937) carried out experiments from which he concluded that the antigenic properties of gonadotrophic extracts are independent of their potency as hormones; and he found that injections of 'purified prolan' into rabbits was not followed by the presence of prolan-antibodies in their sera; and serum giving pronounced antiprolan reactions showed no prolan antibodies. Sulman also gave combined injections of prolan and pig serum to test for hapten properties, with negative results. He concludes that pure prolan is neither an antigen nor a hapten. Werner (1938) adduced another sort of evidence. He found that two pituitary extracts, prepared by different methods from the same source and possessing equal gonadotrophic activity, varied considerably in antigenic potency. From his experiments he concludes that refractoriness induced to gonadotrophic extracts is an immune response to foreign protein associated with the hormone and not necessarily a reaction to the gonadotrophin itself.

Simonnet & Michel (1939) gave repeated injections of gonadotrophic extracts from various sources into different groups of rabbits and then tested their sera for inhibitory properties against several gonadotrophic extracts. Their results (Table 41) show that the inhibitory reactions were almost entirely confined to the species from which the immunizing extracts had been derived. There does not appear to be much evidence of any organ-specificity in the results of their experiments.

TABLE 41. The production of antigonadotrophic sera in rabbits by various gonadotrophic extracts (Simonnet & Michel, 1939)

Nature of gonadotrophic extract used to produce antigonadotrophic rabbit serum	Source of gonadotrophin against which the rabbit serum was tested			
	Human pregnancy urine	Pregnant mare serum	Horse pituitary	Ox pituitary
Human pregnancy urine (placental gonadotrophin)	+	–	–	–
Urine from woman after gonadectomy (pituitary gonadotrophin)	+	+	–	–
Serum of pregnant mare (pituitary gonadotrophin)	–	+	–	–
Horse pituitary	–	+	+	–
Ox pituitary	–	–	–	–

The + sign indicates an inhibition of gonadotrophic action

In other experiments Simonnet & Michel (1940) found that the presence of the pituitary was not essential for the production of antigonadotrophic property in the blood.

Collip & Anderson (1935), in a study of thyrotrophin, say that the responsiveness of an individual to administered hormone varies inversely with the hormone content or production of the subject's own gland, and that for each hormone there is an antagonistic substance. They regard the hormone-antihormone complex as a buffer system, the endocrine stability of the individual depending upon the absolute amount of hormone and antihormone.

In the majority of the experiments quoted above, organ extracts have been used, and these in many instances have been obtained from one species and tested on another. The impossibility of determining in these circumstances how much of a given result is to be attributed to the hormone and how much to the associated impurities renders the value of any result difficult to assess. In experimental work of this kind, until it has become practicable to isolate the trophins in pure form, the only sure methods of studying their effects without the interference of impurities seem to the writer to be by using homologous grafts of pituitary or homologous extracts, or else by joining two individuals in parabiotic union, and usually in these instances, as will be shown later, no immunity is induced.

Of particular interest, because of their importance to the clinician, are some observations by Spence, Scowen & Rowlands (1938), who point out that the prolonged administration of human placental gonadotrophin to human patients does not invoke resistance. Rowlands & Spence (1939) gave gonadotrophin prepared from pregnant mares' serum daily or twice daily for 12 weeks to nine boys with undescended testis. Blood samples were taken before treatment and at intervals later. Before treatment the patients' sera, when tested against the preparation used, showed no antigonadotrophic potency. After about 6 weeks of treatment the sera of most of the patients showed antigonadotrophic properties, which increased throughout the period of observation. None of the testes descended under this treatment. Three months later biweekly injections of human placental gonadotrophin were begun in six of the patients and in three of these the testes descended. In one patient, when the antigonadotrophic activity against the equine gonadotrophin was at a maximum, there was no power to neutralize human pituitary or placental gonadotrophin.

(b) *Will pituitary gonadotrophin produce a specific resistance to placental gonado-trophin, and vice versa?* When trying to answer this question we can depend but little on experimental results attained by injecting extracts obtained from one animal into another, for, as experiments just quoted have shown, immunal reactions to the foreign proteins thus introduced are apt to obscure other less pronounced effects. A few workers have thought they could discern some specific reactions attributable to the organ from which the gonadotrophic extract had been prepared. Zondek & Sulman (1937), as already mentioned, attributed 99·5 per cent of the resistance to gonadotrophin induced in rabbits by human placental gonadotrophin to immunity against foreign protein, the organ from which the gonadotrophin was derived playing little part in the reaction. Nevertheless they believed there was some slight evidence of organ-specificity in the degree of resistance obtained. Fluhmann (1935) gave repeated injections of human placental gonadotrophin to rabbits for a period of 74 days, at the end of which time their blood inhibited the action of human placental gonadotrophin but did not inhibit the action of human pituitary gonadotrophin.

Parkes & Rowlands (1936) submitted rabbits to a course of injection of an extract made from the anterior pituitaries of oxen. Serum from these rabbits injected intravenously into test rabbits inhibited the ovulation-stimulating activity not only of ox-pituitary extracts but of similarly prepared extracts of equine pituitaries and of saline suspensions made from the pituitaries of cows and sheep. Given intravenously to rabbits immediately after mating the serum inhibited the ovulation, which otherwise would have followed in 10 to 12 hours. Furthermore, Parkes & Rowlands found that gonadotrophin prepared from the pituitary did not lead to a protection against chorionic gonadotrophin. Conversely, resistance to chorionic gonadotrophin did not entail resistance to gonadotrophin prepared from the anterior pituitary.

Simonnet & Michel (1939) found that human pituitary gonadotrophin induced in the rabbit a resistance to gonadotrophin prepared from pregnant mare's serum, which is thought by some to be of pituitary origin (p. 80).

It seems possible from these results that there may be some degree of organ-specificity in the immunological reactions induced when gonadotrophic extracts derived from one species are injected into another.

(c) *Do FRH and LH give independent immunal reactions?* A satisfactory answer to this question can hardly be expected until these hormones have been isolated each in a pure condition. Meanwhile available evidence suggests that each of these hormones may perhaps produce its own immunal reactions independently of the other. Rowlands (1938) states that ox-pituitary contains relatively much luteinizing hormone and little FRH, whereas the pituitary of the castrated horse is rich in FRH and poor in LH, and he found that rabbits, if repeatedly injected during a period of 3 or 5 months with extracts of ox-pituitary, produced an antiluteinizing serum. When this serum was given to infantile rats with an extract of castrated-horse pituitary a pure follicle ripening effect was produced without luteinization. These findings were confirmed by Freud (1939), who tested Rowlands' rabbit serum on immature female rats. The rabbit serum when given to rats together with human placental gonadotrophin, or with a gonadotrophic extract of geldings' pituitaries, lessened the increase in size of the ovaries and prevented the luteinization which was caused by the other extracts given alone (Table 42).

TABLE 42.   The production of an antiluteinizing factor in the serum of rabbits after treatment with pituitary gonadotrophin rich in LH, from the ox, as shown by tests on immature female rats (Freud, 1939)

Material injected into immature rats	Weight of ovaries (mg.)	Average number of corpora lutea per ovary	Cornification of vagina
Human chorionic gonadotrophin	16·6	6·6	+
Human chorionic gonadotrophin *plus* Antiluteinizing rabbit serum	9	0	—
Pituitary gonadotrophin from castrated horse	82·4	65	+
Pituitary gonadotrophin from castrated horse *plus* Antiluteinizing rabbit serum	52·8	0	+

(d) *Can resistance to gonadotrophin be induced in an animal by an extract of its own tissues or the tissues of an individual of the same species?* Selye, Collip & Thomson (1934b) implanted rat pituitaries daily into female rats which were 21 days old when the experiment began, for a period of 68 days, at the end of which time the ovaries were of normal or subnormal size. They then gave doses of 250 units of human placental gonadotrophin on thirteen successive days in addition to the pituitary implants, with the result that great enlargement of the ovaries ensued. The experiment seems to show that an irresponsiveness to pituitary gonadotrophin can be induced in young rats by pituitary substance derived from individuals of the same species, and that this irresponsiveness is not accompanied by resistance to a gonadotrophic extract derived from another species. An experiment which agrees with this in some respects and differs in others is recorded by Collip (1937), who gave daily subcutaneous injections of a gonadotrophic extract of sheep's pituitary to lambs and continued the treatment for a period of several months during which the gonadotrophic potency of their sera was tested from time to time on infantile rats. The tests showed an enhanced potency at first; later an antigonadotrophic property was shown in the sera, which disappeared after cessation of the injections of gonadotrophin. The lambs' sera were antagonistic to gonadotrophin obtained from the pituitaries of sheep, cattle and pigs and from the blood of pregnant mares, but not to human placental gonadotrophin.

In contrast with these experiments, in which antigonadotrophic properties were induced in animals by homogenic material, is the following observation in which such material induced no antigonadotrophic effects. Katzman, Wade & Doisy (1937) treated female rats with repeated implantations of one or two rat pituitaries daily during a period of 9 months, at the end of which time the rats still had enlarged ovaries and their plasma augmented the ovarian response of immature rats to pituitary implants. In other words there was no sign of any antihormone in their blood.

Martins (1935) has called attention to the fact that when a spayed female rat is joined in parabiosis with a normal male or female partner for as long as 6, 8 or 15 months, the normal partner at the end of this time still shows signs of persistent stimulation of the gonads, evidence of the formation of antigonadotrophin in this instance being absent throughout. Du Shane, Levine, Pfeiffer & Witschi (1935) have made a similar observation (see also Rowlands & Spence, p. 81).

Viewing these experiments as a whole we feel justified in concluding that resistance to gonadotrophin cannot be induced in an adult by the individual's own gonadotrophin, or by preparations derived from a homogenic individual, unless the proteins have been so altered in the process of extraction as to have become capable of acting as antigens.

When contemplating this problem it has to be remembered that individuals of the same species, unless closely related, may show tissue resistance toward each other. The boundary line of immunological reaction does not necessarily correspond exactly with that which defines a particular species.

Resistance induced by intraperitoneal injection. Reference ought to be made here to experiments which seem to show that some resistance against a given gonadotrophin may be induced by injecting it into the peritoneal cavity. Leonard, Hisaw & Fevold (1935) found that the action on the ovaries of a human chorionic gonadotrophin ('antuitrin-S'), given subcutaneously to immature rats, could be largely prevented by the intraperitoneal injection of a gonadotrophic extract prepared from sheep's pituitaries. Other pituitary or placental gonadotrophins introduced into the peritoneum did not show this inhibitory effect against 'antuitrin-S' given subcutaneously. Collip & Williamson (1936), Freud (1937), Dingemanse & Freud (1939) and Fraenkel-Conrat, Simpson & Evans (1940) also have recorded this phenomenon.

Difference in responses by different Species

All species do not respond in precisely the same manner or degree to the same gonadotrophin. Smith (1935) noted that mice are much more sensitive than rats to FRH, whereas rats give a greater response than mice to extracts which are rich in LH. Both rats and mice, in response to extracts of pregnancy urine, show maturation of follicles and luteinization; monkeys given the same extract show no follicle growth.

Temperature

The responses of the ovary and testis to gonadotrophin are largely influenced by the temperature to which the gonads are exposed. Ovaries of mice grafted into the ear, where they are subjected to cooler surroundings than in their natural position, will produce an enhanced quantity of androgen in response to gonadotrophin. The testes, subjected to the warmth of the abdominal cavity, will cease to respond to normal supplies of FRH by spermatogenesis. Some experiments which demonstrate these altered reactions to gonadotrophin consequent on changes of the local temperature are discussed elsewhere (p. 161). An additional one will be quoted here.

Turner (1938) castrated infantile rats and transplanted their testes into the anterior chamber of the eye. In some instances the testes were grafted into the eye of a female. In this situation in males the testes responded to gonadotrophin by spermatogenesis. In non-spayed females the grafted testes produced sperm heads in a few instances only (6·5 per cent), but in spayed females spermatogenesis, as judged by the presence of sperm heads, occurred in 62 per cent. It was ascertained by tests with a thermocouple that the temperature of the anterior chamber of the eye was slightly lower than that of the scrotum and considerably lower than that of the abdomen (Table 43).

TABLE 43.   Comparative temperatures in different anatomical situations as bearing on spermatogenesis in the transplanted testis of the rat (Turner, 1938)

	Temperature (degrees C.)		
Anatomical locality	Rat I (male)	Rat II (male)	Rat III (female)
Anterior chamber of eye	31·28	29·75	30·91
Interior of scrotum	32·46	31·98	—
Subcutaneous tissue (abdomen)	35·32	35·00	34·96
Subcutaneous tissue (thorax)	35·91	35·62	35·01
Peritoneal cavity	34·90	35·88	36·12
Thoracic cavity	35·71	35·52	35·91

Chapter IV. *Factors which affect the Cytological Structure and Weight of the Anterior Lobe of the Pituitary*

Seasonal changes. The oestrous cycle. Pregnancy, pseudopregnancy and lactation. Gonadal hormones. Gonadotrophin. Sex. Castration, and destruction of seminal epithelium. Nutrition. Age.

SINCE Fichera (1905 a) called attention to the fact that castration is followed by enlargement of the anterior lobe of the pituitary gland with histological changes in its structure, many workers have confirmed and amplified his original observations; and the cytology of the pituitary has become a subject of much importance in endocrinological studies of various kinds. The matter will be discussed briefly here in order to correlate, if possible, the changing cellular constitution of the pituitary with its gonadotrophic functions and with the activities of the reproductive organs.

Normally the glandular cells of the anterior lobe of the pituitary are of three main types, namely chromophobe, eosinophile and basophile, the chromophobe being regarded as the parent of the other two. The relative numbers of these cells vary in different physiological conditions. Besides this changing numerical relationship the cells undergo alterations in volume and other features during different stages of sexual activity. In pregnant or castrated animals, cells with special and distinctive characters occur and are known as 'pregnancy cells' and 'castration cells' respectively. In the embryo chromophobes only are seen (Tuchmann, 1937).

Seasonal Changes

In animals with limited breeding seasons, during periods of sexual quiet when the output of gonadotrophin is at a minimum, small chromophobe cells predominate in the pituitary; a resumption of the sexual function is associated with an increase in the number and dimensions of the chromophiles. Cushing & Goetsch (1913, 1915) observed that sexual inactivity, whether associated with clinical evidence of hypopituitarism in man, or with experimentally produced pituitary deficiency in animals, or during hibernation, is accompanied by a reduction in the number and size of the chromophilic cells in the pars anterior. They confirmed the work of Gemelli (1906), who studied the seasonal changes in the pituitary of the marmot and showed that in this animal at the end of hibernation the pituitary swells, the glandular cells enlarge and the chromophiles stain more deeply.

Moore, Simmons, Wells, Zalesky & Nelson (1934) and Nelson (1936b) have studied the sexual rhythm of the ground squirrel (p. 72) and have correlated the histology of its pituitary with the functional conditions of its reproductive organs throughout the year. This mammal hibernates in the winter and the earliest signs of awakening ovarian function are visible at the beginning of January. A single oestrum occurs when the female first emerges from her winter quarters in April, and parturition takes place in May or June, after which the reproductive organs undergo involution. While the sexual activities are completely in abeyance the anterior lobe of the pituitary is reduced in volume, consists mainly of small chromophobe cells and contains no detectable gonadotrophin. With the gradual arousing of the sexual functions the pituitary increases in weight, the basophile cells, and the eosinophiles too, though to a less degree, increase in number and volume and as time goes on contain larger and more numerous granules. Coinciding with these histological changes in the pituitary is an increase of its gonadotrophic potency. At the end of the breeding season a degranulation of the chromophile cells takes place and eventually they are once more represented by small chromophobes, at which stage there is no detectable gonadotrophin in the pituitary. Kayser & Aron (1938) have observed a similar correspondence between the cytological condition of the pituitary and the reproductive faculties in another hibernating animal, the hamster.

The Oestrous Cycle

Rasmussen (1921) studied the histological changes which take place in the pituitary of the woodchuck (*Marmota monax*) at the time of its emergence from hibernation, when it becomes almost at once sexually active. The chief features of this brief period he found to be a considerable enlargement of the pituitary and an increase in the number and staining capacity of the basophilic cells. Reese (1932) noted in the rat during oestrum, as indicated by vaginal smears, that the majority of the pituitary eosinophiles are deeply packed with readily stained granules, whereas in dioestrum the eosinophiles are stained lightly and show less granulation. Wolfe, Cleveland & Campbell (1932) observed in the dog a pronounced increase in the number of the basophiles during prooestrum and their degranulation during oestrum; and in this animal they noted a decrease in the eosinophiles and an increase of the chromophobes during prooestrum. Cleveland & Wolfe (1933) reported that in the sow as in the dog in prooestrum there is an increase of basophile cells which undergo degranulation during oestrum. They (Wolfe & Cleveland, 1933 a) made differential counts of the rat's pituitary cells at different stages of the oestrous cycle. Their results are shown in Table 44 and are comparable with those obtained by Chadwick in the guinea-pig as shown in Table 45. When comparing the tables it may be remembered that there is no pronounced lutein phase in the rat's oestral cycle.

TABLE 44.   Pituitary cell-counts at different stages of the oestrous cycle of the rat (Wolfe & Cleveland, 1933 a)

Period of cycle	Eosinophiles	Basophiles	Chromophobes
Prooestrus	32·8	5·0	62·2
Oestrus	31·7	6·1	61·2
Metoestrus	33·6	4·2	62·2
Dioestrus	31·8	4·4	63·8

Chadwick (1936) checked the different stages of the oestrous cycle in guinea-pigs and made histological examinations of the pituitary with cell-counts. Basophiles, he found, were at a maximum in size and number during prooestrus, becoming degranulated in late oestrus and metoestrus. Both basophiles and eosinophiles remained scanty during the lutein phase. The chromophobes were at their highest level in dioestrus. These results are set out in Table 45.

TABLE 45.   Pituitary cell-counts at different stages of the oestrous
cycle in the guinea-pig (Chadwick, 1936)

Period of cycle	Eosinophiles	Basophiles	Chromophobes
Early prooestrus	52·0	1·8	46·2
Late prooestrus	52·8	2·1	45·1
Oestrus	53·2	1·4	45·4
Metoestrus	46·7	0·6	52·7
Early dioestrus	46·1	0·7	53·2
Late dioestrus	46·3	0·9	52·8

Summing up, it appears that the characteristic changes in the pituitary during the oestral cycle are (1) an increase of eosinophiles in prooestrus and early oestrus with their subsequent degranulation, (2) an increased basophilia in prooestrus with degranulation beginning in the early stage of oestrus, and (3) an increase of chromophobes for a period after oestrus.

Pregnancy, Pseudopregnancy and Lactation

As the changes in pituitary cytology are the same in all these conditions, they can be considered together. Compte (1898) examined the pituitary in six pregnant women and noted a considerable enlargement of the gland with changes in number and staining reactions of the cells. The average weight of the gland in these women was 1·177 g., compared with a normal weight of between 0·6 and 0·75 g. Erdheim & Stumme (1909) found the average weight of the pituitary in forty nulliparous women to be 0·72 g. During pregnancy the pituitary weight gradually rises, they say, to a little over 1 g. at full term; immediately after parturition a rapid fall occurs, the normal weight being reached 2 months later. Erdheim & Stumme recognized and named the characteristic 'pregnancy cell'. This is described by Haterius (1932 a, b) as a large ovoid cell with an eccentric vesicular nucleus and a clear homogeneous cytoplasm which stains deeply with eosin. He made cytological examinations of the female rat's pituitary at intervals throughout gestation and lactation and found that at about the 3rd, 4th or 5th day after copulation the eosinophiles decrease in number and pregnancy cells begin to appear and gradually increase in number, size and staining quality until the 12th day; they persist throughout gestation and suckling and disappear 3 or 4 days after weaning. If at any time the litters are taken away the pregnancy cells vanish and oestrus is resumed. Pregnancy cells are present, he says, throughout the course of pseudopregnancy. Just before oestrus they diminish in size, become more granular and disappear. In all conditions in which pregnancy cells were seen in the pituitary, corpora lutea were present in the ovary. Cytologically the pituitary of the lactating female is in no way distinguishable, Haterius says, from that of pregnancy. Desclin & Brouha (1931) showed that the cytological changes in the pituitary during pregnancy are not due to the presence of the foetus or to pregnancy changes in the uterus, for they accompany pseudopregnancy after hysterectomy. Wolfe & Cleveland (1933 b), describing the pregnancy cells in the rat's pituitary, state that at an early stage of gestation some of the eosinophile cells become enlarged and their coarse granules disappear; the finer granules which remain take a pale colour with Orange G. Such cells, they say, occur also in pseudopregnancy and are not characteristic of pregnancy, for they are to be seen whenever the ovaries contain active corpora lutea. It will be remembered that active corpora lutea do not develop in the rat in the course of the normal oestrous cycle. Severinghaus (1937) states that changes in the pituitary similar to those of pregnancy may be caused by injections of oestrone, foetal or placental extracts, pregnancy blood or extracts of human pregnancy urine, all of which lead to ovarian luteinization. Severinghaus (1938) says that a feature of the pregnant guinea-pig's pituitary is the large number of degranulated or partially degranulated basophiles rich in mitochondria. Toward the end of gestation a significant increase in eosinophiles occurs and persists after parturition.

Several facts suggest that the pituitary changes which accompany pregnancy may be due largely to progestin, or more certainly perhaps to some influence derived from the corpora lutea. Haterius (1932a) showed that typical pregnancy effects in the pituitary can be induced in normal female mice by repeated implantations of pituitary gland, which cause luteinization of the ovaries. Similar implantations into spayed females or castrated males do not cause these pregnancy changes in the pituitary; whereas in males into which ovaries have been successfully grafted the pregnancy changes can be induced, whether the host has or has not been castrated. Furthermore, it was found by Charipper (1934) that the appearance of 'pregnancy cells' in the pituitary could be brought about, in male as well as in female rats, by repeated injections of an extract prepared from corpora lutea.

Chorionepithelioma. It is of interest to note that histological changes in the pituitary identical with those of pregnancy have been seen in a case of chorionepithelioma of the testicle in a human subject (Entwistle & Hepp, 1935).

Gonadal Hormones

Gonadectomy is followed by enlargement of the anterior lobe of the pituitary and an increase of its gonadotrophic potency; it has also been shown that these effects are due to cutting off the supply of gonadal hormones. The histological effects of the different gonadal hormones on the pituitary will now be discussed.

(a) *Oestrogens.* Haterius & Nelson (1932) castrated male rats and then found that the histological changes in the pituitary which in ordinary circumstances follow castration could be prevented, or if already present could be reversed, by means of ovarian implants. Nelson (1934c) gave oestrin to castrated and cryptorchid rats, and noted the effects of this treatment on the pituitary. There was an almost complete disappearance of 'castration cells' and a great increase in the number of chromophobes; these cells had an enlarged Golgi apparatus and numerous mitochondria, and could be recognized as basophiles which had undergone a change. Some of these degranulated basophiles resembled 'pregnancy cells'. The eosinophiles also had become degranulated, though they were less profoundly affected than the basophiles. Martins (1936) transplanted the pituitaries of rats into their own eyes and twice a week gave the rats subcutaneous injections of 2,000 to 10,000 i.u. of oestrone, similar doses being given also to rats with intact pituitaries. Whether transplanted or not the pituitaries responded in the same way; they became enlarged, the chromophobe cells were much increased in number and the basophiles and eosinophiles were reduced. Wolfe & Chadwick (1936) subjected female rats to daily injections or to a single big dose of oestrin. This treatment was followed by enlargement of the pituitary, the increase in weight of which was proportional to the amount of oestrin used and the duration of the experiment. Histological examination showed (1) degranulation of the basophiles which were converted into chromophobes, (2) degranulation and reduction in number of eosinophiles, and (3) a considerable increase in the number of chromophobes many of which were enlarged.

By the colchicine method, Kuzell & Cutting (1940) showed that oestrone, oestradiol and diethylstilboestrol each caused, after a latent period of 4 or 5 days, a considerable and prolonged increase of mitosis in the anterior lobe of the rat's pituitary. This reaction occurred in normal and spayed rats.

(b) *Androgens.* Unlike oestrogen the androgens do not cause enlargement of the anterior lobe of the pituitary, and, speaking in a general way, they appear to have less influence than oestrogens over this gland. Nevertheless their control of the pituitary is of great biological importance. Lehmann (1927) observed that the characteristic changes which occur in the rat's pituitary after castration can be prevented or abolished by testicular transplants or the injection of watery extracts of testicle. By these means the enhanced weight of the pituitary and the increase in number and size of its chromophilic cells are counteracted. These effects, he says, are not caused by the transplantation of ovaries into male rats. Nelson & Gallagher (1935, 1936) confirmed these results in rats, using different androgens, and they showed that the same effects were produced in the pituitaries of spayed female rats. Male and female rats were gonadectomized and from the next day onward daily doses of androgen were given for a month, when the rats were killed and their pituitaries were examined. In this way Nelson & Gallagher learned that androgens, like oestrogens, prevent the changes which normally follow gonadectomy, and at the same time cause degranulation of the basophiles. In these reactions they noticed a difference between the two sexes, for in the male five times as much androgen was required as was enough in the female to produce an equal effect on the basophiles. Possibly this is one more example of the fact that females appear to be more responsive to androgens than are males, whereas males seem to be more sensitive than females to oestrogen. The androgens used in this experiment were androsterone, androstanediol and androstenedione. It was found by Allanson (1937), also, that the changes which normally follow castration can be counteracted by androgens. Rats were castrated when between 40 and 60 g. in weight and injections of androgen were started a month or more later and continued for 20 days. At the end of this period some of the androgens used had restored completely the pituitary to normal, except for the persistence of an occasional small castration cell. Testosterone, testosterone propionate and androstenedione, in daily doses of 0·5 mg., all had this effect. Androstanedione, trans-androstenediol and transdehydroandrosterone were relatively ineffective in reversing the castration changes in the pituitary. The superior potency of oestrone was shown by the fact that 0·05 mg. daily restored the pituitary after castration to the normal state.

Wolfe & Hamilton (1937*a*, 1939) performed comparative experiments on a number of young female rats weighing between 100 and 125 g. and noted that ten daily injections of 500γ of testosterone acetate caused some degranulation of basophiles unaccompanied by any change in the eosinophiles and chromophobes, or any alteration in the weight of the pituitary. On the other hand 500γ of oestradiol benzoate given daily for 10 days caused complete degranulation of the basophiles, much degranulation of the eosinophiles, an increase in the number of chromophobes and an increase in the weight of the pituitary. The simultaneous administration of testosterone acetate prevented the enlargement of the pituitary which would have been caused by oestradiol if given alone (Table 46).

TABLE 46. The effects of ten daily injections of oestradiol benzoate and testosterone acetate on the pituitaries of female rats (Wolfe & Hamilton, 1937*a*)

Treatment	Dose (γ)	Average weight of pituitaries (mg.)
None	—	6
Oestradiol benzoate	500	10·7
Testosterone acetate	500	5·7
Oestradiol benzoate plus Testosterone acetate	500 } 500	·7·3

McEuen, Selye & Collip (1937) reported that daily doses of testosterone, even though continued for a long while, do not cause enlargement of the pituitary, and Reece & Mixner (1939; see also Reece, 1941) made a similar observation. They gave daily injections of 200γ of testosterone propionate for 15 days to spayed female rats and noted that the treatment did not cause any increase in the size of the pituitary. Wolfe & Hamilton (1939) gave 1 mg. of testosterone propionate daily to female rats for 90 days and at the end of that time compared the histological state of their pituitaries with those of untreated male and female littermates. The pituitaries of the treated females showed an increase of eosinophiles and granular basophiles making a cellular pattern which closely resembled that of their normal untreated brothers and differed widely from that of their normal untreated sisters (Table 47). Selye (1939*a*) has made the same sort of

TABLE 47. The effect of androgen on the histology of the pituitary in
female rats (Wolfe & Hamilton, 1939)

Subject	Weight of pituitary in mg. per 100 g. of bodyweight	Eosino- philes (%)	Granular basophiles (%)	Non- granular basophiles (%)	Chromo- phobes (%)	Castration cells (%)
Untreated male	2·3	49·0	6·1	0·6	43·4	0·9
Testosterone- treated female	3·1	47·3	5·2	1·8	45·1	0·6
Untreated female	4·6	31·8	1·8	3·0	63·4	0

observation in mice; in his experiment daily doses of 5 mg. of testosterone propionate given for 20 days to female mice did not cause degranulation of the chromophile cells.

(c) *Progestin.* Little direct study has been made of the action of progestin on the pituitary. Wolfe & Cleveland (1933 a) found that 'pregnancy cells' are not characteristic of pregnancy, but are seen whenever the ovaries contain active corpora lutea. Charipper (1934) gave fifteen daily injections of a luteal preparation to rats and noted that 'pregnancy cells' appeared in the pituitaries of all the rats so treated. Brooksby (1938) injected 0·5 mg. of progesterone daily into spayed rats and noted that this treatment prevented castration changes in the pituitary and, like oestrin, caused its enlargement. Oestradiol and progesterone, given together, caused greater enlargement than when either was given alone.

Herlant (1939) gave thirteen injections of 0·35 mg. of progestin during 15 days to adult rats, after which their pituitaries showed a reduction in the number and size of eosinophiles with a diminution of their granules, a slight degranulation of basophiles, which otherwise showed little change, and a small increase in the number of chromophobes. To other rats Herlant gave four daily injections of 1 mg. of progestin; these caused an increase in the number and granulation of eosinophiles and a small reduction in the number of chromophobes. Correlating histological with experimental findings, Herlant concludes that the eosinophiles are the source of luteinizing hormone.

Gonadotrophin

Since Lehmann (1928) called attention to the fact that the repeated injection of placental extracts into female rats will bring about changes in the pituitary resembling those which accompany pregnancy, much experimental inquiry has been devoted to the subject. As an outcome of this work it seems probable that many of the effects produced by gonadotrophin on the pituitary are attributable to the gonadal hormones whose production is stimulated. Haterius & Charipper (1931) implanted fresh pituitaries daily into female rats and thereby caused changes in the recipients' pituitaries which resembled those of pregnancy. Similar implants into spayed females or normal males did not result in these changes. If pituitary implants were given to males, whether castrated or not, in which ovaries had been successfully grafted, the 'pregnancy changes' followed in their pituitary glands. Haterius & Charipper concluded that these changes in the pituitary were sequels to the ovarian luteinization induced by the grafts. Karp (1933), also, observed, after the injection of human placental gonadotrophin into rabbits with intact ovaries, changes in the pituitary like those of pregnancy. Wolfe, Phelps & Cleveland (1933) gave to immature female rats injections of an extract of human placenta which was free from oestrin but rich in gonadotrophin. This treatment caused in the pituitary a degranulation of the basophiles and, to a less degree, of the eosinophiles. If the animals were castrated at the beginning of the experiment these changes failed to appear; as they are like those changes produced by androgens and oestrogens it seems likely that they depend upon an enhanced supply of these under the stimulus of gonadotrophin. Severinghaus (1934 a, b) and Nelson (1934 b) have done confirmatory work. Wolfe (1933, 1934) treated immature female rats of 25 to 35 days with two daily subcutaneous injections of 'a relatively crude oestrin-free extract of placenta' for a week. The effects in the pituitary were constant and definite, namely a degranulation of the basophiles and a slight but constant reduction in the number of eosinophiles. Wolfe, Ellison & Rosenfeld (1934) gave extracts of pregnancy urine to normal female rats for a period of 15 days, after which their pituitaries showed a degranulation of basophiles and a less marked degranulation of eosinophiles. Both these types of cell were reduced in number, while the chromophobes were increased. Colloid material was present in the cleft of the pituitary in all the injected rats. The pituitaries were enlarged. If injections of gonadotrophin were continued for an extended period the changes produced at an early stage of the treatment disappeared (Wolfe, 1935 a).

Severinghaus (1934 a, b) treated castrated and non-castrated adult male rats with placental gonadotrophin, i.e. an extract rich in LH. In the non-castrated rats there was an increase in the number of basophiles and a rapid degranulation of their cytoplasm. The pituitaries of the castrated rats, on the other hand, could not be distinguished from those of castrated controls. Severinghaus also submitted adult female rats for periods of 4 to 6 weeks to daily injections of an extract consisting mainly of FRH, and found that the basophiles in the pituitary had undergone degranulation and in some instances degeneration. The acidophiles were perhaps reduced in number, but were otherwise unaltered; the chromophobes were unaffected.

The changes found in the pituitary in these experiments appear to be identical with those produced by oestrogen and androgen, and it seems probable that some and perhaps all were brought about by an increased output of hormones from the gonads in response to the injected gonadotrophic hormones. Such a suggestion is supported by the fact that the changes did not occur in the absence of the gonads.

Sex

Apart from the changes in the pituitary during oestrous cycles in females and the fact that no such cyclic changes have been detected in the male, there appear to be qualitative differences between the pituitaries of the two sexes. Erdheim & Stumme (1909) compared the weights of the pituitaries in a small number of adult men and women and found that the gland was heavier in women than in men (Table 48), as is the case also in the rat (Hatai, 1913 b) and mouse (see pp. 64, 280).

TABLE 48. Comparative weights of pituitary in men and
nulliparous women (Erdheim & Stumme, 1909)

Sex	Age	Number	Average weight of pituitary (g.)
Female	23 to 49	6	0·72
Male	23 to 83	11	0·66

Nelson (1935 a) made cell-counts of the pituitaries of sixty-eight normal male and fifty-six normal female rats and recorded a higher percentage of basophiles and eosinophiles and fewer chromophobes in the male pituitaries (Table 49).

TABLE 49. Differential cell-counts in the pituitaries of male and female
rats expressed as percentages (Nelson, 1935 a)

Sex	Basophiles	Acidophiles	Chromophobes
Male	8·1	49·8	42·1
Female	4·2	34·4	61·4

It has been shown by implantation experiments that the pituitary of the male has a greater gonadotrophic (FRH) potency than that of the female, and there are several reasons for thinking that basophile granules are associated with the production of follicle ripening hormone. The artificial administration of gonadal hormones causes a decrease of the gonadotrophic power of the pituitary associated with degranulation of the basophile cells; and oestrin produces these effects more readily than testosterone. Some facts of human pathology accord with this general view; Fröhlich's syndrome, the symptoms of which are largely due to a deficient supply of gonadotrophin, is found to be associated with a deficiency of basophile cells in the pituitary. Furthermore, the increased gonadotrophic potency of the pituitary which follows castration is accompanied by a pituitary basophilia.

Castration, and Destruction of Seminal epithelium

As mentioned already Fichera (1905a, b) showed that gonadectomy is followed by enlargement of the anterior lobe of the pituitary with histological changes. Since then many workers have studied the phenomenon. Addison (1917) castrated rats and afterwards compared their pituitaries with those of littermate controls. He found that after castration the basophiles increase in size and number. Two months after castration the largest basophiles begin to become vacuolated, and as time goes on the vacuoles increase in size, the nucleus becomes pressed to one side of the cell, and in this way are formed the typical 'signet ring' or 'castration' cells. The acidophiles, he states, are not much affected at first except for a slight reduction in size; later they gradually become degranulated and less readily stained. Seven months after castration their numbers are much reduced. Stein (1933) has reported similar changes in the rat's pituitary following castration.

Severinghaus (1932) did not observe quite the same changes in castrated guinea-pigs although their pituitaries showed an increased gonadotrophic power. His guinea-pigs were 30 days old at the time of gonadectomy and were killed when between 72 and 205 days old. Their pituitaries showed an increase of eosinophiles, but no increase of basophiles, nor were 'castration cells' present. Nelson (1935c) has noticed a difference in the response of the guinea-pig's pituitary to gonadectomy, as compared with the response in other animals, though in his experiments the difference appears to have been one of degree only. He found that the pituitaries of guinea-pigs which had been castrated not less than 2 months previously contained an increased percentage of basophiles, though the increase was not so pronounced as in the rat. Moreover, in the guinea-pig 'castration cells' were not usually seen, though in guinea-pigs which had been castrated a long while previously a few small vacuoles were present in the cytoplasm of the basophiles. From Nelson's observations it seems that the only difference between the responses of the pituitary to gonadectomy in the guinea-pig and the rat concerns the duration of time required for the consequent changes to take place. There are reasons for thinking that the guinea-pig's adrenal has a larger share than that of the rat in producing gonadal hormones; the differences between the effects of castration on the pituitary in the two species may perhaps be accounted for in this way.

In a later work Severinghaus (1938) states that an increase of basophiles follows castration in the rat, guinea-pig, rabbit and monkey. In rats, he says, the increase of basophiles may be seen as early as the end of the 2nd week after castration and at the same time the acidophiles are already reduced in size and number. The same effects of castration on the pituitary have been noticed in man. Biggart (1934) examined the pituitaries of four men after castration and recorded (1) a pronounced increase in the number of basophiles, (2) the appearance of large chromophobes, with transitional stages between these and basophiles, (3) vacuolation and the formation of colloid in the basophiles. The changes in the eosinophiles varied.

Cryptorchidism causes changes in the pituitary like those which follow castration (Martins, Rocha & Silva, 1931), though they take longer to develop. Nelson (1934 *a*, *b*) rendered rats cryptorchid on the 40th day of life. Pituitary basophilia was present 75 days later and vacuolation of the basophilic cells was advanced by the 200th day. At the end of 360 days the pituitary changes resembled those of 'advanced castration'. The basophiles at this time were increased, the acidophiles slightly decreased, there was a more pronounced diminution of chromophobes and many 'castration cells' were present. Nelson (1935 *c*) found that in the guinea-pig, as in the rat, cryptorchidism produced changes in the pituitary similar to those following castration, though taking more time to develop.

Ligation of the vasa efferentia. Van Wagenen (1925) ligated the vasa efferentia of the rat's testis so as to cause degeneration of the seminal epithelium within 21 days. Eight weeks later she found typical 'castration changes' in the pituitary; that is to say, there was an increase in size and number of the basophile cells which were undergoing vacuolation.

X-ray sterilisation. As mentioned earlier (p. 50) destruction of the seminal epithelium of both testes by X-rays is followed by enlargement of the pituitary with histological changes like those brought about by castration.

Thyroidectomy, adrenalectomy and parathyroidectomy. It has long been known that thyroidectomy causes enlargement of the pituitary with cellular changes resembling those which follow gonadectomy. Delille (1909) states that the pituitary becomes enlarged also after adrenalectomy and parathyroidectomy.

Nutrition

The destructive effects on the seminal epithelium which are caused by depriving an animal of vitamin-E have been discussed on an earlier page (p. 68). The testicular deficiency so caused in rats has been shown by Nelson (1933) to be followed by 'castration changes' in the pituitary. In females, however, no castration changes occurred in the pituitary as a consequence of vitamin-E deficiency, which is to be explained by the fact that the ovary is not directly injured by a lack of this vitamin.

Sutton & Brief (1938) reported that in male and female rats suffering from vitamin-A deficiency the basophilic cells of the pituitary are increased, as they are after gonadectomy. The effect is most pronounced in males and may be attributed, probably, to the gonadal injury caused by the avitaminosis.

Age

The chief characteristics of the immature sow's pituitary, according to Cleveland & Wolfe (1933), are (1) a high percentage of granular basophiles such as occurs in the adult during prooestrus and (2) a scarcity of eosinophiles. The latter contain colloid which stains bright yellow with Orange G; many of these cells become amalgamated so as to form large colloidal masses surrounded by eosinophiles.

In old rats the pituitaries appear to be characterized mainly by aberrations from the normal. Wolfe, Bryan & Wright (1938) examined the pituitaries of twenty-six old breeding female rats whose average was 632 days; all of these rats had littered at least once. The pituitaries varied in weight from 10 to 45 mg., the majority being less than 17 mg. In fifteen (58 per cent) of the pituitaries collections of abnormal cells were present. These cells were large with bulky nuclei in which little chromatin was apparent. The nucleoli were very large. The majority of these cells were chromophobes, but in some instances enlarged eosinophiles also were present.

PART II. GONADAL HORMONES

Chapter V. *A General View of the Gonadal Hormones*

Experimental investigation. Chemical structure. Sources, functions and excretion. Bisexual activities. Co-operative and antagonistic effects between different hormones. Factors influencing biological action. Gradients of responsiveness. Reversible and irreversible effects. Actions common to gonadal and adrenal hormones. Influence of gonadal hormones on behaviour and on tissue growth.

THE gonadal hormones are chemical substances which regulate the male and female characters of the body, and control sexual impulse and behaviour. A conspicuous part of their function is to stimulate the growth and activity of the accessory genital organs. In the absence of these hormones the accessory organs of reproduction would not come into existence, or if already present, would not attain full development.

John Hunter (1728–1793) (1837*b*) perceived that the sexual characters depend largely on gonadal activity. He remarks that before puberty there is not much to distinguish the male from the female, and that the special features of the two sexes appear mostly at the time of puberty. He further notes that the female, after the reproductive period has ended, loses some of her feminine attributes and in appearance tends to approach towards the male, or more properly perhaps, he says, towards the hermaphrodite.

The output of gonadal hormones from the gonads and other organs is regulated by the pars anterior of the pituitary gland the activity of which is itself controlled, not only by the gonadal hormones formed in response to pituitary action, but by many additional influences. The complexity of the biological machinery is manifest, and it may be difficult to predict or to observe in detail all the results of disturbing any one part of the endocrine system which is concerned with reproduction.

In recent years our knowledge has rapidly extended, and, while admitting that our comprehension of the subject is still primitive, there are now sufficient ascertained facts to allow a useful discussion.

Experimental Investigation of the Action of Gonadal Hormones

This was impeded in the past by a number of intrinsic difficulties. At the beginning the investigator had to draw conclusions from the effects of removing or transplanting whole organs or portions of them, or else from the results of injecting crude tissue extracts. This form of inquiry continued for many years before any of the sex hormones had been isolated in a pure form. In spite of such a handicap, these earlier inquiries have stood well the trial of time, and the conclusions drawn by the biological pioneers have hardly been shaken by the more satisfactory investigations of to-day which are carried out with the advantage of pure, or at least partly purified, materials.

Although to some extent inaccuracies can be avoided by using chemically pure preparations of the hormones, yet the disturbing fact remains that the tests have to be made on the living animal, which cannot be standardized with the same exactitude as a chemical compound. By using animals of long-established inbred strains and by stabilizing their environment the errors and difficulties due to biological variability can be reduced, though they cannot be eliminated (Emmens, 1939 *a, b, c*). Time and circumstance in these investigations cannot be made uniform for every occasion; and it would be a gross error to regard the sensitive living individual as the equivalent of a purified chemical reagent. It has to be remembered, too, that animals of different species do not all respond in exactly the same way to a given gonadal hormone; although in a broad sense they all respond alike. The differences concern minor biological details, but are not negligible. For example, in a male mouse subjected to frequently repeated doses of oestrone the coagulation glands readily undergo a keratinizing metaplasia which is followed later by a similar change in the seminal vesicles. A rat treated in the same way does not show an identical response; a keratinizing metaplasia is not so easily induced in the rat, and it appears first in the seminal vesicles, the coagulating glands showing the change later or not at all. The different reacting organs in the same species show a varying readiness and degree of response, and the alterations induced in a single organ are not uniform; one part may be affected to an advanced degree while another part shows but little change.

Even individuals of the same inbred strain vary in their reactions to a given hormone. If a number of male mice of one strain be submitted persistently to equal doses of oestrone, some may succumb to hydronephrosis, others to suppuration in the accessory genital glands, and yet others may remain free from these complications and survive for long periods, perhaps to die of a pituitary tumour or mammary cancer. The probability of the last will depend upon the strain from which the mouse has been derived, or, if nursed by a foster-mother, upon the strain to which she belonged. The reactions to a gonadal hormone may vary even among littermates; indeed, the same mouse may show unequal responses to equal doses of a gonadal hormone at different times, amid different environments, or in varying conditions of health.

Seasonal and cyclical influences, the nature of which has not yet been elucidated, may affect the reactions of an individual to a given dose of hormone. Gallagher & Koch (1930) and David (1938) have called attention to a seasonal variation in responses to gonadal hormones, and cyclical variations of shorter duration have been noted by Wade & Doisy (1935 *a*) and Zuckerman (1937 *b*).

The method of administration and other details have to be considered when the effects of giving gonadal hormones are under comparison; some of these details will be mentioned later. Meanwhile the risk of forecasting results without experimental tests may be exemplified by some observations of Emmens & Parkes (1939 *a*). They say that, in tests on the capon's comb, the accessory reproductive organs of the castrated rat or the uterus of the spayed rat or immature rabbit, testosterone given orally is less effective than when injected subcutaneously. On the other hand methyltestosterone given by the mouth is almost as effective as when injected subcutaneously in causing progestational changes in the uterus;

in the other tests methyltestosterone has a greater activity than testosterone when both are given by the mouth. Methyltestosterone, they say, shows more androgenic potency than testosterone when tested in rats, less potency when tested on the capon's comb; and it is more effective in causing progestational changes in the uterus by whatever route it is given.

Chemical structure

Experimental inquiry into the nature and actions of the sex hormones has, because of the drawbacks just mentioned, been largely qualitative. Another obstacle to establishing the subject of sex hormone activity as an exact science is the existing difficulty or impossibility of correlating chemical structure with biological behaviour.

The chemical structure of the three main types of gonadal hormone—oestradiol, testosterone and progesterone—has the same basis, and the divergences from the common pattern which account for the diverse actions of these hormones in the body are seemingly slight (Fig. 1). The difference of biological activity caused by small changes in the chemical structure is exemplified by the following

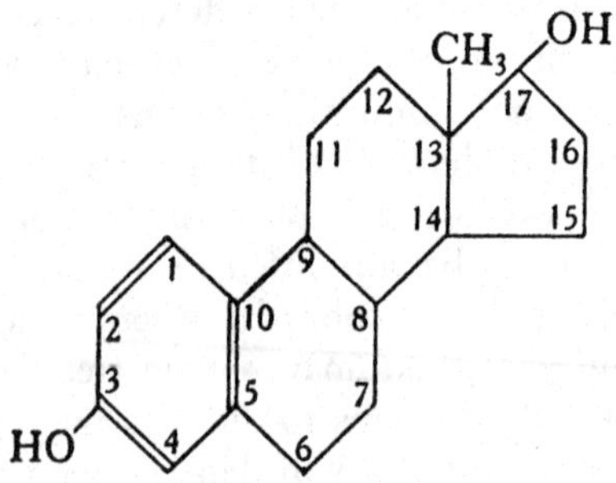

I. Oestradiol 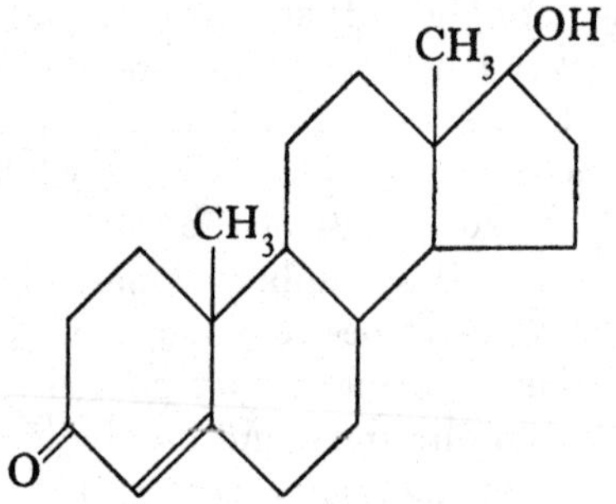   II. Testosterone

Fig. 1a.

example to which Butenandt (1936) called attention: Δ^1-Androstenedione is an oestrogen, Δ^4-Androstenedione is an androgen, and Δ^5-Androstene-3 : 17-diol has both androgenic and oestrogenic properties (Fig. 2). The general similarity in the molecular structure of these hormones seems remarkable in view of their differing effects in the living body.

The biological inactivity of isomers of potent compounds is a feature which may be noticed in this context, though there does not appear to be any instance of two isomers having a different kind of action; that is to say, isomers of oestrogens are not androgenic, nor have isomers of androgen been found to have oestrogenic action.

The physiological activities of the gonadal hormones do not seem to depend on the possession of a sterol nucleus, although this is a characteristic of those which are produced naturally in the body. In the laboratory many compounds of other constitution and yet possessing oestrogenic capacity have been produced (Dodds, 1936, 1937; Dodds & Lawson, 1936). For example, in mice diethylstilboestrol and triphenylethylene (Fig. 6) will cause metaplasia in the accessory genital organs, hernia, cancer of the breast and most if not all the other phenomena which follow the administration of oestrone or oestradiol.

In making this statement one has to remember that an organic compound introduced into the body may there undergo important changes, and the biological effects which follow may result not directly from the chemical substance introduced, but from its metabolic products.

Another feature of the actions of gonadal hormones is the uncertainty of their anatomical limitations. The specific effects which these hormones produce are graded, so that the thresholds of reactivity vary from place to place even in a single organ. Moreover, the activities of each hormonal compound are not closely circumscribed; the reactions produced in the tissues by different hormones overlap (Fig. 2). This overlapping of biological effect is often associated with a similarity in structure. For example, oestradiol and ergosterol have comparable formulae (Fig. 3) and both have specific effects on epiphyseal cartilage and the metabolism of calcium; moreover, ergosterol in very large doses is said

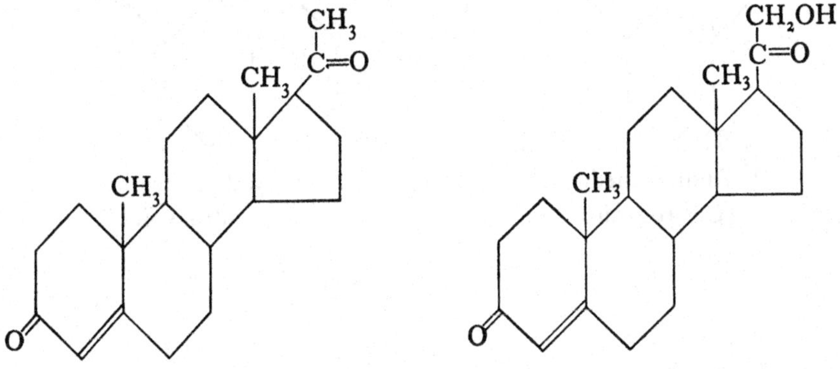

III. Progesterone   IV. Deoxycorticosterone

Fig. 1 b.

to be slightly oestrogenic. But obvious correspondences of this kind between molecular conformation and biological activity are far from being invariable, as will be seen by comparing the natural oestrogens with some of the artificial compounds which produce almost identical effects (Figs. 5, 6). For the present it must be admitted that no constant, predictable, and easily determined features have been recognized as sure guides to the biological action of this kind of compound.

International units. To test the hormonal potency of a substance whose composition is unknown it has been necessary to compare its effects with those produced by a given quantity of a known hormone. For this purpose 'international units' of potency have been agreed upon and have been of great use. They do not in fact represent comparative efficiency for they are based on the minimal amount which will evoke some particular immediate response without regard to the duration of effect and thus are apt to convey erroneous conclusions. For example, suppose a compound X has only half the capacity of oestrone to cause cornification of the mouse's vagina, its potency in terms of the international unit will be one-half that of oestrone. If however, as may be the case, the response aroused by a minimal dose endures four times as long as that caused by the same dose of oestrone, then X may well be regarded, not as less, but as more potent than oestrone. For this and other reasons it seems desirable always when describing an experiment or a clinical case to mention if possible the actual weight of material used instead of its amount as denoted by international units.

Sources, Functions and Excretion

Sources. It is not to be supposed that the gonads are the only source of the gonadal hormones, that the so-called 'male hormones' are produced only by the

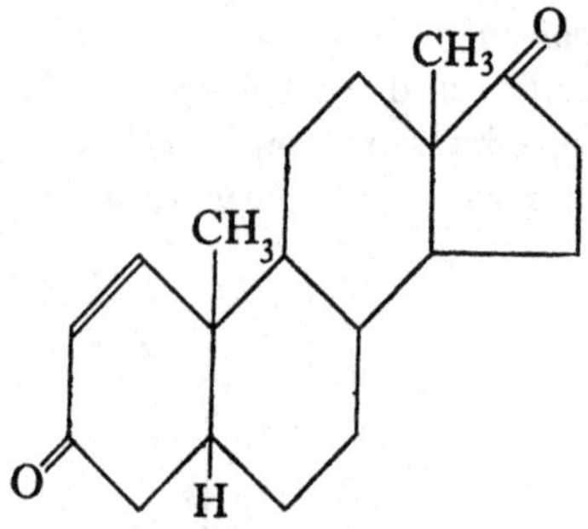

Δ^1-Androstenedione
(Oestrogenic)

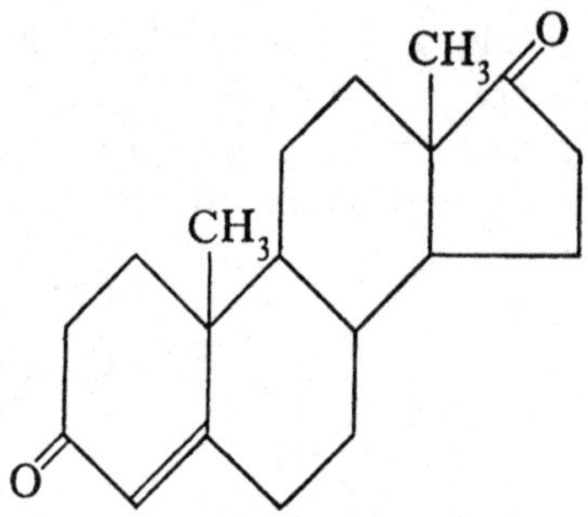

Δ^4-Androstenedione
(Androgenic)

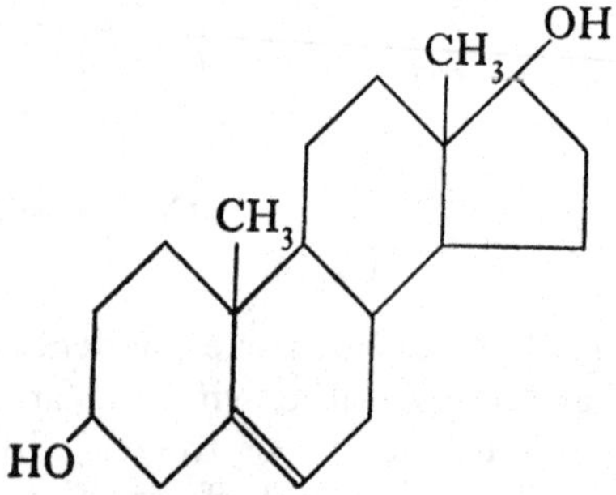

Δ^5-Androstene-3:17-diol
(Oestrogenic and Androgenic)

Fig. 2.

testicle, or that the two kinds of 'female hormone' originate in the ovary alone. The three types of gonadal hormones are elaborated by both sexes, and although the ovary and testis are concerned with their manufacture they do not possess a monopoly. Androgen, oestrogen and progestin, it appears, may all be derived from the testis, ovary, adrenal, placenta and perhaps elsewhere.

Functions. The functions of each of these hormones, too, are not confined to the sex indicated by their names. Structures common to both sexes, for example the embryonic müllerian and wolffian apparatus, the nipples and mammae, the external generative organs, react alike to the same hormone whether the possessor is male or female. Thus in the assay of androgens by applying the material under test to the combs of day-old chicks it has been found that sex has little influence on the responses of the comb (Danby, 1938); other examples will be given later.

A chief role of androgens is to assist the development and maintain the activities of the accessory reproductive organs appropriate to the male, while oestrogens perform a similar part in the development and maintenance of organs appropriate to the female. Progestin controls the relationship between the embryo and the mother; nevertheless, as will be shown later, even its activities are not limited to the female sex. Occasionally a particular character is attributed to one or other

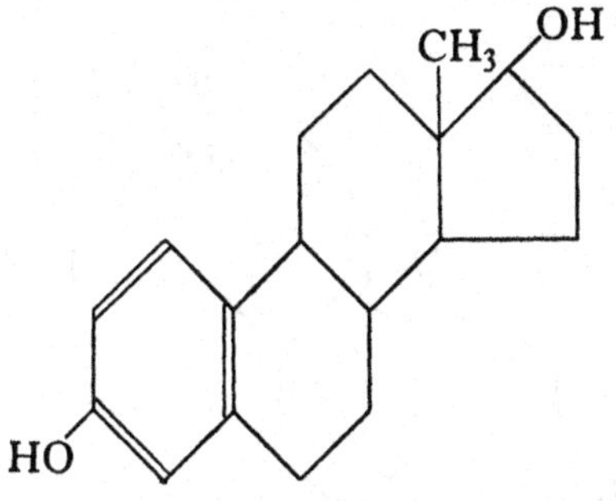

I. Oestradiol

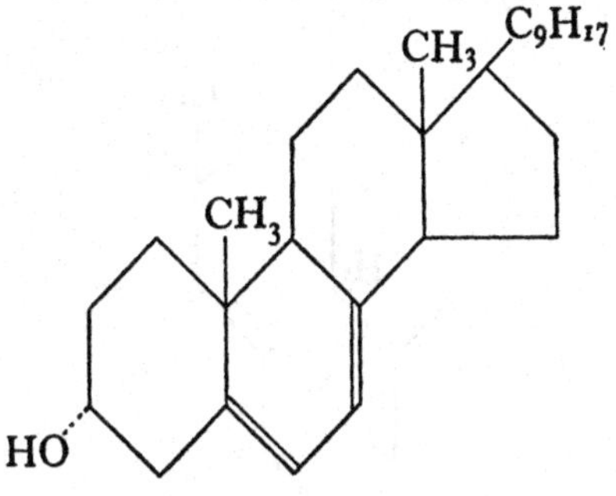

II. Ergosterol

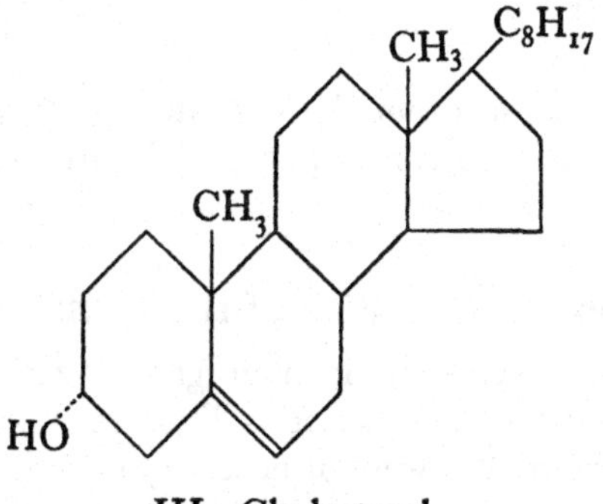

III. Cholesterol

Fig. 3.

sex when it is in fact neutral. Thus it is sometimes stated that removal of the ovary from a hen will cause it to acquire male plumage, whereas it really acquires neutral plumage because the oestrogen is lacking on which the female character of its feathers depended.

The capacity of a cell or a tissue to respond to a gonadal hormone is innate in that cell or tissue and is retained if it is transplanted to another part of the body or to another individual, whether of the same or the opposite sex. Martins & Cardoso (1931) grafted seminal vesicles into the abdominal walls of young male mice. In this unnatural situation the seminal vesicles were found to be healthy, turgescent and secreting 5 months later, showing that they were reacting normally to the gonadal hormones supplied by the host. The principle holds good for transplants of skin in birds and can be seen in those instances in which the differential sexual plumage depends on gonadal hormones (Danforth & Foster, 1929).

Inactivation and excretion.   There is little or no storage of gonadal hormones in the body. They are used, inactivated or excreted nearly as rapidly as they are formed. From this it is clear that their production must be almost or quite continuous during the sexually active periods of life. The hormones are excreted usually in an altered form. Testosterone, which is the chief androgen produced by the testicle, is excreted by the kidneys mainly as the less potent compounds androsterone and *trans*dehydroandrosterone; oestradiol, the main oestrogen of the ovary, is excreted mainly as oestrone, and progesterone is largely eliminated

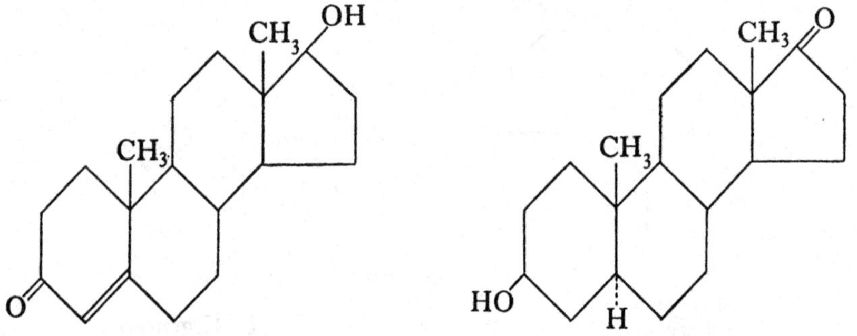

I. Testosterone   II. Androsterone

Fig. 4a.

as pregnanediol, which is almost though not quite biologically inert. Other waste products than these are derived from the gonadal hormones and are excreted in the urine.

So-called Bisexual Activities of the Gonadal Hormones

Attention may be given to some unfortunate generalizations which have arisen from the nomenclature and have caused a little confusion. The terms 'male hormone' and 'female hormone' were of necessity introduced in the earlier days of endocrinology and are responsible for the kind of misunderstanding referred to, which is to regard each of the two main groups of gonadal hormones— androgens and oestrogens—as arising only in the appropriate gonads and as acting only on the organs of one sex. It should be understood that neither type of hormone, in origin or effect, is peculiar to the male or the female.

There is no essential antagonism between the testis and ovary, nor between the individual of one sex and the gonads of the other. This has been shown by transplanting testes into females and ovaries into males (Steinach, 1916; Moore, 1921; Finlay, 1925). The absence in adults of any vital incompatibility between the gonads has been shown also by joining together a male and female in parabiosis so that an exchange of blood takes place between them. After such an operation the gonads of the two individuals are but little affected and may retain their procreative functions (Morpurgo, 1908).

The later classification of gonadal hormones into androgens, oestrogens and progestins is useful and probably essential for a detailed consideration of their respective activities; yet this classification is not always a reliable guide to the effects which will be caused by introducing any one of these hormones into the living organism. Apart from the fact that a hormone of one type may be changed into that of another within the body, there is much overlapping of the potentialities of the different hormones, and it seems that the only sure way to avoid erroneous conclusions is to study the effects of each individual hormone separately, without surmising that because it belongs to one particular group it will therefore exercise exactly those capacities which are commonly accredited to that group.

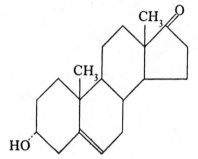

III. Aetiocholane-3(a)-ol-17-one   IV. *Trans*Dehydroandrosterone

Fig. 4b.

(i) *The transformation of one type of gonadal hormone into another in vivo.* An androgen may be converted into an oestrogen in the living body. Steinach, Kun & Peczenik (1936a, b) found that injections of androsterone in normal and castrated rats were followed by an increased output of oestrogen in their urine. In further experiments Steinach & Kun (1937) showed that in man injections of testosterone were followed by an increase of oestrogen in the urine; and the same result though to a less degree was caused by injections of androsterone. Such observations have since been confirmed (Foss, 1939, and others).

In a man with a malignant tumour of the adrenal (Burrows, Cook, Roe & Warren, 1937) the urine was found to contain as much as 3,000 i.u. of oestrogen per litre, and an androgenic activity (comb test) of only 50 to 100 i.u., which is not greatly above the normal. It seems possible that in this case an excess of androgen may have been elaborated and then converted into oestrogen and excreted as such in the urine, just as happens in men receiving large artificial doses of testosterone. With adrenal tumours in women, an excessive excretion of androgen is the rule, though there may also be an abnormally large output of oestrogen.

Schoeller, Schwenk & Hildebrandt (1933) have achieved *in vitro* the reverse process, the conversion of oestrogen into androgen.

It is interesting to note that Hamblen, Pattee & Cuyler (1940), as the result of 835 assays of androgen in the urine of twenty-two women, learned that the administration of oestrogen reduced the output of androgen. All of these women had been suffering from uterine disorders or from discomforts attending the menopause.

(ii) *Formation of oestrogen by the male.* Androgens and oestrogens are both normally produced by each sex, and the difference between male and female in

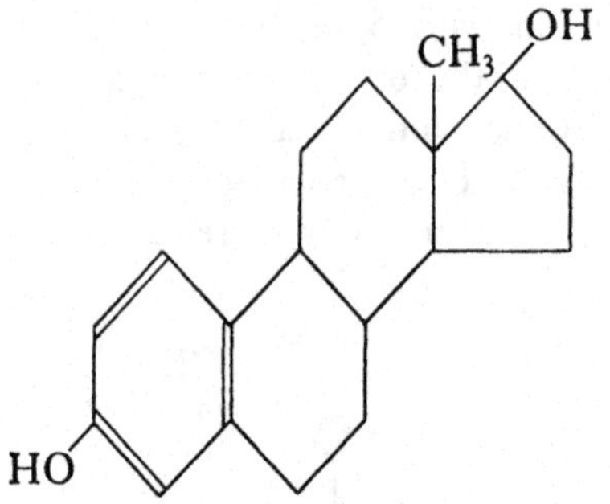

I. Oestradiol

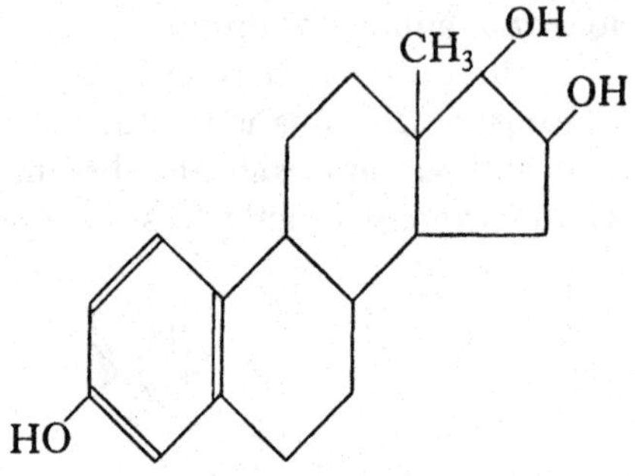

II. Oestriol

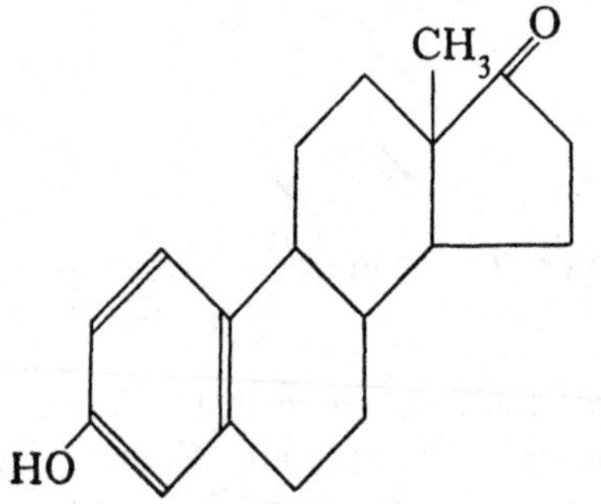

III. Oestrone

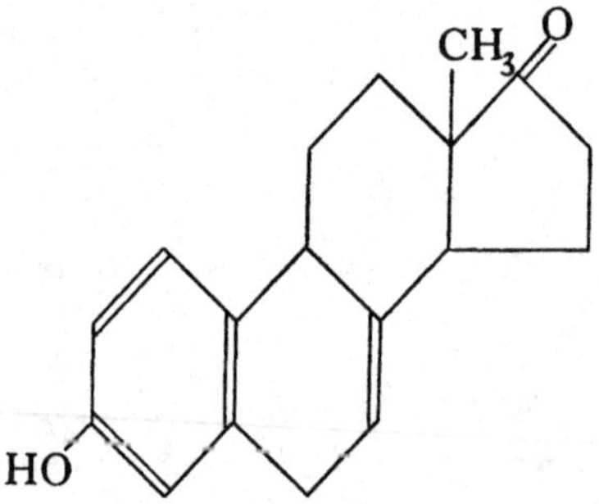

IV. Equilin

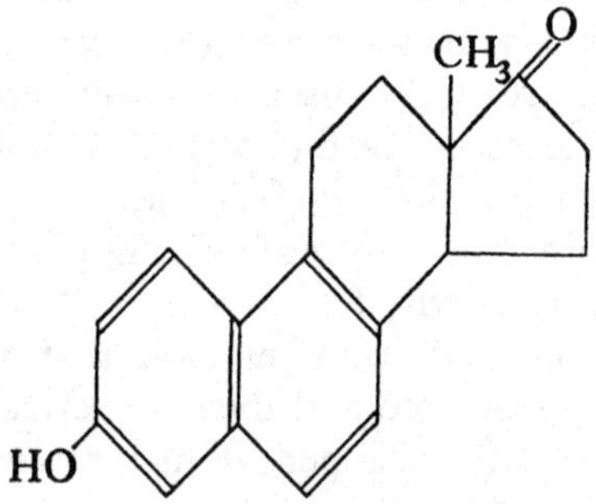

V. Equilenin

Fig. 5. Naturally formed oestrogens

the elaboration of gonadal hormones seems to be, apart from the feminine sexual cycle, one of degree alone. Assays of urine have established that a large excretion of oestrogen by men or of androgen by women is abnormal and usually denotes some pathological disturbance. This is not the case with every species of animal.

Zondek (1934a) found in stallions' urine quantities of oestrogen varying in different horses from 10,000 to 400,000 m.u. per litre, and sometimes exceeding the quantity present in the urine of pregnant mares. In two testes weighing 350 g. removed from a stallion Zondek found 23,100 m.u. of oestrogen, the stallion's testis being thus richer in oestrogen than any other tissue hitherto examined. Geldings and stallion colts excreted relatively small amounts. He concluded that the hormone was formed in the testes. Large concentrations of oestrogen were not found in the urine or testis of the bull. In normal men small amounts of

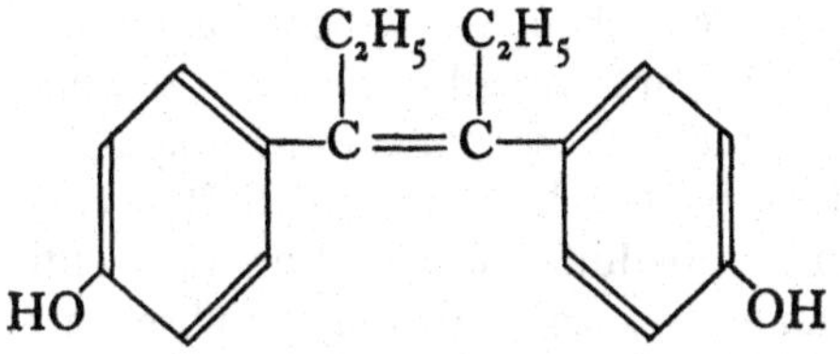

I. Diethylstilboestrol.
(Dodds, Goldberg, Lawson & Robinson, 1938)

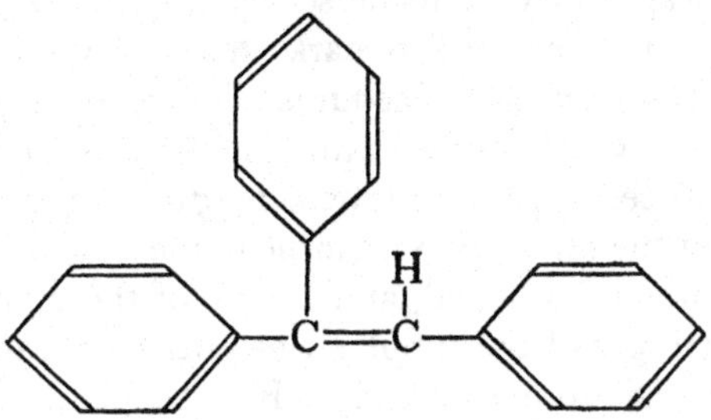

II. Triphenylethylene
(Robson & Schönberg, 1937)

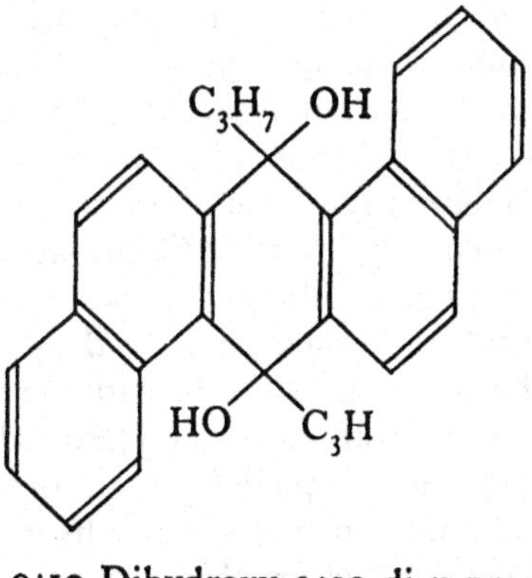

III. 9:10-Dihydroxy-9:10-di-n-propyl-
9:10-dihydro-1:2:5:6-dibenzanthracene
(Cook & Dodds, 1933)

Fig. 6. Synthetic oestrogens

oestrogen are found both in the testis and in the urine (Laqueur, Dingemanse, Hart & de Jongh, 1927; Laqueur & de Jongh, 1928; Fee, Marrian & Parkes, 1929; Borchardt, Dingemanse & Laqueur, 1934; Österreicher, 1934).

(iii) *Formation of progestin by the male.* The bull offers another extreme example of the excretion by a male of a so-called 'female hormone'; Marker, Wittle & Lawson (1938) isolated from the pooled urine of a number of bulls more pregnanediol than was recovered by similar methods from the same amounts of urine excreted by pregnant cows or by pregnant women (p. 394).

(iv) *Formation of androgen by the female.* Steinach & Kun (1931) noticed that in a guinea-pig whose ovaries had been exposed to a sterilizing dose of X-rays the clitoris gradually became converted into a penis-like organ, as happens also in this animal in response to the artificial administration of androgens. The same phenomenon was caused in the female guinea-pig by giving repeated injections of a pituitary extract. In both instances the ovaries had become highly luteinized. Whatever in these instances may have been the hormone which caused masculinization of the external genital organs, the ovary, in part at least, must have been responsible for its presence. Lipschütz (1937a) has verified these experiments on guinea-pigs and remarks that the increased androgenic activity induced in the ovary by the procedures adopted does not depend on histological changes suggestive of gonadal intersexuality, although such a change may occasionally take place. Lipschütz (1936, 1937a) removed one ovary and all but a small remnant of the other from infantile guinea-pigs. In some of these guinea-pigs a penis-like organ developed, and in two of these among twelve there was a definite hypertrophy of the tubular elements of the ovarian medulla, representing apparently a masculine rudiment. But the penile changes occurred also in the absence of these testis-like transformations in the residual ovarian tissue.

Effects comparable with the foregoing were obtained by Hill (1937), who grafted ovaries into the tips of the ears of castrated male mice and thereafter observed signs attributable to the production of androgen by the grafted ovary. Lipschütz (1932) had previously observed that an ovarian graft in a castrated male guinea-pig might in exceptional instances produce enough androgen to maintain the prostate and seminal vesicles in a normal condition, and de Jongh & Korteweg (1935) noted that an ovary grafted from a littermate into a castrated mouse prevented to some degree the atrophy of the accessory reproductive organs which otherwise would have ensued. Bradbury & Gaensbauer (1939) gave daily injections of gonadotrophin, ranging in different animals from 2 to 10 r.u., to female rats from the 6th to the 30th postnatal day. At the end of this period the rats showed a conversion of the clitoris into a penis with the cartilaginous anlage of an os priapi. The foreskin was separate from the glans, which resembled that of a male and was furnished with horny spicules. As in the experiments of Steinach & Kun, and Lipschütz, on guinea-pigs, these results are possibly attributable to luteinization of the ovaries by gonadotrophin.

(v) *The effects of gonadal hormones are organ specific, not sex specific.* The contrasts seen in the results produced by gonadal hormones in the male and the female respectively depend little on the recipient's sex. Broadly speaking the organs which are represented in both sexes react alike in both sexes to the same hormone: for example, the structures derived from the wolffian and müllerian systems, the external genitalia, mammae and nipples, skin and skeleton. The distinction between male and female in regard to their reactions to gonadal hormones seems to lie chiefly in the possession of different tissues competent to react to each type of hormone.

The earliest recorded experiments in endocrinology demonstrated the similarity of response of homologous organs in the two sexes to the same hormonic stimulation. John Hunter (1794), in the course of experiments on grafting, discovered that if the spur of a young cock were transplanted into the leg of a young hen the tissue took root but no spur grew, while the intact spur on the other leg of the donor cock developed as usual. Hunter then grafted the rudimentary spur of a young hen into a young cock, and observed that the transplanted spur grew nearly as fast and to as large a size as the natural spur on the cock's other leg. These experiments he repeated several times, always with the same result.

Much later Loewy (1903) reported that injections of testicular substance into young capons caused the development of male skeletal characters as well as comb growth. Walker (1908 a, b) injected an extract of cocks' testicles daily into two hens. The injections began in February and were continued until October. The hens ceased laying in the middle of March, and while the injections were in progress their combs and wattles grew and became highly coloured. When the injections ceased the combs rapidly shrank to their normal feminine dimensions. It was noticed by Champy & Kritch (1925) that growth and turgidity of the fowl's comb are brought about not only by testicular extracts, but by extracts of ovary also; and Parkes (1937) has pointed out that the hen's comb, like that of the cock, atrophies after gonadectomy and enlarges in response to androgens but not to oestrone. The conclusion is that the ovary of the normal hen produces enough androgen to protect her comb from atrophy. Parkes also found that when an extract of sow's ovary was smeared on the atrophied comb of a capon it caused the comb to enlarge.

Other examples could be given to illustrate this general principle that the effect of a gonadal hormone on a particular organ is not dependent on the sex of the individual in which that organ is situated. The mamma affords a good instance, for its reactions to oestrogen in both male and female have been investigated in detail by many workers and have been found to be alike in both sexes. The action of oestrogen in causing absorption of the pubic bones and the cartilage which unites them, with a consequent separation of the two ventral halves of the bony pelvis, may also be mentioned to illustrate the current theme; in mice these bony and cartilaginous changes are readily caused in males as well as in females by the artificial administration of oestrogen (Burrows, 1935 f). To multiply examples will hardly be necessary in this place; the reader will notice additional instances in the chapters devoted to the effects produced by individual types of gonadal hormone.

(vi) *Similar actions by testis and ovary.* The results of gonadectomy in certain respects may be identical, or nearly so, in both male and female. Such a similarity of effect can be attributed in some instances to the deprivation of a single type of hormone, which is produced by testis and ovary alike. The case of the fowl's comb which has just been mentioned appears to be an example of this.

In adult birds it is easier to remove the testes than the ovary. For this reason the following examples are rather one-sided, being confined to experiments on males; but they seem to illustrate a similarity of action exerted by ovary and testis. When rearing their young the crop glands of pigeons, both male and female, secrete the crop milk with which the young are fed. Castration of male pigeons prevents the hypertrophy and functional activity of the crop gland. Van Oort & Junge (1933) give other examples. They say that the summer plumage of the blackheaded gull, which is common to both sexes, does not appear in males which have been castrated at the time of their first winter plumage. A curious example quoted by them concerns the ratio between rods and cones in the fowl's retina.

In the chick this ratio, they say, is 28 : 100, and in adult cocks and hens the ratios are 48 : 100 and 45 : 100 respectively; after gonadectomy in either sex the ratio remains the same as in the chick.

(vi) *Similarity of responses to different types of gonadal hormone.* (a) *Oestrogen-like action of androgens.* Oestrogens cause premature cornification and opening of the vagina in the rat and mouse. Brouha & Simonnet (1928) observed that four daily subcutaneous injections of an extract of bulls' testes, given to immature rats of 22 days, had similar consequences. The same effects have been produced by several pure androgenic compounds. Butenandt & Kudszus (1935) reported that androstenedione, *trans*dehydroandrosterone, androsterone and testosterone, all of which induce comb growth in capons and stimulate development of the seminal vesicles in rodents and therefore are classified as androgens, when given to normal infantile female rats or mice cause precocious opening of the vagina accompanied by cornification; and Parkes (1935) found that androstanediol and 17-methyl-androstane-17-ol-3-one, both of which are androgens, will cause vaginal corni-fication in the immature rat. Nathanson, Franseen & Sweeney (1938) obtained the same results from the administration of testosterone propionate to immature female rats even after spaying or hypophysectomy. They suggest as an explana-tion that perhaps the immature rat has the capacity of converting testosterone into an oestrogen. Comparable results have been obtained in immature rats by Warren (1935) with androsterone and by Courrier & Cohen-Solal (1937) with testosterone.

Deanesly & Parkes (1937c) called attention to several instances in which androgen and oestrogen produce similar effects. The Sebright bantam cock is hen-feathered; caponization is followed by the growth of 'male' bantam plumage. Small doses of oestrone, they say, will cause such a capon to revert to hen-feathering; so also will daily doses of the androgens *trans*dehydroandrosterone, *trans*androstenediol, androstenedione and testosterone.

Nelson & Merckel (1937), investigating the actions of a number of androgenic compounds, found that daily administrations of testosterone, androsterone, de-hydroandrosterone and androstanedione to adult spayed female rats for a period of 30 days all failed to produce vaginal cornification. On the other hand dehydro-androsterone and *cis*androstenediol caused prolonged cornification of the vagina in normal adult rats; and this effect was induced whether the rats had previously undergone hypophysectomy or not. Deanesly & Parkes (1937c) state that, pro-vided large enough doses are employed, *trans*androstenediol, dehydroandro-sterone, androstenedione and testosterone all cause an increased development of the immature rabbit's uterus like that caused by oestrone.

Courrier & Gros (1938b) gave daily doses of 5 to 10 mg. of testosterone pro-pionate to spayed female cats for periods lasting from 7 to 15 days. The result was a pronounced hypertrophy of the uterus with enlarged uterine glands, the changes resembling those produced by oestrone.

In the monkey the changes which occur in what is referred to as the sex-skin can be induced either by oestrogen or androgen (Hartman, 1940).

Another example of oestrin-like action exerted by androgens was noticed by Wolff (1936a). He injected androsterone in doses varying from 0·5 to 2 mg. into hens' eggs on the 5th day of incubation. In male chicks this treatment caused the development of an ovarian cortex in the left gonad, converting it into an ovo-testis, and led also to the preservation of the proximal segments of the müllerian ducts. Willier, Rawles & Koch (1938) made a similar observation. They injected testosterone, androsterone and dehydroandrosterone into hens' eggs after 43 to 72 hours of incubation. The effect of all these androgens on the female chicks was of a masculinizing nature, that is to say, they caused the ovaries to become like testes in shape with hypertrophy of the medullary tissue. In males, however, they caused the left testis to appear flattened and like an ovary in shape while the oviducts persisted throughout their length. In fact the males closely resembled females. In other words, these androgens had a masculinizing effect on the females but acted like oestrogens on the males.

An observation by Duyvené de Wit (1938, 1940) might be mentioned in connection with this subject. He found that all the three types of gonadal hormones, androgen, oestrogen and progestin, cause enlargement of the ovipositor in the female bitterling (*Rhodeus amarus*).

(*b*) *Androgen-like action of oestrogens.* Although a main function of the androgens is to keep the accessory male generative organs in an active state, this function is not confined to the androgens; oestrogens and progestins take a part. David, Freud & de Jongh (1934) investigated the responses of the prostate, seminal vesicles and preputial glands of young rats to several different oestrogens. The rats were castrated when between 20 and 30 days old and 3 days later were given injections of oestrogen twice a day for 5 days, and were killed on the 6th day. The results are given in Table 50, where it will be seen that the compounds injected caused an increase in weight of the prostate and seminal vesicles as compared with those of castrated control rats. The effects on the preputial glands were not so definite.

Table 50. The effects of various oestrogens on the prostate, seminal vesicles and preputial glands of the castrated rat (David, Freud & de Jongh, 1934)

Hormone	Dose (γ per diem)	Weights of organs expressed in mg. per 100 g. of bodyweight		
		Prostate	Seminal vesicles	Preputial glands
None	—	22⎫ 17⎭ Two series	13⎫ 14⎭ Two series	34
Oestrone	2	44	25	30
Oestrone	10	39	27	48
Oestriol	2	28	18	22
Oestriol	10	48	19	38
Oestradiol	2	25	46	35
Oestradiol	6	75	45	45
Equilin	6	30	27	50
Equilenin	2	25	15	44

Korenchevsky and his colleagues have made comparable observations.

(*c*) *Progesterone-like action of androgens.* In many respects the biological actions of androgens resemble those of progestin; thus, in association with oestrogen, they cause the development of alveoli in the mammae, induce mucification of the vagina and bring about progestational changes in the uterus.

Klein & Parkes (1936a) found that five different androgens, given in doses between 10 and 20 mg., caused progestational changes in the uterus of the immature rabbit corresponding to those induced by 0·25 to 0·5 mg. of progesterone. These compounds were testosterone, methylandrostanediol, methyl*trans*androstenediol, methyldihydrotestosterone and methyltestosterone. To this list other androgens have been added (Klein & Parkes, 1937; Klein, 1938b), including androstenedione, ethyldihydrotestosterone, ethylandrostanediol and others. The progestational action of these compounds on the rabbit's uterus occurs, Klein states, in the absence of the ovary, provided that the rabbit has received appropriate daily doses of oestrin before the androgen (see also p. 319).

McKeown & Zuckerman (1937) caused progestational changes in the rat's uterus by daily injections of 14 i.u. of oestrone given for a period of 4 days and then followed by daily injections of 2 mg. of testosterone for 10 days.

Robson (1937d) reported that testosterone propionate inhibits the action of oxytocin on the rabbit's uterus, an effect which is produced by progesterone also; and Scipiades (1937) states that testosterone propionate, when given to pregnant rats, like progesterone causes a prolongation of gestation, and prevents abortion after spaying. Greene & Burrill (1939) have confirmed this observation, finding that daily doses of 5 mg. of testosterone propionate, 20 mg. of testosterone, or 10 mg. of androstenedione, enabled the continuance of pregnancy to full term in rabbits which had been spayed between the 11th and 14th days of gestation. It may be explained that removal of the ovaries at this stage in the rabbit without further treatment is usually followed by death of the embryos.

A biological action which progesterone has in common with several androgens including testosterone has been described by Shapiro (1936) and Zwarenstein (1937), namely the induction of ovulation in the clawed toad (*Xenopus laevis*), even after hypophysectomy. The administration of oestradiol alone to this amphibian is not followed by ovulation.

(d) *Androgen-like activity of progesterone.*   Steinach & Kun (1931) noticed that, in female guinea-pigs whose ovaries had become highly luteinized in response to repeated injections of gonadotrophin or a sterilizing dose of X-rays, the clitoris gradually became converted into a penis-like organ, as happens also in this animal when subjected to successive doses of androgen. Guyénot, Ponse & Wietrzykowska (1932) have made a comparable observation. They repeatedly gave alkaline extracts of ox-pituitary to female guinea-pigs, with the result that the clitoris in these animals gradually became converted into a penis-like organ with cornified papillae on the glans and a retractable foreskin; these females displayed male behaviour. Their ovaries had undergone extensive luteinization. Papanicolaou & Falk (1934) found that continued treatment of immature guinea-pigs with placental gonadotrophin caused their ovaries to exercise an androgenic function, so that the clitoris enlarged and assumed the form of a penis with well-differentiated prepuce, corpus cavernosum and glans. Bradbury & Gaensbauer (1939) have obtained similar responses to gonadotrophin in immature rats. Subsequent experiments seem to show that these results may be attributed to the androgenic properties of progesterone when produced in large amount and free from the inhibiting action of oestrogen.

Burrows (1936 *f*) showed that progesterone, like androgen, to some extent protects the mouse's prostate and seminal vesicles from the effects of oestrogen, and subsequent observations have demonstrated that progesterone, when given in large enough dosage, exerts an influence on the accessory male organs like that of androgens. Lamar (1937) gave 0·5 to 3 rb.u. of progesterone daily to rats which had been castrated when 25 days old. The injections were begun almost immediately after the castration and were continued daily for 15 days. At the end of this time there were significant increases in the weights of the prostate, coagulating glands, seminal vesicles, preputial glands and periurethral tissue as compared with the weights of these structures in control castrated rats. The most pronounced influence was on the prostate, periurethral tissue and preputial glands, and as regards its effect on these structures 1 mg. of progesterone appeared to be the equivalent of 0·03 mg. of testosterone. Greene, Burrill & Ivy (1939*b*) found that progesterone caused in female rats enlargement of the clitoris, and that in the castrated male rat the prostate and to a less extent the seminal vesicles could be maintained in weight and secretory activity by large daily doses of progesterone. For example, rats were castrated when 19 days old and for the next 20 days were given 3 mg. of progesterone daily; at the end of this period the animals were killed and their prostates and seminal vesicles were compared with those of castrated littermate controls. The androgenic effects of the treatment are shown in Table 51, which represents the outcome of one of their experiments. To maintain these androgenic effects in adult castrated rats 9 mg. or more of progesterone were needed daily.

TABLE 51.   The effect of progesterone on the prostate and seminal vesicles of the immature castrated rat (Greene, Burrill & Ivy, 1939)

Condition	Daily dose of progesterone (mg.)	Interval between castration and autopsy (days)	Total dose (mg.)	Mean weight of ventral prostate (mg.)	Mean weight of seminal vesicles (mg.)
Castrated, untreated controls	0	20	0	11·8	5·0
Castrated and treated with progesterone	3	20	60	69·1	13·2

Yet another effect common to androgens and progestins has been reported by Nelson (1937*a*). It is known that spermatogenesis can be maintained in rats for a considerable period after hypophysectomy, provided that androgens are continually supplied from the time of the operation. Nelson found that progesterone and the progestin *epiallo*pregnaneolone will maintain spermatogenesis in similar circumstances, the latter being more efficient in this respect than progesterone.

In connection with the androgen-like activities of progestins it may be mentioned that pregnanediol has been identified in the urine of female patients with virilism caused by adrenal hyperplasia or carcinoma (Malley & Bradshaw, 1941; Salmon, Geist & Salmon, 1941).

(*e*) *Similarity of action of androgen, oestrogen and progestin.*   In rats and mice the preputial gland, though more largely developed in the male, is functional in both sexes. Van der Woerd (1938) tested the responses of the gland to the three types of gonadal hormone in castrated rats and found that it was stimulated by androgens and oestrogens, and perhaps to a slight extent by progestin. These results will be quoted more fully in connection with the co-operative effects of gonadal hormones (p. 115).

On Cowper's gland, also, androgen and oestrogen act alike; either will cause enlargement and secretory activity of the gland in the castrated rat (Tschopp, 1936).

Androgen, oestrogen and adrenal cortical hormone all cause enlargement of the nipple in the male guinea-pig, according to Jadassohn, Uehlinger & Margot (1938), who obtained this result with the *androgens* testosterone propionate, androsterone, androstenedione and androstanedione; the *oestrogens* oestrone, equilin and equilenin; and the *adrenal cortical hormones* adrenosterone and corticosterone.

Several workers have reported enlargement of the mamma after the administration of androgen, and it has also been observed that androgen, like oestrogen, will induce coloration of the sexual skin in monkeys (Hartman, 1937). Another example of a similar action exerted by the different gonadal hormones is the production of hypospadias in infantile female rats and mice. This result, following the administration of androgens and oestrogens, has been recorded by Hain (1935 a, b), Wiesner (1935), Lacassagne (1936e), Greene & Ivy (1937), and has been confirmed by the writer, who found that hypospadias may be caused in these animals by progesterone also (Burrows, 1939c). Non-descent of the testes, like hypospadias, may be caused by any of the gonadal hormones, androgen, oestrogen or progestin (Burrows, 1939c), and probably by the same action, namely an inhibition of the supply of gonadotrophin by the pituitary.

(*f*) *A single hormone may possess androgenic, oestrogenic and progestational powers.* Emmens & Parkes (1939b) have investigated the biological actions of ethinyltestosterone or pregneninolone, a compound discovered by Inhoffen & Hohlweg (1938), and they find that, although its potency is relatively small, it has androgenic, oestrogenic and progestational capacities, inasmuch as when given in large enough doses it will induce growth of the capon's comb, cornification of the vagina in the spayed rat or mouse, and progestational changes in the uterus of the immature or ovariectomized rabbit following preparatory injections of oestrone.

Courrier & Jost (1939) have shown that pregneninolone will permit nidation of ova and will maintain pregnancy in the spayed rabbit, and Salmon & Salmon (1940) noticed that in addition to this progesterone-like action it has androgenic properties, causing hypertrophy of the fibromuscular components of the uterus and hypertrophy of the clitoris in the rat.

Perhaps a more complete example of the same specific response to different gonadal hormones is that of the bitterling's ovipositor, which according to Duyvené de Wit (1938, 1940) becomes enlarged under the influence of androgen, oestrogen, progestin or adrenal-cortical hormone.

Another character which is shared by all three types of gonadal hormone and adrenal cortical hormones is their capacity to influence the excretion of water, sodium and chlorine.

Thorn & Harrop (1937) tested the influence of various gonadal hormones on the excretion of sodium in a normal male dog. All the hormones tested by them, namely oestradiol, oestrone, progesterone, pregnanediol, testosterone and testosterone propionate, caused a decreased excretion of sodium, progesterone being the most effective in this respect.

(g) *Special effects of androgen on female organs and of oestrogen on male organs.* The term 'bisexual hormonal activity' has been applied to reactions of female organs to 'male' hormones and of male organs to 'female' hormones, or to a common action of androgen and oestrogen on both male and female gonads. To illustrate this kind of 'bisexual' action experiments on sex determination may be quoted; they show that some but not all androgens, given at an early stage of development, may cause intersexualism in both male and female embryos.

Wolff (1935, 1936a) applied androsterone to the chorion of chicks between the 3rd and 5th day of incubation. The embryos were extracted from the eggs and examined between the 15th day and the end of incubation. The results showed that under the influence of androsterone the genetic females were masculinized, while many of the genetic males were feminized. Willier, Rawles & Koch (1938) also have introduced gonadal hormones into fowls' eggs which had been incubated from 43 to 72 hours previously. Their results confirm those of Wolff, except that the androgens they employed did not all act alike. They used testosterone propionate, androsterone and dehydroandrosterone. All of these caused changes in the reproductive organs of the females; the right gonad was enlarged by hypertrophy of the medullary tissue, and both gonads tended to have the shape of testes. In other words the androgens used had a masculinizing effect on genetic females. On the males testosterone propionate did not show any feminizing action. Androsterone and dehydroandrosterone on the other hand caused the left testis to become flattened and ovary-like in shape and to have the structure of an ovotestis. The oviducts persisted entire and were in some instances hypertrophied, the feminizing effect of these androgens being thus pronounced, as in Wolff's experiments.

Noble & Greenberg (1940) treated chameleons (*Anolis carolinensis*) with testosterone propionate and observed that it caused hypertrophy of both male and female genital ducts and keratinization of the cloaca in females and in castrated males.

(h) *Inhibition of the gonadotrophic potency of the pituitary by gonadal hormones.* It is surprising that the most important of all the so-called bisexual activities of the gonadal hormones is not usually included under the 'bisexual' heading, namely the capacity common to androgens, oestrogens and progestin of inhibiting the output of gonadotrophin by the pituitary in both sexes.

Korenchevsky (1937b) has suggested that the gonadal hormones should be classified as follows:

    I.   Purely Male or Female Hormones.
    II. Partly Bisexual Hormones:
          (a) Those having chiefly male effects.
          (b) Those having chiefly female effects.
    III. Bisexual Hormones.

The utilization of such a classification is rendered difficult by the indefinite nature of the boundary lines. Korenchevsky mentions progesterone as the only example of a purely female hormone; yet it has been shown that the formation of pregnanediol is not confined to the female, and that some of the activities of progesterone resemble those of testosterone (pp. 105, 110, 394).

The varied reactions of the gonadal hormones suggest that the biological capacities of each should be studied individually without any preconceived notion that because a given hormone is commonly classed as an androgen it will therefore possess all the properties of testosterone, or that being derived from the ovary a hormone will be quite free from properties usually associated with androgens. It may be thought, too, that the term 'bisexual' as applied to the activities of the gonadal hormones is of little real service and might be discarded with advantage.

(*i*) *The production of androgen and oestrogen by the gonads does not depend directly upon the genetically determined sex of the host.* As the experiments of Steinach & Lipschütz have shown, a testis transplanted into a female is still able to produce androgen, and an ovary in a male will manufacture oestrogen, provided always that the organs are placed in regions of the body which are suitable for their appropriate functions.

(*j*) *The response of a tissue to a gonadal hormone is innate in the tissue and is not dependent on the sex of the host.* Danforth & Foster (1929) performed some experiments on fowls which afford a striking illustration of this principle. They learned first that if skin be transplanted from a day-old chick to one of another colour, the feathers growing in the transplant will retain the characters of the original donor provided that the recipient is of the same sex as the donor of the graft. Thus a black feather graft will remain black although growing in a white fowl, and vice versa. If, however, a graft be made from a chick belonging to a strain in which the cocks and hens have different colouring, the feathers of the graft will assume the colour of those appropriate to the sex of the host; that is to say, the grafts will have retained their capacity to respond in the ordinary way to gonadal hormones.

So-called intersexual conditions. The writer has already criticized the use of the phrase 'bisexual action' in connection with the gonadal hormones. He would ask permission here to disagree with the term 'intersexual' as applied in physiological and clinical literature. For clear thinking straightforward, undeceptive language is required, because false terminology tends to confuse the mind. When 'intersexual' conditions are mentioned the adjective itself suggests the existence of an individual who cannot properly be described either as a male, a female, or a hermaphrodite, whereas in fact the term is intended merely to denote an individual who possesses an excessive proportion of the attributes of the other sex, or one who lacks some of the normal endowments of his own. The expression came into use in the earlier days of endocrinology and has outlived its usefulness. Consider an example. A hen whose ovary has been removed will assume certain male features which may be caused also by inserting testis grafts into an otherwise normal hen (Domm, 1927). After either of these procedures the subject is still a female, and to refer to her as an example of an 'intersex' seems to the writer misleading.

Co-operation of different Gonadal Hormones

In many instances the sexual organs, whether of the male or female, in order to attain a full development and complete functional capacity, require the co-operation of both androgen and oestrogen. Such co-operation might be included in the scope of bisexual activities if such a term were acceptable. Usually, and perhaps always, in these co-operative actions the available quantities of the two types of hormone must bear a more or less strict relationship to each other. Though androgen and oestrogen supplied in certain proportions often act co-operatively, they may be antagonistic to each other, even strongly so, when this proportion is disturbed. Laqueur (1932) and de Jongh (1934b) have noted that while small doses of oestrone will co-operate with androgen in restoring the accessory genital organs of the rat or mouse after castration, large doses of oestrone will inhibit this restorative action.

(i) *Co-operation between androgen and oestrogen.* Freud (1933b) castrated rats when they were between 3 and 5 weeks old and weighed from 25 to 45 g. Three days later injections of gonadal hormones were given twice daily for 4 days and on the 5th day the rats were killed and their seminal vesicles were examined microscopically. By these methods it was found that a testicular androgenic extract and oestrin exercised contrasted effects on the seminal vesicle, the androgen causing hypertrophy and functional activity of the epithelium while oestrin caused hypertrophy of the fibro-muscular components; so that for a full development of the seminal vesicles the action of both androgen and oestrogen was required. De Jongh (1934b) verified these observations, and noted also that doses of androgen insufficient to resuscitate the atrophied epithelium of the seminal vesicle in castrated rats or mice will do so if small doses of oestrone be given at the same time. Furthermore, it was found by David, Freud & de Jongh (1934) that an androgenic extract of testicle does not interfere with the stimulating action of oestrone on the fibro-muscular wall of the rat's seminal vesicle, but that the two hormones mutually enhance the hypertrophic action. This has been confirmed by Overholser & Warren (1935).

Del Castillo & Pinto (1938) found that brief periods of treatment with oestrone cause hypertrophy of the prostate and seminal vesicles in rats and guinea-pigs, that oestrone to some extent hinders the atrophy of these organs after castration, and that the hypertrophy of the accessory male organs induced by testosterone is enhanced by the addition of oestrone and progesterone.

Van der Woerd (1938) has studied the action of gonadal hormones on the preputial gland of the rat; in this animal the gland is functional in both sexes. Male and female rats were deprived of their gonads when between 25 and 28 days old. A fortnight later daily treatment with hormones was begun. The injections were continued for 14 days, after which the rats were killed and their preputial glands weighed. From the subjoined table it will be seen that the glands had been stimulated by androgen and oestrogen, and perhaps in the males by progesterone. In both sexes the glands attained the largest size in those rats which had been given androgen and oestrogen together (Table 52).

TABLE 52.   The effect of gonadal hormones, given singly or in combination, on the rat's preputial gland after gonadectomy (Van der Woerd, 1938)

Hormone	Daily dose (γ)	Mean weight of preputial glands	
		Males (mg.)	Females (mg.)
Controls	—	27	31
Testosterone	200	79	86
Androsterone	200	59	61
Oestrone	10	43	40
Progesterone	200	36	32
Controls	—	32	36
Testosterone plus Oestrone	200 / 10	106	97
Androsterone plus Oestrone	200 / 10	57	67
Progesterone plus Oestrone	200 / 10	39	42

Korenchevsky and his colleagues (1935, 1936, 1937, 1939) also have shown that after gonadectomy, whether in the male or female, the resulting atrophy of the accessory reproductive organs can be allayed to some extent by the gonadal hormone of the opposite sex, and that a supply of both androgen and oestrogen is required for the full development of the accessory generative organs (Tables 53, 54).

TABLE 53.   The effect of gonadal hormones on the uterus and vagina of spayed rats (Korenchevsky, Dennison & Simpson, 1935)

	Weight of uterus (mg.)	Weight of vagina (mg.)
Normal untreated rats	355	233
Spayed untreated rats	33	81
Spayed and treated with androstanediol	123	172
Spayed and treated with oestrone	155	146

TABLE 54.   The effect of daily doses of gonadal hormones on the prostate and seminal vesicle of castrated rats (Korenchevsky, Hall, Burbank & Ross, 1939)

Condition of rat	Weight of prostate (mg.)	Weight of seminal vesicles (mg.)
Normal untreated	853	952
Castrated untreated	65	12
Castrated and treated with oestradiol dipropionate (6γ)	86	51
Castrated and treated with testosterone dipropionate (150γ)	624	528
Castrated and treated with testosterone dipropionate (150γ) and oestradiol dipropionate (6γ)	677	601

Cotte & Noel (1936) have reported that testosterone acetate in daily doses of 2·5 mg. has a restorative effect on the endometrium of the spayed rat.

Another example of co-operation between androgen and oestrogen is seen in the production of vaginal mucification. Normally this condition, which accompanies pregnancy and pseudopregnancy, is caused by the simultaneous action of oestrogen and progesterone in definite proportions. Androgens have been found capable of replacing progesterone in effecting the same result.

The co-operative effects of androgen and oestrogen on the accessory organs of reproduction are not limited to an increase in their weight and histological appearances but extends to their functional activity. This was proved by Kun & Peczenik (1937), using their quantitative test for androgenic activity. This test is made by injecting the hormone in graduated amounts into castrated rats and noting the minimal dose which will lead to ejaculation in response to electrical stimulation of the rump by an interrupted current of 30 volts. By this test they showed that the activity of androgen is greatly increased by the simultaneous administration of oestrone. This hormonal co-operation appears to be applicable to man. Foss (1939), treating a man of 38 who had been castrated, found that copulatory power could be partly restored by testosterone the action of which was considerably increased by the addition of oestradiol benzoate or progesterone.

A zoological example of co-operation between the sex hormones has been described by Champy & Coujard (1939), who have noticed that oestrogen as well as androgen is required for the full development of the clasping digit of the male frog, both these hormones being supplied by the testis.

(ii) *Co-operation between oestrogen and progestin.* (a) *Causation of deciduomata and progestational changes in the uterus.* Perhaps the earliest example of a co-operation between different types of gonadal hormone was that noted by Weichert (1928). Following Loeb's experiments, which had been made on guinea-pigs, Weichert found that uterine trauma in the normal non-pregnant rat did not lead to the development of a deciduoma; if, however, daily doses of an extract of corpora lutea were given immediately after oestrum, injuries of the rat's uterus were followed by the formation of deciduomata. Spayed rats treated with luteal extract did not show this reaction unless they had been submitted previously to a course of oestrin injections. Comparable observations were made by Hisaw & Leonard (1930) and by W. M. Allen (1930), who showed that progestational changes were not induced in the endometrium of normal immature or spayed adult rabbits by injection of oestrin-free luteal extracts unless the animals had been submitted to doses of oestrin for a few days previously. Shelesnyak (1933) proved that deciduomata could be induced in spayed rats by oestrin and extracts of sow's corpora lutea if the two hormones were given successively in the correct relative proportions.

To effect progestational changes in the uterus the two hormones must be supplied successively; that is to say, a preliminary treatment with oestrogen followed by treatment with progesterone supplies the essential condition. If oestrin be given in more than minimal amounts simultaneously with progesterone, even though all the other conditions are favourable, it will prevent progestational changes (Hisaw & Leonard, 1930). Korenchevsky & Hall (1937) state that in the rat a dose of oestrone greater than 1γ will counteract the progestational influence of $1,500\gamma$ of progesterone.

(b) *Mucification of the vagina.* In the course of pregnancy and pseudo-pregnancy the vagina undergoes a characteristic histological transformation in many species. Stratification of the epithelium occurs as in the period prior to oestrus; thereafter the more superficial layers of cells instead of becoming squamous and cornified, as they do in oestrus or under the influence of artificially administered oestrogens, are rounded, swollen and ultimately columnar in shape, and become loaded with mucus-like material. This condition, described as vaginal mucification, is dependent on the co-operative action of oestrogen with progestin or androgen.

Unlike progestational changes in the uterus, which depend on a preliminary subjection to oestrogen followed by progestin, vaginal mucification requires the simultaneous action of these hormones, an essential factor being the relative proportions in which the two hormones are available.

(c) *Development of the mamma.* The co-operative effects of oestrogen and progesterone on the mamma will be discussed in a later chapter.

Vagaries of co-operation. Before ending the present discussion certain reactions may be recalled which might be thought anomalous. They are reminders of the caution required when we are considering the actions of sex hormones, and the difficulty of making correct generalizations in this branch of physiology. For it is not to be assumed that the co-operative effects of the three main types of gonadal hormone—androgen, oestrogen and progestin—always follow clear-cut lines. The fact that a certain androgen shows a particular effect in co-operation with a given oestrogen does not justify the supposition that any androgen will co-operate with any oestrogen to produce a similar result. Korenchevsky, Hall & Ross (1939) found that while oestradiol dipropionate co-operated with testosterone in restoring the accessory generative organs of castrated rats it did not co-operate in this way with testosterone propionate in the doses used, and was antagonistic to androsterone and *trans*dehydroandrosterone.

When the combined effects of two gonadal hormones of different types are under consideration, attention must be paid both to the relative dosages and to the precise nature and identity of the two hormones concerned.

Mutual Antagonisms between Gonadal Hormones

Before discussing the inhibitory effects which one type of gonadal hormone may exercise against another it may be made clear that there is no essential incompatibility between either type of gonad and the sex of the host; nor does the gonad of one sex hinder the survival of a gonad of the opposite sex when the two coexist in the same body. If gonadectomy is performed on littermates and the ovaries are transplanted into the males and the testes into the females, the gonads will grow, and, though they may fail to form mature germinal cells in their new environment, they will produce hormones and by these will affect the sexual characters and behaviour of the host (Steinach, 1916; Lipschütz, 1917; Moore, 1919; Sand, 1919 a, b). The compatibility between the gonads and individuals of the opposite sex was shown in another way by Morpurgo (1908), who joined male and female rats in parabiosis, to resemble Siamese twins, and noted that the individuals retained their sexual characters in spite of the interchange of blood which took place between them. Males in this condition were able to sire normal offspring. Hill (1932) has performed the same experiment with the same results.

Instances have been recorded in animals and man in which a testis and an ovary have coexisted in the same body (Witschi, 1932; Raynaud, Marill & Xicluna, 1939). In such circumstances the gonads are not quite normal in function because they are subjected to heterosexual hormones. An ovary grafted into a male will show maturation of follicles without subsequent luteinization because the pituitary of the male does not supply LH; and the testis in the presence of an ovary may fail to produce mature spermatozoa. Apart from such functional defects there is no incompatibility between the gonads of the two sexes when they are both present in the same individual.

The interplay of the sex hormones is complicated, and there are many ways in which one gonadal hormone may interfere with the production or action of another. At present we are far from being able to explain all these antagonisms. The fact of their existence is important, not only from an academic outlook, but from the practical standpoint of human diagnosis and treatment.

Loeb (1914) seems to have been the first to demonstrate an interference by one hormone with the activity of another. He showed that ovulation is prevented by the presence of corpora lutea the extirpation of which in the guinea-pig is followed by ovulation even during pregnancy. Papanicolaou (1920) made confirmatory observations and found that removal of the corpora lutea from a guinea-pig accelerates the appearance of the next oestrum, and that the continued presence of a corpus luteum inhibits ovulation and oestrus, as shown by vaginal smears.

Before exemplifying antagonistic reactions between the gonadal hormones, it may be recalled that all three types of gonadal hormone can prevent the pituitary from forming adequate supplies of gonadotrophin. By this inhibitory action on the pituitary, testosterone or progesterone given in large enough doses to a female will suppress follicular maturation, ovulation and the oestrous cycle; and so long as the treatment is continued oestral changes in the vaginal epithelium and other consequences of oestrogenic action will not occur, because the production of oestrogen is suspended. This indirect form of antagonism by inhibiting pituitary function has been discussed earlier (p. 51); the present survey will be confined to more direct antagonisms between gonadal hormones when administered at the same time.

The co-operative activities of the gonadal hormones depend upon a correct proportion between the available quantities of each hormone. The counterpart of this principle is that a disproportion between the amounts of each hormone supplied may lead to the display of antagonistic effects.

(i) *Androgen versus oestrogen.* (*a*) *Inhibition of oestrogenic action by androgen.* An interference by testicular hormones with the action of oestrogen was discovered by the Dutch workers (de Jongh, 1933, 1934*b*, 1935*a, b*; David, Freud & de Jongh, 1934). They found that the epithelial changes induced in the prostate and seminal vesicles of the mouse by oestrin could be entirely prevented by the simultaneous administration of androgen. De Jongh (1935*b*) noticed that oestrin causes metaplasia in the prostate more readily in castrated than in non-castrated mice, possibly in consequence of the antagonistic action of the androgen produced in the latter. In castrated mice he ascertained that 3 c.u. of a testicular androgen (Hombreol) were required to counteract the metaplastic activities of 3γ of oestrin. The testicular androgen (testosterone) was more potent than a urinary androgenic extract (androsterone) in preventing this action of oestrin on the prostate. De Jongh (1935 a) discovered also that the metaplastic effects of small doses of oestrogen on the prostate could be prevented by the administration of gonadotrophin, which presumably acts by causing an increased output of androgen by the testis.

Zuckerman & Parkes (1936 a) found that daily injections of 5 mg. of andro-stanediol prevented the changes in the prostate and uterus masculinus which would otherwise be caused by daily doses of 100γ of oestrone in the immature rhesus monkey; and Lacassagne (1937) demonstrated that in the mouse 1,000γ of testosterone given once a week inhibited the action on the prostate of 100γ of oestrone given during the same period.

Rusch (1937) showed that 250γ of testosterone given twice a week quite counteracted the metaplastic effects of simultaneous injections of 25γ of oestrone benzoate on the prostate and seminal vesicles of mice. Doses of 150γ of testo-sterone had some protective effect but did not quite annul the influence of the oestrone benzoate.

In a like manner androgen will prevent some of the effects of oestrogen on the vagina. In spayed mice Robson (1936 c, 1937 b) observed that the vaginal corni-fying effect of 1γ of oestrone given twice daily was inhibited by doses of 400γ of testosterone given at the same time. In causing this effect, he says, testosterone propionate is 10 times as effective as progesterone; and it is 20 times as effective as progesterone in checking the action of oestradiol on the vagina. Courrier & Cohen-Solal (1937) performed similar experiments on spayed rats in which the vaginal cornifying effects of subcutaneous injections of 0·02 mg. of oestrone were prevented by the simultaneous injection of 20 mg. of testosterone, but not by 10 mg. Hain (1937), having determined that the minimal dose of oestrone required to cause vaginal cornification in the spayed rat when given 4 times during 36 hours was 7·5γ, found that this effect could be prevented by the simultaneous injection of testosterone in four doses amounting together to 2·5 mg.

Other examples of an interference with the action of oestrogen by androgen have been recorded. Gardner & Pfeiffer (1938 b), when studying the effects of oestrogen on the skeleton in the mouse, noticed that the changes produced in the bones by oestrin were more readily induced in castrated than in non-castrated mice. The matter was thereon put to an exact test. Injections of hormone were begun when the mice were between 24 and 75 days old and were continued for periods varying from 127 to 232 days. To these mice a weekly dose of 500 or 1,000 i.u. of oestradiol benzoate was given, and into some of them 1·25 or 2·5 mg. of testosterone propionate were injected weekly. These doses of androgen were enough to prevent the changes in the skeleton, including dissolution of the symphysis pubis, which otherwise were brought about by the action of oestradiol in the dosages used.

Skowron (1935) has recorded a protective action of androgen against the aborti-facient action of oestrin on the pregnant rabbit, an effect which may perhaps be one more example of a similarity between the actions of androgen and progestin.

It is possible that androgens interfere with oestrogenic activity in more than one way: thus (1) they may directly inhibit the organ responses to oestrogen, (2) they may check the animal's own supply of oestrogen from the ovary by preventing the output of gonadotrophin from the pituitary, and (3) they may perhaps oppose the action of gonadotrophin on the ovary. Gley & Delor (1937) used two groups of immature female rats. Those of Group I were given 2 mg. of testosterone propionate every other day, while those of Group II received no testosterone. After 7 days all the rats were given equal daily doses of pituitary gonadotrophin for 3 days, the injections of testosterone being continued meanwhile in the rats of Group I. On the 5th day after the first dose of gonadotrophin all the rats were killed and their ovaries were examined. It was found that the ovaries of Group I rats were but little overdeveloped and weighed 150 mg. per 100 g. of bodyweight, compared with 60 mg. which the authors give as the normal ovarian weight at that age. The ovaries of Group II rats which had received no testosterone weighed between 300 and 500 mg. per 100 g. of bodyweight. They conclude that testosterone checks the action of gonadotrophin on the ovary.

(b) *Inhibition of androgenic action by oestrogen.* An important function of naturally produced androgen is to maintain the secretory activity of the prostate and other accessory generative glands of the male. This function is speedily inhibited by relatively small doses of oestrogen, the antagonistic influence of which, when supplied in large enough amount, extends over a wide field of androgenic activity. The naturally formed oestrogens, speaking broadly, are more potent biological agents than are the known androgens, and so a relatively small amount of oestrogen will in most organs annul the activity of a much larger quantity of androgen. This is not an invariable rule. Gley & Delor (1937) observed that daily doses of 1 mg. of oestradiol benzoate were required to neutralize the stimulating activity of daily doses of 200γ of testosterone propionate on the capon's comb. Mühlbock (1938, 1939) determined the relative amounts of different oestrogens required to produce this inhibition. The androgens used were testosterone and androsterone. Testosterone was applied directly to the comb on four successive days, measurements of the combs being made on the fifth day. Androsterone was given by intramuscular injection on four successive days, the results also being ascertained on the fifth day. In testing the inhibitory action of oestrogen against testosterone the two classes of hormone were mixed and applied directly to the comb. In the tests with androsterone the oestrogen was injected simultaneously with the androgen. Some of the results are shown in Table 55.

TABLE 55. Relative amounts of oestrogen sufficient to prevent growth of the capon's comb under the influence of testosterone and androsterone (Mühlbock, 1938)

Duration of experiment (days)	Oestrogen	Total dose of oestrogen (γ)	Androgen	Total dose of androgen (γ)	Method of applying the hormones	Comb growth
5	Oestrone	100	Testosterone	400	Smearing comb	+
5	Oestrone	500	Testosterone	400	Smearing comb	−
5	Oestradiol	500	Testosterone	400	Smearing comb	−
5	Oestriol	500	Testosterone	400	Smearing comb	+
5	Equilenin	500	Testosterone	400	Smearing comb	+
5	Oestradiol benzoate	5,000	Androsterone	100	Injection	−

The − sign indicates that comb growth was arrested.

Oestrone given by injection in doses of between 20 and 1,000γ had but slight effect in preventing the action of androsterone on the capon's comb. Mühlbock also found (1939) that diethylstilboestrol interfered with the action of androsterone on the capon's comb, 2 mg. countering 0·1 mg. of androsterone. Emmens & Bradshaw (1939) have also recorded the fact that oestrogens inhibit the action of androgens on the capon's comb (cf. p. 162).

Korenchevsky & Hall (1938b) noticed an inhibition by androgen of an action exercised by oestrogen on the adrenal. Castration is ordinarily followed by enlargement of the adrenal cortex; androgens cause a return of the adrenals in these circumstances to a normal size. The administration of oestrogen, they say, prevents this restorative action of androgens.

(ii) *Progestin versus oestrogen.* A detailed consideration of the antagonisms between progestin and oestrogen will appear later and a few of the facts only will be mentioned here.

(a) *Inhibition of the action of oestrogen by progestin.* The absence of vaginal cornification during pregnancy, pseudopregnancy and lactation, that is to say at periods when active corpora lutea are present, led to the supposition that luteal hormones might inhibit the cornifying action of oestrogens on the vaginal epithelium. Experiments have shown the supposition to be correct. Progesterone, like androgen, will co-operate with oestrogen in producing vaginal mucification and will prevent the cornification which oestrogen acting alone would produce. The quantity of progestin required for this inhibitory action is relatively large. Desclin & Dessiennes (1940) found that as much as 1 mg. of progesterone given subcutaneously failed to prevent vaginal cornification in spayed mice in response to 0·001 mg. of oestradiol benzoate.

Progestin also counteracts the hypertrophic action of oestrogen on the uterus, and to a slight extent protects the male accessory genital organs from the injurious influence of oestrogens. De Fremery (1937) has reported another instance in which the action of oestrogen is prevented by progesterone. He treated a male macacus rhesus monkey with 1,000 i.u. of oestriol applied on alternate days by unction. This caused reddening of the sexual skin, scrotal swelling and hyperplasia of the mammary glands. If, in addition to oestriol, 1 rb.u. of progesterone were given on alternate days the scrotal swelling did not occur, though the reddening of the sexual skin and development of the breast were not prevented.

If the supply of progesterone is proportionately low, oestrone will increase the tone of the uterus and increase its sensitivity to oxytocin, whereas progestin, if present in large enough amount, annuls this action of oestrone and reduces the tonicity of the uterus (Robson, 1933 a, b, 1935 a). This inhibitory action of progestin against oestrone appears to be important, not only in preventing premature parturition, but also in permitting the uterus to accommodate the growing embryo (Reynolds, 1937). These matters will be discussed later in greater detail.

(b) *Inhibition of the action of progestin by oestrogen.* Although as mentioned earlier both oestrogen and progestin are required for the production of progestational changes and of decidual tissue in the uterus, the two types of hormone must be present in suitable proportions to achieve these effects. Oestrogen when supplied in excess will suppress both the progestational phenomena and the development of a placenta. Robson (1935 a) found in experiments on spayed rabbits that the endometrial reaction to a total dose of 0·75 mg. of progesterone given during a period of 4 days was completely inhibited by the injection of 10γ of oestrone.

(iii) *Progestin versus androgen.* Mühlbock (1938) found that progesterone, like the oestrogens, can counteract the stimulating effect of testosterone on the capon's comb. When smeared together on to the comb, 500γ of progesterone prevented the action of 0·4γ of testosterone. When both were injected simultaneously into the pectoral muscles 500γ of progesterone largely inhibited the action on the comb of 100γ of androsterone.

Factors Influencing the Action of Administered Gonadal Hormones

A. VARIATION OF RESPONSES DUE TO DIFFERENCES IN THE STRUCTURE
    OF THE HORMONE OR TO THE METHOD OF ITS ADMINISTRATION

(i) *The effect of time on the potency of stored material.* When experimental tests with gonadal hormones are undertaken, it is well to remember that their potency may change while they are in storage, and their ability to resist change will depend largely on the method of keeping them. D'Amour (1940) tested several different commercial preparations and found that both oestrogen and progesterone retained their potency in oily solution but gradually lost some of it in water. Starkey, Grauer & Saier (1943) state that androsterone and dehydro-*iso*androsterone when kept in alcohol acquired increased androgenic potency.

(ii) *Division of doses.* Allen, Francis, Robertson, Colgate & Johnston (1924) were investigating the action of follicular fluid obtained from sows' ovaries, and of extracts prepared therefrom, on spayed adult rats. The effects were watched by means of vaginal smears, with cornification of the epithelium as a criterion. During these experiments it became apparent that a dose when given in successive fractions was more effective than the same dose given as a single injection. These results were confirmed by Evans & Burr (1926) and have since been found to hold good for oestrone in a pure form and for other hormones. Laqueur & de Jongh (1929) compared the effects of oestrin on spayed rats when (1) given as a single injection or (2) divided into six portions and injected at intervals during 48 hours. The total volume of injected liquid was the same in all cases, that of the single dose being 0·6 c.c. and of the six divided doses 0·1 c.c. each. Vaginal smears were the criteria of effect. The results (Table 56) show an increased reaction obtained by dividing the dose.

TABLE 56.   The increased effect on the spayed rat's vagina of oestrin
when given in divided doses (Laqueur & de Jongh, 1929)

Number of rats	Total dose of oestrin (γ)	Number of doses	Percentage of positive vaginal smears
11	0·9	6	0
14	3·6	6	64
7	5·4	6	86
15	7·2	6	93
6	7·2	1	0
16	21·6	1	0

Laqueur & de Jongh found that an increased effect from divided doses of oestrin was obtained both with oily and aqueous solutions, and they suggest that the enhanced result probably depends largely on a maintenance of the supply of hormone, that is to say a prolongation of its action. As Parkes has said, the amount of hormone contained in an endocrine gland at any particular moment is usually small compared with that which has to be injected to get a response in a test animal; the reason for which appears to be that there is a continuous output of hormone by the gland with but little storage: conditions which may be difficult to imitate by periodical injections.

Marrian & Parkes (1929b), testing the effects of oestrone on spayed mice, also noted that an enhanced response is obtained by giving divided doses whether an aqueous or an oily medium be used. Deanesly & Parkes (1937a) say that the increased response to oestrone obtained by dividing the dose is not nearly so pronounced in the rabbit as in the rat. In their experiments, as in those of Laqueur & de Jongh, the enhanced effect obtained by dividing the dose was more marked with aqueous than with oily solutions.

To some extent the increased reaction following divisional dosage depends upon the hormone used, the results varying with different compounds. Emmens (1939a) using spayed mice as the test objects found that the total dose of oestrone necessary to cause a positive vaginal response in 50 per cent of the mice could be reduced by 30 per cent when four successive injections were given instead of two. Cornification of the vagina could be obtained with 7γ of oestriol when given in two doses, whereas $0\cdot16\gamma$ sufficed if this was given in four separate portions— a more than fortyfold increase of effectiveness. Some of the results obtained with oestriol are shown in Table 57.

Table 57.   The enhanced oestrogenic effect of oestriol when given in successive divided doses (Emmens, 1939a)

Total dose (γ)	Number of injections	Percentage of test mice in which vaginal cornification was obtained
10	2	70
7	2	50
0·5	4	100
0·18	4	70

(iii) *Site of application.*   (a) *Oral administration.*   Speaking generally, this is not well adapted for accurate experimentation with gonadal hormones on animals. Some of these hormones, though not all, have but little action on the body when given by this route. Marrian & Parkes (1929b) have stated that when aqueous solutions of oestrone were being tested on spayed mice administration by the mouth required 100 times the amount of that given subcutaneously to produce vaginal cornification. Girard, Sandulesco & Fridenson (1933), too, found that much larger doses of oestrone are required to cause vaginal cornification in spayed mice if given orally instead of subcutaneously. Oestrone resists the digestive juices, they say, and its relative ineffectiveness when given by the mouth is probably due to the fact that it is not readily absorbed from the intestinal tract. They found, however, that the rate of absorption from the stomach and intestines could be influenced by several factors, being enhanced by dividing the doses, by increasing the dilution of the solutions in which it is given and by the addition of alcohol. They found that oestradiol given *per os* in three equal fractions to spayed rats is slightly more active in causing cornification of the vagina than an equal dose of oestrone given in an oily solution as a single subcutaneous injection. Oestradiol it may be noted is 3 or 4 times as soluble as oestrone in water and is very soluble in alcohol. Some of the synthetic oestrogenic compounds are readily absorbed from the intestine. Parkes, Dodds & Noble (1938) found that relatively small amounts of ethinyloestradiol and diethylstilboestrol when given by the mouth to rabbits prevented nidation of the ova after fertilization. A daily dose of 0·5 mg. ethinyloestradiol begun 9 to 10 days after mating led to resorption of the ova. A total dose of 20γ given by the mouth was sufficient to prevent pregnancy in rats if the administration were begun on the day following coitus.

Some compounds when given by the mouth may produce biological reactions even more readily than when administered by subcutaneous injection. Emmens (1939b) reported this to be the case with the higher esters of diethylstilboestrol and ethinyloestradiol. He states that esterification prolongs the action of these compounds when given by injection without increasing the degree of their activity, whereas esterification does not cause any prolongation of effect when the compounds are given by the mouth. Inhoffen & Hohlweg (1938) compared the effective doses of different oestrogens when tested on spayed rats by subcutaneous and oral administration, vaginal cornification being the test. Their results show that while oestrone and oestradiol are relatively inert when given by the mouth, oestriol and ethinyloestradiol are highly effective (Table 58).

TABLE 58. Comparative effectiveness of oestrogens when given to spayed rats by mouth or by subcutaneous injection (Inhoffen & Hohlweg, 1938)

	Effective dose when given subcutaneously (γ)	Effective dose when given by mouth (γ)
Oestrone	0·83	60
Oestradiol	0·1	50
Oestriol	10	10
Ethinyloestradiol	0·1	3

Inhoffen & Hohlweg (1938) say that whereas ethinyltestosterone (pregneninolone) when given subcutaneously has only one-third of the activity of progesterone, given orally it has a greater efficiency than progesterone, 4 mg. causing a positive progestational response in the immature rabbit's uterus. Miescher & Tschopp (1938) have shown that when given by the mouth to rats methyltestosterone is more effective than testosterone propionate. Biskind (1940) believes that this fact may be explained by absorption of the compounds from the alimentary canal through different routes, one being conveyed by the blood vessels and the other by the lymphatics. The latter route would avoid immediate subjection of the compound to the liver and so postpone the inactivating action of that organ. To test the matter he inserted pellets of methyltestosterone into the subcutaneous tissue or spleen of castrated rats. In the former case the usual restorative androgenic effects followed, but not in the latter, presumably because on entering the splenic blood vessels the androgen was conveyed directly to the liver and inactivated there.

(b) *Intraperitoneal injection.* When investigating the biological properties of oestrogenic extracts of sows' ovaries, Evans & Burr (1926) saw a pronounced difference between the effects of subcutaneous and intraperitoneal injections. An amount of the substance enough to cause an oestrous vaginal response in rats when given subcutaneously was without obvious oestrogenic effect when injected into the peritoneum. This route of administration is little used at the present day in experiments with gonadal hormones. It will be remembered that gonadotrophins are relatively inactive when given by intraperitoneal injection. In both cases it seems probable that the inactivation of the hormones may be due to their early subjection to the action of the liver when introduced by this route.

(c) *Subcutaneous injection.* This is the method of choice when seeking the general effects of a given hormone, and in many instances too when the action on an individual structure is under inquiry. The method has the advantage of accuracy in dosage, and the disadvantage, when the effects on a single organ are under observation, that a proportion of the dose is wasted because of its wide distribution through the body.

(d) *Direct local application.* This has proved to be of value in many instances when the organ whose reaction is being tested is accessible. The capon's comb is available for this purpose, as shown by Fussgänger (1934). Greenwood & Blyth (1935a) have found that androsterone injected directly into the capon's comb produces a greater response than that elicited by the same dose injected into the pectoral muscles, and Dessau & Freud (1936) report a similar enhanced effect with testosterone and androsterone when these are smeared on the surface of the comb. Deanesly & Parkes (1937a), when testing a preparation for androgenic potency, noted that a total dose of $2 \cdot 5\gamma$ by direct application to the surface of the capon's comb gave as large a response as 500γ given by injection into the pectoral muscles. Emmens & Bradshaw (1939) confirm the increased responsiveness of the capon's comb to gonadal hormones when directly applied. They say that the action on the capon's comb of androgen given by injection is inhibited much more readily by oestrogens if these are applied directly to the comb than if they are injected elsewhere. The ratio of the effective dose when oestrogen is applied directly to the comb to that when it is given by subcutaneous injection at a distant site is for the different compounds used as follows:

Oestrone, 1 : 470; Oestradiol, 1 : 180; Diethylstilboestrol, 1 : 2.

Dessau (1937) states that different ratios are shown between the potencies of androgens when assayed by injection into the pectoral region and by direct application to the comb. Androstenedione, androstanedione and androstanediol have, he says, about the same potency as androsterone when applied directly to the comb. These findings are in contrast with the relative potencies of the compounds as estimated by pectoral injections.

The vagina also is available for the direct application of hormones. Berger (1935) has found that by vaginal application $\frac{1}{2}$ m.u. of oestrone is enough to cause cornification in rats in which more than 6 m.u. are required when given subcutaneously. Lyons & Templeton (1936) also state that the vagina is very sensitive to direct application of oestrogens, cornification being obtained in rats with 0·1 to 0·8 c.c. of the urine of normal women at different stages of the oestrous cycle.

The increased reaction obtained by the direct application of oestrogens to the vagina is not the same for all compounds. Freud (1939) has compared the vaginal reactions to similar doses of various oestrogens when given subcutaneously or by direct application and he finds considerable differences in the results. Oestradiol benzoate, for example, is equally effective in causing vaginal cornification whether given subcutaneously or applied directly to the vagina, whereas stilboestrol is 12 times as effective when given intravaginally as when given by subcutaneous injection (Table 59).

Emmens (1941) says that some oestrogens, including oestradiol and stilboestrol, are much more active in causing vaginal cornification when directly applied to the reacting tissue than when given subcutaneously at a distant part of the body, an extreme example being provided by oestriol, of which 2,000 times the effective vaginal dose is required to cause cornification if the hormone is given by subcutaneous injection. Other oestrogenic compounds given intravaginally bring about cornification only if given in doses equal to those required to produce the same result when given subcutaneously. Emmens suggests that the latter compounds might be termed prooestrogens, and that they are converted into oestrogens only after their absorption and metabolism.

In further experiments Emmens (1942) spayed mice and divided their vaginae into two compartments, making the anterior one to open suprapubically while the posterior compartment retained its normal perineal aperture. By this means the prooestrogens and oestrogens could be readily distinguished, for oestrogens introduced in small amount in the upper sac caused cornification limited to that cavity, whereas prooestrogens applied in the same way brought about cornification in both sacs if they caused cornification at all. The experiment shows that prooestrogens do not become oestrogenic before they have entered the general circulation.

TABLE 59.   Threshold doses of different oestrogens when given subcutaneously and intravaginally (Freud, 1939)

Compound used	Threshold dose		Ratio
	Subcutaneous injection (γ)	Direct application to vagina (γ)	
Stilboestrol	0·37	Less than 0·03	12 : 1
Oestradiol	0·125	Less than 0·03	4 : 1
Oestradiol benzoate	0·25	Less than 0·3	1 : 1
Oestrone	1·5	More than 0·8	2 : 1

'Espinasse (1939a) states that doses of oestrone dissolved in arachis oil, which are too small to cause detectable reactions in the vagina when given subcutaneously in distant regions of the body, will be effective if injected into the perineum. He also learned that with doses large enough to produce a vaginal reaction when given in either of these ways, the greater result, as determined by direct measurement of microscopical sections, is obtained by the perineal injections.

When testing the effects of oestrone on the plumage pigmentation in brown leghorn capons, Greenwood & Blyth (1935 b) found that by local intradermal injection positive results could be obtained by doses too small to cause a response when injected into the pectoral muscles. The distribution of effect when the hormone was given intradermally showed that it spread by diffusion. The breast feathers of the brown leghorn are normally black in the capon and light-coloured in the hen. Oestrone given while the feathers are growing prevents the black pigmentation, and the effect of a single dose is to cause a light bar across the growing feather corresponding in width to the duration of oestrogenic action. Small doses of oestrone made into the skin of the breast after plucking affect only those feathers in the vicinity of the injection, and the sides of the feathers nearest to the site of injection react sooner and more widely than the sides which are farther away ('Espinasse, 1939 b).

The efficacy of a direct application in producing a local reaction has been demonstrated further by Lacassagne & Raynaud (1937), who obtained an epithelial response to testosterone in the castrated rat's seminal vesicle with a dose of only 2·5 γ when this was injected into the vesicle. Another remarkable example of enhancement of efficiency by the direct application of a hormone to the reacting tissue is reported by McGinty, Anderson & McCullough (1939), who injected progesterone into the uterine cavity of immature rabbits which had received previous treatment with oestrone; by means of single doses ranging from 0·5 to 5 γ they obtained progestational responses equal to those induced by 500 γ when given intramuscularly in five divided doses. These results have been confirmed by Haskins (1939), who, moreover, using the same technique got positive progestational reactions with as little as 0·2 c.c. of serum from a pregnant guinea-pig when this amount was introduced into the uterus of an immature rabbit which had been subjected previously to oestrone.

When smeared on the skin the effect produced by oestrogen on the underlying mamma is greater than that caused by subcutaneous injection at some distant site. 'Espinasse's observation on the diffusion of oestrone in the tissues immediately surrounding the original site of injection may be recalled when considering this phenomenon.

(e) *Cutaneous application.* When applied to the skin in suitable solvents (ether, chloroform, benzene, alcohol, or oil) the gonadal hormones readily enter the bloodstream and affect the susceptible organs. Although not suitable for quantitative work, the method has been employed in qualitative tests and in human therapy. The writer has made extensive use of cutaneous applications when testing gonadal hormones for carcinogenic potency in mice. In these experiments the hormone was dissolved in alcohol, chloroform or benzene, one drop of the solution being applied to the skin of the interscapular region twice a week. The hormones so applied are readily absorbed through the skin and produce all their recognized effects.

In man Zondek (1938) has found that oestrone dissolved in 96 per cent alcohol and rubbed into the skin is as effective as the same dose given subcutaneously.

Most often in human therapy oily and fatty solvents have been used for the cutaneous application of hormones. Moore, Lamar & Back (1938) tested a 'face cream', consisting of 0·625 mg. of oestradiol in each ounce and used for cosmetic purposes by women. Applied to the skin of spayed rats this ointment led to cornification of the vagina and distension of the uterus, and in immature male rats caused a considerable reduction in the weights of the testes.

De Fremery (1936) reported that the effect of oestradiol benzoate on the goat's mamma was greater if the hormone was applied by unction directly to the udder that if injected in another part of the body. A comparable observation was made on the rabbit by Lyons & Sako (1940). They used six males, and, starting the experiment when the animals were 2 months old, applied oestrone in oil into the skin immediately around the nipples of the left side, while sesame oil alone was applied in a similar way to the nipples and areolae of the right side. On each occasion one drop was applied to each mamma by a medicine dropper, these treatments being repeated on 5 days a week. In three of the rabbits each dose contained 3 units of oestrone and in the other three 1 unit was used. At the end of 5 weeks the mammae, fifty-three in all, were examined and it was found that the nipples and glands on the treated side were abnormally enlarged; those on the right side, having received no oestrin, were normal.

MacBryde (1939) compared the effects of oestradiol on the human mamma when given by injection and by local application respectively. For the latter he dissolved the oestradiol in an ointment composed of lanolin and soft paraffin. Each day 5 g. of this ointment, containing 25,000 i.u. of oestrogen, was applied to one of the two mammae over a circular area of 10 cm. diameter with the nipple as the centre. He found that unction applied in this way was more effectual than injection in causing development of the mamma, and the breast directly treated showed a greater ultimate response than its fellow of the opposite side.

Speert (1940) performed the same sort of experiment on immature rhesus monkeys. Oestrone dissolved in alcohol (50γ per 1 c.c.) was painted daily around the left nipple. The right mamma was treated in the same way but with alcohol alone. After 50 or 75 days of this treatment the left mammae were enlarged, whereas the right mammae had remained unaffected.

(f) *Intravenous injections.* These have been little used in experimental work on the gonadal hormones. With most of the small animals available for tests the method has no advantage to outweigh the difficulty of using it. The method has been of value in studies of the distribution, inactivation and excretion of the gonadal hormones, but little attention seems to have been given to the relative effectiveness of hormone when given by this route.

(iv) *The nature and volume of the solvent: Injection of undissolved hormones.*
(a) *The nature of the solvent.*   Miescher, Wettstein & Tschopp (1936) when testing the activities of testosterone perceived that the results depended largely on the nature of the solvent used. Given in paraffin testosterone appeared to be inactive. Deanesly & Parkes (1937a) also noticed the important part played by the solvent in this sort of test. Absorption of oestrone from castor oil took place very slowly, and androsterone dissolved in 1 : 2 prophylene glycol, which is miscible with both oil and water, was more than 3 times as effective as when given in olive oil.

The capacity of oil for spreading over the surface of the skin may be a hindrance to its use as a solvent. Klempner, Frank & Hollander (1940) when using young chicks for the assay of androsterone noticed that an oily solution applied to the comb was apt to overrun the adjacent skin so that some of its capacity to affect the comb was dissipated. Using 95 per cent alcohol as the solvent instead of oil the response to a given amount of androsterone was increased by more than 150 per cent.

(b) *The volume of the solvent.* Deanesly & Parkes (1936*b*) drew attention to the fact that the results of a given dose of hormone are largely affected by the volume of the solvent in which it has been administered. When testing the reactions of the prostate and seminal vesicles of the castrated rat to androsterone dissolved in oil they found that by increasing the volume of the oil 5 times the effectiveness of 10 mg. of androsterone was doubled. The reactions to testosterone also were enhanced by increasing the volume of oil in which the doses were given. Freud, Dingemanse & Polak (1937) have made similar observations. In a later paper, Deanesly & Parkes (1937*a*) say that if the volume of oil is increased up to a point where the solution is stable the effectiveness is reduced, the higher activity of the concentrated solution being the result of a delayed absorption of the hormone, which may have crystallized out shortly after injection. Once the volume of oil is enough to produce a stable solution a further increase of the oil may enhance the effectiveness of the contained hormone by slowing the rate of absorption.

The nature of its solvent may influence the effectiveness of a hormone by causing a change in its chemical structure. An instance has been given by Miescher and his colleagues, and will be discussed in connection with the action of androgens. Interesting examples of delayed absorption due to the nature of the solvent have been recorded by Parkes (1942), who found that thyroxine dissolved in cholesterol is not absorbed in weighable amount in the tissues in the course of several months, and 100 mg. tablets of cholesterol containing 15 per cent of adrenalin yielded only about 5 mg. of the latter in a month.

(c) *The injection of undissolved hormones.* Deanesly & Parkes (1937 *a, b*) found that the action of testosterone and oestrone was prolonged if, instead of giving them in solution, they were introduced in the solid form beneath the skin. For example, 2 mg. of oestrone given subcutaneously as a tablet exerted its action on a capon's feathers for a period of between 2 and 3 months instead of for 1 or 2 days only as was the result of giving the same dose dissolved in oil. By weighing the tablet before and after, the amount absorbed in the intervening period could be determined.

(v) *Augmentation of effect by delaying the rate of absorption.* Apart from changing the nature or volume of the solvent there are two ways of delaying the absorption of a compound injected into the tissues. The one is to cause a local inflammation at the site of injection and the other is to give the compound in a less soluble form.

(a) *Delayed absorption due to inflammatory fixation.* When the factors which influence the effectiveness of artificially administered gonadotrophin were being discussed the increase of effect due to the presence of impurities in the injected material was mentioned and was attributed to a slowing in the rate of absorption brought about by a local inflammatory reaction (p. 74). The principle involved seems applicable also to the gonadal hormones, though the evidence available is not complete on this point, because in past experiments bearing on the subject it is not always possible to separate the effect of inflammatory fixation from that of a chemical change in the gonadal hormone itself.

Laqueur and his colleagues (Laqueur, David, Dingemanse & Freud, 1935; Freud, 1935; Freud, Dingemanse & Polak, 1935) observed that the action of testosterone and androstanediol could be considerably enhanced by the addition of certain tissue extracts which in themselves were devoid of androgenic power. Effective extracts of this kind could be obtained from various sources, including bulls' testes, the blood of oxen and cows, liver, adrenal, ovary and wheat germ. If a castrated male rat was given daily for 4 days 0·015 mg. of testosterone together with one of these augmentary extracts or 'X-substances', as they were called, the responses of the seminal vesicles were increased twofold or more compared with those produced by the same dose of testosterone given alone. The action of androsterone was not enhanced by the addition of these extracts.

In these experiments some of the effects may have been attributable to inflammatory fixation and some to esterification of the hormones.

(b) *Delayed absorption due to esterification of the hormones.* Miescher, Wettstein & Tschopp (1936) confirmed the observations of Laqueur and his colleagues which have been quoted above, and discovered that many different fatty acids were activators of testosterone; the most effective in this respect being palmitic, arachidic and propionic. The tests were made as follows: rats weighing between 60 and 80 g. were castrated and 4 weeks later were given daily doses for 10 days of 50γ of testosterone in sesame oil to which 50 mg. of the various fatty acids to be tested were added. The rats were killed on the 11th day and their accessory genital glands were weighed. A few of the results are shown for comparison in Table 60.

TABLE 60.   The action of testosterone on the accessory genital glands of castrated rats as affected by esterification (Miescher, Wettstein & Tschopp, 1936)

	Weights of organs in mg.			
Material of daily injection	Seminal vesicles	Prostate	Cowper's glands	Preputial glands
Sesame oil only	14	41	49	57
Testosterone 50γ	42	66	75	55
Testosterone 50γ plus Benzoic acid 50 mg.	50	102	128	76
Testosterone 50γ plus Palmitic acid 50 mg.	235	210	205	91
Testosterone 50γ plus Arachidic acid 50 mg.	135	182	174	96
Testosterone 50γ plus Propionic acid 50 mg.	117	158	150	70

Callow (1936a) reported that the effectiveness of androsterone could be increased by giving it in the form of benzoate, as is true also of oestrone; and Freud, Dingemanse & Polak (1937) showed that the potency of testosterone is enhanced when combined with acetic or propionic acid.

Deanesly & Parkes (1936b; Parkes, 1936) repeated some of these experiments with confirmatory results. They showed that the addition of palmitic acid to oily solutions of testosterone caused a great increase in the response of the accessory generative organs; but no such enhancement of effect was seen if the testosterone was injected into one side of the rat and the palmitic acid by a separate injection into the other. In some instances the potency of a hormone is decreased by esterification. For example, benzoylation, they say, will certainly decrease the solubility in body fluids of androstanediol and *trans*androstenediol, whereas succinylation will certainly increase the solubility of testosterone. They suggest that the relative inactivity of these compounds may be due to the fact that the benzoates became too insoluble to be absorbed and the succinates too soluble to remain in the body long enough to produce a visible reaction.

David, Dingemanse, Freud & Laqueur (1935), when reporting the increased androgenic activity of testosterone after the addition of 'X-substance', pointed out that this increase was not shown in the response of the capon's comb.

(vi) *Adjuvant or antagonistic actions by two or more gonadal hormones when present at the same time.* Between the three main types of gonadal hormones there are many interactions, and when mixed, as they usually are in nature, the biological effects caused by one hormone will be determined largely by the proportions in which others are present. The matter has been discussed earlier (pp. 115, 118).

(vii) *Duration of the experiment.* When the quantitative effects of a given gonadal hormone are being estimated it is essential to consider their duration. As Miescher, Fischer & Tschopp (1937) have found, if enol esters of testosterone be given in daily doses to castrated rats for 10 consecutive days the results as determined on the 11th day will be less than those obtained by equal amounts of testosterone propionate given in the same way, but they will last much longer. Judged by this standard the enol esters of testosterone are more potent than testosterone propionate. For example, testosterone-3-acetate-17-butyrate given in a single dose of 2 mg. caused the seminal vesicles and prostate at the end of 40 days to weigh 140 and 190 mg. respectively, compared with controls of 14 and 45 mg. The effect of an equal dose of testosterone lasts only about 7 days and that of testosterone propionate about 20 days.

A useful summary of variables affecting the estimation of gonadal hormone activities by assay on living animals has been made by Emmens (1939a).

(viii) *The magnitude of the dosage.* The degree of response of an organ or tissue to a gonadal hormone is not always directly proportional to the amount of hormone supplied. To a certain point the reaction will increase in accord with the dose, but above an optimum dosage a further increase may be followed by a diminished response. For instance, moderate doses of oestrogen will cause an increased development of the mammary ducts, whereas excessive amounts will check extension of the ducts, causing them to be stunted.

(ix) *Different effects produced by continuous or intermittent treatment with the same total dosage.* Lipschütz, Rodriguez & Vargas (1941), when investigating the effects of oestrogen on the guinea-pig's uterus, learned that these effects were much greater if the doses were given 3 times a week regularly than if double or treble the dosage was used in one week and an interval of a week or a fortnight were then allowed to pass without any treatment. This observation applied to the induction of general enlargement of the uterus, adenomatosis, and the formation of polyps and fibromyomata.

B. Variations in the Response to a given Dose of Gonadal Hormone due to Differences in the Recipient Individual or Tissue

(i) *Innate variations in a single species.* Laqueur & de Jongh (1929), during a study of spayed rats and mice in connection with the accurate assay of gonadal hormones, noticed that individual animals did not all respond exactly alike to the same dose of hormone. For example, twenty-four mice were divided into two groups of twelve and each was injected with oestrone 4 times at intervals of a week. The mice of the first group received less than 1 m.u. for each dose, and those of the second group received about 2 m.u. for a dose. Not a single mouse in either group reacted in quite a uniform manner to all the four injections, according to the meticulous grading of responses employed. Even regarding straightforward positive and negative results, only six of the mice gave identical reactions to all the four injections. Three of the mice never gave a positive vaginal smear. These differences of response were not always constant attributes of individual mice in these two groups; in most instances they altered with time. A wide experience however has shown in rare instances, Laqueur & de Jongh say, that discrepancies in reaction have persisted in individual mice. The result appeared to be affected in some instances by the animal's health. In these comparative tests differences of body-weight were found to be negligible factors.

Wider discrepancies are seen when the actions of a given hormone on different inbred strains of mice are compared. It has been shown (Bonser, 1936; and Bonser & Robson, 1940) that the three well-known strains of inbred mice R_{III}, Strong A and CBA do not respond exactly alike to oestrogens. Once a week 3 mg. of triphenylethylene or 50γ of oestradiol dipropionate were given to males of the three strains. In R_{III} mice these doses caused the development of mammary alveoli, mammary cancer, scrotal hernias, cornification of the coagulation gland and degeneration of the adrenal cortex. In the Strong A mice mammary alveoli developed and a few cases of mammary cancer occurred, scrotal hernias formed, but no cornification appeared in the coagulation gland, degeneration of the adrenal cortex was slight or absent, and a striking hyperplasia of the interstitial glandular cells of the testes was caused—a change which was not found in the R_{III} or CBA mice. In the CBA mice no mammary alveoli or cancer developed, no scrotal hernias formed, and hyperplasia of the interstitial gland of the testis and adrenal degeneration were slight or absent; cornification of the coagulation gland occurred as in the R_{III} mice. Some of these results are given in Table 61.

The effect of heredity on oestrogenic neoplasia is exemplified in the CBA females, which show very little liability to mammary cancer under the influence of oestrogen, whereas uterine tumours are readily caused in them, which is not the case with most of the strains subject to cancer of the breast under the same conditions. Miller & Pybus (1942) gave 300 i.u. of oestrone weekly by subcutaneous

TABLE 61.   Varying effects of oestrogen in the males of different inbred
strains of mice (Bonser & Robson, 1940)

Strain	Mammary cancer	Scrotal hernia	Hyperplasia of interstitial glandular tissue of testis	Cornification in accessory genital glands	Lipoid degeneration of adrenal	Enlarge-ment of pituitary
R_{III}	+ +	+ +	−	+ +	+ +	+
Strong A	+	+	+ +	−	+ −	−
CBA	−	−	+	+ +	+ −	+

injection to twenty-six spayed and twenty-eight non-spayed CBA mice. The incidence of uterine tumours—mostly fibromata or fibrosarcomata—was about the same in each group and amounted to 55·5 per cent of the total number of mice employed; mice of other inbred strains have seldom responded to oestrogen by the formation of uterine tumours. Moreover, after prolonged dosage with oestrogen hepatomata occurred with peculiar frequency in CBA males, but not in mice of other strains. These differing effects produced by the same oestrogenic treatment cannot be attributed merely to general factors such as inequality in the rate of metabolism or excretion of oestrogen or to a general difference of susceptibility to its action in the different strains, for the responses of the individual organs varied.

(ii) *Effect of a priming dose.*   When using spayed rats or mice for the assay of oestrogens it is usual after removing the ovaries to give what is called a priming dose of oestrogen. Emmens (1939a) states that the result of an assay will be considerably influenced by withholding this priming dose. He spayed a batch of mice, rested them for a month, and then gave 12γ of oestriol to each in two injections. In three consecutive tests carried out on the same mice in this way positive vaginal responses were obtained in 30, 65 and 75 per cent respectively.

(iii) *Seasonal and cyclical influences.*   Just as Laqueur & de Jongh found unexplained differences in the reaction to oestrogens on the part of individual rats and mice, so Gallagher & Koch (1930) have observed an inconstancy among individual capons in the response of their combs to androgen. The age and weight of the capons and the initial size and shape of the combs did not account for these variations. Gallagher & Koch thought that seasonal influences ought to be considered when these tests are being performed. David (1938) states that the capon's comb is smaller in winter than in the summer and that the application of an androgen will evoke a greater response in the winter than in the summer (p. 97).

Del Castillo & Pinto (1939) have noticed seasonal changes in the reproductiveness of the rats at their institute in Buenos Aires, there being a diminution of births in the autumn and winter. Seasonal variations occur also in the dimensions of the accessory genital organs. The seminal vesicles, they say, have a maximal weight in November, December and January, that is to say in the summer time in the southern hemisphere. After February, which corresponds to autumn in northern latitudes, the weight of the seminal vesicles begins to lessen, reaching a minimum in May. A gradual increase begins in June. Basing their conclusions on the examination of 294 rats, they suggest that seasonal factors cannot be ignored when the effects of artificially administered hormones on rats are being examined.

Del Castillo & Calatroni (1930) discovered what seem to be regular cyclical changes in the responsiveness of the spayed rat's vagina to oestrone. They say that if daily doses of o·1 to o·2 i.u. of oestrone are given to spayed female rats regular cycles of vaginal cornification can be induced. Wade & Doisy (1935 a) gave oestriol (o·65 to 3·0γ) daily to spayed rats for an extended period. During this time they noticed irregular cycles of vaginal cornification. Zuckerman (1937 b) noticed evidence of similar cyclical variations of responsiveness in the monkey's uterus. In a spayed rhesus receiving 100 i.u. of oestrone daily for 1 year phases of uterine bleeding recurred at intervals varying from 5 to 7 weeks.

(iv) *The general condition of the recipient.* The writer has no note of any experiment which bears specifically on this matter, but a general knowledge of biological reactions makes it likely that the response to gonadal hormones, as to other physiological agents, may be affected by the bodily condition of the recipient. As an extraneous and rather special example it may be mentioned that in man the intravenous injection of 10 mg. of thyroxine given to a patient whose basal metabolism is −40 will cause a reaction 7 times as great as when the same dose is given to a patient with a normal basal metabolic rate.

(v) *Hyperaemia of the reacting tissue.* On hypothetical grounds it seems probable that a local hyperaemia would increase the effectiveness of a hormone carried in the circulation, for more hormone would be brought to the hyperaemic reacting tissue in a unit of time than would arrive there in the absence of hyperaemia. However, this suggestion must remain speculative until the subject has been elucidated by experiment.

(vi) *Inflammation.* The only direct investigation into this matter known to the writer is that of Brunelli (1935), who showed that inflammatory agents led to an elective fixation in the inflamed tissue of oestrone given intravenously to rabbits. There is a little indirect experimental support for the suggestion that inflammation may act as an adjuvant to the local action of hormones. We might by this means explain the observation of Wade & Doisy (1935 b) that daily swabbing alone may induce cornification in the spayed rat's vagina. It is known that removal of the ovaries does not entirely abolish the production of oestrogen, and it may be that the small amount circulating in the spayed rat may become sufficiently concentrated under the influence of inflammation to cause a distinct biological response. The finding by Mixner & Turner (1941 a) that the application of turpentine to the nipples and surrounding skin caused a delay of mammary involution after removal of the young might also perhaps be explained in this way.

(vii) *The effect of light and temperature.* Womack, Koch, Domm & Juhn (1931) gave daily injections of an androgenic testicular extract for 10 days to brown leghorn capons. The birds were arranged in three groups. Group I were kept in large pens with free exposure to sunlight and opportunity for exercise; Group II were in small cages with free exposure to sunlight but without facilities for exercise; Group III were kept in small cages in the dark. In these conditions the birds of Group III showed a much greater comb growth than those of Groups I and II. It was concluded that light checked comb growth in response to androgen. Emmens (1939a) was unable to explain variations in the response of spayed mice to oestrone by any seasonal effect. Mice kept in darkness and cold reacted like normal controls. Mice, however, which were exposed to continuous illumination showed a diminished responsiveness. His experiments were done in March. Mice were arranged into four groups of twenty each and for 3 days before and during the tests each group was placed under different conditions of temperature and light. Each mouse received 1 i.u. of oestrone given by two injections. The results are shown in Table 62.

TABLE 62. The effects of light and heat on the vaginal response of spayed mice to oestrone (Emmens, 1939a)

Mean temperature	Lighting conditions	Percentage of positive vaginal responses
25° C.	Normal daylight	55
25° C.	Continual darkness	65
20° C.	Continual illumination by daylight and electric light	30
13° C.	Continual darkness	60

Continual illumination is not a natural environment for mice, and it is conceivable that their health was disturbed by the exposure. However this may be, the experiment shows that continual exposure to light diminished the vaginal response in mice to artificially administered oestrone.

In fowls it seems that moderate heat lessens the response of the comb to androgen. Hain (1938) injected 0·5 mg. of androsterone daily into bantam capons. These were separated into two groups, one being kept at a temperature of 40°–48° F. and the other at 50°–64° F. The ratios of increased height-plus-length of the comb in the two groups at the end of the 5 days were 5·75 : 3·3 mm.

(viii) *Gonadal activity*. This may interfere with the action of gonadal hormones, and for biological assay of these compounds spayed or castrated animals are required. Parkes & Bellerby (1926) showed that the quantity of oestrin needed to produce oestral changes in the vagina of the normal lactating mouse is from 2 to 4 times that which will cause these effects after spaying. Reference to the mutually antagonistic or adjuvant action of different types of gonadal hormone has been made already (p. 122).

(ix) *Sex*. Although it has been stated earlier in this book that the actions of gonadal hormones are organ-specific rather than sex-specific, there are indications that the general responses to them are not always quite of the same degree in the two sexes. Steinach & Holzknecht (1916) spayed young female guinea-pigs and replaced their ovaries by a testicle taken from a brother; in the same way they replaced the testicles of young males by the ovary of a sister. When the animals were fully grown it was found that the grafted testicle had caused the general dimensions of the body in the female to exceed that of a normal male, while the bones were unusually large (Table 63).

TABLE 63. The effect of a grafted testicle on the skeleton and size of the spayed female guinea-pig (Steinach & Holzknecht, 1916)

	Body-weight (g.)	Width between ears (mm.)	Width between zygomata (mm.)	Length of head (mm.)
Normal female	845	22	40	74
Normal male	1,002	31	43	81
Spayed female with grafted testis	1,200	33	48	87

The converse also was observed, that is to say, when his testicles were removed and replaced by an ovary, the male attained smaller dimensions than the normal female, largely because of the restricted growth of the bones (Table 64).

TABLE 64.   The effect of a grafted ovary on the skeleton and size of the castrated male guinea-pig (Steinach & Holzknecht, 1916)

	Body-weight (g.)	Width between ears (mm.)	Width between zygo-mata (mm.)	Head (mm.)	Tibia (mm.)	Femur (mm.)	Humerus (mm.)	Spine (mm.)
Normal male	980	30	43	80	44	36	31	160
Normal female	808	21	40	72	40	33	28	148
Castrated male with grafted ovary	516	19	36	67	38	31	26	136

Lipschütz & Tütso (1925) grafted ovaries into gonadectomized male and female guinea-pigs, and noted that the nipples and mammae of the males enlarged more rapidly than the same structures in the females. This might perhaps be explained by the fact that an ovary growing in a castrated male is exposed continuously to the action of FRH and so may produce more oestrogen than an ovary implanted in a spayed female. Zondek (1936a) states that enlargement of the pituitary is caused more readily by oestrogens in males than in females; and Burrows (1939b) found that when newly born rats were given repeated doses of testosterone, a penis-like organ developed in the females and sometimes attained a larger size than in littermate males treated in the same way.

Several workers have commented on the fact that it may be easier to induce mammary cancer by oestrogens in the male than in the non-breeding female mouse (Suntzeff, Burns, Moskop & Loeb, 1936).

The subject will receive further attention under the heading of acquired resistance (p. 138).

(x) *Age.*   Although there does not appear to be any period in which reactive organs, once their differentiation is complete, will not respond to gonadal hormones, the readiness and degree of their response is not necessarily quite the same throughout life. For example, Hooker (1942) castrated rats at birth and calculated the minimum dose of testosterone needed to stimulate their seminal vesicles at different ages. In normal rats of the colony employed spermatogenesis began when the animals were 40 days old, at which time the vesicles still resembled those of castrated rats. From this time onward the seminal vesicles underwent a rapid development, becoming mature at 60 days. During this period, between the 40th and 60th day of life, it appeared that the seminal vesicles were more responsive to testosterone than they were at an earlier or later age. In castrated rats the least dose of testosterone required to stimulate the seminal vesicles corresponded with the stage of life when in non-castrated rats these organs normally undergo their greatest development, as indicated by the figures quoted below:

Age of castrated rat at time of injection (days)	Minimum effective dose of testosterone required to stimulate the seminal vesicles (γ)
10–20	30
30	25
40–60	5
80	25

Selye & Albert (1942) observed that age influenced some of the rat's responses to oestrogen. Rats were castrated when between 40 and 60 g. in weight. To one batch of these 1 mg. of oestradiol in 0·1 c.c. of peanut oil was administered daily for 10 days; the control batch were given injections of peanut oil only. On the 11th day all the rats were killed and their pituitaries and adrenals were weighed; it was found that the adrenals and pituitaries of the animals which had received oestradiol weighed less than those of the controls, as shown below:

Treatment	Average weight of adrenals (mg.)	Average weight of pituitaries (mg.)
Given oestrogen	16	4
Not given oestrogen	22	5

This effect of oestrogen in reducing the weights of the adrenals and pituitary in the young rat is the reverse of what is known to occur in the adult rat, whose pituitary and adrenals become enlarged under the influence of oestrogen.

In experiments of this kind one has, of course, to bear in mind that a large dose may inhibit growth where a smaller dose would favour it.

(xi)· *Acquired resistance.*   There is some evidence that an individual in the course of time may acquire some degree of resistance to the action of gonadal hormones. Apparently females respond more readily than males to androgens, whereas males are more responsive to oestrogens than females. What seem to be examples of this sex difference in reaction to gonadal hormones are shown in Tables 63, 64, 76 and 80. However this particular phenomenon may be explained, it seems certain that occasionally an individual may acquire in its own lifetime some resistance to excessive supplies of a gonadal hormone when these are administered artificially. The writer (Burrows, 1937 a) has seen resistance to the action of oestrone in a male mouse after prolonged administration of that hormone; the resistance was associated with hyperplasia of the androgen-producing interstitial cells of the testis, and might, perhaps, be attributed to an enhanced output of androgen. If this explanation be correct we seem to have an addition to the known defensive mechanisms of the body, namely the increased production of a therapeutic hormone by tissue newly grown for that special purpose. This kind of testicular response to oestrogens has been described in detail by Burrows (1936 b) and Bonser & Robson (1940).

What appears to be a definite example of adaptation to excessive supplies of a gonadal hormone is provided by experiments of del Castillo & Calatroni (1930), who found that, if between 0·2 and 0·25 r.u. of oestrin are given daily to normal female rats, continued vaginal cornification ensues; but after some months, although the doses are maintained at the same level, intervals of metoestrus occur and gradually normal cycles are resumed. Wade & Doisy (1935 a) made somewhat similar observations. They gave daily doses of oestriol to male and female rats for periods of 113 to 316 days and noticed that with prolonged treatment the effects at first induced by oestriol became less pronounced and in some instances disappeared, showing that some degree of adaptation had come about.

Selye (1940 d) has called attention to what appears to be another example of adaptive resistance. He says that when oestrogen is given to young rats it causes at first a decline in their weights, but growth is resumed later although the treatment is continued. Selye explains this phenomenon as a resistance to oestrogenic action rather than a capacity to inactivate oestrogen.

Emmens (1939d) observed that if oestrone or oestradiol benzoate are given persistently to a brown leghorn cock the comb at first becomes atrophied; after a time, in spite of a continuance of the treatment, the comb may become turgid again.

Bishop & McKeown (1941) investigated the matter in spayed mice. These were divided into two groups each containing thirty-two animals. To Group I 300γ of oestrone were given daily for 3 weeks; meanwhile the animals in Group II received no treatment. At the end of the 3 weeks mice of both groups were given daily doses of 3γ for 26 days. During this time only eight of the thirty-two mice in Group I showed continued vaginal cornification, whereas this was present in every one of the thirty-two control animals which had not been subjected to a preliminary intense dosage.

(xii) *Nutrition.* When discussing the effect of diet on the incidence of spontaneous mammary cancer in mice allusion will be made to the fact that general undernourishment or the absence of certain specific factors from the diet will cause, not only a lowered incidence of cancer, but an inhibition of the oestrous cycles also, as shown by the persistent absence of vaginal cornification. These consequences may be plausibly attributed to a diminished production of oestrogen. Such an explanation does not embrace all the consequences of underfeeding, for excessive cornification of the vagina was early recognized as an effect of vitamin-A deficiency, and at first was described as persistent oestrus; the occurrence is recognized now as part of a general epithelial cornifying metaplasia. The question arises, what effect would an excess of vitamin-A have upon the oestral changes in the vagina? Sherwood, Brend & Roper (1936) investigated the problem. They gave 1,500 or 3,750 i.u. of carotene daily for 15 days to normal rats and examined vaginal smears taken from them at intervals of 8 hours. Every smear, regardless of the oestral phase, contained an excess of nucleated epithelial cells and leucocytes. Normal smears characteristic of the oestral cycle were not obtained until 20 days after the carotene feeding had ceased. The results of this experiment suggest that carotene has some power to inhibit the cornifying influence of oestrogen on the vagina.

(xiii) *Species.* The danger of generalizing on the subject of sex-hormonal activity may be exemplified by the growth of horns in animals. In some species or breeds of animal well-developed horns are the peculiar attribute of the male. This is true of the red deer and of certain breeds of sheep, and in these animals castration at an early age will prevent the growth of horns (Marshall, 1912). In other species, as in the reindeer, well-developed antlers are characteristic of both sexes and do not depend on the gonads (Darwin, 1871). In yet other animals, as in a certain breed of cattle, castration in early life leads to the growth of horns larger than those of the noncastrated animal (Marshall, 1912). The plumage of birds affords further examples of what at first sight might appear anomalous reactions to gonadal hormones (p. 108). Additional instances in which secondary sexual characters depend on genes rather than gonadal influence have been mentioned by Marshall (1910).

The Speed of Reaction to Gonadal Hormones

The swiftness with which organs respond to the gonadal hormones is remarkable. Astwood (1938) found that $0 \cdot 1\gamma$ of oestradiol causes an increase in the weight of the immature rat's uterus of 60 per cent within 6 hours, and Greene & Harris (1940) have noted increases in the weight of the infantile rat's uterus by 25 per cent within 6 hours of a single dose of $1 \cdot 5$ mg. of dehydroandrosterone. According to Astwood the increased weight of the uterus thus suddenly caused by oestrogens is attributable almost entirely to oedema.

Gradients of Responsiveness to Gonadal Hormones

The capacity for reacting to a gonadal hormone is innate in the cells of the re-acting tissue and is retained by them after transplantation to other regions of the body or to another individual. The responsiveness is a graded phenomenon. Each organ or part of an organ has its own special degree of sensitivity, and the threshold dose of a gonadal hormone required to elicit a reaction may vary over a wide range even within a single organ. Pézard (1922) called attention to this phenomenon, having noticed a dissociation of the various male sex characters—spurs, crowing, behaviour—in fowls, different doses of androgenic material being required to evoke these separate characters. Loeb (1928) observed gradients of responsiveness to oestrogen at different levels in the guinea-pig's vagina and uterus, and Moore & Gallagher (1930a), when testing an androgenic extract of bulls' testes on castrated guinea-pigs, noticed different degrees of sensitivity among the different accessory genital organs. Recognizable responses were obtained in the following order, which may be taken as representing sensitivity to the extract: (a) motility of spermatozoa, (b) secretory activity of prostate, (c) secretory activity of seminal vesicles, (d) secretory activity of Cowper's gland, (e) enlargement of epithelium in the vas deferens. While one-twentieth of a rat unit could be detected by the response of the prostate, one-sixth of a rat unit was necessary to cause a recognizable change in the seminal vesicles, Cowper's gland and vas deferens. Burrows (1935c, 1937) has described similar gradients in the action of oestrogens on the accessory genital organs of male mice, and Dessau & Freud (1938) have noticed that when testosterone or androsterone is smeared on the capon's comb the response begins at the base of the comb, gradually spreading from this to the apex of the serrations, which are the last parts to react; the same successive development is seen in the natural development of the cock's comb. The recognition of this graded response is important in testing androgens on the chick's comb by direct application. Frank & Klempner (1937) apply the material to be tested along the base of the comb by a blunt needle and estimate the effects by weighing the comb after removing it by parallel cuts carried down to the bone.

In addition to the examples just cited, a gradation of responsiveness to gonadal hormones has been noticed in the mamma, the external genitalia and surrounding skin, and in the bones.

When the conditions induced by a gonadal hormone regress after curtailment of its supply the order of reversion to normal corresponds with the varying susceptibilities of the different organs or parts; the tissues which respond most readily are usually the last to revert to their former condition. This feature has been studied by the writer (Burrows, 1935c) in the accessory generative organs of the mouse.

It has been noticed that the response of the mamma to oestrogens does not affect all the gland equally; some parts show an earlier or more pronounced reaction than others. Irregularity of the same kind has been seen in the involuting mamma after lactation in the goat (Turner & Reineke, 1936) and perhaps exemplifies the same graded sensitivity of the different parts of the breast to hormonal influences.

The gradients of sensitivity do not necessarily correspond in different species. In the mouse oestrone affects the coagulating gland more readily than the seminal vesicle, whereas in the rat the seminal vesicle is the more sensitive structure; and different hormones, though of the same general type, do not necessarily evoke the same gradients of response.

Different gradients for different hormones. Callow & Deanesly (1935a) compared the effects of the three androgens—testosterone, andradiol and androsterone—on castrated rats. The results showed that while testosterone and andradiol acted equally on the prostate and seminal vesicles, androsterone had a greater effect on the prostate than on the seminal vesicles.

The comparative potencies of androgens have largely been estimated by their influence on the capon's comb. If the gradients of responsiveness to all androgenic compounds remained the same this method could be relied upon; but, as indicated above, the gradients differ. Consequently, the activity of a hormonal compound on the capon's comb being known, it cannot be assumed that its activity on the prostate and other reactive organs are known also. This fact may be illustrated by some experiments by Tschopp (1935). He compared the actions of three androgenic compounds, namely androstenedione, androstanedione and androsterone, on the capon's comb, and on the seminal vesicles and penis of the castrated rat. The results (Table 65) show that while androsterone is more potent than the other two androgens on the capon's comb it has considerably less influence on the seminal vesicles and penis.

TABLE 65. Contrasted effects of different androgens on the male accessory organs of the rat (Tschopp, 1935)

	Amount of one capon unit (γ)	Dose (γ)	Weight of seminal vesicles (mg.)	Weight of penis (mg.)
Androsterone	60	50	11	46
		200	17	65
Androstenedione	100	50	25	85
		200	285	150
Androstanedione	100	50	16	57
		200	51	92

Deanesly & Parkes (1936a) have made similar observations when testing a number of different androgenic compounds on castrated rats. The effects of these compounds on the capon's comb, the prostate and seminal vesicles of castrated rats do not correspond. Some of the compounds used show more potency on the comb than the prostate, others affect the prostate more readily than the comb.

Differences of influence on the prostate and seminal vesicles also were noticed. Testosterone is 10 times as active as androsterone when tested on the seminal vesicles but only between 2 and 5 times as active when tested on the prostate. *Trans*androstenediol is only slightly more active on the prostate than andro- sterone, but 4 to 5 times more active on the seminal vesicles, androsterone having much less influence on the seminal vesicles than on the prostate.

These discordant reactions, in addition to the co-operative and antagonistic effects which the gonadal hormones so often exercise, make biological assays of organ extracts, urine and other body fluids impossible to perform with exact scientific accuracy.

Reversible and Irreversible Effects of Sex Hormones

Many of the changes produced in the living body by the sex hormones endure only so long as the chemical compounds on which they depend are provided in adequate quantity. When the supply of hormone is stopped the tissues return to their former state. A familiar example of a reversible tissue change is afforded by pregnancy, during which the uterus undergoes great hypertrophy, returning to its normal size after parturition. In this instance other factors are involved besides the gonadal hormones, and the same is true apparently of the alterations in the mammary gland connected with lactation and weaning. Less obvious but not less convincing, and dependent directly on the sex hormones alone, are the changes in the uterus and vagina which accompany the oestral cycle.

When tissues have been altered by gonadal hormones it is remarkable how little the lapse of time or the extent of the changes affects the possibility of their return to their former state after the hormonal supply has been re-adjusted. McCullagh & Walsh (1934) found that urinary androgen produced regeneration of the accessory·generative organs in rats which had been castrated 100 days earlier— a long period in the life of a rat.

In the mouse oestrogens will convert the many-pouched coagulating gland possessing a single layer of secretory cells into a simple sac lined with stratified keratinous epithelium. If at this stage oestrogen be withheld the transmuted structure will resume its former histological features and once more will produce a secretion to coagulate the fluid ejected from the seminal vesicle (Burrows, 1935 c). Even the enlargement of the pituitary caused by oestrogen is a reversible condition (Deanesly, 1939).

In some of these reversible reactions the pituitary plays a part. Either androgen or oestrogen persistently given in excess will lead to cessation of spermatogenesis and atrophy of the testes, owing to the power of these hormones to inhibit the supply of gonadotrophin on which the activity of the testes depends. Very soon after stopping the supply of gonadal hormones the pituitary will resume its gonadotrophic functions and so cause the testes to enlarge and resume their normal activities. In other instances the reversible changes do not need the co- operation of the pituitary. For example, cornification of the vagina may be in- duced by oestrogens after the pituitary has been removed, but it will persist only so long as oestrogen is administered.

Reversibility of effect is not an invariable character of the tissue changes induced by gonadal hormones. The conformation of the pelvis and other parts of the skeleton, so far as it depends on the gonadal hormones, is irreversible, and the separation of the pubic bones in mice resulting from oestrogenic action during pregnancy remains after parturition as durable evidence that the mouse has littered. In several species the clitoris can be converted by androgens into a penis-like structure with a well-developed glans, a retractable prepuce and an os penis, and this is another example of a permanent structural transformation caused by a sex hormone; so too is the hypospadias induced in newborn rats and mice by an excess of androgen, oestrogen or progestin. The majority of irreversible changes brought about in the body by gonadal hormones occur during development.

Another kind of irreversible tissue change induced in certain circumstances by gonadal hormones is cancer.

Actions common to Gonadal and Adrenal Cortical Hormones

Before ending this general review of the gonadal hormones reference may be made to some overlapping of activities between them and the hormones of the adrenal cortex. To a large extent progesterone can replace corticosterone and can support life after adrenalectomy, and desoxycorticosterone will perform several of the functions of progesterone on the uterus, vagina and breast. Oestrogen and androgen do not appear able to replace corticosterone in all its actions and have little if any power to maintain life after adrenalectomy, but they have an effect on the excretion of salt and water. Thorn & Harrop (1937) state that oestradiol, oestrone, testosterone, testosterone propionate, pregnanediol and progesterone all have some capacity for reducing the excretion of chloride, sodium and water, progesterone being the most effective in this respect. Their observations were made on a man and on dogs, and are supported by experiments done on immature rats by Zuckerman, Palmer & Bourne (1939). These workers found that oestrogens—like progesterone, testosterone and deoxycorticosterone—lead to retention of NaCl and water. Oestrogens and androgens, they say, cause nitrogen retention too, which progesterone and corticosterone do not.

Another biological capacity which is common to the gonadal and adrenal cortical hormones is their power to cause *enlargement of the bitterling's ovipositor*. Fleischmann & Kann (1938) found that corticosterone has this capacity. They dissolved 3 mg. of corticosterone in 0·3 c.c. of warm ethyl alcohol and added it to 2·7 c.c. of a normal saline solution. After the injection of 0·1 c.c. of this solution into the fish the ovipositor increased in length from 4 to 16 mm. during the next 48 hours. After the addition of 2 c.c. of the solution to a litre of water containing three fish their ovipositors grew from 3, 4 and 4 mm. respectively to 12, 12 and 15 mm. in the next 48 hours.

An overlapping between the biological activities of the gonadal and adrenocortical hormones, it will be seen, takes place in both directions; not only can the gonadal hormones to some extent act as substitutes for the hormones of the adrenal cortex, but the latter are capable of performing some of the functions of the gonadal hormones.

The Influence of Gonadal Hormones on Behaviour

Human conduct, like that of the lower animals, is largely controlled by the sex hormones: a fact which has wide social and legal implications. The mind of a civilized man, however, can often overcome his natural instincts by substituting conditioned reflexes for the reactions of primitive nature, so that the underlying biological urges become obscured, though they are not abolished. In these pages the subject will be considered very briefly and attention will be confined almost entirely to the effects produced by sex hormones on the conduct of lower animals; these are less trammelled by higher faculties and their responses therefore are more easily perceived.

A simple and familiar demonstration of the fact that character and conduct are under gonadal influence is provided by the castration of horses and cattle; the alteration which this operation brings about in the behaviour of the animals is the chief reason why it is done. The conduct of the animal which follows the operation is what may be described as that of the neuter gender.

A different result is caused by the replacement of an animal's own gonads by those of the opposite sex. This procedure has been carried out on rats and guinea-pigs by Steinach (1916), Steinach & Holzknecht (1916), Moore (1919) and Sand (1919 a, b). The sexual behaviour of animals so treated becomes perverted and no longer accords entirely with their innate sex.

Castrated males bearing implanted ovaries show indifference toward females, but when placed with infantile animals of their own species become interested in them, may try to suckle them and will sometimes succeed (Steinach & Holzknecht, 1916). Spayed females bearing transplanted testes behave like males, and when placed in a cage with females in oestrus may attempt to copulate with them.

Moore (1919) and Sand (1919 a, b) showed that male and female gonads can persist in a functional condition in a single individual, and that animals in which such a condition has been brought about by the transplantation of gonads may show bisexual characters. A male, one of whose testes had been replaced successfully by an ovary, behaved to females and young like a male, but to males like a female; his mammae were enlarged and contained milk.

Another way of testing the hormonal control of conduct is to abolish gonadal activity by hypophysectomy, or to stimulate the gonads by giving pituitary extracts. Removal of the pituitary has revealed the fact that some of the functions connected with reproduction are not always entirely regulated by the gonads, because these functions may still be exercised although the gonads have been rendered inactive.

The discovery and isolation of individual gonadal hormones made the investigation of their action on behaviour more decisive, and much experimental work has now been done in this field. No attempt will be made here to review the work comprehensively; a few illustrative examples only will be quoted.

Parental interest. The amount of attention given by the male parent to its offspring varies largely with different animals, and although generally speaking parental instinct is more pronounced in the female, in some species the male will display much solicitude for the young. Experiments have shown that parental interest is largely though not entirely dependent on gonadal activity. Leblond (1938a) says that anxiety for their young is shown by both male and female mice, whether their gonads are or are not present. The main activities relied on in these tests were retrieving the young, licking and cuddling them. (See also Allen & Wiles, 1932, and Leblond & Nelson, 1936, 1937a, b). Leblond (1938a) noticed that the exhibition of maternal solicitude was considerably more pronounced in females which had borne young than in those which had not (Table 66).

TABLE 66.   The effect of littering on maternal behaviour in mice (Leblond, 1938a)

Classification	Age (days)	Percentage of positive maternal response toward young
Before opening of the vagina	23–25	38
Virgins	36–180	56
At 1st lactation	60–148	76
At 2nd and 3rd lactations	116–179	86

Parental behaviour in rats, as in mice, is largely independent of sex hormones, but not to the same degree as in mice (Leblond & Nelson, 1937a). McQueen-Williams (1935) observed that male rats could be induced to display a regard for newborn rats by giving them implants of bovine pituitary tissue, and Riddle, Lahr & Bates (1935) induced motherly behaviour in virgin rats aged between 67 and 81 days by stimulating their ovaries with gonadotrophin and then injecting a prolactic pituitary extract. Before this treatment the rats had shown indifference towards young.

The experiments just quoted seem to show that the degree to which sex hormones govern parental solicitude varies among the different species.

Masculine behaviour. The singing of birds, the rivalry and aggressiveness shown by males generally, the urge to copulate, are all instigated by sex hormones. Usually the male bird ceases to crow or to sing if deprived of the testes and will resume his vocal efforts if given enough androgen, and the same hormone will cause immature birds to behave as if they were adult. Domm (1937) gave daily injections of an extract of bovine pituitary to chicks, beginning on the 2nd day after hatching. This caused precocious male behaviour, including attempts to crow at the age of 9 days. Hamilton (1938a) produced the same results in male chicks by daily injections of 500γ of testosterone propionate given from the 2nd day after hatching, but hen chicks treated in the same way made no attempt to crow though their combs and wattles became enlarged. With injections of dihydroandrosterone Breneman (1939) caused chicks to crow when only 5 days old. Among normal birds the male only, as a rule, expresses himself in song. Shoemaker (1939) treated female canaries by daily doses of 0·076 mg. of testosterone, with the result that they all began to sing, and their songs were indistinguishable from those of normal males. Leonard (1939) and Baldwin, Goldin & Metfessel (1940) have obtained the same results in female canaries by giving them testosterone propionate.

In animals which breed only during part of the year masculine activities are mainly confined, as a rule, to the mating time and disappear in those seasons of the year when the gonads are inactive, just as they disappear after castration. Androgens will cause females, immature males or resting males to behave like mature and active males. Hamilton, in the experiment cited above, noted that his young male chicks under the influence of testosterone became belligerent, and though their spurs had not yet grown they struck at opponents as though their legs were armed. Shoemaker also noticed that the hen canaries when treated with androgen became aggressively quarrelsome. In Domm's experiment the young male chicks only 13 days after hatching attempted to tread, and libido was so strong in them that they had to be isolated to save the lives of their associates.

Results like those above have been observed in mammals. Shapiro (1937) castrated immature rats weighing between 30 and 50 g. and then isolated them for 9 weeks. They were then introduced to females, in which they took no sexual interest. The castrated males were again isolated and given eighteen daily injections of 0·5 mg. of testosterone propionate. When placed among females after this treatment they copulated. Raynaud (1938 a, b) gave injections of testosterone propionate to a pregnant mouse with the result that the female offspring were modified so as to resemble males. One of these when fully grown showed female behaviour, and though there was no patent vagina, accepted the male and encouraged attempts at copulation. She was then given large doses of testosterone, after which her behaviour was quite changed; she was accepted by females and copulated with them as though she were a male. Dantchakoff (1938a) has reported a similar observation. Stone (1940) gave daily injections of 0·62 mg. of testosterone propionate to young rats, beginning when they were between 22 and 25 days old. This accelerated their power to copulate by 20 days, as compared with untreated control rats.

Androgens cause the same effects in man as they do in the lower animals, and will produce precocious virility in boys, restore physical potency to eunuchs, and bring about excessive masculine urges in the normal male adult.

Feminine characteristics and conduct also are dependent on hormonal influence. Female rats, after hypophysectomy, rapidly cease to show any sexual manifestations. Such a state of indifference is largely due to the lack of oestrogen, which is no longer produced by the ovary. This was shown by Ball (1941), who removed the pituitaries from female rats, and a week later gave to each of them 250 i.u. of oestradiol benzoate on five successive days. Under this treatment coitus was accepted by all the five rats employed in the test.

Other examples of the influence of sex hormones on behaviour will be found in the sections devoted to the individual types of hormones. The subject, especially as affecting mankind, is too large to be discussed in these pages. Perhaps enough has been said to stimulate interest in the subject. We are beyond the day when a woman with a beard, the consequence of adrenal hyperplasia, was regarded as a witch, or when the sufferer from brain disease was subjected to chains and punishment. It is surely to be hoped that the victims of sexual infirmity, too, will soon be treated with understanding and kindness instead of persecution and contempt.

Social problems. Biological studies suggest that war and some of the other difficulties which beset human communities may have their origin in the impulses for reproduction and maintenance of the home (Heape, 1931). The mystery of migration, too, may perhaps depend ultimately upon urges brought into action by the gonadal hormones, as suggested by Marshall (1910) in his *Physiology of Reproduction* and supported by Rowan's observations (see also Bullough, 1943).

Anaesthetic Properties of Gonadal Hormones

A notable property of the gonadal hormones is their power to induce general anaesthesia. Selye (1941, 1942) who was the first to observe this property, found that it was common to several hormonal steroids, including testosterone propionate, androstenediol, oestradiol, oestrone, oestriol, dihydroequilin, equilenin, stilboestrol, progesterone, pregnanediol, *allo*pregnanediol, pregneninolone and deoxycorticosterone acetate. The anaesthesia induced by these compounds was not preceded by a period of excitement, and was deep enough to allow prolonged abdominal operations. Females seemed to be more susceptible than males to this anaesthetic action.

The Influence of Gonadal Hormones on Tissue Growth, with special reference to Cancer

It may be convenient here to review the influence which gonadal hormones have upon the growth of tissues in the body, and to recall some of the circumstances which affect this kind of influence. The subject has considerable interest in connection with embryology and cancer. Unhappily our study of the matter is confined to the living organism, where there is such a complexity of factors that it is hard to isolate and identify details. We have to remember that when a certain effect follows the introduction of an organic chemical compound into the body, that effect may be caused, not by the compound as introduced, but by some product of its metabolism. The capacity of living organisms to adapt themselves to foreign agents or to neutralize them by buffering or other means is a property which may make it difficult to analyse results in the kind of physiology to which this essay is devoted. Biological differences, even between individuals of the same species and sex, provide another obstacle, one which renders accurate quantitative work almost impossible. These and other complications often prevent the actions of gonadal hormones on the growth of tissues from being stated in precise scientific terms. Nevertheless, an attempt may be made to formulate some of the observed effects of gonadal hormones, regarded as a group, on the growth of tissues in the living organism. These effects may be tabulated as follows:

    I. Inhibition of growth.

    II. Stimulation of growth.

        A.   Controlled growth.

        B.   Uncontrolled growth or cancer.

        C.   Imperfectly controlled growth, including innocent tumours.

I. THE INHIBITION OF GROWTH BY GONADAL HORMONES

There is more than one way in which a gonadal hormone may retard the growth of a particular tissue. (i) It may so act on the anterior lobe of the pituitary as to check the production of appropriate trophins. Thus, any of the three main types of gonadal hormone will check the formation of FRH by the pituitary and so will arrest the growth of ovarian follicles. (ii) In other situations the growth of a particular tissue seems, though this is uncertain at present, to be inhibited by the direct action of a gonadal hormone. For example, oestrogen arrests the growth of epiphyseal cartilage, causing bony union between the epiphyses and shafts of the bones, a fact which largely explains why women are smaller than men. (iii) Yet a third manner in which a gonadal hormone may impede growth is by interfering with the growth-promoting action of another hormone. In this way oestrogen can prevent the development of placental tissue under the influence of progesterone.

II. The Promotion of Growth by Gonadal Hormones

A. *Controlled growth*

(i) *Cellular differentiation and function.* A characteristic feature of normal tissue growth in response to a gonadal hormone is its adaptation to a particular purpose, in attaining which cellular multiplication, differentiation and function follow each other in orderly succession and appropriate degree. The fact that differentiation and function are imposed upon growth is of vital importance. The writer has noticed that when tissues composed of undifferentiated cells growing from different sources come into contact they unite. Such union of primitive cells is seen in the healing of wounds, the coalescence of tissues during embryological development, and the amalgamation of malignant papillomata growing from opposite points within glandular ducts and acini. When two masses of undifferentiated cells have met and become combined in this way their morphological fate will depend upon the intervention of differentiation and on the form which this takes. In cancer, as a rule, the results of any differentiation which takes place are negligible, but in other kinds of growth differentiation is of fundamental importance. In the healing of an incised wound the granulation cells growing from the edges meet, amalgamate and, becoming converted largely into collagen-producing connective tissue, cause a tough and permanent union. The temporary amalgamation of undifferentiated cells and their subsequent permanent junction or disjunction according to the nature and degree of their differentiation is an important architectural process in embryology. The terminal part of the floor of the penile urethra is formed by the union of the lateral walls of the urethral groove. Normally the edges of these walls after their conjunction become replaced by ordinary firm connective tissue. Gonadal hormones, if supplied early enough and in sufficient quantity, can anticipate this transformation and cause the two uniting epithelial edges to become keratinized, so that union is impossible and hypospadias is brought about. Another example is provided by the relationship between the prepuce and glans penis. These at first are united by two or three layers of relatively undifferentiated epithelial cells and in the absence of a sufficient supply of androgen will remain so. Androgen causes this primitive epithelium to differentiate into a stratified squamous cornified epithelium and so brings about separation of the foreskin from the glans. An essential feature of this differentiation is that it affects first and most the oldest cells, that is to say, those which are farthest from the basement membrane.

It is not only cornification which leads to separation of previously united un-differentiated cells; the formation of mucus or other secretion has the same effect and seems to explain the opening up of the primitive glands and their ducts, which at first consist of solid rods of epithelium.

There are reasons for suspecting that malformations other than hypospadias may be caused by mistimed epithelial differentiation in consequence of a defective supply of gonadal hormone—for example, the partial closure of the vagina which sometimes occurs in children and elderly women.

(ii) *Specificity of action*. The gonadal hormones are not general stimulators of cell proliferation or differentiation. Their normal action on the tissues is specific and is confined almost and perhaps entirely to structures concerned directly or indirectly with sexual reproduction. This specificity is independent of the sex of the individual; homologous structures in the two sexes react alike.

(iii) *Responsive and irresponsive tissues*. Different cells and tissues vary in the degree and nature of their reactions to the different types of gonadal hormone, and the capacity to react is innate in each individual cell or group of cells, is un-affected by their neighbours and is retained after transplantation into another part of the body or into another individual, whether of the same or of the opposite sex.

(iv) *Gradients of reactivity*. The readiness with which different tissues respond to a given gonadal hormone shows wide divergences and the threshold of reactivity may vary through a large range within a single organ. Moreover, the reactions may vary not only in degree but in character also in different parts of a single anatomical system. For example, the persistent administration of oestrogen to a mouse leads to a graded supercalcification and enlargement of the bones, those most affected being the femora and spine. This process is accompanied by a decalcification and gradual disappearance of the symphyseal ends of the pubic and ischial bones.

(v) *Reversible and irreversible effects*. Most of the effects of gonadal hormones on cellular growth and differentiation in adult life are reversible and are maintained only so long as the hormone is supplied in adequate amount; on cessation of the supply the tissues return toward their former state. It might be noted here that the enlargement of an organ under the influence of a gonadal hormone does not necessarily denote that cellular proliferation has occurred. The increased size may be caused mainly or entirely by hypertrophy of the cells already present and by intercellular turgor. In some instances, of course, the enlargement is a result of organized cellular multiplication; hypertrophy of the penis under the influence of androgen, or extended ramification of the mammary ducts under the stimulus of oestrogen may be mentioned as examples.

The growth effects of gonadal hormones are not always reversible. Permanent changes are induced by the hormones, especially during the period of development; union of the epiphyses with the shafts of bones under the influence of oestrogen is one instance, and separation of the prepuce from the glans penis by androgen is another. In early embryological life the examples of irreversible changes are numerous in connection with the development of the accessory reproductive organs. Cancer is another kind of irreversible change caused by gonadal hormones.

(vi) *Adaptational resistance to gonadal hormones.* There is some evidence that with a continued exposure to a particular gonadal hormone the individual may acquire some power of resistance to its action. This resistance is not attributable to the formation of antihormones. Sometimes it may be accounted for, the author believes, by an increased capacity to produce an antagonistic hormone (Burrows, 1937a). The persistent administration of oestrogen to a mouse is apt to cause hypertrophy of the androgen-producing interstitial tissue of the testicle, and when this happens some of the ordinary reactions to oestrogen become less pronounced. Probably there are other adaptive mechanisms.

(vii) *Inactivation and excretion.* It seems certain that the effect of a given quantity of gonadal hormone will depend largely on the rate at which it is inactivated and excreted. The liver is the chief inactivator and hepatic disorder has been shown by experiment to accentuate the effects of artificially administered oestrogen. The effect of renal lesions in retarding the excretion of a gonadal hormone and so enhancing its effect does not seem to have been recorded.

(viii) *The pituitary gonadal relationship.* The formation of gonadal hormones is controlled normally by the pituitary. Any one of these hormones by its action on the pituitary checks the output therefrom of gonadotrophin, and so automatically curtails its own production. The property of inhibiting the output of gonadotrophin being common to all the gonadal hormones it follows that any one of them can inhibit the output of the others. It seems likely that the prevention of mammary cancer in mice by giving them testosterone (Lacassagne, 1939c) is brought about by checking the production of oestrogen in this way.

(ix) *Mutual antagonisms between different gonadal hormones.* Apart from their action on the pituitary, the gonadal hormones in some circumstances show mutual inhibitory effects, so that the growth stimulation by one gonadal hormone on a particular tissue may be prevented by a gonadal hormone of another type. Inhibition by progesterone of the enlargement of the chicken's comb under the influence of androgen is an example.

(x) *Co-operation between gonadal hormones.* The co-operative effects of different types of gonadal hormone are as important as their mutual antagonisms. For this co-operation the two hormones must in some instances act simultaneously and in others successively. Thus it has been found that for full maintenance of the accessory male genital organs after castration, both androgen and oestrogen are required at the same time. In this case the oestrogen must be supplied in relatively small quantities. Given in excess oestrogen will largely counteract the restorative effects of androgen on the atrophied accessory organs. For progestational changes in the uterus a preliminary subjection to oestrogen followed by progesterone is required. Oestrogen given at the same time as progesterone will inhibit the progestational development.

(xi) *The influence of sex on the responsiveness to gonadal hormones.* The reactions of homologous tissues to a gonadal hormone do not differ in character between the two sexes, but there is evidence that the degree of reactivity is not identical in the two sexes; males appear more responsive than females to oestrogen and less responsive to androgen. In females these comparative sensitivities are reversed, and the effect of androgen in them is greater than in males. An example of these differences of reaction in the two sexes is seen if the gonads of littermates are interchanged so that testes are replaced by ovaries and ovaries by testes. The ultimate effect of such an exchange of gonads between the two sexes is that the testicle-bearing female grows to a larger size than the normal intact male, while the ovary-bearing male does not attain the dimensions of a normal female. Experiments with pure hormones also reveal a different sensitivity of the two sexes to the gonadal hormones.

(xii) *The effect of circumstances on the growth promoting and retarding effects of gonadal hormones* is too complicated and not yet well enough understood to be discussed in detail. It must suffice to say that the kind and degree of effect produced by a gonadal hormone will be influenced by many circumstances, including dosage, species, age, season, nutrition, general health and perhaps other conditions.

B.   *Uncontrolled growth*

(a) *Non-specific cancer.*   Although the specific nature of the tissue responses has been insisted on, the gonadal hormones appear to have some capacity for inducing non-specific cancer. Subcutaneous sarcomata have been induced in mice and rats by injections of oestrogen; in mice the same result has followed injections of testosterone (Lacassagne, 1939b). Extensive trials have not shown any capacity on the part of these hormones to cause cancer of the skin.

(b) *Specific cancer.*   Apart from subcutaneous sarcomata, the known malignant activities of the gonadal hormones are confined to the breast, uterus, testis, bones and the white cells of the blood. These lesions will be discussed later.

C.   *Causation of Non-Malignant Tumours by Gonadal Hormones*

Little will be said here about innocent tumours induced by gonadal hormones. So far as is known at present they appear to be caused only by oestrogen and include fibromyomata of the uterus, ovarian cysts, chromophobe swellings of the pituitary and perhaps mammary adenomata. They will receive further reference when the effects of oestrogen on the individual organs are under consideration.

The Influence of Gonadal Hormones on the Determination of Sex

In some lower animals it is known that the determination of sex may be governed by external circumstances as well as by genic influences, whereas in mammals sex appears to depend entirely on genes. Gonadal hormones, especially when supplied in the early stages of development, may cause great changes in the reproductive organs of mammals so that the individual may seem to have undergone sex reversal (Foote, 1940; Witschi, 1942). In these seeming examples of sex reversal, whether induced by gonadal hormones or by abnormal environment, the original sex as determined by genes remains unaffected; only the manifestations of sex have been changed. Consequently, if a vertebrate whose sex has been apparently quite reversed is mated with a normal animal, their progeny will all be the same sex, because the sex chromosomes of the two mates will be identical (Crew, 1921; Wells, Huxley & Wells, 1935).

PART III. ANDROGENS

Chapter VI. *Androgens*

Introductory remarks. General review of biological action. Gradients of influence. Sources, inactivation and excretion.

Introductory Remarks

THE term androgen is used here as a collective title for compounds which resemble testosterone in biological action. The chief function of these hormones is to stimulate the development and activity of the accessory male reproductive organs. On the general dimensions of the body androgens have little effect, the smaller size of the female in mankind and in many other mammals being due to the action of oestrogen rather than to a deficiency of androgen. On the skeleton and general conformation of the body the androgens appear to exercise some influence, as seen, for example, in pathological cases of virilism in women, a condition which is accompanied by an excessive excretion of androgen in the urine. When this disease affects young girls the shape of the body tends in time to become masculine in character; whereas with defective gonadal development in young males the contours of the body may become distinctly feminine, the hips being broad compared with the shoulders. The falsetto voice and hairless face of the eunuch who has been castrated in childhood may be attributed also to a deficiency of androgen; for the testis is the chief though not the only source of this kind of hormone. The distribution of hair on the body in men, the extent of the facial sinuses, and the many curious sexual adornments and peculiarities which the males of different species acquire, especially in the mating and breeding season, also depend for their appearance and maintenance on an adequate supply of androgen. The crest of the newt, the cock's comb, the rugose and swollen clasping digits of the male frog, and the bright coloration of some fishes in the breeding season are examples. At present it will be unnecessary to discuss these curiosities at length, and attention will be concentrated on the more constant and vital activities of the androgens in mammals, including man.

General Review of the Biological Action of Androgens

The results of castration. The more obvious effects of castration on the body have been recognized since ancient times. Such knowledge, however, has been accompanied by many false ideas. It has been thought, not only by the ill educated, that the eunuch is physically and mentally inferior to the normal male. As will be shown in the course of this discussion, the size of the body is not determined by androgens, and there is proof that castration does not have any bad influence on intellectual power. So far as primitive impulses are concerned, experiments on animals have shown that androgens cause an inclination to pugnacious assertiveness, a behaviour which among men has not infrequently been admired, as evidence of a strong character rather than as a defect of self-control. Though the eunuch lacks the primitive urge which leads to battle, he may be perfectly well able to hold his own when confronted with an enemy. Justinian's famous general, Narses, is an instance of a eunuch possessed of outstanding military capacity and general intelligence; and as regards intellectual power he is far from unique among men who have been deprived of their gonads.

The tendency to look down upon the eunuch seems to be instinctive in the human mind, for it has survived through the ages and the supposed disadvantages of castration in man have been constantly exaggerated. An example of this kind of prejudice is seen in the history of the surgical treatment of enlarged prostate by castration. At a time when non-malignant enlargement of the prostate was a fatal disease, it was discovered that removal of the testes would in most cases provide a cure; almost immediately a rumour spread through the world and was accepted that the operation caused mental debility. In spite of the absence of proof, in spite of proof that the rumour was untrue, the idea became a belief and old men preferred to die a miserable and lingering death rather than be cured at such an imaginary risk. Even to-day this erroneous notion prevails (p. 299).

The influence of androgens. The first experimental demonstration of hormonal action by the testes, though its significance remained obscure for a long while, was made by John Hunter when he showed that the rudimentary spur of a hen would grow into a good masculine spur if transplanted into the leg of a cock, whereas the small spur of a young cock would undergo no such development when implanted into the leg of a hen. Furthermore, Hunter directed attention to facts which indicated to him that the accessory genital organs of the male were dependent for their development, and probably for their existence, on the presence of the testicles. With the exception of a few sporadic observations, many years passed after Hunter's death before his experimental methods were applied once more to the problems of sexual anatomy and physiology. Berthold (1849) proved that the effects of castration in fowls could be mitigated by testicular grafts. Brown-Séquard (1889), believing that testicular extracts had a beneficial influence on decrepit women and that the feebleness of old men might be partly the result of testicular failure, performed tests upon himself. At the age of 72 years he gave himself subcutaneous injections of extracts prepared from the testes of dogs and guinea-pigs on 15, 16, 17, 24, 29 and 30 May. The injections caused pain, but he felt himself more resistant to fatigue after the second injection onward, and noticed an improvement of intestinal tone, together with an increased bodily and mental vigour. These effects, however, were not very striking and he thought that possibly they were the consequence of autosuggestion. Variot (1889) gave similar testicular extracts by injection to male patients; the results were the same as those obtained by Brown-Séquard on himself, and Variot, like the latter, believed they might be attributed to autosuggestion. Loewy (1903) found that repeated injections of testicular substance into young capons caused masculine development of the skeleton and comb. Walker (1908 a, b) prepared fresh saline extracts of cocks' testes, using two volumes of saline solution to one of testis, and injected 0·5 c.c. of this extract every day into each of two hens, beginning on 1 February and ending on 4 October of the same year. The hens were between 2 and 4 years old at the beginning of the experiment. They ceased laying in the middle of March and their combs and wattles grew to about 8 times their normal size and became highly coloured; the hens became pugnacious and anxious to attack cocks when introduced to them. When the injections were finally withheld the combs and wattles shrank, regaining their normal dimensions by 9 November, and the hens began to lay again in the following January.

Pézard (1911) made an extract of the cryptorchid testes of pigs. With this material he treated two black orpington capons, giving them intraperitoneal injections twice a week. The result was that the atrophied combs and wattles grew, becoming thick, turgid and erectile, and the two birds crowed and showed combative ardour like normal males. The treatment lasted from January till May, when it was discontinued. The birds now ceased crowing, they lost their sexual instincts, their combs shrank, and their general appearance reverted to that of the ordinary capon.

With the results obtained by these pioneers to give encouragement, a rapidly growing volume of research has been done, some of which will be referred to in the following pages.

Molecular structure of androgens. Biological experiments involving castration, the transplantation of testes, or the injection of testicular extracts had already laid the foundations of our knowledge of the testicular hormones before their chemical characters had been identified. A new era in these studies began when McGee (1927) isolated from bulls' testes a relatively pure product having the potencies of an androgen. Shortly afterward Loewe & Voss (1928) demonstrated the presence of androgen in urine. Since then rapid advances have been made in the chemistry of these compounds. The first androgen to be prepared in a chemically pure form was androsterone, which Butenandt (1931) isolated from human urine. This discovery made it possible to investigate the biological properties of an androgen under exact conditions. A further advance was made when Ruzicka and his colleagues (1934) synthesized androsterone from cholesterol. In the following year the more potent androgen, testosterone, was obtained in crystalline form from fresh testicular tissue by Laqueur and his collaborators (David, Dingemanse, Freud & Laqueur, 1935). Soon this compound too had been prepared artificially from cholesterol by Ruzicka & Wettstein (1935) and Butenandt & Hanisch (1935 a). Since then many different artificial androgens have been prepared and made available for study.

Reference has been made in an earlier chapter to a few details concerning the molecular structure of the androgens, and no more will be said on the subject except to recall the fact that it has not yet been found possible to correlate exactly the peculiar biological activities of these compounds with their molecular constitution, nor can all the androgens be demarcated with precision from the oestrogens and progestins, for their fields of activity overlap.

The chief androgen derived from the testis is testosterone, which is converted, apparently by the liver, into less active compounds including androsterone and its isomer aetiocholane-3(α)-ol-17-one (Callow & Callow, 1939), which are excreted in the urine. Other androgens are formed in the adrenal, including *trans*-dehydroandrosterone. Formulae of the chief naturally formed androgens are shown in Figs. 4a, b (pp. 102, 103).

Biological tests for androgen. The chief tests in common use are founded on the effects produced on (1) the growth of the capon's comb, and (2) the development and functional activity of the prostate and seminal vesicles of the castrated mouse or rat. Among other methods which have been used are the spermatozoa motility test (Moore, 1928), ejaculation test (Moore & Gallagher, 1930a), and cytological tests based on cellular reactions in the prostate (Moore, Price & Gallagher, 1930), seminal vesicles (Moore, Hughes & Gallagher, 1930), vas deferens and Cowper's gland (Moore, 1930). The bitterling test (Kleiner, Weisman & Mishkind, 1936) also may be mentioned. In routine assays, the changes of weight effected in the prostate and seminal vesicles of castrated mice or rats are the indicators now increasingly used; the capon test alone cannot be regarded as a reliable criterion of androgenic potency in mammals.

When the androgenic activity of a particular compound has to be estimated for therapeutical purposes it may be better, for reasons which will be stated later, to observe the reaction of each particular organ concerned, rather than to rely on a general estimate of androgenic activity based on the response of a single structure.

International unit of androgen. Because it is often necessary to estimate the hormonic activity of tissues and tissue extracts without knowing the hormone concerned, a unit based on the degree of the responses caused under standard conditions has been found convenient. In this way it is made possible to state different degrees of potency in approximately accurate terms even though the substance to which that activity is attributable remains unknown. The international unit of androgenic activity as established by the Health Organization of the League of Nations in 1935 is the activity of 100γ of a standard preparation of crystalline androsterone as measured by its effect on the capon's comb.

Gradients and Variations of Androgenic Action

Gradients of responsiveness. Usually the effects of a given androgen are not distributed equally throughout the organism.

Pézard (1922) called attention to a dissociation of the various male characters in fowls, having found that the threshold dose of male hormone is not the same for all manifestations of sex, for example, the growth of spurs, masculine psychology and crowing. Similar variations in responsiveness are seen in the human species; for an excess of androgen sufficient to cause hirsutism in a woman may be compatible with the continuance of oestral cycles, which would be inhibited by a more plentiful supply of androgen. In a single organ different grades of responsiveness can be detected. Champy & Kritch (1925) noticed that the formation of 'muco-elastic tissue' in the comb first appears in chicks at the base of the comb and gradually progresses toward the apex. After castration this special condition of the comb disappears, its retrogression beginning at the apex and spreading to the base, which is the last region to become atrophied. Dessau & Freud (1936) observed a similar gradient of effect when solutions of pure crystalline testosterone or androsterone were smeared directly on to the capon's comb. Although the preparations were applied uniformly over the whole comb the development of 'muco-elastic tissue' proceeded from the base of the comb to the apex of the serrations in the same way as in normal adolescence.

Variations in the distribution of responses to different androgens. Another feature is the varying degree of effect produced by different androgens on the same organ or by the same androgen on different organs. One androgenic compound may affect the prostate more readily than the seminal vesicle, while another affects the seminal vesicle more readily than the prostate. Moore & Gallagher (1930a), using an extract of bulls' testes, found that the dose required in the rat to prevent castration atrophy of the seminal vesicles was 3 times that producing the same effect in the prostate; and Moore (1935b) noticed that after castration retrogressive changes could be recognized sooner in the seminal vesicles than in the prostate. Moreover, it has been found that the action of a given androgen on the prostate and seminal vesicles cannot be foretold by knowing its activity on the capon's comb. An appreciation of these facts enabled Dingemanse, Freud & Laqueur (1935) to show that androgen prepared from the testis (testosterone) is a different substance from the androgen prepared from urine (androsterone). For the experiment they used rats which had been castrated when 3 weeks old and now, at 6 months of age, weighed about 260 g. These were divided into three groups, one being kept for control. To each rat of the other two groups daily injections were given of the same number of capon units of the two preparations under comparison. All the rats were killed at the end of 28 days and the weights of their seminal vesicles compared. The results show that doses of the two androgens which have equal effects on the capon's comb differ considerably in their effects on the seminal vesicles (Table 67).

TABLE 67. Difference of effect between testosterone and androsterone on the castrated rat's seminal vesicle, when given in the same dosage as measured in capon units (Dingemanse, Freud & Laqueur, 1935)

Treatment	Weight of seminal vesicles at the end of treatment (mg.)
I. Untreated	8
II. Injections of urinary androgen (androsterone)	113
III. Injections of testicular androgen (testosterone)	538

Callow & Deanesly (1935b) found that while 1 mg. of androsterone given daily was enough to maintain the prostate in castrated rats of 140 to 160 g., 2 mg. of androsterone given daily in the same manner were required to maintain the seminal vesicles. Further, they noted a species difference in responsiveness to androsterone, for in castrated mice 1 and 2 mg. daily failed to maintain completely either the prostate or the seminal vesicles.

Tschopp (1935), studying the properties of androstenedione, found that this substance is much more active than androsterone in causing growth of the penis and seminal vesicles in the young castrated rat, but is less active in its effect on the capon's comb (Table 68).

TABLE 68. Comparison of effects of different androgens (Tschopp, 1935)

	Capon unit (γ)	Weight of seminal vesicles after dosage of 20 mg. (mg.)	Weight of penis after dosage of 20 mg. (mg.)
Androsterone	60	17	65
Androstenedione	100	285	150

Similar observations have been made by Deanesly & Parkes (1936 *a, c*), who examined the biological actions of eleven different androgens on the accessory reproductive organs of young rats and also on the capon's comb. The effects of these various compounds on the seminal vesicle, prostate and comb respectively did not correspond, some having a greater effect on the comb than on the prostate, while others stimulated the prostate more readily than the comb. Compared with androsterone, testosterone was between $2\frac{1}{2}$ to 5 times as effective on the prostate and about 10 times as effective on the seminal vesicles. Observations of this sort suggest that when the androgenic potency of any particular compound is mentioned, the organs and animals on which it was tested should be stated.

Additional examples, illustrating differences of biological action among the large number of androgenic compounds now known, have been mentioned in connection with the so-called bisexual activities of gonadal hormones and need not be quoted again here.

Variation of responsiveness to androgen in the same species. Gallagher & Koch (1930), when testing an androgenic extract of bulls' testes on the combs of brown leghorn capons, noticed that there were pronounced individual variations in responsiveness among the birds. These variations, they say, could not be accounted for by the age or weights of the capons, or by the initial size or shape of the combs. They suggest that, when performing assays by the capon test, seasonal factors must be taken into consideration. David (1938) also has drawn attention to this matter. He remarks that the capon's comb is smaller in the winter than in the summer and that the response of the comb to an androgen will be accentuated in the winter.

The Sources of Androgen within the Body

A general consideration of the sources of androgen which are available for the individual must include the food supply. In the absence of evidence that this source plays a part in physiology or pathology, no further attention will be given to the subject in the present discussion, which will be confined to the sources of androgen within the body.

(i) *The testis.* General experience of the effects of castration have indicated the male gonads as organs for the elaboration of effective androgen, and the production by the testis of a more potent androgen from androgens of less potency has been proved by experiment in the laboratory. Danby (1938, 1940) perfused blood through bulls' testicles and observed that if testosterone were added to the perfused blood there was no increased androgenic potency of the emergent blood as compared with that with which the perfusion started. If, however, dehydroandrosterone, androstenedione or androstenediol were added to the blood its perfusion through the testis led to a greatly increased androgenic potency of the emergent blood, as shown by testing it on the capon's comb. The mere transfusion of the testis with blood led to the appearance of androgen in the effluent, showing, apparently, that the formation of androgen takes place at least to some extent in the testis.

It may be asked in what part of the testis the elaboration of potent androgen occurs. Ancel & Bouin (1903 *a*, 1904 *a, b*) appear to have been among the first to investigate this problem experimentally. They recognized the fact, well known to veterinary surgeons, that cryptorchid domestic animals may preserve their masculine characters to an awkward degree despite the absence of spermatogenesis, the tubular epithelium in their testes being small in amount and infantile in character while the interstitial glandular tissue is well developed; *a priori*, therefore, it seemed likely that masculine characters depend on the latter. They observed that in dogs, horses and pigs, retention of the testes within the abdomen led to degeneration of the seminal epithelium without immediate atrophy of the accessory genital organs or degeneration of the interstitial tissue. In guinea-pigs and rabbits Ancel & Bouin caused complete disappearance of the seminal epithelium by ligating the efferent ducts of the testicles or by causing sclerosis of the epididymis by injections, and they found that the animals so treated retained their masculinity. This experiment appeared to prove that the hormones responsible for male characters are derived from the interstitial glandular elements of the testicle. Richon & Jeandelize (1903) caused atrophy of the seminal epithelium in rabbits by ligating the vasa deferentia at the age of $1\frac{1}{2}$ months. Littermates were castrated at the same age. When examined 4 months later the external genitalia were infantile in the castrated rabbits but normally developed in those with seminal atrophy caused by ligation of the vasa deferentia. In these rabbits there was no atrophy of the interstitial glandular tissue.

Pézard (1920) found that extracts of the cryptorchid testes of swine stimulated growth of the capon's comb; and in the absence of activity in the seminal epithelium of these retained testes it is again reasonable to conclude that the effective androgen was derived from the interstitial tissue.

Aron (1921, 1925) considered the problem in a different way. In the crested newt (*Triton cristatus*) there is no interstitial tissue in the testis. At the approach of the mating season, he states, some of the seminal tubules near the hilum of the testis undergo a transformation, resembling luteinization as seen in the ovary, and form a glandular body distinct from the seminiferous part of the testis. The development of this new glandular tissue coincides with the appearance of matrimonial adornments. As soon as the mating season ends, both the newly formed gland and these adornments retrogress. Destruction by galvano-cautery of this accessory glandular tissue of the testis, without castration, causes a disappearance of the secondary sexual characters, he says, just as pronounced as that which follows castration. Courrier (1921) found that in the stickleback (*Gasterosteus aculeatus*), when spermatogenesis is at its height, there is a deficiency of interstitial cells in the testis and an absence of the special characters which accompany the mating season, namely skin coloration, the secretion of nest-building mucus by the kidney, and mating. Only after spermatogenesis is completed (in May) does the mating season begin, preceded and accompanied by the appearance of abundant, actively secreting interstitial tissue in the testis and culminating in the growth of sexual adornments and the secretion by the kidney of nest-building mucus. The strict correspondence of these manifestations with the development of the interstitial tissue, and their complete independence of spermatogenesis, seem to indicate clearly that the effective androgen is derived from the former. Courrier's observations on the stickleback have been confirmed in detail by Aron (1925).

Witschi, Levine & Hill (1932) exposed rats to X-rays so as to cause degeneration of the seminal epithelium without injuring the interstitial tissue which is
much more resistant. After this procedure the prostate and other accessory
genital glands became hypertrophied, a fact which suggests that destruction of
the seminal eipthelium had not led to a reduction of the output of androgen from
the testis. Johnston (1934) repeated these experiments on rats, using X-rays in
some instances and radium in others. His results were similar to those of Witschi,
Levine & Hill. Johnston also made the interesting observation that if rats were
rendered cryptorchid by operation so as to impair the seminal epithelium, and
were then given pituitary gonadotrophin, their prostates increased in size by
nearly 200 per cent. This result showed that the pituitary hormone had stimulated
the production of androgens by the testis; and seeing that the seminal epithelium
was degenerate, the increase, almost certainly it seems, must have been derived
from the interstitial glandular cells. Moore (1932) records that a cryptorchid
testis confined in the abdomen was found to be secreting as much androgen as
two normal testes, though its fresh weight was only 2·8 per cent of the normal
testicular mass and it contained no seminal cells.

Further evidence of the part played by the interstitial tissue in the formation
of androgens has been produced by Benoit (1938b), who removed the greater part
of each testis in a number of cocks, and in several instances he noticed that,
although sufficient time had elapsed for regeneration so that the testes had become
as large or larger than the normal, yet the comb remained atrophied. Microscopical examination of these 'regenerated' testes, he says, has shown an ample
development of tubules and seminal epithelium, even with spermatozoa, but an
almost entire absence of interstitial tissue. Some details have been recorded of
the post-mortem examination of a man of 63 who had atrophy of both testicles,
the dimensions of the right being 15 × 10 mm. and of the left 25 × 10 mm.
Microscopical examination of these testicles showed them to be devoid of tubules,
while the Leydig cells were not only abundant but were apparently in a state of
active secretion. In this case there was no diminution in size of the prostate or
other accessory genital organs; the absence of seminal epithelium in fact had not
entailed any consequences which could be attributed to an inadequate supply of
androgen.

The converse of these observations has been recorded. Moore & Samuels
(1931), in testing the effects on rats of a diet deficient in vitamin-B complex,
found that under this regimen the prostate and seminal vesicles involuted although
the production of motile and apparently normal spermatozoa continued. The
interstitial glandular cells in these animals were atrophic. If to rats in this condition an androgen was given the seminal vesicles and prostate were restored to an
active secreting state within a period of 6 to 12 days. Buchheim (1932) also found
that an active seminal epithelium failed to maintain the accessory genital organs
if the interstitial glandular cells were atrophied. He gave injections of pitch dissolved in olive oil to rabbits, guinea-pigs, rats and mice. In all these animals the
treatment caused atrophy of the interstitial cells of the testis and involution of the
accessory generative organs, although the seminal epithelium appeared unaffected and continued to produce spermatozoa.

Among the diseases of man are rare tumours of the interstitial tissue of the testis which produce excessive quantities of androgen, so that when they occur in young boys they cause premature masculinity. Stewart, Bell & Roehlke (1936) have recorded an example in a boy 5 years old. This child had pubic and axillary hair, an enlarged penis 10 cm. in length, and other signs of premature masculine development, unaccompanied however by any sexual manifestations. The affected testis was excised, after which most of the abnormal phenomena became less pronounced.

The foregoing experiments and observations seem sufficient proof that the interstitial tissue of the testicle is the chief normal source of effective androgen.

Observations on men have shown that the testes are not the only source of androgens, for these are present in the urine of eunuchs and women. Callow & Callow (1940) found that the total amount of androgen excreted by a eunuch of 20 years was about the same as that excreted by normal men; there was, however, a pronounced contrast between the nature of the androgens so excreted, the androsterone and aetiocholane-ol-one excreted by normal men being largely replaced in the eunuch's urine by *trans*dehydroandrosterone, derived perhaps from the adrenal cortex (Table 69).

TABLE 69.   Excretion of androgens by eunuch and normal men
(Callow & Callow, 1940)

	Mixed urine of normal men (mg. per litre)	Urine of eunuch (mg. per litre)
*Trans*dehydroandrosterone	0·2	2·0
Androsterone	1·6	0·5
Aetiocholane-3(α)-ol-17-one	1·4	0·9

Hoskins & Webster (1940) collected the urine from two men who had been castrated 10 years and 35 years previously and estimated the total androgen content to be the equivalent of 6 i.u. per diem by the capon-comb test. They too found a qualitative difference between the androgens excreted by these eunuchs and those excreted by normal men, though their results differ from those of Callow & Callow (Table 70).

TABLE 70.   Excretion of androgens by normal and castrated men
(Hoskins & Webster, 1940)

	Normal men (mg. per diem)	Eunuch (mg. per diem)
Androsterone	2·9	0·45
*Trans*Dehydroandrosterone	2·0	0·10

(ii) *The ovary.* Lipschütz (1932) noted that an ovarian graft in a castrated male guinea-pig occasionally maintains the seminal vesicles and prostate in a normal condition, though this does not usually happen. In other experiments Lipschütz (1933, 1935, 1937a) partially spayed twelve guinea-pigs, removing one ovary and leaving in place only a minute fragment of the other. A year later in six of these guinea-pigs the clitoris had become replaced by a structure resembling the penis of the male. Whatever the explanation may be, it appeared that the ovarian abnormality caused by the operation had led to the production of enough androgen to bring about an obvious degree of masculinization.

Hill & Gardner (1936; see also Hill, 1937; Hill & Strong, 1940) performed experiments on mice which clearly indicate that the ovary may be a source of androgen. Before discussing this work, and as an introduction to it, reference should be made to Moore's observations. Moore (1935 b) made a detailed study of cryptorchidism in rats, guinea-pigs, dogs, rabbits and sheep by pushing the testes into the abdomen, which is possible in these animals, and closing the inguinal canal. He found that within a week of such a procedure a complete disorganization of the seminal tubules may ensue. This condition is reversible and a rapid recovery follows replacement of the testis in the scrotum. Moore discovered that the testicular changes caused by cryptorchidism might be attributed, as Crew (1922) had suggested, to the higher temperature to which the testis is subjected when in the abdomen. The difference of temperature between the interiors of the scrotum and abdomen in these animals varied from $1\cdot5°$ to $7°$ C. Incidentally it may be remarked that the difference in sixteen boys was found by Harrenstein (1928) to lie between $3\cdot3°$ and $5°$ C.

Hill & Gardner grafted mouse ovaries into the tips of the ears of castrated male mice. As a result of these grafts the accessory generative organs, instead of undergoing the usual post-castration atrophy, were maintained in several instances in full functional activity. Among thirteen mice so treated, the prostate and seminal vesicles were fully maintained and the functions of the coagulation gland were preserved in six; in two the seminal vesicles and prostate were about one-half the normal size, and in the remaining five the typical consequences of castration ensued. Examination of the successful grafts showed the ovary to consist of clusters of lutein-like cells. Ovaries grafted into the abdomen of males had no androgenic effect. The temperature of the mouse's ear was found to be approximately the same as that of its scrotum, that is to say $5°$ to $6°$ C. below that of the abdomen. The results of these aural grafts were considerably affected by the temperature of the environment. If the mice were kept in a temperature of $22°$ C. evidence was always forthcoming of the production of androgen by the ovarian grafts, but castration changes occurred if the mice were kept in an atmosphere of $33°$ C. By reversing the atmospheric conditions from warm to cold, or from cold to warm, the maintenance or atrophy of the accessory reproductive organs could be regulated, though no change could be detected in the grafted ovaries to correspond with their varying function.

In a later paper Hill (1941) says that an ovary successfully implanted into a mouse's ear continues to produce androgen for about 12 months, and in castrated males enough androgen is formed to maintain the prostate and seminal vesicles in a normal state for 6 months. He could not correlate this androgenic potency with any histological changes in the ovary.

Deanesly (1938 b) repeated some of Hill's experiments, using rats instead of mice. She found that most of the successfully grafted ovaries showed androgenic activity by causing growth and secretory activity in the prostate. But the maintenance effects on the seminal vesicles were either much less pronounced or absent, from which she concluded that the androgen produced by the grafted ovary was probably not the same as that produced by the normal testicle, that is to say testosterone, for this affects the prostate and seminal vesicles of the rat in about equal degree, which is not the case with all androgens.

Pfeiffer (1937) found that if rats were placed in warm surroundings so as to maintain their body temperature at 39° C. for 15½ hours out of the 24, changes similar to those caused by cryptorchidism took place in their testes and there was some reduction of androgen formation.

Further evidence of the production of androgen by the ovary is provided by observations on the fowl's comb, for this structure will respond by an increase in size not only to testicular extracts but to ovarian extracts also (Champy & Kritch, 1925). Moreover, the hen's comb like that of the cock atrophies after gonadectomy and becomes enlarged and turgid in response to androgens but not to oestrogens (Parkes, 1937; Domm, 1937; Dorfman & Greulich, 1937)—from which it seems clear that the hen's ovary produces enough androgen to protect her comb from atrophy. It has been found also that an extract of sow's ovary will stimulate growth of the capon's comb (Parkes, 1937); and it is a matter of common observance that the hen's comb is enlarged during the egg-laying phase, when the ovary is in an active functional condition, and that it shrinks and becomes flaccid when egg-laying ceases. Another example of the production of androgen by the ovary is afforded by some experiments by Domm (1937), who gave daily injections of a gonadotrophic hormone prepared from sheep's pituitary to recently hatched chicks. This treatment caused premature and excessive growth with turgidity of the comb, and to a less degree of the wattles and ear-lobes, associated with great enlargement of the medulla of the ovary. If the ovary were removed, no comb growth was induced. Apparently no functional activity was caused in the right vestigial gonad of the chicks by the administered gonadotrophic hormone. Noble & Wurm (1940) have seen additional evidence of androgen production by the ovary during their studies of the black-crowned night heron (*Nycticorax nycticorax hoactti*). Both sexes of this heron, they say, undergo similar pigmentary changes at the onset of the breeding season. Gonadectomy prevents these changes. Testosterone propionate causes the breeding plumage to appear in immature herons and in gonadectomized adults of either sex. Oestrogens do not have this effect. The conclusion is that the ovary in these birds produces androgen at the onset of the breeding season. Another example of the same sort is mentioned by Witschi & Fugo (1940), who say that during the breeding season in starlings of either sex the bill becomes yellow, and that this feature can be produced in female starlings at any time of the year by androgen, but not by oestrogen. A comparable observation has been made by Kirschbaum & Pfeiffer (1941) on the sparrow. The male sparrow's bill, they say, becomes black in the breeding season and that of the female sparrow darkens at the same time. These effects can be artificially produced in either sex by applying small doses of testosterone propionate to the base of the bill during periods of sexual inactivity. Apparently as the mating season approaches the female produces enough androgen to cause darkening of her bill.

Human pathology has shown that the ovary in some circumstances may produce relatively large amounts of androgen, for in certain cases of ovarian tumour (arrhenoma) pronounced virilism occurs and there can be little doubt that it is the consequence of androgen formed in the affected gonad.

(iii) *The adrenals.* Lespinasse (1924) seems to have been the first to demonstrate by experiment the androgenic capacity of adrenals. He implanted adrenal tissue into the pectoral muscles of a young cockerel and thereby induced precocious male development and behaviour.

For many years clinicians have recognized a close connection between the adrenals and the sexual organs in man. Deficiency of the adrenal cortex is accompanied in men by sexual impotence accompanied by atrophy of the testes, prostate and other accessory glands, and in women by atrophy of the ovary with degeneration of the ova and an absence of corpora lutea (Grollman, 1936). On the other hand hypertrophy of the adrenal cortex, and some adrenal tumours also, when occurring in females or in boys before the age of puberty, often cause virilism, and in such cases an excess of androgen has been detected in the urine (Simpson, de Fremery & Macbeth, 1936, and others), and compounds of an androgenic nature have been isolated in a pure form from mammalian adrenals and from the urine of female patients who were the subjects of adrenal tumours.

From adrenal material Reichstein (1936) obtained and identified adrenosterone, which has about one-fifth the capacity of androsterone for causing growth of the comb in capons, and Pfiffner & North (1940) extracted from the adrenals of cattle an isomer of deoxycorticosterone, namely 17-β-hydroxyprogesterone, which has an androgenic potency about equal with that of androsterone; it had no progestational effect in the rabbit when given in a dose of 5 mg. Callow (1936c; see also Crooke & Callow, 1939) isolated *trans*dehydroandrosterone from the urine of a 6-year-old girl who had an adrenal tumour. Burrows, Cook, Roe & Warren (1937) identified $\Delta^{3:5}$-androstadiene-17-one in the urine of a man with adrenal carcinoma, and Butler & Marrian (1938) isolated *iso*androsterone from the urine of patients with symptoms of virilism. From the urine of a woman with an adrenal tumour Wolfe, Fieser & Friedgood (1941) isolated five different 17-ketosteroids, namely androsterone, 3α-hydroxyaetiocholane-17-one, *trans*dehydroandrosterone, $\Delta^{3:5}$-androstadiene-17-one, and 3α-hydroxyandrostene-17-one.

Riddle (1937) observed sixteen pigeons in which no gonads were discovered, and yet these birds showed normal male behaviour and therefore, it seems, androgen must have been formed somewhere in the body: he suggests the adrenals.

Callow (1938) has shown that oophorectomy in women, although it may reduce the urinary excretion of androgen, does not entirely prevent it, and Hamblen, Ross, Cuyler, Baptist & Ashley (1939), using a colorimetric method of estimation, discovered relatively high titres of androgen in the urine after oophorectomy as well as after the menopause. Hirschmann (1939) isolated two androgens, namely androsterone and *trans*dehydroandrosterone, from the bulked urine of nine women who had undergone double oophorectomy and were still below the age of the menopause.

Price (1936) observed that if a rat is castrated when 1 week old the development of its prostate and seminal vesicles continues until the rat is between 30 and 40 days old, when signs of atrophy begin to appear, and she suggests that in the intervening period a considerable supply of androgen must have been forth-coming; it is reasonable to suspect that the adrenals are the source.

Various experiments on animals have shown by other means that the adrenals have a sexual function. Moszkowska (1935) worked with castrated guinea-pigs, and regarded the cornified denticles which are part of the normal male guinea-pig's penis as indicators of the presence or absence of androgenic activity, for these little structures disappear and the penis as a whole undergoes atrophy in the absence of stimulation by androgen. Moszkowska noticed that if a castrated guinea-pig were treated daily with injections of an alkaline extract of bovine pituitary the atrophy of the penis and its denticles was distinctly delayed. If the injections were given immediately after castration the penis and its adnexa continued for 18 days or so to increase in size and the prostate and seminal vesicles underwent a much reduced involution as compared with those of controls. It seemed clear that androgen was being produced which could to some extent sustain the accessory reproductive organs of the male. In these guinea-pigs, after castration, the adrenals and thyroids were much enlarged.

Davidson & Moon (1936) gave adrenocorticotrophic extracts from sheep's pituitaries to normal and castrated immature male rats. The preparations were free from growth, thyrotrophic and gonadotrophic hormones, though some lactogenic hormone was present. Seven rats were castrated 21 days after birth. Injections of adrenocorticotrophin were begun on the next day and continued daily for 9 days. The rats were then killed and their adrenals and accessory reproductive glands were weighed. The results show, not only an increase in the size of the adrenals, but a considerable increase also in the accessory genital organs as compared with those of untreated castrated rats (Table 71).

TABLE 71. The effect of adrenocorticotrophin on the accessory genital organs of castrated rats (Davidson & Moon, 1936)

Total dose of adrenocorticotrophin (mg.)	Average weight of adrenals (mg.)	Average weight of accessory genital glands (mg.)
297	54·9	130·8
414	69·4	100·9
603	105·5	113·5
0 (controls)	22·4	54·8

Davidson (1937) states that adrenocorticotrophin has no effect on the accessory reproductive organs of rats whose adrenals have been removed previously. The tabulated results suggest that the adrenals under the influence of adreno-corticotrophin produce androgen.

Burrill & Greene (1939, 1940b) noticed, as Price had done, that castration of immature rats does not lead to rapid atrophy of the prostate as it does in the adult; if, however, the adrenals also are removed atrophy of the prostate follows at once. They castrated rats when they were 16 days old and from some of these they removed the adrenals 5 days later, and they adrenalectomized also some non-castrated rats when 21 days old. All were killed when 26 days old and their ventral prostatic glands examined. Results like those shown in Table 72 were obtained only with immature rats; castration later in life always caused atrophy of the prostate.

TABLE 72. The effects of castration, adrenalectomy and of both combined
on the prostates of immature rats (Burrill & Greene, 1940b)

Treatment	Condition of prostate
Castration	Partially maintained
Adrenalectomy	Normal
Adrenalectomy and castration	Atrophic

(iv) *The placenta.* An increased amount of androgen has been noticed in
the urine during gestation. In pregnant mice the adrenal *x*-zone disappears—a
change which is caused by androgen but not by oestrogen or progesterone. It
is suspected, though not proved, that this and other androgenic effects occurring
in pregnancy may be due to androgen formed in the placenta.

(v) *Extraneous sources.* Some androgens are highly effective when taken by
the mouth, and, as with oestrogens, we are not in a position to ignore entirely the
possibility of androgenic reactions being produced by food or drink. A curious
example of androgenic action being naturally induced by a hormone formed outside the body of the reacting individual is that on which the bitterling test for
androgen is founded. Kleiner, Weisman & Mishkind (1936) have shown that
in this little fish the ovipositor grows at the egg-laying season in response to
androgen discharged into the water by the male.

Continual production of androgen and lack of storage. The rapidity with which
the effects of castration in the sexually active adult become manifest show that if
androgen is stored at all in the body in normal circumstances the amount so held
is small; apparently, during active sexual life, the supply is incessant or nearly so.
Such a conclusion has been tested and confirmed by Moore (1932), using his
spermatozoon motility test as a guide.

One notable example of the storage of androgen in special circumstances has
been recorded by Sweet & Hoskins (1941), who found that the pigmented deposit
of fat (the so-called hibernating gland) of the woodchuck during its winter sleep
contains a relatively large concentration of androgen: as large in fact as that found
in the bull's testis, which is otherwise about the richest source of androgen
known.

Inactivation and Excretion of Androgen

Experiments have shown that androgens, like oestrogens, are rapidly inactivated
in the body and excreted in the urine as compounds with reduced androgenic
potency. After subcutaneous injections of testosterone, or its acetate, propionate
and benzoate, into castrated dogs, Kochakian (1939) was able to recover from the
urine only 0·3 per cent of the androgenic potency of the compounds injected, as
tested on the capon's comb.

The fact that androgens recognized in the urine are not necessarily those which
are active in the body was first suggested by Dingemanse, Freud & Laqueur (1935),
who found that androgen extracted from male urine had considerably less action
on the seminal vesicles of castrated rats than was shown by equal doses, estimated
in capon units, of androgen extracted from testes. It is now recognized that the
effective androgen formed by the testis, namely testosterone, is not excreted in
the urine as such; before excretion it is converted into other less active compounds, including androsterone (Butenandt, 1931), aetiocholane-3(α)-ol-17-one

(Butenandt & Dannenbaum, 1934; Callow & Callow, 1938, 1939), and perhaps *trans*dehydroandrosterone and some form of oestrogen.

Sites of inactivation of androgens. Neither the inactivation of testosterone, nor its conversion into less active androgen, depends on the gonads. Dorfman (1940) gave 60 mg. of testosterone daily by the mouth for 5 days to a man of 40 who had been castrated 20 years earlier. The urine during this period was collected and about 15 mg. of androsterone was obtained from it, testosterone being absent as in the urine of normal men.

Danby (1940) found that if a dog's liver or cow's kidney is perfused with blood containing androgen the latter is inactivated. The capacity of the liver to inactivate androgen has been confirmed by Biskind (1940), who used castrated rats, displacing the spleens of some of them into the subcutaneous region without dividing the splenic vessels. He then implanted pellets of methyltestosterone subcutaneously into the intact rats and into the displaced spleens of the others. When this had been done the splenic vessels in some of the rats were ligated. As a result the rats carrying subcutaneous pellets, or pellets implanted in the spleens whose vessels had been tied, showed the usual androgenic effects in their accessory generative organs. In contrast with these, the rats carrying pellets of methyltestosterone in otherwise intact spleens showed no response to the androgen. Biskind's interpretation of this result is that the androgen in the latter group was conveyed directly to the liver and inactivated there, whereas in the former two groups the androgen passed directly into the general circulation. He was able to draw the same conclusion from a similar experiment done on female rats (Biskind, 1941 a). Pellets of testosterone propionate were implanted in their spleens, and in some instances the splenic vessels were tied, in others they were left intact. If the vessels were tied the testosterone inhibited oestrus, if the splenic vessels were intact the oestral cycles continued, presumably because in this case the dissolved testosterone passed directly to the liver, where it was inactivated.

Burrill & Greene (1940a) have performed a comparable experiment. They castrated rats 20 to 23 days old, and in one group sutured half of a testis to the gastrosplenic mesentery and in the other group implanted half of a testis into the subcutaneous tissues. The animals were killed 1 or 2 months later, at which time the prostates and seminal vesicles of the rats with subcutaneous grafts were between 5 and 6 times as large as those in the rats bearing mesenteric grafts. The result seems to be explained, as in Biskind's experiment, by the direct conveyance of the absorbed androgen from the mesenteric grafts to the liver and its inactivation there.

The same workers (Burrill & Greene, 1942b) have later made more direct tests. They inserted pellets of testosterone into the spleens of castrated rats and noted that the accessory genital structures were very much less affected thereby than by similar doses lodged in the subcutaneous tissues. It was further noted that pellets of methyltestosterone lodged in the spleen caused some enlargement of the genital accessories, though not so much as that produced by subcutaneous doses, from which it seems that testosterone is inactivated more readily than methyltestosterone by the liver.

As already stated the chief metabolic products of testosterone, apart from oestrogen and inert compounds, appear to be androsterone and aetiocholane-3(α)-ol-17-one. Callow (1939) gave 100 mg. of testosterone propionate daily to a man. While he was under this treatment the androgenic activity of his urine was 4 times that determined during a later period when he was not under treatment. From this it can be concluded that the excess of androgen in the urine was derived from the testosterone which had been given. Analysis showed that the chief androgens excreted in the urine were androsterone and aetiocholane-3(α)-ol-17-one and they were present in about equal quantities (Table 73).

TABLE 73. Excretion of androgens by a man under daily doses of
100 mg. of testosterone propionate (Callow, 1939)

	Androsterone (mg. per litre)	Aetiocholane-3(α)-ol-17-one (mg. per litre)
Normal men	1·2	1·4
Men under treatment with testosterone	8	7·7

Dorfman (1940, 1941) also has shown that the administration of testosterone propionate is followed in men by an increased excretion of androgen in the urine. In such cases he has isolated from the urine androsterone and two of its isomers, namely aetiocholanol-3(α)-17-one and aetio*allo*cholanol-3(β)-17-one.

The conversion of androgen into oestrogen in the living body. Another metabolic product of androgens is oestrogen, perhaps in the form of oestrone.

Steinach, Kun & Peczenik (1936b) found that the administration of androsterone to rats was followed by an increased excretion of oestrogen in the urine. Further, they proved that this transformation did not take place in the testes, for it occurred in castrated as well as in normal rats. Steinach & Kun (1937) reported similar results in man following the administration of either testosterone propionate or androsterone. They estimated the oestrogen content of the urine of six men of varying ages. To five of the men testosterone propionate in doses of 50 mg. was then given 3 times a week until each had received twenty doses. The sixth man was given 20 mg. of androsterone 4 times a week. The oestrogen content of the urine was estimated at intervals and was found to rise steadily during the treatment. Before receiving androgen the oestrogen content lay between 0 and 36 r.u. per litre of urine. With the continued injection of androgen the urinary output of oestrogen steadily rose and with cessation of the injections declined. The largest amounts registered were 1,200 r.u. per litre after testosterone propionate, and 210 r.u. after androsterone. The exact nature of the oestrogen excreted was not determined. An increased output of oestrogen in men following the administration of androgen has been observed also by Callow (1938), Foss (1939) and others.

Callow, Callow & Emmens (1939) examined the urine of two men with enlarged prostate, one man with acromegaly and three eunuchs, all of whom were treated with testosterone propionate, and they found that during this treatment there was a largely increased excretion in each case of both androgen and oestrogen (Table 74).

TABLE 74.   Increased excretion of androgen and oestrogen in men under treatment with testosterone propionate (Callow, Callow & Emmens, 1939)

Condition	17-Ketosteroids by colour test (mg. per diem)	Androgen by capon test (i.u. per diem)	Oestrogen (i.u. per diem)
Enlarged prostate			
Before treatment	—	16	<4
After ,,	—	50	34 approx.
Acromegaly			
Before treatment	9·6	12	27
After ,,	39·9	126	621
Castrated			
Before treatment	7·4	16·3	7
After ,,	12·4	50·2	36·2

In women a decrease of the urinary excretion of androgen has followed the administration of oestrogen. Hamblen, Pattee & Cuyler (1940) estimated by Oesting's colorimetric method the androgenic titres of the urine of ten women who had passed the menopause; the women were then treated with oestrogen, after which the amount of androgen in the urine was found to have become less. The same sequel to treatment with oestrogen was seen in seven women who were the subjects of 'hypoovarianism' (Table 75).

TABLE 75.   Decreased excretion of androgen in women after the administration of oestrogen (Hamblen, Pattee & Cuyler, 1940)

Condition of subject	Number	Excretion of androgen per diem		Percentage reduction of androgen excretion
		Before treatment (i.u.)	After treatment (i.u.)	
Past the menopause	10	84	60	28
'Hypoovarian'	7	70	43	40

An observation by Parkes (1935) may be recalled in connection with the foregoing remarks. He noticed that androstanediol provokes vaginal cornification in normal but not in spayed rats, a fact which suggests that in this instance the ovary may play some part in the conversion of an androgen into an oestrogen.

Excretion of androgen otherwise than by the kidney.   Apparently no great attention has been given to the share, if any, which the alimentary canal takes in the elimination of androgens.

During lactation androgen may be excreted in the milk, and if given to the lactating mother may pass to the infant in sufficient amount to exercise recognizable biological effects (Raynaud, 1938).

Excretion of Androgen by Boys and Men

(i) *Before puberty*. Androgens affect embryological development and are needed therefore at an early stage of intrauterine life. This matter will be discussed later. To exemplify the production of androgen by the gonads before birth we can quote an experiment by Koch (1931), who extracted more androgen per gram from the testes of foetal bulls than from those of calves or adult bulls. This does not mean that the elaboration of androgen is independent of age. The estimation of androgen excreted in the urine, in spite of drawbacks inherent in this method of investigation, has yielded much information on the production of androgen in the body and has given also a good basis for both academic conclusions and human therapy. In man the output of androgen is small in the early years of life, rises at puberty, reaches a maximum during active sexual life, and declines in old age.

Bourg (1930b) observed that ten daily injections of man's urine into immature rats caused hypertrophy of the accessory sexual structures whereas no such effect was produced by the urine of children. Using the capon-comb test as a criterion, Womack & Koch (1932) were unable to detect any androgen in the urine of boys under 10 years old, whereas the urine of boys of 14 years and upward apparently contained as much as that of adult men.

By improved methods of assay Dorfman, Greulich & Solomon (1937) estimated the output of androgen in eighteen boys of various ages and found a notable rise in the amount at the onset of puberty (Table 76).

TABLE 76.   The urinary excretion of androgen and oestrogen in boys of various ages (Dorfman, Greulich & Solomon, 1937)

Age in years	Average output of androgen (i.u. per diem)	Average output of oestrogen (i.u. per diem)
7–8	1·1	Less than 9
9–10	2·0	Less than 5
11–12	2·0	12·5
13–14	6·3	17·5
14–15	11·7	38·0
15–16	16·6	43·2
16–17	16·2	68

From this table it will be seen that the increase of androgen in the urine at puberty is accompanied by an increase in the amount of oestrogen. This may be accounted for, wholly or in part, by the capacity of males to convert androgen into oestrogen, as mentioned earlier. Nathanson, Towne & Aub (1939) also, using a colorimetric method of assay and making several tests with each individual, have shown that an increased excretion of androgen accompanies the onset of puberty in boys (Table 77).

TABLE 77.   The urinary excretion of androgen in boys of various ages (Nathanson, Towne & Aub, 1939)

Age in years	Average output of androgen (i.u. per diem)
$6\frac{3}{12}$	6·5
$8\frac{1}{12}$	8·6
$10\frac{1}{12}$	17·7
$12\frac{9}{12}$	22·6

(ii) *After puberty.* The increased excretion of androgen by men during sexual maturity, with a subsequent decline in later years, has been shown by Kochakian (1936), Dingemanse, Borchardt & Laqueur (1937), Hansen, McCahey & Soloway (1936) (Tables 78, 79).

From table 79 it will be seen that there are wide variations in the different assays, and it has been noticed by several workers that even in the same individual the urinary content of androgen may differ much from time to time

(Koch, 1937; Callow, 1938; Hamblen, Ross *et al.*, 1939). McCullagh (1939), who has called special attention to this variability, found the average output of androgen per diem in twenty normal men was 40·5 or 35·1 i.u. as estimated by two different methods.

TABLE 78. Excretion of androgen in young adult and in elderly men
(Kochakian, 1936)

Number of men	Age in years	Average excretion of androgen (c.u. per diem)
12	21–29	12–16
5	50–76	2–3

TABLE 79. Excretion of androgen by males at various ages
(Dingemanse, Borchardt & Laqueur, 1937)

Age in years	Average output of androgen (i.u. per diem)	Extremes
5–15	20·5	8–50
20–34	99·3	15–170
59–67	20·3	5–40

It will be noted that urinary assays of androgen differ when made by separate observers. No doubt such discrepancies will vanish when methods of assay have become standardized and more care is taken in collecting the urine.

Excretion of Androgen by Young Girls and Women

As already stated the formation and excretion of androgens is not confined to the male sex. These hormones may be excreted by normal adult women in approximately the same amounts as by men (Womach & Koch, 1932; Callow, 1936b; Koch, 1938). Children excrete smaller quantities.

Dorfman, Greulich & Solomon (1937) assayed the androgen content of the urine of young girls, and their results show that, as with boys, a rise occurs on the approach of puberty (Table 80).

TABLE 80. The urinary output of androgen in young girls
(Dorfman, Greulich & Solomon, 1937)

Average age in years	Average output of androgen (i.u. per diem)	Average output of oestrogen (i.u. per diem)
7–8	2·0	—
9–10	0·7	10·0
10–11	2·2	5·6
11–12	3·0	<5·0
12–13	7·0	20·0

Nathanson, Towne & Aub (1939), though their estimates exceed those of Dorfman and his colleagues, show a similar rise in the output as puberty approaches. They used a colorimetric method of assay (Table 81).

Dingemanse, Borchardt & Laqueur (1937) found a considerable increase in the output of androgen in the urine of girls after puberty (Table 82).

TABLE 81.   The urinary output of androgen in young girls
(Nathanson, Towne & Aub, 1939)

Age in years	Average output of androgen (i.u. per diem)
$7\frac{7}{12}$	5·5
$8\frac{8}{12}$	7·7
$10\frac{4}{12}$	7·2
$11\frac{6}{12}$	14·9
$11\frac{8}{12}$	14·8
$12\frac{5}{12}$	14·7

TABLE 82.   The urinary excretion of androgen in young girls and
women (Dingemanse, Borchardt & Laqueur, 1937)

Age in years	Average output of androgen		Extremes (i.u. per litre)
	(i.u. per litre)	(i.u. per diem)	
5	11	8·3	10–12
20–26	42	40·3	15–65

From the quantities mentioned in previous pages it will be seen that there is no very great difference in the amount of androgen excreted by men and women. The relative output of 17-ketosteroids in adults of the two sexes have been determined by Callow, Callow, Emmens & Stroud (1939), who say that the difference is not significant (Table 83).

TABLE 83.   Excretion of androgens by adult men and women
(Callow, Callow, Emmens & Stroud 1939)

	Mean value of 17-ketosteroids excreted in the urine per diem (mg.)
Men	9·05
Women	6·75

Some variation in the urinary excretion of androgen has been noticed during the menstrual cycle in women. Hamblen, Cuyler & Baptist (1940), using Oesting's colorimetric method, estimated the androgen output in nine women during the menstrual cycle and found that there was a significant decrease during the bleeding stage in seven of the women. The average titre in the entire group was 72 i.u. per diem during the non-bleeding phase and 50 i.u. per diem in the bleeding phase.

After the menopause the output of androgen falls. Dingemanse, Borchardt & Laqueur (1937) found that the androgen output in women between 20 and 26 years of age amounted to 42 i.u. per litre of urine with extremes of 15–65, whereas in old women the amounts ranged from 7 to 11 i.u. per litre.

There appears to be little if any difference in kind between the androgens excreted by normal men and women. Callow & Callow (1939) assayed the bulked urine collected from a number of healthy women, and isolated therefrom three different androgens in amounts which are shown in Table 84 together with assays made by them (Callow & Callow, 1940) on the urine of a eunuch and on the urine of normal men.

TABLE 84.   Excretion of androgens by a eunuch and by normal
          men and women (Callow & Callow, 1940)

Source of urine	Androsterone (mg. per litre)	Aetiocholane-3(α)-ol-17-one (mg. per litre)	Transdehydro-androsterone (mg. per litre)
Eunuch	0·5	0·9	2·0
Normal men	1·6	1·4	0·2
Normal women	1·3	1·3	0·2

Callow & Callow state that both androsterone and aetiocholane-3(α)-ol-17-one
are derived from testosterone when this is given by injection, though it is not
necessary to infer from this that testosterone is their natural precursor in women.
The ovary certainly is not the only source of androgen in women, for removal of
both ovaries does not abolish the output of androgen in the urine (Parkes, 1937).
Hirschmann (1939) isolated androsterone and aetiocholane-3(α)-ol-17-one from
the bulked urine of nine spayed women, all under the menopausal age, in amounts
only a little less than those obtained from the urine of normal women. *Trans-*
dehydroandrosterone was also recovered from this urine.

Hamblen, Ross, Cuyler, Baptist & Ashley (1939), using a modification of
Oesting's colorimetric method, have noticed some apparent correlation between
the output of androgen and the complexion of the skin in women, the amounts
being larger in brunettes than in blondes (Table 85). The administration of
oestrogen, they found, caused a decrease of the androgen excretion; androgens
caused an increase. In five women who had undergone oophorectomy the output
of androgen in the urine was not reduced. From this they conclude that most of
the androgen excreted by women is not derived from the ovary.

TABLE 85.   Excretion of androgen by blondes and brunettes
          (Hamblen, Ross, Cuyler, Baptist & Ashley, 1939)

Complexion	International units of androgen excreted in 24 hours
(1) Blonde	14
(2) Medium blonde	15
(3) Medium brunette	21
(4) Medium brunette	22
(5) Marked brunette	34

Factors other than Age and Sex in the Production,
Inactivation and Excretion of androgen

The normal production, inactivation or excretion of androgen may be disturbed
by several circumstances.

(i) *Pituitary influence.*   The pituitary controls the production of androgen
because it is the source of trophins on which the formation of androgen depends.
At present we have little knowledge of any part played by the pituitary in the
excessive production of androgen under natural conditions by the adrenal, ovary,
testis or any other organ. We are better informed about circumstances in which
the output of trophins and consequently of androgens may be reduced. This
matter has been dealt with already (p. 32) and no more will be said about it here
except to remind the reader that the output of gonadotrophins by the pituitary is
affected not only by the gonadal hormones but by various other agencies, in-

cluding diet, general health, poisons, environment and psychological and physical stimuli.

(ii) *Absence or disease of the androgen-producing organs.* An excessive production and excretion of androgen is a recognized result of certain tumours of the testicle, ovary and adrenal. In some instances, well appreciated in the case of the adrenal, a simple hyperplasia of the androgen-producing tissue exists. A possible connection between these lesions and the pituitary cannot be dismissed from the mind, but until the connection, if any, can be traced we have to regard these disorders as though they had arisen from local causes alone. A simple hyperplasia and overaction of the interstitial tissue of the testis, or of the androgenic medullary elements of the ovary, are less firmly established as pathological entities. If in fact they occur the matter may be one of sociological interest, for there can be little doubt that androgens influence behaviour, and their excessive production might cause erratic conduct.

A diminished output of effective androgen may be associated with testicular or adrenal defects.

The excretion of androgen by cryptorchids, eunuchoids and eunuchs. Castration, incomplete descent of the testes and underdevelopment of the testes in eunuchoid conditions are all accompanied by quantitative changes in the formation of gonadal hormones and their excretion in the urine.

The effects of cryptorchidism on the output of androgen by the testicles has been studied in rats by Nelson (1937a). He castrated rats or made them cryptorchid when between 39 and 44 days old, and recorded the results 300 or 510 days later. The reduced weights of the seminal vesicles and prostate suggest that a long continued, slow decline of the output of effective androgen followed replacement of the testes in the abdomen (Table 86).

TABLE 86. The effects of castration and cryptorchidism on the bodyweight and accessory genital organs of the rat (Nelson, 1937a)

Rats	Experimental period (days)	Body-weight (g.)	Weight of seminal vesicles (g.)	Weight of ventral prostate (g.)
Normal	300	406	1·375	0·462
Cryptorchid	300	357	0·125	0·394
Castrated	300	328	0·019	0·033
Normal	510	427	1·284	0·484
Cryptorchid	510	343	0·057	0·109
Castrated	510	323	0·021	0·035

Hanes & Hooker (1937), by a different method, also found a reduced formation of androgen by cryptorchid testes. They assayed testes from adult boars for their content of androgen by the capon-comb test. Two separate batches of material gave these results:

Testes	Weight of testicular material containing 1 c.u. of androgen	
	Experiment I (g.)	Experiment II (g.)
Normal	38·7	27·0
Cryptorchid	86·7	53·5

Kenyon, Gallagher, Peterson, Dorfman & Koch (1937) assayed the urine for androgen and oestrogen in a number of boys and men and found that a pronounced decrease in the excretion of both androgen and oestrogen accompanied an absence or morbid condition of the testicles (Table 87). It will be noted that cryptorchidism is eventually accompanied by a reduced output of effective androgen. The fact that masculine traits and the condition of the male accessory organs, both in animals and man, may be maintained for a considerable time though the testes are absent from the scrotum is well recognized. Nevertheless many different observations have proved that a reduced formation of effective androgen is a feature of prolonged cryptorchidism.

TABLE 87. The effects of testicular activity on the output of gonadal hormones in the urine (Kenyon, Gallagher, Peterson, Dorfman & Koch, 1937)

Number	Condition	Ages	Average excretion of androgen (c.u. per diem)	Average excretion of oestrogen (γ per diem)
4	Normal men	26–35	40	10
3	Cryptorchid men	13–26	24·5	2·7
8	Eunuchoid men	20–36	20	1·8
3	Men with gynaecomastia	15–24	15	10·7
2	Castrated men	21–56	2·25 (per litre)	3·25 (per litre)
4	Normal women	23–34	28	27

McCullagh & Lilga (1940) have published assays of urine from men with testicular or adrenal defects, and their results, shown in Table 88, are comparable with those of Kenyon and his colleagues.

TABLE 88. Excretion of androgen in the urine of men in various abnormal conditions (McCullagh & Lilga, 1940)

Condition	Number of men	Average amount of androgen per litre (i.u.)	Extremes
Normal men (23–37 years)	20	37·8	18–86
Eunuchoids	12	9·5	0–31
Castrated	5	6·4	0–18
Addison's disease	3	25·8	10–45

Callow, Callow & Emmens (1940) compared the excretion of 17-ketosteroid and oestrogen in eunuchs and normal men and found that although the output of 17-ketosteroids by the eunuchs was somewhat low it was not outside the range of variations which occur in normal men (Table 89).

Note. When considering the excretion of androgen, as shown in this and other tables, it is to be remembered that the figures do not always refer to identical subjects. (1) In some experiments androgen excretion has been estimated by biological tests, usually on the capon's comb; that is to say, the figures set out represent the *effective* androgen content of the urine as tested on comb growth. In this case the results would almost certainly be different if tested on prostate or other accessory generative organs, and in any event they indicate not the *quantity* of androgen present but its *effectiveness*; and the different androgens vary greatly in their efficacy. (2) In other tables the output of androgen is represented by the total 17-ketosteroids present in the urine and the figures given do not signify the androgenic potency of the urine, nor do they represent the total androgen present. However, experience has shown that either method will give useful information as to the androgen production by individuals. One might add that a statement of the total amount of androgen excreted in 24 hours is more informative than a report of the amount contained in one litre of urine.

TABLE 89.   Excretion of gonadal hormones by castrated men
(Callow, Callow & Emmens, 1940)

| | Average amounts excreted in urine | |
Subjects	17-Ketosteroids (mg. per diem)	Oestrogen (i.u. per litre)
Eunuchs	7·77	5·8
Normal men	9·05	10–30

The qualitative changes in the excretion of androgen after castration are more striking than the quantitative changes. Callow & Callow (1940) collected and assayed a large amount of urine from a eunuch. They found that, as compared with the output of normal men, the eunuch's urine contained less androsterone and aetiocholane-3(α)-ol-17-one, and more *trans*dehydroandrosterone, which is a less potent androgen (Table 84, p. 172). It will be seen that the total weight of androgenous substance excreted has not been much, if at all, diminished by castration, but that the androgens excreted by the eunuch are less potent, regarded as a whole, than those present in the urine of normal men.

(iii) *Impaired activity of the liver.*   Yet another cause for an excess of available androgen in the living body may perhaps be a failure of the individual to inactivate or excrete it. It has been shown that testicular androgens are inactivated by the liver (p. 166). Talbot (1939) found that by causing cirrhosis of the liver, the inactivation of oestrogen is to some extent prevented. The supposition that a defective action of the liver perhaps may allow an excessive concentration of androgen in the blood seems not unreasonable in the light of these facts.

Chapter VII. *The Action of Androgen on the Reproductive Organs before their Complete Differentiation*

Ovary, testis. Freemartins. Accessory generative organs.

Ovary, Testis. Freemartins

MANY years before any androgenic compound was identified and made available in pure form for experiment much had been learned about the biological properties of these hormones by noting the results of castration, transplantation of testes or the injection of testicular extracts. A study of the conditions in which freemartins were produced supplied additional knowledge, and inquiries into the changes in the gonads and accessory genital organs of animals which have a limited breeding season also gave valuable information. Later work with the pure hormones has confirmed and extended the knowledge thus acquired.

Permanent and transient effects. Some of the effects of androgen on the reproductive organs are permanent, others are reversible and endure only while androgen is being supplied in adequate quantity. The permanent effects are produced when growth and sexual development are still incomplete. This does not mean that lasting changes can be brought about only during embryological life, for many organs do not acquire their permanent form and character until long after birth. The gonads become fully differentiated early in embryological existence and afterwards remain relatively immune from permanent injury by androgen. The accessory generative organs complete their development much later and meanwhile their growth and form may be permanently interfered with by androgen. For example, the larynx and the bony pelvis do not acquire their ultimate masculine or feminine conformations until puberty and after; earlier in life they may become permanently deformed under androgenic influence. A few structures, of which the clitoris is an example, appear to remain, even in the adult, susceptible to some degree of permanent metamorphosis when submitted to an excess of androgen, and this seems true also of the facial dermis; for a beard once grown is slow to disappear after the chief sources of androgen have been removed.

Different ways in which androgens act. Androgens affect the reproductive organs in more than one way: (1) they act on the pituitary so as to check the output of gonadotrophin or to alter the relative amounts of FRH and LH which it produces, and by this means they influence the formation of androgen within the organism; (2) they stimulate the growth and functional activity of structures associated specifically with the male sex; (3) they diminish, modify or enhance some of the activities of oestrogen and progestin; and (4) to some extent they can fulfil the functions of progesterone.

THE ACTION OF ANDROGEN ON THE GONADS

Before discussing the effects produced by androgen on the testis and ovary some details of the growth and differentiation of these organs in natural circumstances may be recalled. Experiments on the larvae of axolotls (Humphrey, 1928) have shown that if, before gonads have appeared, a strip of mesoderm containing gonadal primordia is removed and grafted into the body cavity of an embryo host, the grafted tissue may yet produce a gonad. The experiment seems to prove that, in the axolotl, the determination towards gonadal formation lies in the primordial tissues themselves.

The primitive gonad is a bisexual organ, containing both male and female elements. In the testis the androgen-producing tissue becomes active at a date which precedes by a considerable interval the time when the female gonad becomes physiologically active.

Ancel & Bouin (1903 b) found that interstitial glandular cells were already well developed in the testes of embryo pigs of 30 mm., before the sexual cords have acquired lumina. In fact, they say, the testis at this period is made up chiefly of interstitial cells which contain numerous granules of secretion (see also Bouin & Ancel, 1903 a, b). Allen (1904) and Whitehead (1904) observed a rapid growth of interstitial cells in the testes of embryo pigs of 25 mm. These cells, which are abundantly distributed throughout the testis, are only occasionally if ever seen in the ovary at this early stage. After a length of 35 mm. has been reached by the pig embryo, the interstitial tissue of the testis, according to Allen, undergoes involution, the cells being reduced almost to the state of naked nuclei. They come into activity again when the embryos have reached 20 cm. in length and remain so until full term, at which time they are the chief feature of sections of the testis seen under the microscope. In the rabbit's ovary Allen found no interstitial cells until 45 days after birth. In the male calf Lillie & Bascom (1922) found interstitial cells very early in embryological life, at the beginning of differentiation of the testis, whereas in the calf's ovary they do not appear until near the time of birth. Willier (1932) states that the first decisive change towards sexual differentiation is seen in the medulla. In the male, the rete testis continues to proliferate, remains compact and shows the formation of tubules, while the cortex disappears. In the female proliferation of the rete ceases early, the cortex becomes thickened and the germ cells multiply. According to Burns (1928), during the earliest stage of sexual differentiation the germ cells in the male are distributed evenly through the gonad, there being no distinguishable aggregation of them toward the surface, whereas in the female they become arranged about the periphery so as to form a distinct cortex. Pronounced cytological differences also, he says, are seen in the germ cells of the two sexes at this early stage. Dantchakoff (1937 b) observed in the gonads of male embryo guinea-pigs, at a very early stage of development, special glandular cells which are free in the mesenchyme and resemble closely the Leydig cells of the adult testis. She believes that the function of these cells is to produce androgen, and that they are responsible for the development of the male accessory organs. She found no such cells in the gonads of female embryos. Whatever the significance and function of these glandular cells may be, it seems probable for several reasons that the production of androgen by the male in effective quantity occurs at an early stage of embryonic existence.

We may summarize the foregoing discussion by stating that when a gonad first can be distinguished in the embryo it appears morphologically as a bisexual organ, containing a medulla which represents the male component and a cortex which is the female element. In the process of development one or other of these sexual elements becomes suppressed, while the other continues to flourish so that the sex of the embryo becomes recognizable. Whether the primitive, bisexual gonad will be biased to the male or to the female side will depend in vertebrates upon the genetical constitution of the individual. Cytological appearances, as just mentioned, suggest that in certain mammals gonadal hormones are produced much earlier in the male embryo than in the female, and in later pages it will appear that a development of male structures in the female mammal is a more frequent occurrence than the development of female structures in the male.

We may now consider the direct effects produced by androgen upon the male and female gonads; and in the discussion of this matter we shall utilize much first-class experimental work which was performed before pure androgens had become available, and the outcome of which is attributable by inference only to the action of these hormones.

(a) *The effects of androgen on the developing ovary.* It may clarify the subsequent exposition if we suggest that the direct effect of androgen upon the ovary will depend largely upon the stage of gonadal differentiation which has been already attained.

(i) *The effects of testicular tissue on the embryonic ovary.* Minoura (1921) grafted small pieces of testis on to the chorio-allantoic membrane of developing chicks after they had been incubated for periods varying from 2 to 16 days. In some of the females so treated there was a persistence of the right gonad, which normally remains rudimentary in hens, and a modification in a masculine direction of the left gonad. Both Greenwood (1925) and Kemp (1925) repeated this experiment on a large scale without being able to confirm Minoura's observations. Their methods differed in one respect from Minoura's, inasmuch as none of their grafts were inserted before the 7th day of incubation, whereas some of Minoura's were introduced as early as the 2nd day; and it seems that the propensity of the female gonad to undergo a masculine change under the influence of testicular hormones diminishes as sexual differentiation of the gonads proceeds (see Willier & Yuh, 1929).

Burns (1928) states that in larval axolotls (*Amblystoma punctatum*) the gonads remain without sexual differentiation even when large enough to be recognized and transplanted with the aid of a binocular microscope. He carried out a series of transplantations of these undifferentiated gonads, including part of the mesonephros with the gonad, into older larvae in which sexual differentiation had already occurred. The hosts were killed at the end of periods varying from 30 to 72 days and their own and the grafted gonads were then examined. Among sixteen such individuals available for examination, including eight males and eight females, the hosts' gonads were normal in all but two, which showed some evidence of bisexual structure. On the other hand in seven of the grafted gonads there was a mixture of male and female characters, showing that the structure of the grafted undifferentiated gonads had been materially changed by influences exerted by the host. When, however, gonads which had already undergone differentiation were transplanted they did not undergo any modification so as to show bisexual features.

That there is little if any incompatibility between testis and ovary in the adult animal has been proved by transplantation experiments carried out by Steinach (1916) on rats and guinea-pigs, Moore (1921) on guinea-pigs, and Finlay (1925) on fowls, and also by the observation of examples of true hermaphroditism in mammals (Witschi, 1932). Wells (1937b) made a detailed examination of two hermaphrodite squirrels; in each animal all the reproductive organs of both sexes were present except that the external genitalia were male and the prostate was absent. The ovotestes were situated in the abdomen. Eighty per cent of the volume of the medullary part of these gonads consisted of seminiferous tubules and interstitial tissue. Spermatogonia and primary spermatocytes were present. The ovarian tissue, which formed a cortex to the gonad, showed the presence of primordial ova and follicles with granulosa, but no follicles with antra and no corpora lutea were seen. Gonadotrophin from pregnant mare's serum caused a simultaneous formation of spermatozoa, graafian follicles and corpora lutea. Raynaud, Marill & Xicluna (1939) have described a case of human hermaphroditism in which the right gonad was an ovary and the left a testis. The psyche was female, the mammae and uterus were well developed and the ovary was functional. There was no vaginal opening, and hypospadias was present. The testis showed the presence of scanty spermatids.

The foregoing observations seem to show that the differentiated gonads, whether ovary or testis, are little influenced directly by the gonadal hormones. There is plenty of evidence, however, that the incompletely differentiated ovary may be profoundly affected by androgen supplied from the testis; and it seems that the degree of effect may depend on the stage of differentiation attained before the excessive supply of androgen has begun.

One way to study the influence of naturally produced hormones on the gonads of the other sex is provided by symbiosis, that is to say the sharing of a common blood supply by individuals of different sexes; and three kinds of symbiosis are available for these inquiries, namely (i) normal symbiosis between the mother and the embryo, (ii) the occasional but natural symbiosis in which male and female foetuses share a common placenta—a participation which produces the freemartin—and (iii) artificial symbiosis in which two individuals are joined together by operation so that their bloods can intermingle. As we are concerned at present with the action of androgen only, we need not discuss here the symbiotic relationships between the maternal ovary and the embryo gonad.

Freemartins. In the freemartin there are many features which bear upon the present discussion. From ancient times it has been recognized that when a cow gives birth to two calves, one a bull and the other a cow, the bull will grow into a normal male and the cow will often though not always be a freemartin, that is to say her reproductive organs, including the gonads, tend in a varying degree toward the male type. On the other hand, when both the twins are of the same sex, whether male or female, they are normal. John Hunter observed the condition in the horse, ass, sheep and cow, but not in man, dog or cat. He described the naked eye anatomy, and remarked that the freemartin had the general appearance of a spayed heifer or ox. In connection with this it may be mentioned that Marsman (1937), investigating the output of gonadal hormones by freemartins, finds that they produce negligible quantities of androgen and oestrogen, being thus from the hormonal standpoint on an equality with the ox. William Harvey (1578–1657) had noted in *Equidae* and in those cloven-footed animals which occasionally produce twins, that in some instances each of the twins had its own separate placenta, whereas in others the twins shared a single placenta. Lillie (1917), in a classical paper, showed that the freemartin is zygotically a female which has become modified by the sex hormones of the male twin owing to fusion of the two chorions and the consequent interchange of blood between the two individuals. By comparing the number of corpora lutea present with the number of foetuses, Lillie further showed that the freemartin is not the result of mono-zygotic twinning. In every one of eighty-one instances in which a single foetus was present there was only a single corpus luteum. In every one of twenty-two pregnancies in which twins were present, one being a freemartin, there were two corpora lutea; so that one may conclude that the freemartin is the result of a secondary fusion of the two initially separate placentae with anastomosis of their blood vessels. Chapin (1917) studied the gonads of embryo bovine twins 7·5 cm. in length. In the male twin seminiferous tubules were already visible and were separated by connective tissue and interstitial glandular cells, and the organ was enclosed in a tunica albuginea. In the female twin the gonads were small com-pared with the gonads of the male twin, being 2·07 mm. in length compared with 3·5 mm. Each ovary consisted of a medulla surrounded by a tunica albuginea; very few germ cells were present and no interstitial glandular cells were seen. In a freemartin 21 days after birth the gonads resembled those of a male in the absence of cortex, the presence of a tunica albuginea, and the structure of the rete. Some of the sex cords were like medullary cords, others being like seminiferous tubules. The position of the rete was that in the normal ovary. Willier (1921), examining a series of freemartins, noticed considerable differences in the grade of transformation of the female gonad toward that of the male.

Lillie (1932) states that fusion of the two blastodermic vesicles of twin cattle may precede sexual differentiation of the gonads, so that hormones produced by the developing testis would be available from the very commencement of their formation and at a time preceding the suppression of the male components of the primitive ovary.

A notable feature of the influence exercised by the gonad of the male over that of his twin sister is the permanence of the effect, for the adult freemartin does not have periods of oestrus, nor does she display any sexual interest in the males of her species, nor do they pay any sexual attention to her. The ovary has been so much altered by its early subjection to androgens that it does not acquire functional activity, and the adult animal has the characters of one from which the gonads have been removed in early life.

A condition is seen occasionally in the human species which might be regarded as the equivalent of the freemartin. The patients genetically are females, but their gonads have become so modified as to be functionless. The adult subject is feminine in appearance, has no sexual urge and does not menstruate. In such cases it seems probable that the ovaries, before their complete differentiation, have been subjected to an excess of androgen, though this has not been derived from a male twin. A case which seems to be an example of this condition has been reported, with illustrations, by Weisman (1941).

Artificially produced symbiosis. A symbiotic condition, comparable physiologically with that naturally present in twins which share a single placenta, can be brought about by uniting two individuals to each other by operation. When such a union is effected during the embryonic stage it is followed by gonadal abnormalities, which to some extent vary according to the degree of differentiation already attained at the time when the union is made. Burns (1925) succeeded in joining together, side by side, larval axolotls (*Amblystoma punctatum*) before their gonads had undergone sexual differentiation, a change which occurs relatively late in these creatures. Eighty of his parabiotic pairs survived until their sex could be identified by histological examination of the gonads, and among them the members of each pair were without exception of the same sex; forty-four of the pairs were both males, and the remaining thirty-six were both females. No intersexual abnormalities were seen, every pair consisting of apparently perfect males or perfect females. Clearly in this experiment a reversal of gonadal sex, as distinct from genetic sex, was induced in a large number of the individuals. In another experiment of the same kind Burns (1930, 1931) used larvae of *Amblystoma tigrinum*. In this series no instance of complete reversal of sex was found, but a large proportion of the pairs had gonads with combined male and female characters. In a total of 57 there were 16 pairs of males, 12 pairs of females and 29 pairs consisting of a male and female combination. All but one of these bisexual combinations showed some degree of intersexuality:

In 23 the male caused modification of the female gonad.
In 4 the female caused modification of the male gonad.
In 1 equally balanced changes were present in the gonads of both individuals.
In 1 no gonadal changes had occurred, the parabiotic pair consisting of a normal male and a normal female.

In the most advanced examples the change almost amounted to the transformation of an ovary into a testis.

Witschi & McCurdy (1929) performed the same sort of experiment on frogs and salamanders. In parabiotic frog larvae, they say, the two sexes at first develop independently. Later the male partner becomes predominant and causes a modification of the female gonad toward that of the male. They also used larval salamanders (*Triturus torosus*) and the results were comparable with those already rendered familiar by study of the freemartin. Of twenty-two pairs joined together immediately preceding the onset of muscular movements, eleven eventually consisted either of two males or two females, and eleven consisted of a male and female combination. Among the latter the male partner was without exception predominant, the gonads of the female being very much smaller than those of the male and containing few or no germ cells. Ovocytes were never seen, and the medullary cords were compact and resembled those of the testis. In four of the female partners a very few spermatogonia were present. These four cases might be regarded as confirming Burns' observation of sex reversal. Witschi & McCurdy noticed in male and female combinations that the male gonads also were modified to some extent, that is to say the germ cells were reduced in number. Although this change was pronounced in certain instances, they did not find any example in which spermatogonia were entirely absent.

(ii) *The effect of pure androgen on the embryonic ovary.* Wolff (1935) dissolved androsterone in oil (20 mg. in 1 c.c.) and deposited a few drops on the chorion of embryonic chicks between the 3rd and 5th day of incubation. The embryos were extracted between the 15th day and the end of incubation. His results are shown in Table 90.

TABLE 90.   The effects of androsterone on the gonads of embryo chicks (Wolff, 1935)

Dose of androsterone per egg	Sexual condition of chicks on 15th day of incubation		
	Males	Females	Intersexes
0·27 mg.	12	13	2
0·35 to 1 mg.	8	0	42

The intersexes formed a nearly continuous series between females and males. In the least pronounced examples the left gonad had the structure of an ovary, but the right gonad was more developed than in normal females and contained toward the hilum a mass of tissue with traces of testicular cords and also gonocytes. In the most pronounced examples the gonads were of nearly equal size, the left being slightly the larger; both resembled hypertrophied testes and histologically had the structure of testes, though the left gonad showed an ovarian cortex with oogonia, which was the only character by which the embryos could be recognized as intersexual.

Dantchakoff (1937 *a*, *b*) noted the effect of testosterone propionate on the ovaries of embryonic chicks and guinea-pigs. In chicks so treated the abnormalities produced were slight; there were no obvious changes in the rudimentary right gonad, and the left gonad was flattened and limp. In guinea-pigs there was a pronounced hypertrophy of the ovarian medulla with a well-developed rete. Raynaud (1937 *a*, *b*, 1938 *a*, *b*) injected testosterone propionate into pregnant mice and noted the effects on the young. In the ovary there was a large development of rete with the formation of numerous anastomosing tubules which entered the ovary by the hilum and penetrated to the central zone where they formed large lacunae. Otherwise the ovary was unaffected. Greene & Ivy (1937; see also Greene, Burrill & Ivy, 1938 *a*, *b*) treated pregnant rats with repeated doses of testosterone and androsterone and noted similar changes in the ovaries of the young. Burns (1939*a*) obtained the same results by injecting testosterone propionate into newly born opossums. Hamilton & Gardner (1937) gave daily subcutaneous injections of 500γ of testosterone propionate to rats during the last third of pregnancy. The female offspring, when examined at maturity, showed normal, functioning ovaries containing follicles and corpora lutea.

Burns (1939*b*) injected 25γ of testosterone propionate into larval axolotls (*Amblystoma punctatum*) every 4 or 5 days, beginning when they were 60 days old and continuing for 45 days, so that each animal received during this period a total dose of 250γ. At the age of 60 days the gonads in this axolotl have already undergone some degree of differentiation in normal circumstances, and therefore complete reversal of sex by the action of androgen would not be expected. At the end of the treatment the testes were found to be normal, whereas in the ovaries almost all degrees of intersexuality were seen, as happened after parabiosis in the experiments mentioned a few pages earlier.

The relatively slight effects produced by androgen on the ovary in birds and mammals in most of the experiments quoted can probably be explained by the advanced state of gonadal differentiation already reached when the androgenic treatment was begun. In the experiments of Minoura (1921), Burns (1925, 1928), Witschi & McCurdy (1929) and Wolff (1935), in which the treatment was begun before sexual differentiation of the gonads had been completed, the effects were more profound and somewhat different in character.

It is difficult to get a perfectly clear understanding of the effect of androgen on the undifferentiated gonads of lower vertebrates, because the different species do not appear to respond alike. Foote (1940) found that by adding testosterone propionate to water containing larvae of *Amblystoma maculatum*, the testes in most of the males were converted into ovaries, a result which was caused also by oestrone or oestradiol propionate. Witschi (1942) applying the same treatment to larval frogs obtained the opposite effect; the gonads in all the test animals treated with testosterone or androsterone developed as testes.

(b) *The action of androgen on the ovaries after their differentiation.* Experiments reported by Selye (1941 a) suggest that ovarian atrophy under the influence of androgen may be attributable not only to an inhibition of the formation of gonadotrophin as discussed on page 55, but to a direct inhibitory influence on the ovary itself. From the 2nd day of life onward for a fortnight female rats were each given daily injections of 1 mg. of testosterone propionate; one ovary was then removed from three of the animals. After this 100 i.u. of chorionic gonadotrophin were given daily for 6 days, the injections of testosterone propionate being continued in some. All the animals were killed on the 7th day after the treatment with gonadotrophin had been begun. The average weights of the ovaries in the three groups at the end of the experiment were as follows:

Ovaries after treatment with testosterone only   6 mg.
    ,,   ,,   ,,   testosterone plus gonadotrophin   18 mg.
    ,,   ,,   ,,   gonadotrophin only   90 mg.

(c) *The effects of androgen on the testicle.* Androgens appear to have little *direct* effect on the testicle at any stage of its development. It may be that in early embryonic life they take a part in suppressing female rudiments existent at that time in the male gonad. The testes like the ovaries pass through a sexually indifferent stage when both medulla and cortex are represented. The cortex, which in the female will eventually form the germinal tissue of the ovary, disappears as a distinct structure at an early stage from the male gonad, and it may be that this change is effected partly or entirely by androgen. Evidence has been adduced already that the effects on the ovary are much more profound if androgens are supplied to the embryo before sexual differentiation of the gonad has occurred. The same may be true, perhaps, of their influence on the testicle;

however, there is little definite evidence that androgens by *direct* action modify in any way the male gonad during the early stages of development.

As to the *indirect* action of androgen on the testis by inhibiting the supply of gonadotrophin from the pituitary there can be no doubt; given in excess to young males androgens prevent the testes from attaining their full size and from arriving at the stage of spermatogenesis. The undersized immature testes that are the result do not descend into the scrotum, so long as the administration of androgen is continued; if they had already descended before treatment they may return into the abdomen provided that the inguinal passages remain patent as they do in many species—rats and mice for example. This phenomenon may be observed in young mice if androgens are given in sufficient amount soon after the testes have descended. In older mice the indirect action of androgen on the testes, through the pituitary, is not so quickly observed, and requires a longer period and perhaps larger doses of androgen for its production.

Although androgens suppress the output of gonadotrophin they do not prevent the pituitary from acting synergistically with chorionic gonadotrophin, to which the testes will respond in spite of the continued administration of testosterone (Selye, 1941 *a*) (see p. 74).

Sex reversal. This term is sometimes used in describing the action of gonadal hormones. It is, of course, not applicable to a mere alteration of the outward manifestations of sex. But the sex may be regarded as reversed if the sexual character of the gonads has been changed, as when an ovary becomes a functional testis. Nevertheless in such an instance the genic constitution of the individual continues unaltered, so that an ovary which has been converted into a testis producing spermatozoa and is able to fertilize ova will fail to impart to them any genic constitution other than that with which it was originally endowed. Consequently the offspring will all be of one sex (Crew, 1921). In the complete sense of the term, therefore, it may be doubted whether sex reversal in vertebrates has been achieved in the laboratory.

Accessory Generative Organs

THE ACTION OF ANDROGEN ON THE DEVELOPMENT OF THE ACCESSORY GENERATIVE ORGANS

(i) *The müllerian and wolffian systems.* We can regard the wolffian system as concerned with the male share in reproduction and the müllerian system with the female share. It will be remembered that the epididymis, vas deferens and ejaculatory duct, together perhaps with the seminal vesicle, are derived from the wolffian system, while the oviduct and uterus are survivals of the müllerian system. In the normal adult male the müllerian structures have disappeared except for vestiges of the cranial and caudal ends, which are represented in man by the sessile hydatid of Morgagni and the uterus masculinus respectively. In the female the epoöphoron persists as a surviving remnant of the cranial end of the wolffian duct and may be regarded as the homologue of the vasa efferentia testis and epididymis. As will be seen, the chief effect of androgen on these structures is to preserve those which arise from the wolffian system and to bring about their full development. If enough androgen is available, this result will be produced in females as well as in males.

The effect of androgen on the müllerian system is not so clearly defined, and it may be left an open question for the present whether the failure of müllerian structures to complete their development in the male is the consequence of a suppressive action of androgen or of a deficiency of oestrogen. Experimental results appear ambiguous on this point; in some instances the administration of androgen seems to have caused a persistence of müllerian structures in the male, and in others to have suppressed them. In females the development of the oviducts has been stimulated by androgen (Witschi & Fugo, 1940).

In his essay on the freemartin Lillie (1917) showed that the anatomical peculiarities of this condition could be attributed to the influence on the female of hormones derived from the male twin, and it is interesting to note that John Hunter, more than a century earlier, had recorded that he could not find any oviducts in a freemartin ass which he dissected, although the horns of the uterus seemed normal. In two of three freemartins which he examined vasa deferentia were present and were associated at their urethral ends with small seminal vesicles; in one case he mentions that the vasa deferentia communicated with the vagina near the opening of the urethra.

Steinach (1916), noting the results of transplanting testes into spayed female guinea-pigs, concluded that the maintenance of the wolffian ducts depended upon the testis. Minoura (1921) came to the same conclusion. He grafted small pieces of testis on to the chorio-allantoic membrane of developing chicks between the 2nd and 16th days of incubation, and he noted in a number of instances that the wolffian ducts persisted in many of the females so treated, whereas controls grafted with pieces of other organs (liver, spleen, thyroid, thymus) did not show this persistence. Greenwood (1925) repeated these experiments with this difference that the grafts were inserted on the 8th day of incubation. The chicks were removed from the egg-shells between the 14th and 17th day of incubation and examined. Among twenty female chicks in which the testis grafts had survived no intersexual conditions were found. His results do not confirm those of Minoura; and he states that in untreated female chicks the wolffian ducts are still present at the time of hatching. Kemp (1925) also repeated Minoura's experiment, inserting the testicular grafts between the 7th and 10th day of incubation, and like Greenwood he failed to confirm Minoura's work. In some of Minoura's experiments the grafts were inserted at an earlier stage of incubation and possibly this fact may account for some of the discrepancies. When this kind of experiment is being considered it must be remembered that exposure of an embryo to adverse circumstances at any particular stage of its existence is apt to interfere with the developmental processes which normally take place at that stage. Willier & Yuh (1929) found that if hens' eggs were chilled on the 8th day of incubation, at which time differentiation of the wolffian and müllerian structures normally occurs, regression of the müllerian ducts in males, or of the right müllerian duct in females, was retarded or prevented. Even the appropriate differentiation of the gonads may be interfered with in this way, as Witschi showed by exposing 5-week-old larvae of *Rana sylvatica* to an abnormally high temperature. In these circumstances the development of male gonads was little affected, but in the females the formation of normal ovaries was prevented; they became differentiated into organs resembling testes in their structure.

However, there can be no doubt that androgens cause survival and increased growth of the wolffian ducts and the structures derived from them.

Wolff (1935) applied androsterone to the chorion of embryo chicks on the 3rd, 4th or 5th day of incubation and examined the embryos on the last few days of incubation. When the dose of androsterone exceeded 0·35 mg. all the females and apparently many of the males showed an intersexual condition. In the females which had minor sexual modifications the right müllerian duct was larger than normal, whereas in the individuals with more pronounced sexual modifications the müllerian ducts had almost or entirely disappeared.

Raynaud (1937 a, b) gave injections of testosterone propionate, in doses varying from 5 to 10 mg., to pregnant mice. This treatment frequently, though not always, caused abortion. Of two litters borne by mothers who had been given testosterone on the 13th day of gestation, seven of the mice were killed on the day of birth and submitted to detailed histological investigation. Among the seven, there were one normal male, one normal female, one female with persistent wolffian derivatives on one side only and four with bilateral abnormalities involving the wolffian system. The following abnormalities were observed in one of these mice. The medulla of the ovary showed a well-formed rete with numerous anastomosing tubules which emerged from the hilum of the ovary and united with a structure like an epididymis from which proceeded a wolffian duct running parallel with the müllerian duct and comparable in size with a normal vas deferens. The müllerian system also was affected, for the two horns of the uterus, though large, united to form a single rudimentary tube. Another female mouse, whose mother had received 10 mg. of testosterone propionate 7 days before parturition, was killed 30 days after birth, when the müllerian ducts, though rather shorter than those of a female of the same age, were otherwise normal. The wolffian system was preserved, as in the normal male, and epididymis, vas deferens and ejaculatory ducts were all present; at the lower end of each wolffian duct was an evagination forming a gland which Raynaud regards as a seminal vesicle. The same author (Raynaud, 1938c) found that an epididymis was present in the female mice only if their mothers had received the injections of androgen at least 6 days before parturition. Effects similar to the above were induced by dehydroandrosterone.

Dantchakoff (1937a, 1938a) injected testosterone into the amniotic cavity of embryo guinea-pigs at about the 21st day of intrauterine life, and gave further injections after birth. The results were similar to those obtained by Raynaud in mice. Normally in the female adult guinea-pig the only remnants of the wolffian system are some small fragmentary canals representing perhaps the ductuli efferentes and epididymis of the male. In female guinea-pigs treated before the 45th day of embryonic life with testosterone the wolffian system was preserved, and a massive rete testis was present together with epididymis, vas deferens and ejaculatory duct. The müllerian structures were well developed but the ends of the uterine horns and the ovaries, instead of being near the kidneys, were lying beside the bladder.

Greene & Ivy (1937) injected testosterone into three pregnant rats toward the end of pregnancy, and examined the offspring when adult; and they found that development of the müllerian structures had been inhibited while that of the wolffian structures had been stimulated. In a female rat of this series, whose mother had been given 2·5 mg. of testosterone on the 12th day of pregnancy, oviducts were absent and the uterus was represented by a very rudimentary structure dorsal to the bladder. A rudimentary vas deferens was present. Greene, Burrill & Ivy (1938b) repeated the experiment on a larger scale. They gave injections of testosterone or testosterone propionate to 152 rats at various stages of pregnancy. In ninety-seven of these the foetuses died, the remainder had living litters or were delivered by caesarian section. The females of nineteen such pregnancies were examined after death. In these, although the mothers had received doses of testosterone propionate ranging from 5 to 9 mg., the horns of the uterus and oviducts were preserved. A vas deferens, parallel with the uterus, was present bilaterally in fifteen and unilaterally in ten. In some the complete wolffian system was represented from gonad to urethra. In three of these females when adult there persisted ductuli efferentes, epididymis with head and tail, vas deferens and male accessory organs, together with normal oviducts, uterus and upper part of vagina. The same workers (1938a) gave androsterone in divided doses, the totals varying from 40 to 280 mg., to pregnant rats. Examination of twenty-three of the female offspring showed that the derivatives of the müllerian system were well developed. In ten of these females wolffian ducts also were present parallel with and adjacent to the uterus (bilaterally in seven and unilaterally in three). Continuous with the rete of the ovary in these cases were efferent tubules which communicated with an epididymis from which issued a vas deferens opening into the urethra in the same position as in the male. Raynaud (1937 a, b, 1938 a, b, c) in a fine series of papers has described in detail the müllerian and wolffian systems of mice and their transformations by androgen. In normal untreated male mice at birth, he says, the epididymis and vas deferens are well formed, while Müller's ducts have disappeared entirely except at their lower extremities, where they fuse and form a uterus masculinus. In the normal untreated female at birth the oviducts, uterine horns and the body of the uterus are well formed; the wolffian ducts have degenerated, vestiges of the upper and lower ends only surviving. The upper vestige consists of the ovarian rete and ductuli efferentes, which however do not intercommunicate as they do in the normal male. The lower vestige of the wolffian duct is represented by two tubes which fuse caudally with a mass of cells at the junction of the lower end of the uterus with the upper end of the vagina.

Burns (1939 a, c) experimented with the young opossum (*Didelphys virginiana*). At birth, he says, this animal is almost undifferentiated sexually; wolffian ducts are present but müllerian ducts do not appear till several days after birth. He injected the young on the 1st or 2nd day after birth with testosterone propionate in doses of 50, 75 or 100γ. This treatment caused in the females the presence of large rete canals in the ovary, which were often dilated and were directly connected with an epididymis which was continued into a vas deferens, the whole wolffian system being thus preserved. In males the müllerian duct derivatives were greatly enlarged; in females also testosterone propionate stimulated this growth of the müllerian structures.

In further experiments Burns (1939b) found that testosterone propionate, given to larval axolotls when their gonads had already undergone some sexual differentiation, caused hypertrophy and differentiation of the wolffian ducts in both sexes while the müllerian ducts were suppressed. In males and females the cloaca and its glandular derivatives were hypertrophied.

An unexpected result of this kind of experiment has been reported by Willier, Rawles & Koch (1938), who introduced various androgens, including testosterone, androsterone and dehydroandrosterone, into fowls' eggs when they had been incubated for periods varying from 43 to 72 hours. The chicks were examined after 16, 17 or 18 days of incubation. In the females all the androgens used had a masculinizing effect, causing the right ovary to become enlarged by hypertrophy of the medullary tissue and causing both gonads to resemble testes in shape. In the males testosterone and dehydroandrosterone caused persistence of the müllerian ducts throughout their whole length, and these were in some instances hypertrophied. The left testis contained ovarian tissue. These androgens, in fact, had acted as androgens on females but to some extent like oestrogens on males.

Selye (1940c) has reported a curious change produced by androgen on the oviducts of newborn rats. To these he gave 1 mg. of testosterone propionate by intraperitoneal injection daily for a fortnight, after which the daily doses were increased to 3 mg. and were continued for another fortnight. The rats were then killed. The oviducts had been transformed into gelatinous masses, their texture resembling that of Wharton's jelly.

(ii) *The seminal vesicles, prostate and coagulating glands.* John Hunter believed that the penis, urethra and all parts connected with them, including the seminal vesicles, were dependent for their existence on the testes. He noted that the prostate became atrophic after castration and that the seminal vesicles in eunuchs were much diminished in size, whereas if one testis only were removed, the other being normal, no atrophy of the seminal vesicles would ensue. The first scientific confirmation of his views on the development of the accessory reproductive organs seems to be that of Lillie (1917). He noted, like Hunter, the presence and persistence of wolffian ducts (vasa deferentia) and seminal vesicles in the freemartin, and showed that their existence in the female twin is attributable to testicular influence derived from the male twin with which the freemartin shared a common blood supply.

Korenchevsky (1935) treated spayed rats for 3 weeks with androsterone and diol derivatives of androsterone; and when dissecting the rats at the end of this period he found well-developed glands, which are not seen in the normal female rat, attached to the proximal end of the urethra at the base of the bladder. The largest of these glands was $6 \times 4 \cdot 2 \times 2 \cdot 2$ mm. Further inquiry (Korenchevsky & Dennison, 1936a) showed the presence of atrophic glandular vestiges in the same position in a considerable percentage of normal female rats. They concluded that these vestigial glands were the morphological counterparts of the male prostate. Under the influence of androsterone the vestiges in the female become hypertrophied and functionally active, and in this condition resemble histologically the prostate of the normal male.

Greene & Ivy (1937) likewise found, in female rats whose mothers had been given testosterone during pregnancy, paired glandular structures attached to the proximal end of the urethra at the base of the bladder. They agree with Korenchevsky in regarding these as representing the prostate of the male. Other glands resembling seminal vesicles were present also. Similar findings have been recorded by Greene, Burrill & Ivy (1938 a, b) in the female offspring of rats which had been given androsterone during pregnancy (see also Price, 1939).

In female mice whose mothers had been given 10 mg. of testosterone 7 days before parturition, besides epididymis, vas deferens, ejaculatory duct and colliculus, Raynaud (1937b, 1938a) was able to identify prostate, coagulating gland and seminal vesicle. Raynaud's observations have been confirmed by Turner, Haffen & Struett (1939). Dantchakoff (1937b) has made a similar observation in female guinea-pigs which had been treated during embryonic life by direct injection of testosterone into the amniotic cavity between the 35th and 45th day of gestation. Burns (1939a) has recorded the development of a prostate in the female opossum under the influence of testosterone propionate administered from the first day of birth.

The male genital structures—epididymis, vas deferens, wolffian duct, seminal vesicle—which have been preserved in the female by giving androgens during development do not disappear after the administration of androgen has been stopped. They persist afterward, as Raynaud (1938c) discovered, resembling the same atrophic organs in the castrated male.

(iii) *The vagina.* Androgens, given to pregnant mice, rats and guinea-pigs, will affect the female young *in utero* so as to hinder the complete separation of the urethra from the vagina, in consequence of which the perineal part of the vagina fails to develop and the cranial segment opens into the urethra, either by a single median opening or by two lateral apertures. This vaginal deformity has been produced experimentally by Dantchakoff (1938a), Greene & Ivy (1937) and others; and the writer has on many occasions observed the phenomenon in rats and mice whose mothers had been given androgen toward the end of pregnancy. The malformation of the vagina thus produced by androgens is permanent and the animals so affected are apt to die when adult from pyometra. A similar absence of the perineal part of the vagina has been noted in freemartins (Lillie, 1917) and is attributable in them also to an excessive supply of androgen during early embryonic life. Odd as it may appear this condition is normal in the female spotted hyaena (*Hyaena crocuta*), so that until puberty it is difficult to distinguish female from male. Matthews (1941), who has made a special study of the animal, says that her vagina opens into the urethra and has no external orifice of its own. In the young of this species the clitoris is almost identical with the penis of the male, except that it is slightly smaller. The glans is free from the well-developed prepuce and is furnished with keratinous spines as in the male. At this stage the external urethral meatus is a slit only 2 to 3 mm. long. At puberty the whole clitoris enlarges and the meatus expands and extends down the ventral part of the clitoris so as to cause a moderate degree of hypospadias, the meatus attaining a length of 1·5 cm. Copulation occurs by the urinary meatus and the young are born through the urethra, which becomes sufficiently dilated for the purpose. The newborn cubs are in a rather advanced state of development, being relatively large and able to walk; their eyes are open and the incisor and canine teeth are already erupted.

Investigation of the urine to discover the nature and relative amounts of gonadal hormones excreted during gestation in this animal might be of great interest. Matthews remarks that the ovaries of all the females he examined showed a high degree of luteinization, so that the greater part of the ovarian weight was attributable to lutein tissue.

Premature opening of the vagina. Androgens given to young female rats and mice after birth show another effect, namely premature opening and keratinization of the vagina. This was first recorded by Butenandt & Kudszus (1935), who gave various androgens to rats varying in age from 18 to 25 days. Their results are shown in Table 91.

TABLE 91.   Keratinization and opening of the vagina in immature rats under the influence of androgens (Butenandt & Kudszus, 1935)

Androgen	Dose (mg.)	Age at opening of vagina (days)
Androstenedione	6	24–26
Dehydroandrosterone	6	28–30
Testosterone	2·4	27
Androsterone	6	46–54
Untreated controls	—	45–57

This consequence of androgens, which has been confirmed by Deanesly & Parkes (1936a), Nathanson, Franseen & Sweeney (1938), Rubinstein, Abarbanel & Nader (1938), and others, was unexpected because these hormones do not cause vaginal keratinization in rats and mice after puberty (Warren, 1935).

Chapter VIII. *The Action of Androgen on the Reproductive Organs after their Complete Differentiation*

The Gonads

Ovary and Testis.

THE writer has already discussed the action of androgens on individual organs before their complete differentiation. The purpose of this separation is to distinguish between the early, often permanent, effects, and the usually transient, reversible effects which androgens induce at a later stage in the life of the reproductive organs. The distinction between the two groups is not quite logical, for there is no sharply defined boundary between them. Nevertheless, in connection with human pathology and treatment—which are in the mind of the writer throughout this essay—the distinction is important; and any overlapping of thought or repetition of narrative which the arrangement entails will, the writer hopes, be forgiven.

For a long while after birth, and in some instances throughout life, the sexual characters have not assumed a fixed and final form, and some of the responses to androgen, even in late postnatal existence, are irreversible and permanent, or nearly so.

I. THE OVARY

(a) *Gonadotrophic action.* The first effect of androgen on the ovary of the immature animal, and perhaps on that of the adult, is an acceleration of follicular maturation. A comparable initial effect is exercised by androgens on the testis as shown by a favourable influence on spermatogenesis (p. 198).

Shapiro (1936) reported that ovulation could be induced in the clawed toad (*Xenopus laevis*) by various androgens as well as progesterone; and adrenal cortical extract also had this effect. The reaction was obtained even though the pituitary had been previously removed. Ovulation was not obtained in Shapiro's experiments with oestrogens; the doses of these mentioned in his paper are very large —10 mg. of oestrone and 8 mg. of oestradiol. Dantchakoff (1938b) observed the formation of multiple follicular cysts in the ovaries of young guinea-pigs which had been subjected to testosterone. The condition, she says, resembles that produced by FRH, namely the simultaneous maturation of many follicles which, failing to rupture or to undergo luteinization, remain as cysts. Salmon (1938b) gave single injections of 1 to 5 mg. of testosterone propionate or androstenediol in sesame oil to twelve female rats 17–30 days old. To six female rats of the same age he gave sesame oil only. Opening of the vagina within 72 hours occurred in all the rats treated with androgen, and their ovaries showed follicular maturation between 60 and 72 hours after injection, and corpora lutea between 96 and 192 hours after injection. In the control rats, which had received sesame oil only, the vagina remained closed and the ovaries did not show evidence of stimulation. Nathanson, Franseen & Sweeney (1938) performed a similar experiment on rats with the same consequences.

Starkey & Leathem (1938) used 148 mice, keeping some as littermate controls. When the mice were between 18 and 21 days old they were given single sub-cutaneous doses of 0·5, 1, 1·5 or 2 mg. of testosterone propionate, and were killed 48, 72, 96 or 120 hours later. Post-mortem examinations showed the following effects: 48 hours after the injection pronounced follicular activity was present, reaching a maximum at 72 hours; the uterus was hypertrophied, reaching its largest dimensions at 72 hours; mucification of the vagina was present at 96 hours. Large follicles were still seen at 120 hours but there were no corpora lutea. These sequels were most pronounced after the largest doses and were least with the doses of 0·5 mg.

The gonadotrophic effect as disclosed by the experiments just quoted seems to be exerted only upon the follicles. Selye (1939a) gave daily injections of 5 mg. of testosterone propionate to adult mice of the *dba* strain for 20 days, at the end of which period the ovaries, though smaller than those of controls, consisted of large follicles, follicular cysts and a few cystic corpora lutea; the interstitial tissue of the ovary was atrophied. In later experiments (1940b) he found that the atrophic influence of oestradiol on the ovary and the testis could be prevented to some extent by testosterone; an observation which might possibly be relevant to the present discussion.

Aschheim & Varangot (1939) noted accelerated maturation of follicles in the ovaries of adult rats which had been given 1 mg. of testosterone propionate daily for varying periods. The ovaries were hyperaemic, and follicles in all stages of development were present with mitoses in the granulosa. The interstitial cells were atrophic and their nuclei displayed the 'cartwheel' appearance charac-teristically present after hypophysectomy. As will become apparent as this dis-cussion proceeds, it is not easy to reconcile the foregoing evidence of an early and temporary follicle-stimulating effect of androgens, with other evidence which will be mentioned of an opposite effect. Perhaps the age of the animal, the duration of treatment or the size of the dose of androgen may affect the outcome of an experiment of this sort (v. p. 212).

(b) *Enlargement and maintenance of corpora lutea.* Several workers have noticed that androgens influence luteinization in the ovary. Korenchevsky, Dennison & Hall (1937) gave 0·5 and 1·5 mg. of testosterone propionate daily to rats for a period of 21 days. After the latter doses the ovaries were larger by 29 per cent than those of untreated rats, and contained more numerous and larger corpora lutea (Table 92).

TABLE 92.   The effect on the rat's ovary of brief treatment with testo-sterone propionate (Korenchevsky, Dennison & Hall, 1937)

Daily dose	Duration of treatment (days)	Mean weight of ovaries (mg.)
None (controls)	—	75
0·5 mg.	21	74
1·5 mg.	21	97

McKeown & Zuckerman (1937) have reported the enhanced development of corpora lutea in rats treated with testosterone propionate given daily in doses of

0·2 mg. for a period of 10 days, and Nelson & Merckel (1937) have seen the same result in rats following daily injections of 2 mg. of testosterone.

Wolfe & Hamilton (1937 b) showed that this effect of testosterone propionate on the corpora lutea is largely dependent on the phase of the oestrous cycle during which the injections of testosterone are first given. Forty adult rats were given 2 mg. of testosterone propionate daily for 10 days and were killed on the day after the last injection. Their oestral condition was ascertained by vaginal smears. The results showed that if the treatment was begun during dioestrus, little if any enlargement of the corpora lutea ensued, whereas considerable enlargement followed if the first injections were given during oestrus (Table 93). For comparable results see Schilling & Laqueur (1942).

TABLE 93. The effect of testosterone propionate on corpora lutea in the rat (Wolfe & Hamilton, 1937 b)

Phase of oestrous cycle when treatment began	Duration of treatment (days)	Mean diameter of corpora lutea (mm.)
No treatment (controls)	—	0·93
Oestrus	10	1·5
Dioestrus	10	1·0

Freed, Greenhill & Soskin (1938) believe that the kind of effect produced by androgen on the ovary may vary considerably with different doses. Normal adult female rats were given daily subcutaneous injections of testosterone propionate for 16 days; they were killed on the 17th day and the ovaries and uterus were weighed. The rats which had received 1 mg. daily showed signs of an increased luteinization; their ovaries contained numerous large corpora lutea and progestational changes were present in the uterus. The rats which had received 0·05 mg. per day had atrophic ovaries. Their results may be summarized as follows:

Small doses of testosterone propionate caused ovarian atrophy, diminished output of oestrogen, anoestrus.

Large doses of testosterone propionate caused large corpora lutea, diminished output of oestrogen, anoestrus.

In both cases the production of FRH appeared to be prevented. The relative weights of ovaries and uterus in these rats are given in Table 94. The enlargement of the uterus after the bigger doses of testosterone is explained by the presence of progestational changes.

TABLE 94. Relative effects of large and small doses of testosterone propionate on the rat's ovaries and uterus (Freed, Greenhill & Soskin, 1938)

Daily dose (mg.)	Number of rats	Average weight of ovaries (mg.)	Average weight of uterus (mg.)
1	7	56	570
0·1	8	32	298
0·05	8	24	262
None	6	74	370

(c) *Atrophy of the ovary.* This is a constant effect of androgens when these are supplied in sufficient quantity and for a long enough time. Mazer & Mazer (1939) gave 0·5 mg. of testosterone propionate to ten rats weighing between 30 and 40 g.; they continued the injections 3 times a week during a period of 102 days, and killed the rats on the day after the last injection. They also treated twenty adult rats in the same way for 62 to 68 days. Oestrus was suppressed in all. The ovaries in both classes were much reduced in size, the development of follicles was arrested and corpora lutea were absent (Table 95).

TABLE 95.   The effect on the rat's ovary of prolonged treatment with testosterone propionate (Mazer & Mazer, 1939)

Rats used	Treatment	Number	Duration of treatment (days)	Mean body-weight (g.)	Mean weight of ovaries (mg.)
Infantile	None	8	—	149	45
Infantile	Testosterone propionate	10	102	145	10
Adult	None	6	—	175	61
Adult	Testosterone propionate	20	62–68	181	20

Groher (1938) gave daily an androgenic preparation ('testiglandol') to 10-day-old mice, and noted a gradually increasing atrophy of the ovary with atresia of follicles. The same effects on the ovaries of adult guinea-pigs were noticed by Boling & Hamilton (1939) after daily injections of 4 mg. of testosterone propionate.

(d) *Arrest of ovulation.*   In contrast with the slightly stimulating action of androgen on the ovary, which has been mentioned earlier, are the more pronounced and more readily appreciated inhibitory actions. Among these appears to be the capacity to prevent the release of ova from ripe graafian follicles. Geist, Gaines & Salmon (1940) treated with testosterone propionate two women who had normal oestral cycles. One received 925 mg. during a period of 31 days, and the other received 1,225 mg. during 15 days. Laparotomy at the end of these periods revealed that ovulation had not occurred in either individual, although it is supposed that ovulation in women in normal circumstances occurs spontaneously and almost with regularity. In some animals, as in man, rupture of follicles with discharge of ova spontaneously occurs when maturation is complete. In other species, including the rabbit, some extraneous stimulus is required to induce ovulation. Mating has this effect and so, too, have several other sensory impulses, such as those provided, for example, by the presence at close quarters of a possible mate. These impulses lead to ovulation by causing the pituitary to release an additional amount of gonadotrophin into the blood stream; and ovulation may be induced in a rabbit at any time during its prolonged oestral state by injecting luteinizing gonadotrophin into an ear vein. We know that androgens check the output of LH from the pituitary, and it seems probable, therefore, that the inhibition of ovulation by androgen is effected by a withholding of LH. Gray & Lawson (1939) have produced experimental evidence bearing directly on this matter. Nine female rabbits were given intramuscular injections of 10 mg. of testosterone propionate daily for 22 days, and three rabbits were kept as untreated controls. At the end of this period urinary gonadotrophin was given intravenously to all the rabbits except three of those which were being treated with testosterone; and 3 days later one tube and ovary were removed. Treatment with testosterone was continued. After an interval of 15 days from the injection of gonadotrophin the rabbits were killed, and as shown in Table 96 all those which had received gonadotrophin showed ovulation whether they had received testosterone or not, whereas in the three which had been given testosterone without gonadotrophin the ovaries showed no sign of activity. Testosterone, even in the presence of gonadotrophin, appears to have prevented the formation of corpora lutea in these rabbits.

TABLE 96.   The effect of testosterone propionate on ovulation in the rabbit with and without artificially administered gonadotrophin (Gray & Lawson, 1939)

Number of rabbits	Testosterone pro-pionate given	Gonadotrophin given	Condition of ovaries	
			At laparotomy (25th day)	At autopsy (37th day)
6	+	+	Multiple ovulations	Minute, pale corpora lutea
3	+	−	Small, pale ovaries	Small, pale ovaries
3	−	+	Multiple ovulations	Large corpora lutea

Burdick (1940) has produced evidence which suggests that the inhibition of ovulation by testosterone may not be an immediate effect in every species, though the prolonged administration of androgen undoubtedly has this result. Among mice receiving 0·5 mg. of testosterone propionate daily several were seen by Burdick which had ovulated during the course of the injections, and the question arises whether in these instances ovulation occurred because of or in spite of the injections. In a mouse which had been mated and then treated with testosterone propionate, four unicellular ova were found in the ovarian end of the oviduct on the 6th day. A similar result was seen in another mouse in the same circumstances. Apparently the mice had ovulated on the 5th or 6th day after mating.

(e) *Arrest of oestrus.* Ihrke & D'Amour (1931) gave daily injections of 6 c.u. of an androgenic extract of bulls' testes to normal female rats and noticed that oestrus was suspended throughout the period of this treatment. Normal cycles reappeared soon after the injections were withheld. Furthermore, they found that rats in persistent anoestrus caused by the injections of testis hormone came into oestrus within 2 or 3 days if gonadotrophin were given in addition to the testicular extract. Moore & Price (1932) performed a similar experiment on rats with the same results. Robson (1936c) gave 0·1 mg. of testosterone twice daily by injection to mature mice having regular oestral cycles. During the 2 weeks in which this treatment was continued the mice remained in anoestrus; and when the injections were stopped oestrus was resumed. No mating occurred during the period of dioestrus.

(f) *Arrest of menstruation.* Because they arrest oestrous cycles it is not surprising that androgens given to primates will prevent menstruation. Zuckerman (1937a) found that 25 mg. of testosterone propionate given twice a week to normal mature rhesus monkeys stopped the oestrous cycles for as long as the treatment was continued. The reproductive system was not permanently disordered in these monkeys and menstruation restarted when the injections were withheld. Papanicolaou, Ripley & Shorr (1938) gave intramuscular injections of 50 mg. testosterone propionate every other day to a girl of 18 with the conse-

quence that menstruation was arrested; it was resumed 28 days after the last injection. Loeser (1938) and others have recorded the arrest of menstruation in
women under treatment with testosterone propionate.

That the effects of testosterone on the ovaries and on oestrus are largely indirect and attributable to arrest of the output of gonadotrophin from the pituitary
has already been pointed out (p. 113).

(g) *Formation of ovarian cysts.* The presence of follicular cysts, as reported
by Dantchakoff (1938b) and Selye (1939a), following the administration of
androgens has been quoted (pp. 191, 192). Champy (1937) has seen a different
kind of ovarian cyst in guinea-pigs which had been given injections of 'lipoides
testiculaires'. These cysts, he says, are formed by invaginations of the germinal
epithelium to form mucoid cysts in which papillomata may develop. The same
condition, he says, may be produced by oestrone.

Simple ovarian cysts lined by flat cells and derived apparently from follicles
have not been uncommon among mice treated by the writer for prolonged periods
with testosterone.

Korenchevsky & Hall (1940) have mentioned the presence of numerous ovarian
cysts in the ovaries of rats which had been treated by them with testosterone and
oestradiol at the same time.

II. The Testis

(a) *Atrophy and arrest of spermatogenesis.* The most obvious effect of an excessive supply of androgen on the testicle in postnatal life is atrophy. Moore
(1930), in calling attention to this effect, attributed it to a reduction in the output
of gonadotrophin from the pituitary, and this explanation is generally accepted.
The injury inflicted is grossly manifested by a reduction in the size of the testes;
at the same time the accessory reproductive organs are hypertrophied through
the direct action on them of the excess of androgen (Table 111). This double
effect, together with a comparison of the potencies of different androgens, is
shown by an experiment of Bottomley & Folley (1938a). They daily injected
2 mg. of various androgens dissolved in sesame oil into immature male guinea-
pigs weighing about 175 g. for a period of 30 days. Controls were given injections
of sesame oil alone. At the end of the treatment the animals were killed and their
organs weighed. The results almost consistently showed a reduced weight of the
testes and an increased weight of the seminal vesicles and prostate under the
influence of androgens (Table 97). It may be noted that the table reveals no

TABLE 97. The effects of various androgens on the testes and accessory
generative organs of immature guinea-pigs (Bottomley & Folley, 1938a)

	Weights of organs (expressed as mg. per 100 g. of bodyweight)		
	Testes	Seminal vesicles	Prostate
Sesame oil	279·9	100·9	49·7
Androstenedione	88·9	154·8	117·4
Androsterone	193·3	142·4	68·6
transDehydroandrosterone	258·4	142·2	71·3
Dihydrotestosterone	49·9	204·4	100·8
17-Methyltestosterone	45·2	237·3	136·4
Testosterone propionate	74·9	306·7	99·1

strict correlation between the effects on the testes and those on the seminal vesicles and prostate.

Experiment seems to show that androgens impede testicular function most readily in the young. Moore & Price (1932), using an androgenic extract of bulls' testes, found that the daily administration of 6 c.u. for 20 days caused little or no change in the testicles of adult rats; in young rats between 34 and 54 days old the same treatment caused arrest of spermatogenesis, degeneration of the germinal epithelium and a reduction in the size of the testes. Moore & Price (1937) repeated the experiment using androsterone instead of testicular extract and again they found that the deleterious effect of the injections on the rats' testes depended largely on the age of the rat at the time of treatment. Daily injections of androsterone varying from 0·25 to 6·0 mg. given for 20 days to immature rats caused a large reduction in the weights of their testes as compared with those of untreated controls, whereas daily doses of 6 mg. given to adult rats for the same period had no clearly adverse effect on the testes. A similar differential effect on the testes of young and old rats respectively was produced by testosterone propionate (Moore & Price, 1938).

The atrophic effect of androgens on the testis, however, is not confined to immature animals, although more readily produced in them. This was shown by Korenchevsky & Hall (1939), who gave daily doses of androgen to old rats. The androgens used were androsterone, dehydroandrosterone, testosterone and testosterone propionate. All these preparations, which were given in doses of 0·75 to 7·5 mg. daily, caused a reduced weight of the testes.

As already mentioned the inhibitory action of an excess of androgens on the functions of the testes has been attributed to the capacity which androgens possess for curtailing the supply of gonadotrophin from the pituitary. This effect has been demonstrated by an experiment performed by Bottomley & Folley (1938b), in which they found that the testicular atrophy which ordinarily follows the administration of androgen in excess can be prevented by giving gonadotrophin at the same time. To one group of guinea-pigs they gave daily doses of androgen alone, and to another group they gave similar doses of androgen together with gonadotrophin. After 32 days of this treatment the testicles were atrophic in the guinea-pigs which had been treated with androgen alone, but were not atrophic in those which had received simultaneous doses of gonadotrophin (Table 98).

TABLE 98. Prevention by gonadotrophin of testicular atrophy in the guinea-pig under the influence of androgen (Bottomley & Folley, 1938b)

Daily treatment	Duration (days)	Number of guinea-pigs	Mean weight of testes at end of experiment (mg.)
2 mg. Testosterone propionate	32	5	247·0
2 mg. Testosterone propionate plus Gonadotrophin	32	5	1,581·8

It will be understood that the atrophic changes caused in the testes by androgen are not permanent; they last only so long as the excessive supply of androgen continues.

In this sort of experiment it has to be remembered that androgens prevent descent of the testes in young animals, and in older animals may lead to their retraction within the abdomen; and, as is well known, the mere retention of a testis within the abdomen will cause a destruction of the seminal epithelium and a considerable reduction in the size of the testis.

(b) *Gonadotrophic action of androgen.*   Inhibition of the gonadotrophic activity of the pituitary with consequent gonadal atrophy and arrest of spermatogenesis is not the only effect which androgens have on the testis. Their immediate and direct action appears to be gonadotrophic, that is to say they cause at first an increase in the size and functional activity of the testes. As already shown they have a comparable effect on the ovaries. Several experimenters have observed that if androgens are given to an animal when its pituitary is removed, spermato-genesis and the formation of motile spermatozoa will be maintained for a while, which does not happen after hypophysectomy unaccompanied by the admini-stration of androgen. From such experiments it appears that androgens exert a gonadotrophic influence only during the period which immediately follows re-moval of the pituitary; they fail to restore functional activity if a long enough time has elapsed after hypophysectomy to allow atrophy of the testis to become esta-blished. This gonadotrophic action appears to be exercised only on the spermato-genic functions of the testicle; the interstitial glandular cells are unaffected and undergo atrophy after removal of the pituitary whether androgens are or are not given.

Walsh, Cuyler & McCullagh (1933, 1934) removed the pituitaries of rats and noted that atrophy of the gonads followed, being almost complete at the end of 20 days. Daily injections of 9 c.u. of androsterone prevented this atrophy so that after 20 days the gonads could not be distinguished from those of a normal rat. Nelson & Gallagher (1936) confirmed this result and showed that normal sperma-togenesis could be maintained in the hypophysectomized rat for as long as 40 days by the daily administration of androsterone (Table 99). Males treated in this manner, they say, have sired normal litters, though the degenerative changes in the interstitial cells consequent on hypophysectomy were not prevented by the administration of androsterone.

TABLE 99.   The effect of daily doses of androsterone in maintaining spermato-genesis in rats after hypophysectomy (Nelson & Gallagher, 1936)

Condition	Daily dose of androsterone (c.u.)	Duration of observation after hypophysectomy (days)	Average weight of testes (mg.)	Motility of sperm
Untreated controls	0	—	2,432	Good
Hypophysectomized	0	22	517	No sperm
Hypophysectomized	10	22	801	Fair
Hypophysectomized	14	22	1,031	Good
Hypophysectomized	20	22	1,609	Good
Hypophysectomized	20	40	1,586	Good
Hypophysectomized	0	40	423	No sperm

Nelson & Merckel obtained comparable results in rats (1937) and in mice (1938). They removed the pituitaries of adult mice and then injected 8γ of various androgens daily for 20 days, beginning on the 2nd day after the operation. The mice were then killed, and their organs were weighed and examined (Tables 100, 101). Atrophy of the seminal epithelium had been prevented by the androgens, though these had not prevented atrophy of the interstitial cells.

TABLE 100.   The effects of androgens on spermatogenesis in hypophysectomized mice (Nelson & Merckel, 1938)

| | Organ weights (mg.) | | | |
Treatment	Testes	Seminal vesicles (full)	Prostate	Sperm motility
None	211	243	48	+
Hypophysectomy only	31	17	12	No sperm
Hypophysectomy *plus* Androstanedione	198	149	41	+
Hypophysectomy *plus* Androstanediol	175	263	52	+
Hypophysectomy *plus* Testosterone propionate	168	579	85	+

TABLE 101.   The effects of androgens on spermatogenesis in the hypophysectomized rat (Nelson & Merckel, 1937)

| | | | Weights of organs in mg. | | | |
Treatment	Androgen	Daily dose (mg.)	Seminal vesicles	Pro-state	Testes	Presence of spermato-genesis
None	None	.	963	295	2,432	+
Hypophysectomy	None	.	.	54	527	−
Hypophysectomy	Testosterone	1	1,815	472	1,243	+
Hypophysectomy	Androsterone	1	958	277	1,785	+
Hypophysectomy	Dehydroandrosterone	1	627	276	1,789	+
Hypophysectomy	Androstanedione	1	1,297	415	2,117	+
Hypophysectomy	*cis*Androstenediol	1	497	215	1,790	+

The experiments on rats showed that the capacity of the various compounds for maintaining the size of the testis and its spermatogenic functions after hypophysectomy was not related to their androgenic activity as determined by their actions on the prostate and seminal vesicles, for, among the androgens used testosterone had about the weakest influence in preserving the weight of the testis and spermatogenesis (Table 102).

TABLE 102.   Relative capacities of different androgens for maintaining spermatogenesis in the hypophysectomized rat (Nelson & Merckel, 1937)

(The numbers indicate the relative androgenic values of the substances tested)

Androgen	Androgenic potency as shown by maintenance of accessory reproductive organs after hypophysectomy	Capacity to maintain spermatogenesis after hypophysectomy
Testosterone	1	6
Androstanedione	2	1
Androstenedione	3	2
Androsterone	4	5
Dehydroandrosterone	5	4
*cis*Androstenediol	6	3
*trans*Androstenediol	7	7

Gaarenstroom & Freud (1938) removed the pituitaries from seventeen adult rats and thereafter gave them daily subcutaneous injections of 0·6 mg. of testosterone propionate. Under this treatment atrophy of the testes did not occur and spermatogenesis was maintained. They suggest that pituitary gonadotrophin may perhaps act only upon the interstitial glandular tissue of the testis and that a secondary hormone is required to stimulate mitogenesis in the tubular epithelium.

Hamilton & Leonard (1938) made a still more extended inquiry. They removed the pituitaries from forty-one rats weighing between 110 and 150 g. and at the same time manipulated one of the testes of each rat into the abdomen and took steps to prevent its redescent. The rats were divided into two groups. The rats of one group received 0·5 mg. of testosterone propionate each day, the others being kept as untreated controls. At periods varying from 13 to 44 days after the operation the rats were killed and their testes examined. It was found that not only the scrotal testes but the cryptorchid testes also had to some degree maintained their weight under the influence of testosterone (Table 103).

TABLE 103.　The effect of testosterone propionate on the cryptorchid testes of rats after hypophysectomy (Hamilton & Leonard, 1938)

Treatment	Position of testes	Average weights of testes (mg.)
Testosterone propionate (0·5 mg. daily)	Scrotal	620
None	Scrotal	320
Testosterone propionate (0·5 mg. daily)	Abdominal	182
None	Abdominal	126·5

Hamilton & Leonard found that if the treatment with testosterone was not begun until an interval of 9 or more days had elapsed after hypophysectomy, spermatogenesis was not renewed.

Cutuly, McCullagh & Cutuly (1937a) removed the pituitaries from a series of adult male rats of the same strain, and again found that by injecting androgen spermatogenesis could be maintained although atrophy of the interstitial cells was not prevented. Androsterone or testosterone in daily doses of 1·5 mg. caused complete maintenance of most of the tubules, so that at the end of 18 days they still contained intact seminal epithelium and spermatozoa, although varying degrees of damage could be seen in all the testes examined. It was noted particularly that the testes remained in the scrota, which had not become atrophied. In rats after hypophysectomy, unless some androgen is given, the scrota shrink and the testes recede into the abdomen, and Cutuly and his colleagues believe that the action of androgen in maintaining spermatogenesis after removal of the pituitary is largely the result of keeping the testes in the scrotum. They are of opinion that spermatogenesis in the rat is, to some extent at least, independent of any direct hormonal stimulation. Some of their results are shown in Table 104.

TABLE 104.　The effects of androgens on the testes and accessory generative organs of rats after hypophysectomy (Cutuly, McCullagh & Cutuly, 1937a)

Surgical treatment	Hormone (1·5 mg. daily)	Testes	Seminal vesicles	Ventral prostate	Coagulating gland
Hypophysectomy	Androsterone	1,570	431	239	139
Hypophysectomy	None	540	96	34	42
None	None	2,410	745	325	185
Hypophysectomy	Testosterone	1,370	906	251	144
Hypophysectomy	None	560	65	25	38
None	None	2,320	668	331	148

Weights of organs in mg.

Chu (1940) experimented with pigeons and found that whereas testosterone caused atrophy of the testes and arrested spermatogenesis in adult normal birds it had an opposite effect after hypophysectomy, for in this condition it maintained spermatogenesis and to some degree the weight of the testes; though these were reduced in size they were more than 5 times as heavy as those of hypophysectomized controls which had received no androgen.

Wells (1936b) demonstrated the gonadotrophic action of androsterone by giving the hormone to immature ground squirrels or to adult squirrels at a period of the year (18 August to 6 November) when they are not sexually active. He gave 1·5 mg. of androsterone daily by subcutaneous injection during 20 to 31 days to thirteen immature and three adult squirrels. Examination on the day after the last injection revealed that precocious spermatogenesis had been induced in both the immature and non-rutting adults. Moore (1937), also, has reported that androgen will cause an increase in the size of the testes of ground squirrels if given during a period when in normal circumstances these organs are quite inactive.

Maintenance of spermatogenesis by androgen after hypophysectomy has been observed in other species than those already mentioned, namely the rabbit (Greep, 1939) and guinea-pig (Cutuly, 1941).

The gonadotrophic action of androgen in men whose pituitary functions are impaired has been noted by many clinicians. Hamilton (1937) reported the case of a patient aged 27 who had the characters produced by castration in childhood, including a high-pitched voice, wide hips, female distribution of hair, a penis only 2·5 cm. long with incomplete separation of the prepuce from the glans. The scrotum was flat and the testes impalpable. He was given subcutaneous and intramuscular doses of testosterone acetate 3 times a week until a total of 550 mg. had been given. Later he was given 20 mg. of testosterone propionate twice a week. Under this treatment the scrotum became enlarged and pigmented, and within it both the testes became enlarged and palpable. Vest & Howard (1938), Villaret, Justin-Besançon & Rubens-Duval (1938) and Spence (1940) have also recorded an increase in the size of the testicles in hypogonadal subjects under treatment with androgen. A possible accessory cause of this androgen-induced testicular enlargement, apart from hypertrophy and multiplication of cells, may be oedema of the organ or what has been described as such (p. 204).

When considering these 'gonadotrophic' effects of androgen in subjects who have been deprived of their pituitaries or whose pituitaries are not in full activity, the suggestion made by Cutuly and his colleagues that the effect may be explained partly by the action of androgen on the scrotum must be borne in mind. Nevertheless, such action cannot be regarded as a complete explanation of the results, for these have been noticed, as already mentioned, in rats with testes rendered cryptorchid by artificial means (Hamilton & Leonard, 1938) and in pigeons whose testes are permanently in the abdomen (Chu, 1940); moreover, in some of the examples which have been observed in man the testicles, though atrophic, were already in the scrotum before the administration of androgen was begun. Again as Nelson and his colleagues have shown, the capacity of different androgens to maintain spermatogenesis after hypophysectomy is not proportional to their potency as measured by their action on the accessory reproductive organs.

Selye (1940b) states that the testicular and ovarian atrophy, which ordinarily follows the administration of oestradiol through its effect in checking the supply of gonadotrophin, may be prevented to some degree by testosterone. This effect, it seems, might also be attributed to a gonadotrophic effect exercised by the androgen.

Androgen has been found to stimulate spermatogenesis in fish. Bullough (1942a) reported that when given to minnows (*Phoxinus laevis* L.) after the spawning season testosterone propionate may cause abundant mitosis in the seminal epithelium with a copious production of spermatozoa.

(c) *Sperm motility*. This appears to depend upon the secretions of the accessory genital glands (Walker, 1910), and as these depend upon a supply of androgen, the motility of spermatozoa also depends on such a supply. This fact has been used as the base of a delicate test for androgen (Moore & Gallagher, 1930a). Guinea-pigs are used, their testes having been removed after isolation of the epididymis which is left in the scrotum.

(d) *Descent of the testes*. Before considering the effect of androgen on the descent of the testes into the scrotum it may be helpful to take a general survey of the matter first.

The descent of the testis may be regarded as consisting of two stages: (1) the transit from its original site near the kidney to the abdominal opening of the inguinal canal, and (2) its passage along the inguinal canal to the scrotum.

The first stage of the movement is not yet understood well enough for discussion. Investigating the second stage Engle (1932b) experimented with immature macaques. In these monkeys, when weighing between 2,500 and 4,400 g., the testes lie in the upper part of the inguinal canal. The internal ring at this age is closed and so the testes cannot be retracted into the abdomen. They can be moved a short distance along the inguinal canal but cannot be forced into the scrotum. When the monkeys were treated with pituitary or chorionic gonadotrophic extracts the testes gradually enlarged until at the end of 3 or 4 weeks they had doubled in size. While they enlarged they also became more mobile, so that they could be moved a further distance along the inguinal canal, and after 14 days of treatment they had reached the level of the base of the penis or the upper end of the scrotum, which by this time had increased in volume. As the treatment continued so did the progress of the testes, until they occupied the scrotum. Engle noted that chorionic gonadotrophic extract was more active in causing these changes than were pituitary extracts. Zondek (1935) gave daily injections of chorionic gonadotrophin to immature rats for 8 to 14 days and noted that the testes increased in size and descended into prematurely enlarged scrota, the accessory reproductive glands being also precociously well developed. The changes of weight induced in the testes and accessory genital glands, including the prostate, seminal vesicle and coagulating gland, are shown in Table 105.

Moore & Price (1932) gave gonadotrophin to immature rats and observed that, though spermatogenesis was not expedited, there was an increase ranging from 30 to 90 per cent in the weight of the testes and of 2,076 to 5,000 per cent in the weight of the seminal vesicles. Baker & Johnson (1936) gave daily injections of 5 r.u. of prolan to twelve ground squirrels at a time of the year when these animals are sexually inactive. These injections caused enlargement and partial descent of the testes. Spermatogenesis was induced in all, and the seminal vesicles, prostate and Cowper's glands enlarged and began to produce secretion.

TABLE 105. Increased weights of testes and accessory genital glands induced by gonadotrophin in immature rats (Zondek, 1935)

Daily dose of gonadotrophin (r.u.)	Mean weight of testes (mg.)	Mean weight of accessory genital glands (mg.)
0 (controls)	355	470
100	390	840
1,000	605	1,170

Greep, Fevold & Hisaw (1936; see also Greep, 1937) compared the effects on hypophysectomized rats of two gonadotrophic extracts, one of which consisted mainly of FRH and the other of LH. FRH caused an increase in weight of the testis and proliferation of the tubular epithelium but failed to induce any change in the interstitial testicular tissue or the accessory organs. LH also caused enlargement of the testes but did not cause proliferation of the tubular epithelium; it caused a pronounced increase of the interstitial cells and enlargement of the scrotum and accessory reproductive glands. Both FRH and LH caused descent of the testis, though the latter was the more influential in this respect. The effect of gonadotrophin on the testis and its descent into the scrotum has been demonstrated in another way by Cutuly & Cutuly (1938). Rats weighing between 50 and 150 g. were joined in parabiosis and one of each pair was hypophysectomized. This was followed in the rat whose pituitary had been removed by atrophy of the testes and their retraction into the abdomen. If now the normal partner were castrated so as to increase the gonadotrophic output of his pituitary, the testes and entire genital tract of the hypophysectomized rat gradually resumed a normal condition, the scrotum expanded and the testes descended.

These various experiments show that gonadotrophins cause descent of the testes into the scrotum; they also show that this descent is preceded by an increase in the size of the testes, an increased output of androgen by them and a consequent enlargement of the accessory genital structures including the scrotum.

A deprival of gonadotrophin by hypophysectomy leads to a reversal of these phenomena: the accessory reproductive organs become atrophic, the testes shrink, become flabby and, in animals in which this is a mechanical possibility, are retracted into the abdomen.

The size and consistency of the testis as factors in its descent. When gonadotrophins are lacking the testes become flat, flabby and much reduced in size; with a good supply of these hormones the testes enlarge and become rounded and firm, and their size and consistence perhaps may be contributory factors in the descent into the scrotum. Martins (1939) opened the tunica albuginea of rats weighing between 150 and 200 g., removed the testicular contents and replaced them by paraffin models of about the same size and shape as normal testes. The artificial testes were then returned to the scrotum. In five rats treated in this way the pellets remained within the tunica albuginea. Between 4 and 5 weeks after the operation the artificial testes lay either in a suprascrotal position in the canal or in the abdomen and they could not be manipulated into the scrotum by abdominal compression. Two of these rats, which had been operated on 11 months before, were given 5 mg. of testosterone propionate in 1 c.c. of oil, the dose being repeated at the end of 5 days. On the 10th day three of the four artificial testes had descended fully into the scrotum. The fourth was prevented from descending by adhesions. The experiments suggest that androgen, by causing enlargement of the scrotum and spermatic cord, and perhaps of the canal also, plays an active part in descent of the testes.

Rost (1933) likens the progress of the testis along the infundibular canal to the motion of a cherry stone projected by squeezing it between the thumb and finger, a propulsion which cannot be imparted in the same way to a soft flabby body. However, the testes are not turgid in all species at the time of their transit into the scrotum; in the rabbit they are small, soft and flabby at the time of their descent.

An extended search by the writer has not discovered any of the abdominal viscera herniated into the scrotum in mice unless they have been treated with oestrogen, although the relative dimensions of the structures and the passage-way to the scrotum in the dead body seem to offer no mechanical hindrance to their travelling in that direction. In the living animal, even when deeply anaesthetized, attempts by the writer to manipulate abdominal viscera into the scrotum have failed.

The writer at one time thought that perhaps some substance might diffuse from the testis and cause a relaxation of the cremaster muscle in its immediate neighbourhood, but he found that such a hypothesis, though attractive, is not supported by experimental tests (Burrows, 1934c, 1936a); and it appears to be eliminated by the experiments of Martins with artificial testes made of paraffin.

In connection with the size and consistence of the testis in relation with its descent into the scrotum there is a factor to which perhaps insufficient attention has been given in the past. As observed in the mouse or rat the undescended testes of the immature animal enlarge as puberty approaches. This enlargement is partly due to hypertrophy of the seminal epithelium and an increase in the calibre of the seminal tubules; an additional cause is the collection of fluid in the intertubular spaces. The latter phenomenon has been investigated by Van Os (1936), who describes the fluid as a homogeneous substance which is stained pink with eosin, purple with Dominici stain, yellow with van Gieson, usually blue but sometimes red with Mallory, and violet with Weigert's fibrin stain. It appears, he says, to be a fluid containing coagulated protein. He has noticed this interstitial fluid in the testes of rats, mice, guinea-pigs and rabbits. Indeed it is perhaps the most striking difference between the retained testes of early post-natal life in these animals and the mature organ after its descent into the scrotum. The fluid begins to collect in the mouse and rat before descent occurs and it seems to the writer that it may be an example of the local retention of water such as occurs, sometimes to a pronounced degree, in other organs of the body under the influence of gonadal hormones. To some extent it may be responsible for the enlargement of the testes which has been noticed by Spence (1940) and others in hypogonadal human patients who have been treated with androgen.

The influence of androgen in descent of the testis. A sufficiently well-developed inguinal canal and scrotum and a long enough spermatic cord are clearly essential conditions for the descent of the testis. They result from the action of androgen, and this may account for the fact that in some circumstances the artificial administration of androgen will assist testicular descent.

(*a*) *Descent of the testis assisted by androgen.* Cutuly, McCullagh & Cutuly (1937*a*) observed that daily doses of androsterone or testosterone prevented, in the rat, shrinkage of the scrotum and recession of the testes into the abdomen after hypophysectomy. Hamilton (1938*b*) discovered that premature descent of the testes could be induced in the rhesus monkey by testosterone. In this monkey (*Macacus rhesus*) the testes, he states, are in the scrotum at birth, shortly after which they return to the abdomen, where they remain until the onset of sexual maturity at the age of 4 or 5 years, when they again occupy the scrotum. Daily injections of 5 mg. or more of testosterone propionate caused descent of the testes in immature rhesus monkeys on about the 15th day of treatment. Enlargement of the accessory organs preceded descent of the testes which, having reached the scrotum, remained there.

Martins' (1939) experiment, in which artificial testes made of paraffin were caused to descend into the scrota by testosterone propionate, is another illustration of the part taken by androgen in descent of the testes. Some observations by Biddulph (1939) are of interest here. He gave 2γ of testosterone propionate daily to rats from the day of birth till they were 31 days old, to other rats he gave daily doses of 10γ in the same conditions, and control rats received no androgen. He noted that in the animals which had been under the influence of the smaller doses of testosterone the testes descended sooner than in the untreated controls, and that the larger doses of testosterone delayed testicular descent. He further noted that the development of the scrotum was more advanced in the rats which had received the smaller doses of androgen. Hypothetically such results might be explained on the supposition that the output of gonadotrophin by the pituitary was not inhibited in proportion to the different dosages.

The results of these experiments seem to be of significance for human therapy. In the treatment of boys for non-descent of the testis gonadotrophin, especially that prepared from pregnancy urine, has usually been employed. It may be that the addition of androgen in small doses would be beneficial in some of these cases by ensuring an increased development of the structures concerned. It is conceivable that occasionally the testes, because of their long retention in the abdomen or inguinal canal, or for other reasons, are unable to produce enough androgen in response to gonadotrophic stimulation. Commonly an increased development of the external genitalia provides evidence of a sufficient androgenic influence on boys who are treated with gonadotrophin for incomplete descent of the testis.

(*b*) *Arrest of testicular descent by androgen.* The more usual result of giving androgen to young animals is atrophy of the testes with an arrest of their descent. This effect, which has been recorded by Cutuly & Cutuly (1938), Burns (1939*a*) and Burrows (1939*c*), is common to all the gonadal hormones—androgens, oestrogens and progesterone alike—and may be attributed to the fact that all these hormones inhibit the supply of gonadotrophin from the pituitary and so lead to atrophy of the testes (Cutuly & Cutuly, 1938).

To sum up, we may say that the conditions usually required for descent of the testis through the inguinal canal to the scrotum seem to be (1) a testicle which is large and firm enough, (2) a canal and scrotum sufficiently yielding and voluminous to permit the passage and reception of the organ, (3) a spermatic cord the length of which will allow the testis to reach the scrotum. Condition (1) is induced by gonadotrophin and perhaps to some extent by androgen, conditions (2) and (3) depend on androgen produced by the testis itself in response to stimulation by gonadotrophin.

The gubernaculum probably plays little part other than that of a guide in the absence of which the testis may miss its correct destination. The writer (Burrows, 1936c) divided the gubernacula in mice and pressed the testes into the abdomen. The absence of gubernacula did not prevent the displaced testes from descending, but they occasionally found their way into the wrong compartment of the scrotum.

Once they have reached the scrotum an additional factor in retaining the testes there is the cooler temperature to which they are subjected in this situation. A lowered temperature encourages spermatogenesis (Moore, 1924, 1935b) and causes an increase in the volume, consistence and weight of both testis and epididymis (p. 161).

Chapter IX. *The Action of Androgen on the Accessory Generative Organs*

Uterus. Vagina. Oviducts. Prostate and coagulating gland. Seminal vesicle. Vas deferens. Epididymis. Epoöphoron. Cowper's gland. Preputial gland. Scrotum. Perineum. Penis. Clitoris. Nipples and mammae.

The Uterus

(a) *The stroma.* The first effect of androgen on the uterus seems to be *oedema*, the result in this respect resembling that produced by oestrogen. The response occurs swiftly; Greene & Harris (1940) found a considerable increase in weight of the infantile rat's uterus within 6 hours of a subcutaneous injection of 1·5 mg. of dehydroandrosterone.

Several observers have reported *hypertrophy of the uterine musculo-fibrous stroma* under the influence of androgen. The reaction occurs in normal adult females (Korenchevsky & Hall, 1937), after spaying (Korenchevsky, Dennison & Simpson, 1935; Korenchevsky, Dennison & Brovsin, 1936; Deanesly & Parkes, 1937c) and during lactation (Brooksby, 1938). Aschheim & Varangot (1939) gave 1 mg. of testosterone propionate daily to spayed and non-spayed rats and found that the changes induced in the uterus were the same in both cases. They describe the uterus as being very vascular, with oedema of the circular layer of myometrium.

In most of the experiments just quoted the hormone used was testosterone, and it may be that not every androgen will produce exactly the same consequences as those described. Korenchevsky & Hall (1940) found that, though androsterone caused enlargement of the uterus in the normal rat, it showed this effect less when given to a rat which had been spayed. Some of the effects of testosterone and testosterone propionate on the rat's uterus are shown in Table 106.

TABLE 106.   The action of androgen on the rat's uterus and vagina
(Korenchevsky & Hall, 1937)

Condition of rat	Androgen	Duration of treatment (days)	Daily dose (mg.)	Average weight of organs in mg.		Progestational changes in uterus	Vaginal mucification
				Uterus	Vagina		
Normal	None	—	—	365	252	—	—
Spayed	None	—	—	35	131	—	—
Spayed	Testosterone	21	1·5	140	251	—	+
Spayed	Testosterone propionate	21	1·5	251	320	+	+ +
Normal	Testosterone propionate	21	1·5	690	444	+	+ +

(b) *The endometrium.* The action of testosterone on the epithelium of the uterus cannot be stated in precise terms, because competent workers have observed widely differing results. In some instances, which will be quoted presently in connection with the arrest of menstruation by androgen, atrophy of the endometrium has been recorded. In other experiments androgens have brought about effects like those produced by oestrogen or by progestin.

Changes resembling those caused by oestrogen were found in an intersexual lesser shrew (*Sorex minutus*) which was examined by Brambell & Hall (1935). In this animal the male organs were normal and spermatozoa were present. No ovarian tissue was discovered. Oviducts were present and patent, the vagina was well developed cephalad but tapered distally and opened into the vasa deferentia near their junction with the urethra. The uterus was large, distended with fluid and showed cystic glandular hyperplasia. A similar condition after the administration of androgen has been recorded by Nathanson, Franseen & Sweeney (1938), who gave single doses ranging from 2·5 to 10 mg. of testosterone propionate to immature female rats. Within 130 hours of the injection the ovaries were enlarged and showed stimulation of the follicles with the subsequent formation of corpora lutea. In these rats the endometrium displayed glandular hyperplasia with mitoses followed later by secretory activity. These features were not present in control animals. Phelps, Burch & Ellison (1938) gave daily subcutaneous injections of 1 mg. of testosterone propionate to adult spayed guinea-pigs for periods of 26 or 34 days; at the end of these periods there was an advanced degree of cystic hyperplasia of the endometrium with hypertrophy of the uterine stroma. The endometrial changes resembled those caused by oestrogen except perhaps that the cystic appearance induced by the androgen was more pronounced. Aschheim & Varangot (1939) have reported the occurrence of cystic glandular hyperplasia in the uterus of spayed as well as of normal rats after daily doses of 1 mg. of testosterone propionate.

The influence of androgen on endometrioma. Wilson (1940) has reported the case of a woman of 28 with a rapidly growing endometrioma in the recto-vaginal septum, involving both rectum and vagina. Intramuscular injections of 50 mg. of testosterone propionate were given every 2nd or 3rd day for 3 months and thereafter every 3rd or 4th day. During one year of this treatment menstruation was suppressed, the pelvic mass became reduced to one-quarter of its former size and the patient remained free from pain. At the end of the year the treatment was relinquished because it had produced hirsutism and other signs of virilism. After the treatment had ceased the pain gradually returned, and the tumour grew.

Progestational changes in the endometrium. In some of their effects on the uterus androgens resemble progesterone. Klein & Parkes (1936b, 1937; see also Emmens & Parkes, 1939a) gave 5γ of oestrone on alternate days for a period of 6 days to immature rabbits weighing between 500 and 800 g. At the end of this period daily doses of 10 to 20 mg. of androgen were given for 5 days. By these experiments it was shown that testosterone, methylandrostanediol, methyl-*trans*androstenediol, methyldihydrotestosterone and methyltestosterone all caused some degree of progestational response in the uterus, their activity in this respect being about one-twentieth that of progesterone. Methyltestosterone was almost as active when given by the mouth as when given by injection, and by whichever way administered was more potent than testosterone in causing progestational changes in the rabbit's uterus. Klein & Parkes noted also that the ovaries are not essential for these progestational reactions, for they occur in rabbits which have been spayed. Robson (1937d) produced progestational changes in the uterus of adult spayed rabbits by daily doses of 9 to 10 mg. of testosterone propionate given after a week's treatment with oestrone. Korenchevsky (1937 b) showed that testosterone propionate had considerably more influence than testosterone in causing progestational changes in the uterus of the spayed rat.

McKeown & Zuckerman (1937) gave o·2 mg. of testosterone propionate daily for 9 days to rats having normal oestrous cycles. At the end of this period there were large corpora lutea in all and progestational changes were present in the uterus. Similar endometrial changes were not induced in spayed rats treated with testosterone propionate in the same way for 9 days after a preliminary 4-day course of oestrone injections, and McKeown & Zuckerman regard the progestational changes induced by testosterone propionate as an indirect effect caused through the corpora lutea. This conclusion is not accepted by Aschheim & Varangot (1939), who obtained a different result in rats, nor is it easy to reconcile with the observations of some other experimenters. It seems possible however that corpora lutea if already present might co-operate with androgen in the causation of a progestational endometrium in the absence of any previous administration of oestrogen. Brooksby (1938) gave 500γ of testosterone propionate daily for 10 days to lactating rats—in which active corpora lutea are normally present—and at the end of this time a progestational condition of the endometrium was found, though the epithelium was lower and the ridges of epithelium not so pronounced as in normal progestation. Spayed lactating rats treated in the same way showed an endometrium of the dioestrous type.

Mazer & Mazer (1940) obtained results with testosterone propionate which seem to show that it may cause progestational changes without any preparatory supply of oestrogen. They gave 2 mg. of testosterone propionate daily to adult spayed rats and after varying intervals the rats were killed and examined. The effects produced on the uterus may be summarized briefly as follows:

Days of treatment	State of endometrium
3	Little change
8	Slight progestational changes
15	Complete progestational condition
23–29	Same as after 15 days

Noble (1939) caused progestational changes in the rat's uterus by 70 mg. of testosterone propionate given in the course of 28 days after hypophysectomy and removal of the ovaries, which suggests that the action is direct.

Co-operative effects of androgen and oestrogen on the uterus. Apparently the uterus, like other accessory sexual organs whether in the male or female, requires for its well-being a balanced supply of both androgen and oestrogen.

(c) *Motility of the uterus.* The induction of progestational changes in the uterus is not the only effect which some androgens have in common with progesterone. Robson (1937 d) gave daily injections of o·01 mg. of oestrone to adult spayed rabbits for 7 to 8 days and after this gave them injections of androgen twice a day for 4 days and on the 5th day tested their uterine reactions to pituitrin and oxytocin. He found that testosterone propionate in doses of 9 to 12 mg. given in this way inhibited the normal response of the uterine muscle to intravenous injections of o·1 unit of oxytocin. The response to pituitrin was also prevented. Progestational changes were not essential for these inhibitory actions, which are the same in character as those caused by progesterone. Testosterone has also been found to inhibit contractions of the fallopian tube (Geist, Mintz & Salmon, 1939).

(d) *Suppression of oestrus.* Although in biological literature the term oestrus is often used as though synonymous with vaginal cornification, it will be convenient to consider it here in connection with the effect of androgen on the uterus, so that the discussion of oestrus, uterine bleeding and menstruation can be regarded in a co-ordinated fashion.

Ihrke & D'Amour (1931) reported that when female rats were given daily injections of an androgenic extract of bulls' testes they remained in dioestrus as long as the treatment was continued. Soon after the injections had been stopped normal oestrous cycles were resumed. These results obtained with a testicular extract have been reproduced with pure androgens. Robson (1936c) gave 0·1 mg. of testosterone dissolved in oil twice daily to mice which previously had shown normal oestrous cycles. The treatment was continued for a fortnight or longer and during this time the mice remained in dioestrus, the cycles being resumed after the injections had been stopped. Injections of oil alone in control mice caused no interference with the periodical vaginal changes. Browman (1937) obtained a similar result in rats with androsterone and testosterone. The daily doses of testosterone ranged from 0·5 to 3 mg., and those of androsterone were 3, 4 and 5 mg. All the rats so treated were in dioestrus by the 3rd day after the first injection and remained in that condition throughout the 16-day period of treatment. After the injections had been stopped vaginal cornification appeared within 4 to 8 days. Nelson & Merckel (1937) also reported the arrest of oestrus in rats by daily doses of 2 mg. of testosterone, and there is little doubt that such a reaction is not confined to particular species.

The phenomenon may be explained partly by the fact that androgens interfere with the output of gonadotrophin by the pituitary. Such an explanation is not entirely adequate, for Robson (1936c) has demonstrated that androgen can prevent oestrogen from causing vaginal cornification in animals whose ovaries have been removed. He gave to spayed mice doses of oestrogen which were known to cause cornification of their vaginae. If sufficiently large doses of testosterone were given at the same time, cornification was prevented. It seems, therefore, that there are two mechanisms by which androgen may prevent cornification of the vagina in the adult rat or mouse, namely (1) an inhibition of the gonadotrophic functions of the pituitary so that ovarian activity is arrested, and (2) a direct antagonism of androgen against the influence of oestrogen on the vagina.

(e) *Arrest of menstruation.* As mentioned elsewhere, menstruation occurs only in the primates. Hartman (1937) and Zuckerman (1937a) have both found that menstruation in monkeys may be arrested by androgens. The latter gave 25 mg. of testosterone twice a week to a rhesus monkey (*Macaca mulatta*) weighing 5·2 kg. This treatment was continued for 210 days, during which no uterine bleeding took place; previously menstrual cycles had recurred regularly at intervals varying between 23 and 30 days. At the end of the 210 days the monkey was killed. Her ovaries were very small and contained no corpora lutea or large follicles, thus manifesting a deficiency of gonadotrophin. Loeser (1938) states that 50 mg. of testosterone propionate given to a woman by intramuscular injection every other day suppressed the menses; curettage, he says, revealed a 'completely atrophic' endometrium. Geist, Salmon, Gaines & Walter (1940) say that testosterone propionate, given 3 times a week so that 500 mg. or more are received in a month, will suppress menstruation in a woman and will cause atrophy of the endometrium so that it resembles that after the menopause (cf. p. 207).

(*f*) *Inhibition of uterine bleeding.* If a female animal is subjected to a continued dosage with oestrogen or progestin and after a while these hormones are suddenly withheld, uterine bleeding will follow, as it does also and for the same reason in primates after removal of the ovaries during active sexual life. Hartman (1937) states that the uterine haemorrhage which follows removal of the ovaries, or which follows the sudden stoppage of a course of oestrone injections, can be prevented in the monkey by giving adequate doses of testosterone. He gave 100 r.u. of oestrone daily to a normal adult female monkey for 5 days and then stopped the injections. Uterine bleeding began on the 5th day and continued for the next 11 days. Later he gave the same doses of oestrone for 5 days, and then injected 10 mg. of testosterone propionate daily for the next 13 days, during which no uterine bleeding occurred. In another experiment, using the same method and doses, by continuing the injections of testosterone propionate for 63 days, uterine bleeding was prevented for the whole of this time. Zuckerman (1939) and Abarbanel (1940) have done similar experiments on monkeys with the same results as those of Hartman. The prevention of bleeding from the uterus by testosterone may be associated with its progesterone-like activity, for progesterone has the same effect.

(*g*) *The formation of deciduomata and endometrial moles.* Testosterone propionate, given in large enough doses, will not only produce progestational changes in the endometrium, but it will act like progesterone in causing deciduomata. McKeown & Zuckerman (1937; see also Parkes & Zuckerman, 1938) gave daily injections of 14 i.u. of oestrone to six female rats for 4 days and thereafter gave them 2 mg. of testosterone propionate daily for 10 days. On the 6th day of the testosterone injections both uterine horns were injured by transfixion with a needle. On the day after the last dose of testosterone the rats were killed. The uterine horns in all were found to be swollen and in one a well-developed deciduoma was present. In the others the endometrium was progestational in character. Brooksby (1938) found that testosterone propionate in daily doses of 500γ, though causing a progestational condition in the rat's uterus, did not cause deciduoma in response to injury. He further found that with this dosage, testosterone propionate inhibited neither the formation of deciduomata in response to progesterone nor the formation of endometrial moles in response to oestrogen.

(*h*) *The growth of fibromyomata of the uterus.* Experiments on animals show that fibromyomata of the uterus are the results of an excessive supply of oestrogen or an excessive susceptibility to its influence. Human experience has proved that removal of the ovaries, or their lapse into quietude at the menopause, is followed by a cessation of growth of these tumours, and indeed a reduction of their size, and it seemed possible that their formation might be prevented or their continued growth checked by the administration of androgen. Lipschütz, Vargas & Ruz (1939) have demonstrated by experiment such an inhibitory action of androgen. They found that whereas uterine fibroids are readily induced in guinea-pigs by oestrogen, their formation can be checked by androgen. Testosterone prevented the formation of uterine fibroids in spayed guinea-pigs under the influence of oestradiol monobenzoate when the proportion of testosterone to oestradiol was in a ratio of 22:1.

Turpault (1937) treated with testosterone twenty-one women all of whom had uterine fibroids and had not reached the menopause. Daily intramuscular injections of 5 to 10 mg. of testosterone were given during 3 weeks in each month. Six of the patients did not return and were not available for an assessment of the results. In the remaining fifteen a diminution or cessation of haemorrhage, a shrinkage of the tumours and a lessening of the discomforts, were secured in eleven, the remaining four showing no benefit.

Against its possible advantages there are many objections to the prolonged treatment of women with androgen, and the results mentioned above are quoted merely because of their academic interest.

The Vagina

(a) *Before puberty*. It is noteworthy that androgens do not have the same action upon the vagina of the infant and that of the adult. In the former they induce cornification of the vagina, whereas in the adult, with few exceptions, they do not.

Premature opening and cornification of the vagina. Butenandt & Kudszus (1935), while investigating the effects of androgens on immature rats, observed that premature opening of the vagina with cornification of its epithelium was caused. They used rats 18–25 days old and injected the androgen to be tested subcutaneously twice a day on three successive days. Some of their results appear below (Table 107).

TABLE 107.   The effect of androgens on vaginal opening in immature rats (Butenandt & Kudszus, 1935)

Hormone	Total dose (mg.)	Age at opening of the vagina (days)
Testosterone	2·4	27
Androstenedione	6	24–26
Dehydroandrosterone	6	28–30
Androsterone	6	46–54
None (controls)	—	45–57

This action of androgen on the vagina of the immature rat has been confirmed. Salmon (1938b) gave single doses of 1–5 mg. of testosterone propionate or androstenediol to rats aged 27 to 30 days. In the animals so treated opening of the vagina occurred within 72 hours; between 60 and 70 hours after the injections the ovaries showed follicular stimulation like that caused by pituitary gonadotrophin, and when examined between 96 and 192 hours later well-developed corpora lutea were present. In control rats of the same age the vaginae were still closed. The same results were obtained in rats by Nathanson, Franseen & Sweeney (1938), using single injections of testosterone propionate (2·5–10 mg.). Opening of the vagina with cornification occurred within 96 hours of the injections, and 24 hours later the ovaries were enlarged and showed that stimulation of the follicles had taken place; corpora lutea developed later (v. p. 191).

In considering these results the gonadotrophic action of androgen on the ovary and testis will come to mind, and it might have appeared plausible to attribute the premature opening of the vagina caused by androgen to the gonadotrophic influence which these exert on the ovaries. But Andersen & Kennedy (1933) had already proved that opening of the vagina is not entirely dependent on ovarian activity, for though delayed it is not prevented by spaying; furthermore, Rubenstein, Abarbanel & Nader (1938) found that testosterone propionate caused opening of the vagina in immature spayed rats, so that the ovary does not appear to be essential for the reaction. Even the pituitary is not required for the phenomenon. Nathanson, Franseen & Sweeney (1938) showed that the reaction still takes place in the absence of the ovaries, or of the pituitary, or of the ovaries and the pituitary. In hypophysectomized immature and adult rats Noble (1939), also, found that testosterone caused opening of the vagina with cornification. We know that the androgens used in the experiments just quoted do not cause vaginal cornification in the adult rat whether intact or spayed provided that the pituitary is present and capable of function; and it seems justifiable to suspect that this reaction to androgen both in immature intact rats and hypophysectomized adults is due to the same cause, namely an absence or deficiency of gonadotrophin.

(b) *After puberty. The prevention of post-castration atrophy.* When androgens are given to a spayed animal the consequent vaginal atrophy can often be partly prevented, or if already present can be overcome to a considerable extent, by the administration of androgen. Korenchevsky and his colleagues (Korenchevsky & Dennison, 1936b; Korenchevsky & Hall, 1937) gave various androgens daily for a period of 21 days to spayed rats and found that all the hormones under test protected the vagina against atrophy to some extent. A few of their findings are represented below (Table 108).

TABLE 108. The action of androgens on the spayed rat's vagina
(Korenchevsky & Hall, 1937)

Hormone used	Daily dose (mg.)	Prevention of post-castration atrophy		Weight of vagina (mg.)
		Epithelium	Stroma	
Δ^5-Androstenediol	0·2–0·6	+	+ (slight)	174
Δ^4-Androstenedione	3·0	+ (slight)	+	224
Testosterone	1·5	+ (slight)	+	251
Testosterone propionate	1·5	+	+	320
*Trans*dehydroandrosterone	1·0–4·0	+ (very slight)	+ (very slight)	163
Uninjected spayed controls	—	—	—	131
Intact controls	—	—	—	252

NOTE. The + sign in the table denotes 'prevention'.

Nelson & Merckel (1937) reported that post-castration changes in the vagina of the spayed rat could be largely prevented by daily injections of testosterone (o·5 to 1 mg.), androsterone (1·5 mg.), dehydroandrosterone (1 mg.) or andro-stanedione (1 mg.).

Hypertrophy of the fibromuscular stroma of the vagina. As with the other accessory reproductive organs of the female androgens cause hypertrophy of the vaginal stroma. This effect is reported in the paper by Korenchevsky & Hall (1937) to which reference has been made (Table 108). In some instances the hypertrophy resembles that which accompanies pregnancy, the calibre of the vagina being enlarged and its walls folded so as to accommodate the enlargement.

Mucification of the vaginal epithelium. One of the effects which many androgens have in common with progesterone is mucification of the vaginal epithelium. This change, for which the co-operation of oestrogen usually is required, is induced in spayed rats by androsterone, testosterone, testosterone propionate, *trans*dehydroandrosterone, androstenediol, androstenedione and androstanediol (Nelson & Gallagher, 1936; Korenchevsky & Dennison, 1936b; Korenchevsky & Hall, 1937; Noble, 1939; MacDonald & Robson, 1939). Experiments by the last named provide an example. Into mature mice which had been spayed a fortnight or more beforehand they injected the artificial oestrogen triphenyl ethylene alone or together with progesterone or testosterone propionate. Four injections were given during 48 hours and the mice were killed at varying intervals within the next 5 days. The results show that testosterone, like progesterone, co-operates with oestrogen to cause mucification of the vaginal epithelium (Table 109).

TABLE 109. Mucification of the vagina induced by the co-operation of oestrogen and androgen (MacDonald & Robson, 1939)

Hormones used	Total doses (mg.)	Number of mice	Percentage with vaginal mucification
Triphenylethylene	0·4–0·5	46	10·7
Triphenylethylene *plus* Progesterone	0·4–0·5 / 1	13	46
Triphenylethylene *plus* Testosterone propionate	0·4–0·5 / 0·25–0·5	14	79

Vaginal cornification. Some androgens produce cornification instead of mucification of the vaginal epithelium in normal adult rats. This is the case with *cis*androstenediol and dehydroandrosterone given in doses of 1 mg. daily (Nelson & Merckel, 1937). After removal of the pituitary, vaginal cornification instead of mucification may be caused by other androgens in adult animals and has been noted in these circumstances after the administration of dehydroandrosterone and *cis*androstenediol (Nelson & Merckel, 1937), testosterone propionate (Parkes & Zuckerman, 1938), testosterone (Noble, 1939) (see p. 108).

Freud (1938) has pointed out that the formation of a many-layered stratified epithelium in the vagina is a separate phenomenon from the cornification or mucification of the superficial epithelial cells and is attributable to a different cause, which he designates mitogenetic. He suggests that gonadal hormones may be grouped as follows according to their action on the vaginal epithelium (see also p. 304):

A. *Mitogenetic (i.e. producing a many-layered stratified lining).*
Oestrone, androstenediol, androsterone, dehydroandrosterone (*none of which inhibits cornification*).

B. *Mitogenetic and Mucifying.*

 Testosterone, testosterone propionate, methyltestosterone, Δ^4-androstenedione, androstanediol. (*These induce a many-layered stratified epithelium but lead to mucification instead of cornification of the superficial layers.*)

C. *Non-mitogenetic and Mucifying.*

 Progesterone, pregnanediol. (*These inhibit cornification and cause mucification, but they do not bring about a many-layered stratified lining.*)

 Stratification of vaginal epithelium. Freud's classification seems to fit the facts, and it brings out an effect which is common to androgen and oestrogen, though more pronounced with the latter, namely the formation of a many-layered stratified lining to the vagina, the superficial cells of which will become cornified unless cornification is prevented by androgen or progestin, under the influence of which they will become mucified.

 Inhibition of cornification. As mentioned in connection with Freud's suggested classification of gonadal hormones, many of the androgens, like progesterone, inhibit vaginal cornification, though relatively large amounts may be required to do so (Robson, 1936 c).

The Oviducts

 (a) *Stimulating action.* When describing the action of androgens in the embryo examples were quoted to show that these hormones may have a stimulating action on the development of the oviducts. This effect has been particularly noticeable in birds but has been recorded also in the opossum. In birds the effect is not confined to embryonic life, for Witschi & Fugo (1940) have noted hypertrophy of the oviduct after the injection of androgen into the adult hen during the non-breeding season.

 (b) *Repressive action.* In contrast with these observations are others which reveal a suppressive action by androgen on the oviducts, which have been found absent in freemartins and in mice subjected to androgen during embryonic life. Whether these discrepant results are due to the different stages of development at which the androgens have acted, to different dosages, or to some other cause is a problem which awaits elucidation (p. 184).

The Prostate and Coagulating Gland

Camus & Gley (1896, 1897) noticed that the guinea-pig's seminal fluid coagulates rapidly after its discharge, and they found that this coagulation is caused by a specific enzyme formed by the glands adjoining the seminal vesicles. Walker (1910) inquiring further into the matter discovered that the specific coagulating enzyme is secreted by a special 'coagulating gland' which in the guinea-pig and some other rodents is distinct from the prostate. The reader will bear in mind that when mentioning the prostate in connection with the action of hormones many writers do not distinguish between the prostate and coagulating gland, but refer to both as prostate.

 Moore, Gallagher & Koch (1929; see also Moore & Gallagher, 1930 b) injected lipoid extracts of bulls' testes into castrated guinea-pigs and by this means maintained the prostate, coagulating glands and seminal vesicles in functional activity.

In ordinary circumstances these glands become atrophied in the absence of the testes.

By a different method Martins & Rocha (1931) showed that prostatic hypertrophy could be caused by an excessive production of androgen by the testes. They joined two male rats in parabiosis and castrated one of them; the result was that the prostate of the non-castrated rat became hypertrophied because its testicles were stimulated to produce an excess of androgen owing to the increased supply of pituitary gonadotrophin derived from the castrated partner. The increased output of gonadotrophin by a castrated animal's pituitary has already been mentioned (p. 45).

Korenchevsky, Dennison & Kohn-Speyer (1932) castrated rats when they were between 22 and 29 days old. About 2 months later the rats were given injections of androsterone twice a day for 7 days, at the end of which they were killed and their prostates were weighed. The results are shown in Table 110.

TABLE 110. The effect of androsterone on the prostate of the castrated rat (Korenchevsky, Dennison & Kohn-Speyer, 1932)

Daily dose of androsterone (c.u.)	Mean weight of prostate per 200 g. of bodyweight (mg.)
None	60
1	95
2·5	115
5	154
7	212

Freud (1933b) performed a similar experiment showing that either a testicular extract or androsterone will cause growth and activity in the prostate and other accessory sexual organs after castration.

Walsh, Cuyler & McCullagh (1933, 1934) removed the pituitaries of rats, thus causing atrophy of the testes and accessory genitalia. In such rats they found that prostatic atrophy could be prevented by the daily injection of '9 bird units' of androsterone.

Callow & Deanesly (1935b) showed that the prostate, coagulating and ampullary glands could all be maintained in a functional state after castration in rats, mice and guinea-pigs by a daily administration of androsterone. Itho & Kon (1935) gave 50 mg. of androgen extracted from human urine every day to young dogs, with the result that the accessory organs of generation became much enlarged as compared with the same organs in littermate controls (see Table 111).

Bottomley & Folley (1938a) gave 2 mg. daily of various androgens dissolved in sesame oil for 30 days to immature guinea-pigs, with the result that the testes were diminished in weight, owing to inhibition of the gonadotrophic function of the pituitary, while the seminal vesicles and prostate were enlarged as compared with controls (Table 97).

Zuckerman & Parkes (1938) treated a castrated rhesus monkey weighing 6·72 kg. with injections of testosterone propionate. At first the dosage was 5 mg. given every other day, and later 17·5 mg. were given once a week. At the end of 91 days the prostate, seminal vesicles and Cowper's glands were of normal size and in full secretory activity.

For this stimulating action of androgen on the prostate it is not essential that the age of puberty should have been attained. Price (1936) found that hypertrophy with precocious secretory activity could be induced by androgen in rats at the age of 12 days.

The generalization that the actions of androgen, and of other gonadal hormones, are organ-specific and not sex-specific is exemplified by the reactions of prostatic tissue. Korenchevsky & Dennison (1936a; Korenchevsky, 1937a) tested a number of androgens on spayed female rats and noted that every one of the compounds caused an enlargement with secretory activity in the vestigial prostates of these animals. In normal female rats these glands resemble histologically the prostates of castrated males, consisting of diminutive tubules lined with low cubical epithelium (Marx, 1932). After treating female rats with effective doses of androgen these vestigial glands become clearly visible to the naked eye, and histologically resemble the prostate of the normal male rat except that secretion in most of the tubules in the treated female is less than in the normal male. The androgens found to have these stimulating effects on the female prostate included androsterone, *trans*dehydroandrosterone, androstanediol, testosterone, testosterone propionate, Δ^4-androstenedione and Δ^5-androstenediol.

Moore & McLellan (1938) made a microscopical study of surgically enucleated prostatic tissue from five men who before operation had been treated with intramuscular injections of testosterone propionate in doses varying from 285 to 1,125 mg. during periods of 12 to 95 days before the operation. They were unable to recognize any results of the treatment in the sections examined, and concluded that either benign hypertrophy in man is an irreversible condition or its presence is not dependent only upon a senile diminution in the supply of androgen.

Attention may be directed to what at first sight might appear an anomalous consequence of giving certain androgens. Korenchevsky & Hall (1939) found that some old rats treated with androsterone and dehydroandrosterone showed a decrease instead of an increase in the weights of their seminal vesicles, prostate, preputial glands and penis. Testosterone and testosterone propionate, on the other hand, caused an increased weight of these organs. This odd-seeming result they explain by the inhibitory effect of androgen on the output of gonadotrophin by the pituitary, with a resultant arrest of formation of androgen by the testes. Seeing that testosterone is considerably more potent than androsterone or dehydroandrosterone the explanation appears sound, provided that the inhibitory influence of androsterone and dehydroandrosterone on the gonadotrophic activity of the pituitary is relatively in excess of their stimulating influence on the accessory genital organs.

These various experiments show that the main effect of androgen on the prostate and coagulating glands is confined to the epithelial structures, which become hypertrophied and functionally active. Any increase of stroma induced is, it seems, merely that required to support the hypertrophied epithelium and its secretory products. These effects are in contrast with those produced in the accessory male generative organs by oestrogen, which causes atrophy and eventually metaplasia of the epithelium together with hypertrophy of the stroma. Apparently small doses of oestrogen may co-operate with androgen in maintaining the prostate and other accessory generative organs after castration, whereas larger doses of oestrogen are antagonistic to androgenic action. Androgen in excess will protect the prostate from the epithelial atrophy and metaplasia which would be caused by oestrogen in its absence.

These results may be summarized by stating that androgens cause enlargement and secretory activity of the prostatic epithelium. This effect is produced whether the individual is (1) normal, (2) immature, (3) castrated, (4) hypophysectomized, (5) male or (6) female. Furthermore, androgen will protect the prostate and coagulating gland from the deleterious effects of an excess of oestrogen.

The effect of androgen on the production of acid phosphatase by the prostate. Kutscher & Wolbergs (1935) discovered that the normal human adult prostate contains a large amount of a phosphatase for whose activity the optimum pH is 4·9 which, it may be noted, is about the pH of the vagina during oestrus in women. Gutman & Gutman (1938a) found but little acid phosphatase in the kidney, liver, duodenum, vertebrae, testis or Cowper's gland, whereas it was present in high concentration in the prostate and preputial gland. The amount detectable in the prostate increased at puberty; at 4 years of age 1·5 King-Armstrong units were found per g. of fresh tissue as compared with 73 units in a boy of 13 years. A high acid phosphate concentration has been found in the osseous metastases of a malignant prostate (Gutman, Sproul & Gutman, 1936; Huggins, Stevens & Hodges, 1941) and in the serum of patients with prostatic cancer (Gutman & Gutman, 1938b; Huggins & Hodges, 1941; Huggins, Scott & Hodges, 1941). In such cases a fall of the serum acid phosphatase has been noted to follow castration or the administration of oestrogen (Huggins, Scott & Hodges, 1941), whereas androgen caused a sharp rise of the serum content of acid phosphatase in three cases of prostatic cancer (Huggins & Hodges, 1941).

Huggins, Masina, Eichelberger and Wharton (1939) devised a method for collecting the prostatic secretion of dogs and by direct assay found the secretion of acid phosphate was arrested by castration and resumed once more if the dogs were given testosterone propionate. A comparable observation has been made with the rhesus monkey (Gutman, 1942).

The treatment of non-malignant enlargement of the prostate in man by castration. The dependence of the gland upon the testes enabled surgeons to cure benign enlargement of the prostate by castration. This successful operation was discarded largely because of an unwarranted rumour that mental deterioration was apt to follow. The effect of castration, and of gonadal hormones, upon the secretory activity of the prostate in old dogs has been studied experimentally by Huggins & Clark (1940).

The Seminal Vesicle

John Hunter called attention to the atrophy of the seminal vesicles which follows castration. Most of the modern experimenters when inquiring into the effects of androgen on the prostate and coagulating glands have noted also their effects on the seminal vesicles. The results are identical in kind; androgens favour the development and functional activity of the epithelium of the seminal vesicle, but have little if any detectable action upon the stroma except so far as this is needed for the support of the hypertrophied epithelium and the secretion produced by its activity. Examples of the effect of androgen on the weight of the seminal vesicles are given in Tables 97, 100, 101. As with the prostate and coagulating glands, androgen, when available in sufficient amount, will protect the seminal vesicles from the epithelial atrophy and metaplasia caused by oestrogen when unbalanced by androgen.

The response of the seminal vesicles to androgen, though similar in kind, is usually not equal in degree to that of the prostate (including with this the coagulating gland). Callow & Deanesly (1935a) showed that androsterone had a smaller effect on the seminal vesicle than on the prostate; 1 mg. daily maintained in an active state the prostate of the adult castrated rat, whereas to maintain the seminal vesicles a daily dose of 2 mg. was required. Moore & Price (1937, 1938) have confirmed this observation.

The Vas Deferens and Epididymis

(a) *Histological effects.* Vatna (1930) studied the results of castration and of an androgenic extract of bulls' testes on the vas deferens of the rat. Normally the vas is lined by a tall, single layer of columnar ciliated epithelium. After castration the cilia disappear, the epithelial cells become small and rounded, cease to secrete, and form a stratified layer many cells thick. The muscle also becomes reduced in size. After castration, Vatna points out, these changes may be prevented, or, if they have already occurred, may be reversed by means of testicular extract. These observations on the rat were confirmed by Moore (1932). Itho & Kon (1935) made observations on puppies, and their results, though few, are interesting. Four littermate puppies were used and when they were a month old two of them were given daily doses of 50 mg. of an androgenic extract prepared from human urine. One treated and one control puppy were killed 3 weeks after, and the other two 6 weeks after the experiment began. A summary of the results is shown in Table 111.

TABLE 111.   The effect of daily doses of 50 mg. of androsterone on the weights of the testes and accessory generative organs of young dogs (Itho & Kon, 1935)

	Testes (mg.)	Epididymis (mg.)	Vasa deferentia (mg.)	Seminal vesicles (mg.)	Prostate (mg.)
1. Treated	75	265	182	610	910
2. Untreated	122	107	47	298	397
3. Treated	152	476	230	1,370	3,220
4. Untreated	344	220	64	435	570

It will be seen from the table that the administration of androsterone had caused a diminution in the size of the testes accompanied by an enhanced development of the accessory genital structure, including the epididymis and vas deferens.

(b) *Functional effects.* The action of gonadal hormones on the motility of the vas deferens of the rhesus monkey was tested by Martins & Valle (1938) using *in vitro* methods. They found that testosterone inhibited contraction of the vas while oestrone stimulated contraction, as also did castration. Their results may be compared with those of progesterone on the uterus and ureters.

The Epoöphoron

Except the vestigial prostate found in some female animals, few homologues of the male reproductive organs remain in the adult female apart from those composing the external genitalia. The only persistent relic of the wolffian body in the female is the epoöphoron, which represents the cranial end of the wolffian apparatus, corresponding with the vasa efferentia testis and epididymis of the male. Its response to androgens is the same as that of the homologous structures in the male. Granel (1939) gave three injections, each of 5 mg. of testosterone propionate, at 5-day intervals to adult female guinea-pigs. This treatment caused changes in the epoöphoron in the absence of noticeable reactions elsewhere; the epoöphoron was hypertrophied, its tubules being longer and more convoluted than normal. Under the microscope its sections resembled those of the epididymis and the epithelium had assumed a glandular character. Using Dustin's colchicine technique on guinea-pigs which were under treatment with testosterone, Granel found that mitosis was conspicuous in the epoöphoron, though not so pronounced as in the uterine glands and ovarian follicles—which, however, are normally the sites of germinative activity which the epoöphoron is not.

Cowper's Gland

Like the other accessory generative organs, Cowper's glands become atrophic after castration (Moore, 1932). This atrophy can be largely prevented by extracts of testicle or by pure androgens. Heller (1930, 1932), using rats and guinea-pigs, studied in detail the effects of castration on Cowper's gland; this becomes reduced in size, the tubules and acini merge together, the epithelial cells become atrophic and cease to secrete. The daily administration of a testicular extract for 30 days to a rat which had been castrated 8 months previously caused a return of the glandular epithelium to the normal size. Tschopp (1936) castrated rats and gave them ten daily injections of various androgens, including testosterone, androsterone, androstanediol and androstenedione, and found that in every case castration atrophy was largely prevented. He made the further observation that Cowper's glands, and some other accessory generative organs in the castrated male, may be stimulated into growth and activity by appropriate doses of oestrogen. This effect is amplified when androgen and oestrogen are both supplied (Table 112).

TABLE 112. The effect of androgen and oestrogen on the accessory generative organs in the castrated rat (Tschopp, 1936)

Hormone given	Daily dose	Number of rats	Seminal vesicles	Prostate	Cowper's glands	Preputial glands
None	0	3	10	29	33	30
Testosterone	50γ	4	28	53	71	64
Oestrone	20 r.u.	3	38	57	73	43
Testosterone plus Oestrone	50γ 20 r.u.	3	69	98	174	84

An experiment which showed the stimulating effect of androgen on Cowper's gland in the castrated monkey has been mentioned earlier (p. 216).

The Preputial Gland

In rats and mice these glands are well developed in both sexes, and their activity is stimulated by androgens and oestrogens. After castration they become atrophic. Voss (1931) found that this post-castration atrophy in the mouse could be overcome by means of testis transplants or by the injection of androgen. Freud (1933 a) showed that a testicular extract (Hombreol) restored the preputial glands, which had become atrophied following castration in young male rats, and caused an increased weight of these glands in normal male rats. A still greater increase was obtained when some oestrone was given in addition to the androgenic preparation.

Tschopp (1936) made a similar observation on rats and found that hypertrophy of the preputial glands could be produced not only by a number of different androgens, but by oestrogens as well, and oestrogen and androgen co-operated in this action (Table 112).

Korenchevsky, Dennison & Brovsin (1936) gave daily doses of 167γ of testosterone to gonadectomized male and female rats and so brought their preputial glands up to and above their original weight. This they say is an exceptional result; enlargement of other accessory generative glands in castrated animals is caused by androgen, but the glands do not as a rule regain their normal weight.

Bulliard & Ravina (1937) made a similar observation after the use of testosterone propionate. Van der Woerd (1938) obtained comparable results in gonadectomized male and female rats. The rats were castrated or spayed when 25 to 28 days old. After an interval of 14 days they were given daily injections of 200γ of testosterone or androsterone for a period of 14 days, at the end of which they were killed and the preputial glands were weighed (Table 113). Nelson & Merckel (1937) observed that either cisandrostenediol or dehydroandrosterone given daily to normal or hypophysectomized female rats caused pronounced enlargement of the preputial glands accompanied by cornification of the vagina and enlargement of the clitoris.

TABLE 113. The effect of androgen and oestrogen on the rat's preputial gland after gonadectomy (Van der Woerd, 1938)

Hormone	Daily dose (γ)	Average weight of preputial glands in mg.	
		Males	Females
None	0	27	31
Oestrone	10	43	40
Androsterone	200	59	61
Testosterone	200	79	86
Oestrone plus Androsterone	10 200	57	67
Oestrone plus Testosterone	10 200	106	97

The stimulating effect of androgen on the preputial gland of the female was demonstrated also by Mazer & Mazer (1940). They gave 2 mg. of testosterone propionate daily for 29 days to adult spayed rats and weighed their preputial glands at the end of this period. Not only were the preputial glands of the treated rats greatly enlarged as compared with those of untreated spayed rats, but they were twice as large as those of normal untreated females (Table 114).

TABLE 114.   The effect of testosterone propionate on the preputial glands
of the spayed rat (Mazer & Mazer, 1940)

Rats	Treatment	Average weight of preputial glands at the end of 29 days (mg.)
Normal	None	80
Spayed	Sesame oil	50
Spayed	2 mg. daily of testosterone in sesame oil	182

Noble (1939) performed a similar experiment with the same sort of result.

External Genitalia

The external genitalia of the two sexes are based in the main upon a single model,
the components of which show essentially the same reactions to androgen in male
and female. For brevity, therefore, the two sexes will be considered together in
the present discussion.

(i) *Scrotum and perineum.* The dependence of the scrotum on the testes for its
full development is demonstrated by the atrophy which it undergoes after
castration. Wells (1937 a) says that the weight of the scrotum depends mainly on
the cremaster muscle, and he showed that the scrotal atrophy which follows
castration is not due merely to the absence of the testicle, for the average weight
of the whole scrotum after complete castration is less than the weight of half a
scrotum from which one testis has been removed while the other testis remains
intact; though unilateral castration does lead to some atrophy of the empty half
of the scrotum. These observations were carried out on the ground squirrel
(*Citellus*). Hamilton (1936, 1937) showed that the development and maintenance
of the rat's scrotum are dependent on androgen. In the white rat the ventro-
caudal portion of the scrotum, he says, is wrinkled and reddish yellow in colour.
In ten castrated and seven hypophysectomized adult rats the scrotum became
atrophied, lost its pouch-like form and its pigment, and eventually resembled the
perineal skin of a female rat, becoming closely attached around the perineum and
base of the tail. Testosterone acetate prevented these changes. Furthermore,
twenty immature male rats given daily injections of testosterone for 19 days,
beginning when they were 14 days old, had scrota $1\frac{1}{2}$ to 3 times as large as those
of non-injected littermates.

According to Nelson (1937a) the rat's scrotum is particularly sensitive to
testosterone, for a daily dose of 0·25 mg. maintains the scrotum just as well as
3 mg., whereas most of the accessory genital organs show a greater response to
3 mg. than to 0·25 mg.

Enlargement of the scrotum has often been noticed in boys after treatment with
gonadotrophin for undescended testicle. In these instances the scrotal enlargement is due to an increased output of androgen in response to gonadotrophic
stimulation. In eunuchs and men with testicular atrophy scrotal enlargement
occurs as a direct response to treatment with testosterone (Vest & Howard, 1938).

Androgens have well-marked effects upon the perineum of the female also.
Hall (1938) examined the action of testosterone propionate on normal female rats
and on others whose ovaries had been removed during infancy. The injections of

androgen were continued for the last 3 weeks of life. Normally, she says, the vaginal introitus is covered by skin with a few hairs and a squamous keratinized epithelium beneath which are numerous sebaceous glands. After treatment with testosterone this skin is thickened, the sebaceous glands are more numerous and are greatly hypertrophied.

When androgens are given at an earlier stage of growth their effects on the perineum of the female are pronounced. Hamilton & Gardner (1937) gave 500γ of testosterone propionate subcutaneously to pregnant rats during the last third of pregnancy. The external genitalia of the female offspring of these rats were masculine in appearance. The clitoris was large and resembled a penis, there was no vaginal opening and the perineum was extensive and scrotum-like. Raynaud (1938a) performed a similar experiment on mice. During the last few days of pregnancy the mothers were given subcutaneous injections of testosterone propionate. As soon as they were born the young were separated from their mothers and received no more testosterone. When adult these mice showed an absence of the perineal part of the vagina, the ano-genital distance was greater than normal and a kind of scrotum was present.

The writer has seen in adult female rats and mice, which had been treated prenatally in this way, a substantial perineum, with a pronounced median raphe extending as a raised cutaneous ridge on to the caudal surface of the penis, as the organ may be called in its transformed state in these androgen-treated females. The lower end of the vagina was absent and the condition produced had a remarkable resemblance to that of a castrated male.

The motor function of the scrotum. The writer is not aware of any published work directly showing the influence, if any, of androgen on the muscular activity of the scrotum. Andrews (1940) has studied the effects of castration on this function in rats. A thread was stitched to the scrotum and attached to a recording lever. The rat was anaesthetized and immersed in a water bath with its head out. The temperature of the bath was 38° C. at the beginning and was gradually lowered to 20° C. and then raised again to its original level, the changes throughout taking place at the rate of about 1° C. per minute. With normal adult rats cold caused contraction and heat caused relaxation of the scrotum; but with rats which had been castrated when 5 weeks old and were tested 5 weeks or more after the operation there was but little contraction in response to a lowered temperature. If, however, a rat were tested in the same way within a few hours after castration, the scrotum reacted as in the intact animal. With normal infantile rats there was little scrotal response to cold; an increased responsiveness coincided approximately with the arrival of puberty.

(ii) *Penis and clitoris.* Long before the discovery of hormones the penis had been noticed to remain undersized in men castrated early in life. A similar effect follows castration in other animals (Richon & Jeandelize, 1903). In the free-martin, which is genetically a female modified by the sexual hormones of her twin brother, the clitoris, as John Hunter had shown, is much enlarged. These early observations suggested that the development of the penis depends on some influence derived from the gonads.

Numerous experiments on animals and clinical observations on man have since proved that the growth and development of the penis is dependent on androgen.

In the male an excessive supply of androgen causes hypertrophy of the penis, a deficiency of androgen leads to atrophy of this structure.

Korenchevsky, Dennison & Kohn-Speyer (1932) castrated rats when they were between 22 and 29 days old, and gave them two daily injections of androsterone for 7 days, at the end of which period the rats were killed and their organs weighed. It will be seen that androsterone prevented the atrophy of the penis which followed castration in the uninjected control rats (Table 115).

TABLE 115. The effect of androsterone on the penis of the castrated rat
(Korenchevsky, Dennison & Kohn-Speyer, 1932)

Daily dose of androsterone (c.u.)	Average weight of penis expressed as mg. per 200 g. of bodyweight
None (controls)	72
1	86
2·5	104
5	108
7·5	131

Tschopp (1935) has reported comparable results, which illustrate also how androgens do not exert their specific actions on all the responsive tissues alike. He castrated rats weighing between 70 and 80 g. and used them 4 weeks later, when they were given daily doses of different androgens for 20 days, after which they were killed and the organs weighed. The results show, not only that androgens counteract the tendency to post-castration atrophy, but that the effect of each androgen on the penis is not strictly correlated with its effect on the capon's comb (Table 116).

TABLE 116. Comparison of the effects of different androgens on the
rat's penis and capon's comb (Tschopp, 1935)

Androgen	Amount of one capon unit (γ)	Daily dose (γ)	Weight of penis (mg.)	Weight of seminal vesicles (mg.)
Androsterone	60	200	65	17
Androstanedione	100	200	92	51
Androstenedione	100	200	150	285

Clinical observations on man have proved that androgen in excess causes hypertrophy of the penis, and that a deficiency of androgen prevents its full development. Hypertrophy of the penis is a recognized consequence of treating boys for undescended testicle with gonadotrophin and so causing an increased supply of androgen. And in hypogonadism, in which the external genital organs remain infantile after the age of puberty, the administration of androgen will cause their enlargement.

In the female most of the constituents of the penis are present in rudimentary form, and in some animals, including rats and mice, the urethra opens normally in the female at the tip of a small replica of the male penis. Under the influence of androgen this female penis can be changed into a structure almost, and perhaps quite, indistinguishable from the penis of a male. The first observation of this fact was made on female guinea-pigs which had been spayed and grafted with the testis of a littermate.

Steinach (1916) performed a number of such experiments on guinea-pigs. In the following year Lipschütz (1917), when examining one of Steinach's guinea-pigs which had been spayed and grafted with the testis of a littermate, noticed that its clitoris had developed into an organ resembling a hypospadiac penis, carrying two horny styles like those of a normal male. This consequence of inserting testicular grafts into females has been confirmed by Sand (1919a, b) in guinea-pigs and rats, Moore (1921) in guinea-pigs, Romeis (1923) in the guinea-pig and dog. Occasionally the development of an organ resembling a penis occurs spontaneously in the female guinea-pig (Lipschütz, 1917).

The same abnormality in some degree has been induced in the females of various animals by the administration of androgen. Wiesner (1935) caused the phenomenon in rats with androsterone, Dantchakoff (1937a, 1938a) in the guinea-pig with testosterone, Korenchevsky & Hall (1937) and Korenchevsky, Dennison & Eldridge (1937) in rats with testosterone, testosterone propionate, androstenedione and androstenediol, Zuckerman (1937a) with testosterone propionate in the ape, and Burns (1939a) with testosterone propionate in the opossum.

The conversion of the female clitoris into an organ like the penis by the action of androgens is not limited to a general hypertrophy of tissues which otherwise retain their feminine characters. The enlargement may be accompanied by structural and functional alterations of such a pronounced character that the female is enabled to copulate with other females (Raynaud, 1938a, b).

It will be convenient to discuss separately and in succession the changes produced by androgen on the different components of the penis and clitoris.

(a) *The relationship between the prepuce and the glans.* In many species (e.g. rabbit, cat) if a male is castrated when young the prepuce remains adherent to the glans penis in spite of the fact that the balanopreputial epithelial partition is present and well developed (Richon & Jeandelize, 1903; Retterer & Lelièvre, 1912, 1913). This epithelial partition has been described by Retterer & Lelièvre (1913) as an invagination. R. H. Hunter (1935), when studying the human foetus, came to the conclusion that the fold is not an ingrowth. At first, he says, the glans penis is naked and the prepuce gradually extends over it, remaining adherent to it. In whichever way it forms the balanopreputial fold consists at an early stage of a layer, only a few cells thick, of embryonic epidermis. At a period of development which is not the same for all species, this tissue becomes changed into a stratified, squamous epithelium. Later the cells of the middle layer of this epithelium, that is to say the cells farthest from the basement, undergo keratinization and the keratinized cells no longer adhere to each other, so that the layer of epithelial cells covering the glans becomes gradually separated from the layer of cells lining the foreskin (unpublished observations by the author). In the human child this process is usually completed or nearly completed at birth. During the circumcision of young babies areas in which the separation has not been quite accomplished are occasionally apparent, so that a little bleeding is caused when the prepuce is freed from the glans. In some animals the separation occurs only at a more advanced stage of life, and it has been noticed in such cases that an early castration will prevent the separation from taking place. According to Retterer & Lelièvre (1913) the prepuce remains adherent to the glans as a rule in the ox, and in the castrated cat also if the gonads are removed soon enough after birth. The part played by androgen in the separation of the prepuce from the glans has been indicated by Raynaud & Lacassagne (1937). In the mouse, they say, the separation between prepuce and glans is not completed until about the 46th day after birth, and the separation does not take place normally in females nor in castrated males. If, however, testosterone be given to normal females or castrated males, separation of the balanopreputial fold will ensue as in the normal male. The same result follows the administration of androgen in rats and cats. Hall (1938) has made similar observations in rats. The writer has confirmed these findings in rabbits, rats and mice. Zuckerman (1937a) has reported that 25 mg. of testosterone propionate given twice a week to a female monkey of 5·2 kg. caused hypertrophy of the clitoris and a separation between the prepuce and glans.

(b) *The influence of androgen on the development of the glans penis and glans clitoridis.* Wiesner (1935) gave daily injections of androsterone to female rats from 4 to 7 days old and killed them 3 weeks later. They showed enlargement of the clitoris, the presence of cavernous tissue, and hypertrophy of the epithelium of the glans. According to Hall (1938) in the clitoris of the normal adult female rat the exposed part of the glans is covered by a thin layer of squamous epithelium with small, dark nuclei. The glans contains no cavernous tissue. In rats deprived of their ovaries when young the prepuce remains fused with the glans. In the normal male rat the glans penis is free from the prepuce and is armed with numerous small keratinous horns which are not present normally in the female. Hall (1938) gave androgen for a period of 30 days to normal adult female rats and to adult female rats which had been spayed when 3 weeks old. The changes produced in the glans clitoridis by this treatment, apart from an increase in size and a full separation from it of the prepuce, were as follows: (a) a conversion of the thin layer of atrophic squamous cells normally covering the exposed part of the glans into a thick, well-developed, keratinized epithelium, (b) the formation of keratinized papillae on the normally smooth clitoris, (c) an increase in the size of the blood lacunae in the stroma, and (d) an increased development of the muscular and fibrous components, the nuclei of the muscle cells being larger and the fibres thicker than in untreated females. These effects were produced by testosterone propionate, which was the most potent androgen used in the experiment, and by androstenedione and androstenediol. The simultaneous administration of oestrone or oestrone and progesterone together did not prevent these masculinizing effects of the androgens. Oestrone alone or combined with progesterone had no apparent effect on the clitoris. Dantchakoff (1938a) gave a single injection of testosterone into the amniotic cavity of a female guinea-pig embryo 22 days after insemination. After birth five injections of 2 mg. each were given during the first 2 weeks of life. When 2½ months old there was no external sign that this animal was a female. It behaved as a male, copulating with normal females, and the structural details of its penis were similar to those of a normal male.

(c) *Development of an os clitoridis.* In the normal females of some species a small os clitoridis has been observed. Retterer & Neuville (1913) have recorded its occurrence in an adult lioness. The bone was relatively minute, its dimensions being 2 × 0·4 × 0·2 mm. Most female cats, they say, have a fibrous cord in the clitoris representing an os penis; in one, aged 3 years, they found by microscopical examination a small os clitoridis measuring 0·5 × 0·05 × 0·01 mm. In male rats and mice an os penis is well developed, but is absent or at most vestigial in the females.

Wiesner (1935) observed the formation of cartilage in the clitoris of female rats which had been treated when between 4 and 7 days old with androsterone; and Greene, Burrill & Ivy (1938b) and Turner, Haffen & Struett (1939) have noted the development of bone in the clitorides of rats whose mothers had been given injections of testosterone propionate during gestation. The writer has observed the same phenomenon in female rats and mice which have been subjected to testosterone or testosterone propionate before or immediately after birth. An os clitoridis developed artificially in this way, like other male organs induced by androgen in the female, does not subsequently disappear.

(d) *The urethra; hypertrophy and hypospadias.* It seems that the penile urethra was primarily designed for the passage of semen and is therefore to be regarded as a male reproductive organ. Sir Richard Owen in a footnote to the works of John Hunter remarked that in *Ornithorhynchus* and *Echidna* the urine escapes by the cloaca, the urethra serving only for the passage of semen.

The influence of the testis over the structure and functions of the urethra was recognized by John Hunter, who pointed out that in the castrated animal the periurethral glands are small, flabby, tough, and contain little secretion, whereas in the normal male they are large, soft and full of secretion. The normal condition is mainly due to a sufficient supply of testicular androgen, to which the urethral stroma appears very responsive. Freud & Laqueur (1934) noticed that an enhanced growth of the periurethral tissues was evoked in immature male rats by doses of androgen which were too small to cause any obvious reaction in the seminal vesicles.

A general hypertrophy of the urethra involving all its constituents is induced by androgen in normal males; and in castrated males androgen can restore the atrophied tissues toward normality. In females different results are produced according to the amount or potency of the androgen employed or the stage of development at which it is given. (1) With large amounts of androgen given prenatally to rats or mice the penile homologue in the female develops into an organ resembling in structural detail the penis of the male, and the urethra has its full share in this transformation; (2) after smaller doses given before birth or during the first few postnatal days—before differentiation of the external genitalia has become finally established—there occurs a defect in the ventral wall of the distal part of the urethra, so that *hypospadias,** as it is called in human pathology, is produced; (3) given at a later stage, when structural differentiation has become complete, the smaller effective doses of androgen act in the same manner as the larger doses on the female urethra; that is to say, they cause a general urethral hypertrophy, as in males.

* In the female mouse and rat the urethra does not normally open at the level of the vaginal outlet as in women, but at the tip of a column of tissue like a small penis.

Lipschütz (1917) seems to have been the first to record the artificial production of hypospadias; he discovered it in a spayed guinea-pig into which the testis of a littermate had been grafted. Wiesner (1935) injected an androgenic compound ('Androkinin' or 'Proviron') into young rats on the 4th, 9th and 13th postnatal days. In five of the females so treated hypospadias was caused. The urethral cleft appeared between the 7th and 11th days after birth. Hain (1935 b) found that the administration of an androgenic testicular preparation ('Hombreol') in doses varying from 15 to 100 c.u. into pregnant rats towards the end of gestation caused hypospadias in the female offspring. In a second paper (1936) she reported that a total of 5 mg. of androstanediol given in divided doses at hourly intervals on the 20th day of gestation caused hypospadias in all the female members of the litter. Greene & Ivy (1937) have observed a similar deformity in the offspring after the administration of testosterone to pregnant rats. The writer has seen hypospadias in mice as well as in rats as the consequence of testosterone given to the pregnant mothers or directly to the offspring soon after birth. It is of interest to note that this urethral deformity may be caused by oestrogen (Hain, 1935 a; Lacassagne, 1936 e) and progesterone (Burrows, 1939 c) as well as by androgen.

According to Ruth (1934) the urethra in the male rat is completed only a short while before birth. On the 20th day of gestation the penile part of the urethra in this animal is completed only in the proximal two-thirds, being closed ventrally by a thin wall of flattened epithelium only two cells thick. Androgen given just before birth, or within the first few postnatal days, appears to prevent the completion of this ventral wall and perhaps to reopen the terminal portion that has already formed.

Hypertrophy of the clitoris and hypospadias may occur spontaneously in animals. Lipschütz (1927 b) observed a number of female guinea-pigs in which the clitoris was replaced by a structure like a penis. The ovaries in these cases were normal in appearance and when grafted into males produced the usual effects on the mammae. Removal of the ovaries did not lead to any change in the abnormal clitorides of these females. The deformity in these guinea-pigs was hereditary. The writer has seen instances of spontaneous hypospadias in untreated female mice.

(e) *The musculature of the penis.* The muscles of the penis are particularly sensitive to gonadal hormones. Castration, or the administration of oestrogen, will cause them to undergo rapid atrophy, so that they become wasted, pallid and toneless. After castration these muscles can be restored almost to a normal condition by the administration of androgen. The matter has been studied in detail by Wainman & Shipounoff (1941), who castrated rats per abdomen when 60 days old, and killed them 6 weeks later. To some of these rats 500γ of testosterone propionate were given daily for 35 days before death. In the castrated but otherwise untreated animals the bulbocavernosus, ischiocavernosus and levator ani muscles were all atrophic, the most conspicuous histological feature being a decrease in the width of the muscle fibres. These changes had not occurred in the rats which, though castrated, had been treated with testosterone. In non-castrated rats, testosterone caused hypertrophy of the perineal muscles; and these structures became greatly increased in size in immature spayed females under the influence of testosterone propionate. The results are given on p. 229.

Condition of rat	Average weight of perineal muscles (mg.)
Castrated	279
Normal	1,211
Normal and given testosterone propionate	1,942

Action of Androgen on the Nipple and Mamma

(i) *The Nipple.* If given in the embryonic stage of life androgens retard the development of the nipples or suppress their appearance. Hamilton & Gardner (1937) examined ten adult female rats whose mothers had received daily subcutaneous injections of 500γ of testosterone propionate during the last third of pregnancy. In these rats some of the nipples were absent and others were markedly underdeveloped though the mammary glands were nearly normal. Raynaud (1938c) made a similar observation on mice whose mothers had been treated with testosterone propionate during the last few days of pregnancy.

The effects of androgen on the nipples in postnatal life, it seems, are different; at this period instead of arresting or retarding their development, androgen may encourage their growth. Bottomley & Folley (1938 *a, c*) showed that the enlargement of the nipples in the male guinea-pig before puberty depends on the testes, after removal of which they ceased to grow; these workers also found that the administration of androgen to immature guinea-pigs provoked a more rapid enlargement of the nipples than occurred in controls. They estimated the rate of growth of the young male guinea-pigs' nipples to be isometric with the rate of body-growth. Using this fact to check their results they discovered that several androgens, including Δ^5-*trans*androstenediol, testosterone, Δ^5-*trans*dehydroandrosterone and 17-methyltestosterone, promoted growth of the nipples, the first of these being the most potent. *Cis*androsterone, *cis*androstanediol, dihydrotestosterone and Δ^4-androstenedione did not have this effect. The rate of enlargement of the nipples caused by testosterone propionate was, they noted, greater in non-castrated than in recently castrated guinea-pigs. Jadassohn, Uehlinger & Margot (1938) also found that the nipples of the male guinea-pig became enlarged under the influence of testosterone propionate, androsterone, androstenedione and androstanedione. A similar effect, they remark, is caused by oestrone, equilin, equilenin, adrenosterone and corticosterone. Noble (1939) spayed a number of adult rats and 10 days later started treating them with testosterone propionate, giving 70 mg. during 28 days. This caused hypertrophy of the nipples and mammae. In hypophysectomized rats whether spayed or not spayed similar treatment with testosterone propionate caused hypertrophy of the nipples with atrophy of the mammae. The action of this androgen on the nipples seems therefore, in contrast with its action on the mammae, to be direct and independent of pituitary or ovarian influence.

(ii) *Growth of mammary ducts.* Androgens in moderate amount appear to stimulate development of mammary ducts. It may be, as with oestrogens, that larger quantities have a stunting effect. However, it is not yet easy from experimental records to define clearly the action of androgen on the breast. With the exception of Astwood, Geschickter & Rausch (1937), who experimented with immature male and female rats, and Van Wagenen & Folley (1939 *a, b*), who used spayed immature monkeys, most investigators have noted some extension of the mammary duct system under the influence of androgen whether in mature or immature, castrated or non-castrated, male or female animals. Such an effect has been recorded by Van Heuverswyn, Folley & Gardner (1939) and Lewis, Turner & Gomez (1939) in the mouse, Nelson & Gallagher (1936) and Noble (1939) in the rat, Bottomley & Folley (1938c) in the guinea-pig, and Folley, Guthkelch & Zuckerman (1939) in the monkey. In nearly all these instances the duct proliferation is reported as slight.

Possibly some androgens are more effective than others in stimulating growth of the mammary ducts and acini. Nelson & Gallagher (1936) gave 1 mg. of androstanediol daily to spayed rats for 30 days; these doses caused a complete development of the ducts, some acini also being present.

Among the androgens which have been observed to cause extension of the mammary ducts are testosterone and testosterone propionate (0·2 to 2 mg. daily to spayed rats), Δ^5-transandrostenediol, 17-methyltestosterone, and Δ^5-trans-dehydroandrosterone (2 mg. daily to spayed guinea-pigs). Nelson & Gallagher (1936) found that androsterone given in daily doses of 1 mg. to spayed rats for 30 days had no influence on the mamma.

Lewis, Turner & Gomez (1939) induced slight arborization of the mammary ducts in mice by 0·1 mg. of testosterone propionate given daily for 25 days. With daily doses of 0·2 mg. during the same period a considerable growth of the mammary ducts was obtained. Van Heuverswyn, Folley & Gardner (1939) gave subcutaneous injections of various androgens to young castrated and non-castrated mice. The injections were given every other day for 16 days, at the end of which the mice were killed. Examination of the mammae showed an extension of the ducts. This effect was produced by androstenedione, testosterone, dehydro-androsterone, cisandrosterone, Δ^5-transandrostenediol. When considering such results it has to be remembered that the reaction of a responsive tissue to a hormone may depend largely on the dosage (v. p. 332).

(iii) *Growth of alveoli.* As to the action of androgen in causing the development of mammary alveoli there appears to be no doubt; the effect has been recorded by McEuen, Selye & Collip (1936b), Collip, McEuen & Selye (1936), Nelson & Merckel (1937), and Reece & Mixner (1939), in addition to the workers whose names have been mentioned above. The last-named workers spayed thirty-six mature rats and removed one mamma from each. Thereafter one of every pair of rats was given 200γ of testosterone propionate daily for 15 days by subcutaneous injection, while its fellow remained untreated. On the 16th day the rats were all killed and those which had received testosterone propionate showed, in contrast to their fellows, an extensive alveolar system with mammary secretion. Their pituitaries showed an increased prolactin content of 40 per cent, as tested on the pigeon's crop gland.

G. L. Laqueur (1943) gave testosterone propionate in varying doses to young adult virgin rats and found that if the treatment was started during oestrus it caused proliferation of the mammary acini together with hypertrophy of the corpora lutea, enlargement of the uterus and mucification of the vagina. Laqueur suggests that the primary effect of androgen in this experiment had been to cause functional activity of the corpora lutea and a consequent formation of progesterone.

(iv) *Formation of secretion other than milk.* In addition to these effects on the mamma many inquirers have noticed that androgens provoke secretion in the mammary gland. This occurrence has been observed by Selye, McEuen & Collip (1936), and Reece & Mixner (1939), in male and spayed female rats, and by Van Wagenen & Folley (1939 *a*, *b*) in the immature spayed monkey. The secretion is not milk and its production cannot be described as lactation.

(v) *Inhibition of lactation.*   Androgens, like oestrogens, will suppress lactation, possibly by checking the output of lactogen from the pituitary. Robson (1937*c*) found that daily injections of o·1 mg. of testosterone propionate stopped lactation in the mouse, though suckling and maternal solicitude were not interrupted. Similar doses of androsterone had but little inhibitory effect on lactation as judged by the growth of the young. Folley & Kon (1938) have recorded the arrest of lactation in rats, in which testosterone propionate in daily doses of o·4 mg. per 100 g. of bodyweight prevented the production of enough milk to nourish the young. Daily doses of androsterone (o·3 mg. per 100 g. of bodyweight) had no obvious effect on lactation in their experiments. The arrest of lactation by testosterone has been observed in women. By means of testosterone propionate Kurzrok & O'Connell (1938) arrested lactation in all but one of twenty-one women. The dosage varied from 50 to 100 mg. given in divided doses of 10 to 25 mg. each. Usually the treatment was begun on the 3rd day of the puerperium or later. In the single instance in which the treatment failed the administration of testosterone had been started on the 1st day of the puerperium. Portes, Dalsace & Wallich (1939) say that one, two or three injections of 10 mg. of testosterone acetate were enough to stop milk production in each of eight women submitted to the treatment. Beilly & Solomon (1940) have recorded the results obtained in 108 women and state that 25 mg. of testosterone given 3 times at intervals of 12 hours completely suppressed lactation in 66 per cent and caused partial suppression in 29 per cent of the cases.

Prevention of mammary cancer.   Attempts to prevent cancer of the breast by giving androgen will be discussed later (p. 367).

The influence of the pituitary in the mammary responses to androgen.   Apparently the influence of androgen on the mamma, unlike that on the nipple, is indirect, working through the pituitary. McEuen, Selye & Collip (1937) removed the pituitaries from rats varying in age from 38 to 42 days old and thereafter gave them daily subcutaneous injections of 200γ of testosterone. The rats were killed after 19 days of this treatment and examined. No sign of stimulation of the mammae was found. In another group of hypophysectomized rats treated in the same way with testosterone but given a pituitary extract in addition, the mammae had undergone progressive development. Noble (1939) found that 2 mg. of testosterone propionate given daily to spayed rats caused extension of the mammary ducts and the development of alveoli, but these results were not produced if hypophysectomy had been done previously. Reece & Leonard (1942; see also Leonard & Reece, 1942) spayed and hypophysectomized adult rats and then began to give daily doses of 300γ of testosterone alone to some of the rats and combined with pituitary extract to the others. These doses were continued for 15 days, at the end of which the rats were killed. In those which had received testosterone propionate alone the mammae had involuted, whereas in those which had been given in addition pituitary extract the breasts showed a well-developed alveolar system.

Chapter X. *The Action of Androgen on Tissues and Organs other than those already dealt with*

Adrenal. Kidney. Liver. Pancreas. Thyroid. Parathyroid. Thymus. Salivary gland. Skin. Skeleton, bodyweight and muscular system. Fat. Production of Sarcoma.

The Adrenal

A SEXUAL dimorphism in the adrenals has long been recognized, and there is little doubt that the differences between male and female adrenals depend mainly on the action upon them of gonadal hormones.

(a) *Weight of adrenals.* After removal of the testes the adrenals become enlarged. Korenchevsky (1930) castrated rats, usually between the 23rd and 33rd day of life, and after intervals of 77 or 182 days weighed their adrenals and compared them with those of littermate controls. In another series, instead of castrating the rats he made them cryptorchid. It will be seen in Table 117 that castration was followed by enlargement of the adrenals, and that cryptorchidism had no such effect, the adrenals in this condition being slightly smaller than those of intact control animals.

TABLE 117. The effects of castration and cryptorchidism on the weight of the rat's adrenal (Korenchevsky, 1930)

	Intact controls	Cryptorchids	Castrates
Average weight of adrenals in mg. per 200 g. of bodyweight	37·4	34·5	46·3

The increase in the size of the adrenals following castration in the rat appears to be confined to the cortex. Excision of the ovaries has no such effect and the post-castration enlargement can be attributed almost with certainty to the deprivation of androgen caused by removal of the testes.

Several experimenters have shown that androgens cause a diminution of the weight of the adrenals, this effect being more readily produced in young than in old rats (Korenchevsky & Hall, 1939). The effect is seen in castrated and non-castrated males, and spayed and non-spayed females. Korenchevsky, Hall & Ross (1939) castrated male rats when between 22 and 25 days old and thereafter gave them injections of androsterone or testosterone propionate on 5 days a week in amounts totalling 7·5 mg. in a week; in addition they gave to some of the rats oestradiol dipropionate on 3 days a week, totalling 0·090 mg. in the 7 days. The results (Table 118) show that castration without further interference is followed by enlargement of the adrenals and that this enlargement can be prevented by giving androgen; it can be seen also that the action of androgen in reducing the weight of the adrenals can be prevented by oestrogen. In similar circumstances Korenchevsky, Dennison & Simpson (1935) obtained comparable effects on the rat's adrenal with androstanediol.

Mazer & Mazer (1940) did the same kind of experiment on female rats. Ordinarily the size of the adrenal is little if at all affected by removal of the

442

442

4422

4422

44222

442222

4422222

ovaries. As will be seen (Table 119) the administration of testosterone propionate in daily doses of 2 mg. to spayed rats was followed at the end of 8 days by a considerable reduction of the weight of the adrenals.

TABLE 118. The effect of androgen on the weight of the adrenals in the castrated rat (Korenchevsky, Hall & Ross, 1939)

Condition of rat	Average weight of adrenals (mg.)
Normal controls	53
Castrated	74
Castrated and given androsterone	50
Castrated and given androsterone *plus* oestradiol	78
Castrated and given testosterone propionate *plus* oestradiol dipropionate	77

TABLE 119. The effect of testosterone propionate on the adrenals of spayed rats (Mazer & Mazer, 1940)

Condition of rats	Average weight of adrenals (mg.)
Intact and untreated controls	40
Spayed	35
Spayed and given testosterone propionate	27

We shall now have to consider in detail the effects of androgen on the different anatomical components of the adrenal.

(b) *The demarcation zone.* Hall & Korenchevsky (1937, 1938) have given the name 'demarcation zone' to that part of the rat's adrenal where the glomerular and fasciculate zones meet. The demarcation zone, they say, consists of 1 to 4 rows of cells which are smaller than those of the zona fasciculata and zona glomerulosa, and contain few if any vacuoles. The zone, they say, is most conspicuous in sections stained with Scarlet Red, the cells being almost free from lipoid granules. After castration, they find, this zone disappears because its cells become vacuolated and impregnated with lipoid granules. The administration of androgen is followed by a reappearance of the demarcation zone in castrated rats, a sequel which they have observed after the use of androsterone, testosterone, androstanediol, androstenedione and *trans*dehydroandrosterone.

(c) *The X zone.* The X zone is a striking feature of the adrenal in the immature mouse and in man and some other species in early life (p. 385). It is formed by that part of the cortex which abuts on the medulla, and consists of cells which are stained more deeply by eosin and contain less cytoplasm than those of the remainder of the cortex. In natural circumstances this zone disappears in male mice when they are between 30 and 40 days old; in the female it persists longer but disappears rapidly if the mouse becomes pregnant. If males are castrated when young the X zone persists.

Martin (1930) reported that 'male hormone' caused a disappearance of the X zone in mice. This fact has been established by later experiments in which pure androgens were used. Deanesly & Parkes (1937c) castrated mice when they were 21 days old and thereafter gave to some of them daily doses of androgen. By this experiment they verified (a) the persistence of the X zone in castrated mice, and (*b*) its disappearance under the influence of androgens. The compounds and daily doses which showed this effect were testosterone (80γ), androstanediol (160γ), and *trans*androstenediol (250γ). Starkey & Schmidt (1938) found that daily injections of 75γ or 100γ of testosterone caused a disappearance of the *X* zone within 6 days in adolescent and adult nulliparous female mice whether intact or spayed, and in young normal or adult castrated males. Confirmatory observations on mice have been recorded by Tolenaar (1939) and Howard (1940).

(*d*) *Zona reticularis and zona fasciculata.* Although the rat differs from the mouse in not possessing a conspicuous *X* zone during infancy, the changes undergone by the adrenal of the adult rat under the influence of androgen seem to be identical in character with those induced in mice. According to Hall & Korenchevsky (1938) the reticular and fasciculate zones in the rat become enlarged after castration chiefly because of increased lipoid in the cytoplasm (Table 120). The administration of androgens, including androsterone, testosterone, androstanediol, androstenedione and *trans*dehydroandrosterone to castrated rats, caused a return of the hypertrophied fasciculate and reticular cells to normal with a reduction of lipoid granules in their cytoplasm.

(*e*) *Zona glomerulosa.* In Table 120 there is no evidence that the glomerular cells were affected by castration, and very little has been said about their response to androgen. Selye (1939*a*) has noted that 5 mg. of testosterone propionate given daily for 20 days to adult mice of the *dba* strain caused a diminution in the size of all three components of the adrenal cortex, including the glomerular zone, the shrinkage being caused by a decreased volume of cytoplasm.

TABLE 120. The effect of castration on the components of the rat's adrenal (Hall & Korenchevsky, 1938)

Condition of rat	Average weight of adrenals (mg.)	Width of structures			
		Glomerulosa (μ)	Fasciculata (μ)	Reticularis (μ)	Medulla (μ)
Normal	53	67	585	342	1,350
Castrated	73	67	668	415	1,459

(*f*) *The medulla.* In Table 120 it will be noticed that the medulla in the castrated rat's adrenal is larger than normal. In experiments on mice the adrenal medulla has been reported to be reduced in size after castration. Whether these are chance results or not may be laid aside for the moment. Little has been said about any effect on the medulla by most of those who have studied the effects of castration and of gonadal hormones on the adrenal. Cramer & Horning (1937*b*) say that testosterone propionate given in 1 mg. doses 3 times a week to castrated male mice or to sexually mature young females, whether spayed or not, causes an increase of the medulla together with a disappearance of the *X* zone. Absence of male hormone, they say, is accompanied by a diminution in the number of medullary cells fully charged with adrenaline. They state that the *X* zone which appears after castration consists mainly of medullary cells which have been temporarily inhibited from forming adrenaline. By fixation with osmic acid vapour the adrenaline of the medullary cells is rendered visible as osmophil granules. Apparently in the normal male mouse all the medullary cells are fully charged with adrenalin. After castration the adrenalin granules are diminished in number or completely disappear from the peripheral part of the medulla.

To avoid confusion of thought it must be borne in mind that Cramer & Horning include in the X zone the peripheral part of the medulla deprived of adrenalin together with a zone of cells derived from the cortex. Most writers, we believe, would define the X zone as consisting of cortex only.

The Kidney

Korenchevsky & Dennison (1934), while investigating the effects of androgen on various organs in male rats, observed that the kidneys were increased in weight. The rats were castrated when between 21 and 30 days old. After an interval they were given daily doses of 1·5 c.u. of an unspecified androgen for 21 days and then killed. Under this treatment an increase of 13 per cent in the average weight of their kidneys was noted, comparison being made with castrated but otherwise untreated controls. The addition of 180 i.u. of oestrone to each dose of androgen caused an enhanced kidney-weight of 28 per cent. The same effect of androgen on the kidney has been observed in normal and spayed female rats (Korenchevsky & Ross, 1940).

Selye (1939 a, b) reported an enlargement of the mouse's kidney under the influence of androgen and defined certain histological changes which accompanied the enlargement. He gave 0·5 mg. of testosterone propionate daily for 10 days, and then 1 mg. for 6 days, the mice being killed on the 17th day. The increased size of the kidneys he found was due to hypertrophy of the epithelial cells of the proximal and distal convoluted tubules. No change was seen in the glomeruli but the normally thin layer of epithelium covering the parietal surface of Bowman's capsule was hypertrophied, the cells being cylindrical or cuboidal instead of flat. Crabtree (1940) has reported a difference between the kidneys of normal male and female mice. The difference concerns the histology of Bowman's capsule. As usually described both the parietal and visceral layers are lined by flattened epithelium, but this description is not quite correct, for in some of the capsules, while the glomerulus is covered by a thin, flat epithelium, the parietal layer of the capsule, either wholly or in that part which is contiguous with the tubules, is lined by high cuboidal cells. In normal adult males 89 per cent, and in adult females 13 per cent, of the capsules are of this type, Crabtree says. In young males and females he gives the proportions as 25 per cent and 13 per cent respectively. Pfeiffer, Emmel & Gardner (1940) have reported an increased kidney-weight in mice following treatment with androgen and in these circumstances the extension of tubular epithelium into the parietal lamina of Bowman's capsule is, they say, often seen. Testosterone propionate alone caused an increased kidney-weight up to 35 per cent above the normal. The addition of oestradiol alone caused a slight increase, and when given together with testosterone caused the kidneys to weigh more than if the androgen had been given alone.

Korenchevsky & Ross (1940) have noticed another change in the kidneys of rats under treatment with androgens, namely an increase in the lumina of the tubules in addition to hypertrophy of their epithelium.

The effect of androgen on the excretion of water and electrolytes.   References have been made on other pages tó resemblances between the biological actions of gonadal hormones and adrenal cortical hormones. Although androgens do not display this kind of similarity very forcibly yet it can be traced. Unlike proge-sterone, androgens have little power to maintain life after adrenalectomy. Gaunt & Hays (1938) tested their capacity in this respect on adrenalectomized ferrets, to which 5 or 10 mg. of testosterone propionate were given daily. The animals so treated survived for 13, 12, 8 and 11 days respectively, thus showing a slight prolongation of life, for without treatment ferrets die, on the average, in 6 days after adrenalectomy.

There can be no doubt as to the influence of androgen in causing an accumula-tion of water in special localities in the body. Berdnikoff & Champy (1934) showed that the cock's comb has a much higher water content than the capon's comb. In the uterus, too, androgens cause a retention of water. Selye & Fried-man (1940) gave daily injections of 3 mg. of testosterone propionate to rats which had been spayed the day before the treatment began. At the end of 9 days the uteri were slit longitudinally, and 5 days later they all showed gelatinous endo-metrial tumours at the site of trauma. These tumours recall histologically the cock's comb, the swollen sex-skin of the monkey, and the myxoma-like condition surrounding the terminal ends of the vasa deferentia in mice which have been subjected to oestrogen. Retention of water seems to be a leading feature of these conditions.

The effects of androgen on the metabolism of water and electrolytes, as shown by analysis of urine, have been investigated by Kenyon, Knowlton, Sandiford, Koch & Lotwin (1940). They gave intramuscular injections of 25 mg. of testo-sterone propionate twice daily to men and women. These injections were followed by an increase in the bodyweight and a reduction in the urinary output of nitrogen, sodium, potassium, chloride, creatine, inorganic phosphorus and water. These effects were reversed on stopping the doses of testosterone. The androgen did not affect the metabolic rate, respiratory quotient, pulse rate or blood pressure.

The Liver

Korenchevsky & Dennison (1934) gave 1·5 c.u. of a testicular androgenic pre-paration daily to castrated male rats for 21 days. At the end of this time their livers were larger than those of castrated control rats. Expressed as mg. per 200 g. of bodyweight, the ratio was 814 : 684. The injection of 180 i.u. of oestrogen simultaneously with the androgen did not affect the result. In a later experiment Korenchevsky, Hall, Burbank & Cohen (1941) gave testosterone propionate to male and female rats in a dosage of 7·5 mg. per week. Under this treatment the weight of the liver in both males and females was increased (Table 121).

Korenchevsky (1941) has reported cytological changes in the hepatic cells caused by androgens. These changes, he states, are recognized by the number and size of basophile granules which are present in the cells of the liver and can be demonstrated by selective staining. The granules, he says, are coloured by Millon's reagent, are digested by pepsin and hydrochloric acid, are numerous and large in animals well fed with proteins or aminoacids, and small and few in animals which have been starved or fed on a diet rich in fat. They are more numerous, it appears, in males than in females. Gonadectomy, whether in the male or female, causes a reduction in the size and number of the granules and of

TABLE 121. The effect of testosterone on the weight of the liver and abdominal fat in the rat (Korenchevsky, Hall, Burbank & Cohen, 1941)

Condition of rat	Sex	Amount of testosterone propionate given per week (mg.)	Weight of liver per 200 g. of bodyweight (g.)	Weight of abdominal fat per 200 g. of bodyweight (g.)
Normal	Male	0	5·95	7
Castrated	Male	0	5·51	9
Castrated	Male	7·5	6·40	9
Normal	Female	0	6·51	14
Spayed	Female	0	5·63	14
Spayed	Female	7·5	6·75	12

the total weight of the liver also. In such a condition, androgens, Korenchevsky says, will cause a return of the granules to normal in size and number, and at the same time will increase the weight of the liver.

Selye (1939 a) found that 5 mg. of testosterone propionate given daily to mice for 20 days did not alter the weight of the liver. The two experiments are hardly comparable because the dosage in Selye's experiment is much above that used by Korenchevsky.

Using large doses (50 mg.) Gaunt, Remington & Edelmann (1939) did not find any increase of the liver glycogen in rats after the injection of testosterone propionate.

The Pancreas

Rathery & Turiaf (1938) castrated 5 guinea-pigs weighing between 500 and 600 g. and killed them at intervals of 12 to 79 days later. In all they found a hyperplasia of the islets of Langerhans. To another batch of guinea-pigs they gave daily injections of 2·5 mg. of testosterone propionate for 10 days and killed them on the day after the last dose. No hyperplasia of the islets of Langerhans was present in these. They say that testosterone reduced the blood sugar in animals whether they had been depancreatized previously or not. Cornil, Paillas & Rosanoff (1938) gave daily subcutaneous injections of 10 mg. of testosterone propionate for 14 days to each of four dogs weighing between 4 and 5 kg. At the end of this time pancreatic lesions were found in all, the islets of Langerhans having almost completely disappeared. No changes were found in the rest of the pancreas.

The Thyroid

The effects of castration and cryptorchidism on the weights of various organs were investigated by Korenchevsky (1930). He castrated rats or made them cryptorchid soon after weaning and after considerable intervals killed them and compared their organs with those taken from intact controls. He found that both cryptorchidism and castration caused diminution of the thyroid in about an equal degree (Table 122).

The reverse of this experiment was done by Benoit & Aron (1934), who studied the effect of thyroidectomy on the testes. Ducks and cocks were deprived of their thyroids, laparotomy being performed at the same time to record the size of their testes. From the time of the operation the testes shrank rapidly so that at the end of 20 days they had lost nine-tenths of their volume and had ceased to form spermatozoa. Benoit (1937) in later experiments found that thyroid feeding or

TABLE 122. The effect of castration and cryptorchidism on the weights of the thyroid, adrenal and pituitary glands, the penis, prostate and seminal vesicles (Korenchevsky, 1930)

(Weights of organs are expressed in mg. per 200 g. of bodyweight.)

Condition of rat	Adrenal	Thyroid	Pituitary	Prostate and seminal vesicles	Penis
Normal	47·4	27·1	8·5	1,386	222
Cryptorchid	48·6	23·1	8·7	1,062	238
Castrate	64·4	22·7	11·5	76	71

injections of thyroxin stimulated into activity the gonads of immature male ducks, and that pituitary implants from thyroidectomized ducks which had been subjected to abnormal illumination showed little gonadotrophic effect on mice as compared with implants taken from ducks similarly illuminated but with their thyroids intact.

The close association between the thyroid and the gonads which the foregoing observations reveal appears to depend on the pituitary. Kippen & Loeb (1936) say that gonadectomy performed on male or female guinea-pigs causes an increased production of thyrotrophin by the pituitary, which in turn leads to proliferative activity in the thyroid. This consequence, they noted, was less obvious when immature guinea-pigs were used for the experiment.

The effect of a pure androgen on the thyroid has been examined by Selye (1939a), who gave 5 mg. of testosterone propionate daily to adult mice for 20 days. This treatment caused some enlargement of the thyroid, in which the epithelium was increased in height, while there was a reduction in the amount of colloid present. Nathanson, Brues, & Rawson (1940) tested the effect of androgen on the thyroid by the colchicine method. They gave single doses of testosterone propionate to young rats weighing between 40 and 48 g. Twelve hours later, or after longer intervals, colchicine was given (0·05 mg. per 50 g. bodyweight). The rats were killed 12 hours after this injection. Microscopical specimens of the thyroids showed an increase of mitosis together with increased functional activity in the thyroid epithelium.

The Parathyroid

The only reference we have to the action of androgen on the parathyroids is that contained in the paper of Nathanson, Brues & Rawson to which we have just referred. They found that the rat's parathyroid like the thyroid showed increased mitosis and functional activity after a single dose of testosterone propionate.

The Thymus

Chiodi (1938, 1939) castrated rats which were 80 days old, and later compared their thymus glands with those of non-castrated rats of the same age. The experiment showed that castration led to an increase in the weight of the thymus as compared with that of a normal rat, the average weights of the gland in the two groups which showed the greatest difference being 445 mg. in the castrates and 240 mg. in the controls.

Androgens cause a reduction in the size of the thymus. Korenchevsky and his colleagues (Korenchevsky, Dennison & Kohn-Speyer, 1932; Korenchevsky & Dennison, 1934; Korenchevsky, Dennison & Eldridge, 1937; Korenchevsky, Hall & Ross, 1939) tested the effect of androsterone, *trans*dehydroandrosterone, testosterone and testosterone propionate on rats and found that all these hormones accelerated involution of the thymus—a property which they share with oestrogens. This effect has been recorded also by Schacher, Browne & Selye (1937) and Reinhardt & Wainman (1942).

The Submaxillary Salivary Gland

Lacassagne (1940) has described a sexual dimorphism of the submaxillary salivary gland in the mouse. The gland consists of (1) *acini* composed of cells with a cytoplasm which is coloured faintly blue in sections stained with eosin and haematoxylin, (2) a straight duct connecting these acini with (3) *secretory tubules* whose epithelial cells contain acidophil granules; these secretory tubules terminate in (4) the excretory ducts, which have a similar character. The glandular acini, Lacassagne says, are developed to a comparable extent in males and females, whether spayed or not spayed. This part of the gland appears to be sexually neutral. In the male the secretory tubules predominate over the acini, in the female the acini predominate. Females treated with androgen tend to have glands resembling those of the male, and the salivary glands of the male under treatment with oestrogen tend to conform to the female type. The salivary glands of a normal male mouse, he says, weigh 65 mg., and those of a female 48 mg., and the average diameter of the secretory tubules in the male is 47μ as against 37μ in the female. Castration or prolonged treatment with oestrogen much reduces the weight of the submaxillary glands and the diameter of their tubules in the male. This change is reversible and two doses of $2 \cdot 5$ mg. of testosterone propionate quickly restore the weight of the glands and the diameter of the tubules to normal, and if injections of androgen are continued the submaxillary glands will become hypertrophied. In these respects the rat's salivary glands resemble those of the mouse.

The Skin

(a) *General effects on skin; hairiness, acne.* The influence of gonadal hormones on the skin seems to have had relatively little attention from laboratory workers, and for our knowledge of the subject we have to depend largely on clinical experience. Simple observation tells us that in normal man the texture of the skin and hair, and the distribution of the latter, are different in the two sexes. Knowledge gained from individuals in abnormal conditions—from men who produce too little androgen because they have been castrated or are the subjects of hypogonadism from some other cause including old age, and women who manifest virilism because they are subjected to an excess of androgen—suggests that androgens have a stimulating effect on the growth of epidermal structures, especially in certain regions of the body, including the face. The administration of androgen to a woman for a therapeutic purpose has been known to cause an excessive growth of hair.

Youths, at the approach of manhood, are apt to suffer from acne, an inflammatory condition of the hair follicles which accompanies the stronger growth of hair at this period, especially on the face; and it has been noted that in women with hypertrophy or neoplasia of the adrenals, who consequently acquire a growth of coarse hair on the face accompanied by an increased excretion of androgen in the urine, like youths, and probably for the same reason, are subject to acne. The dependence of acne on a relative excess of androgen seems to be proved by the frequency with which this skin condition has accompanied the use of androgen for therapeutic purposes (Hamilton, 1937; Vest & Howard, 1938; Kenyon, 1938, and others).

Loeb & Haven (1929) examined the mitotic activity of the epidermis in guinea-pigs and they found that it was less in females than in males; and in females it varied at different times in the oestrous cycle, being highest at the end of the period, when the supplies of oestrogen and progestin are least. The rate of cutaneous proliferation was found in their experiments to rise during pregnancy to that of the male. The experimental results of Loeb & Haven suggest that the normal proliferative epidermal activity appropriate to each sex may depend upon a proper balance between androgen and oestrogen. If this be correct facial hairiness might be caused either by an excess of androgen or by a deficiency of oestrogen. In this connection may be quoted an inquiry by Lawrence & Werthessen (1940), who made assays of the urine of eight female patients with acne and eight normal women. A 48-hour specimen was used in each instance, collected between the 16th and 20th day of the menstrual cycle. The assays revealed no excess of androgen but a considerable reduction of oestrogen in the patients with acne (Table 123).

TABLE 123. Urinary output of gonadal hormones in normal women and in women with acne (Lawrence & Werthessen, 1940)

Condition of donor	Androgen ('units')	Oestrogen ('units')	Ratio of androgen to oestrogen ('units')
Normal	178	75	2·4 : 1
With acne	158	24	6·6 : 1

Laboratory experience seems to show that the influence of oestrogen in retarding cutaneous growth is more significant than the capacity of androgen to accelerate it (p. 376). Emmens (1942) reported that in rats, although the hair grows more rapidly in males than in females, its speed of growth is unaffected by the administration of androgen.

(b) *Pigmentation.* In brunettes with virilism an increased pigmentation of the skin is not uncommon; an excess of androgen may be suspected of causing this Hamilton & Hubert (1938) have recorded that after castration a man failed to become tanned on exposure to strong sunlight. He was given injections of testosterone propionate and under this treatment responded to sunlight by a normal increase of pigmentation. On ceasing the injections the tanning disappeared, but returned when the supply of testosterone propionate was renewed. Hamblen,

Ross, Cuyler, Baptist & Ashley (1939), by assays of urine, found indications that the amount of androgen excreted by a woman may be correlated in some degree with her complexion, for brunettes excreted more androgen than blondes in the small number of individuals on whom the tests were made (p. 172, Table 85).

The influence of gonadal hormones on pigmentation has been investigated by tissue culture. Hamilton (1940) grew explants of dorsal skin obtained from 6-day-old chick embryos of New Hampshire red or Rhode Island red fowls. Normally, he says, such explants occasionally produce melanophores and rarely red pigment cells also. The explants were bisected, one half of each being cultivated in normal medium and the other in a medium to which gonadal hormones had been added in concentrations varying from 20 to 300γ per 1 c.c. From Table 124 it will be seen that the explants subjected to either oestrogen or androgen showed an increased production of pigment, but so too did some of those subjected to sesame oil alone.

TABLE 124. Production of pigment in explants of chicken skin under the influence of gonadal hormones (Hamilton, 1940)

Additions to culture medium of hormone in sesame oil	Number of embryos	Number of explants	Percentage of explants containing red pigment cells	
			Treated	Control
Testosterone	13	30	60	0
Oestradiol dipropionate	47	148	37·8	6·0
Oestradiol monobenzoate	10	19	63·2	10·0
Sesame oil only	37	73	31·5	2·6

In monkeys some of the cutaneous pigmentary changes which manifest sexual activity can be induced either by androgen or oestrogen (Hartman, 1940).

(c) *Effects of androgen on cutaneous appendages other than hair.* Among the epidermal structures which may become especially well developed in males through the influence of androgen are antlers, teeth and tusks, the spurs and comb of cocks and a large number of the temporary or permanent cutaneous manifestations of the male sex. Many of these, including some of the most striking marriage adornments of birds, amphibia and fishes, are temporary, and appear only during the mating season when the testes are active. Not all the gay plumage which distinguishes male birds can be attributed to testicular androgen; in some species it represents a neutral condition, the distinctive female plumage being caused by oestrogen. In these species a cock's characteristic feathering persists after castration, and a spayed female will assume 'male plumage'.

The Skeleton, Bodyweight, and Muscular System
The Skeleton

Yarrell (1827) seems to have been the first to observe that the gonads affect the shape of the skeleton. His observations were on birds and will be discussed later (Chap. xix). Godard (1859) reported two male patients with, as he believed, congenital absence of the testes. Their general conformation tended toward that of the female, he said, as in males castrated in infancy. His observations might be hardly worth quoting if they had not stimulated Poncet (1877) to investigate the matter by experiments on animals.

Before referring in detail to Poncet's work it may be remarked that when the effects of gonadal hormones on the skeleton are discussed three matters need separate consideration, namely (a) the size of the skeleton as a whole and of its individual bones, (b) the texture of the bones and (c) the distinctive features of the masculine and feminine bodily contours, which depend largely on the shape of the pelvis in addition to the characteristic distribution of fat.

(a) *The size of the bones.* The length of a bone is governed by the activity of its epiphyseal tissues. When the epiphysis joins the shaft by osseous union, growth in length at that end of the bone ceases. It will be shown (Chap. XIX) that oestrogen, in addition to its action on the texture of bone, accelerates osseous union between the epiphyses and shafts and so curtails the growth of bones. This seems to be the chief reason why in man and many other mammals the male is larger than the female.

Poncet's experiments may be considered in the light of these facts. He used littermate male rabbits of the same weight and kept them in identical conditions. One of the rabbits was castrated when 3 months old, the other being kept as a control. Three months later the rabbits were killed and their skeletons were examined. Poncet found that the whole skeletal framework of the castrated rabbit was larger than that of his normal brother. The compact tissue of the bones was a little increased and the medullary cavities were enlarged. The individual bones were longer, stronger and straighter than in the intact rabbit, these changes being most noticeable in the femur, tibia, fibula and ilium, though not confined to these bones. The head of the castrated animal was longer than that of the control. Richon & Jeandelize (1905) repeated Poncet's experiment on rabbits with confirmatory results. The heads of the castrated rabbits were longer by 1 to 6·5 mm. than those of the controls, and the diameters of the face were increased.

Launois & Roy (1903) described the condition of a man of 27 years who had suffered from enteric fever when 16 years old. His testes were not palpable and his penis was very small. He was tall, with limbs which were slender and disproportionately long; he had a small thorax and pelvis, a feminine voice and hairless face. Radiographic examination showed the persistence of epiphyseal cartilages in the metacarpals and phalanges. Launois & Roy refer to another comparable case.

Bouin & Ancel (1906; see also Ancel & Bouin, 1906), with their usual fine enterprise, tested the effects of testicular extracts on the growth of the skeleton. They had already shown that male gonadal hormone is produced by the interstitial glandular cells of the testis and that these cells remain functional in ectopic testes in the absence of spermatogenesis. They obtained ectopic testes from various animals, extracted them with glycerine and water, and tested the extracts so prepared on male guinea-pigs 2 to 4 weeks old. These were divided into three groups, namely (a) normal untreated, (b) castrated and otherwise untreated, (c) castrated and injected with testicular extract. These last received 1 c.c. of extract diluted with 3 c.c. of water, the injections being given every 2nd day regularly for a period of 9 months, at the end of which the guinea-pigs were killed. The results are shown in Table 125, and measurements of the penis and seminal vesicles are included as evidence that the extracts used did in fact contain effective amounts of androgen.

TABLE 125. The effects of testicular extract on the growth of bones in the castrated male guinea-pig (Bouin & Ancel, 1906)

Lengths of structures in mm.

Condition of guinea-pig	Femur	Tibia	Nasal bones	Penis	Seminal vesicles
Normal, untreated	43·7	45·7	19·3	37·3	58·3
Castrated, untreated	44·8	47·3	22·5	20·7	13·0
Castrated, given testicular extract	43·5	46·0	20·7	32·0	35·7

The figures given in the table show a restraining influence of testicular extract on the growth of bone.

Steinach (1916; see also Steinach & Holzknecht, 1916) examined the effects on the skeleton of transplanting whole testes into spayed guinea-pigs. For comparison castrated males, and normal males and females were kept. The operations were done within a day or two of birth, littermates being used. The results (Table 126) appear to be at variance with those previously described, for the bones in Steinach's experiment are smaller in the castrated than in the non-castrated males, and are much enlarged in the spayed females bearing testicular grafts. Moore (1919, 1922) repeated this kind of experiment and obtained the same result as Steinach.

TABLE 126. The effects of gonadal grafts from the opposite sex on the conformation of the head and bodyweight in the male and female guinea-pig after gonadectomy (Steinach, 1916)

Subject	Body-weight (g.)	Ear distance (mm.)	Zygomatic distance (mm.)	Length of head (mm.)
Normal male	998	30	44	80
Castrated male	947	27	43	78
Castrated male with ovarian graft	685	21	39	69
Normal female	836	22	41	73
Spayed female with testicular graft	1,155	32	48	85

The results recorded by Bouin & Ancel seem to show that androgens to some extent curtailed the lengths of the bones, for castration led to an increased length and testicular extract prevented this effect. But it has to be remembered that they were using crude extracts the androgenic influence of which on the bones may have been overcome by other substances present.

Other experiments, besides those of Steinach & Moore, have shown that castration does not invariably lead to an increased length of bone (Pomerat & Coe, 1941, see also Table 129).

We are faced here by what seems a conflict of evidence. According to some reports castration has been followed by an increase in the length of the bones, and according to others the opposite effect has been produced. At present the discrepancy, which has been noted in man as well as in the lower animals, cannot be explained. When the effect of castration on the deposition of fat is under discussion it will be noticed that castration may be followed sometimes by thinness and sometimes by fatness. Whatever cause may lead to these varying results, we may conclude that androgens have little influence on the longitudinal growth of bone except in so far as they may counteract the influence of oestrogen on these structures.

In man there is direct evidence that androgens may favour body-growth. Webster & Hoskins (1940) treated with testosterone propionate eight boys who showed signs of hypogonadism. In each case the rate of growth, estimated in cm. per 100 days, was ascertained before treatment. Thereafter they were given testosterone propionate, the dosage ranging from 75 to 125 mg. per week. While receiving androgen the boys showed an acceleration of growth from 1·36 to 3·6 cm. per 100 days. No evidence of epiphyseal closure occurred during the period of treatment.

(b) *The texture of the bones.* Gardner & Pfeiffer (1938a) discovered that if male or female mice are subjected for prolonged periods to large weekly doses of oestrone benzoate or equilin benzoate a hypercalcification of the general skeleton ensues together with resorption of the pubic bones and their interosseous cartilage. They found further (1938b) that these changes can be prevented by giving testosterone propionate, 1·25 mg. of which will inhibit the action of 0·5 mg. of oestradiol on the bones of mice, the doses of each hormone being given once a week.

(c) *Conformation of the skeleton.* So far as concerns the distinctive shape of the skeleton in the two sexes it is certain that this depends, as Yarrell showed with birds more than 100 years ago, on gonadal activity. In men who have been castrated before puberty or have suffered from hypogonadism in the early years of life the shape and dimensions of the pelvic and shoulder girdles are apt to conform to the female type, that is to say the pelvis is wide and the shoulders are relatively narrow. As one example among many may be quoted a patient whose case was reported by Broster (1941). A man of 32 had never undergone the changes of puberty. His testes were atrophied and the external genitalia undeveloped. The distribution of his hair was like that in the female, and he had never had to shave. His height was 5 ft. 5 in. and the contours of his body were feminine, the width of his pelvis being 31 in. and that of the shoulders 30 in. A photograph which illustrates well the feminine form of the body in a male whose testes had been destroyed in early life accompanies an article by Eidelsberg & Ornstein (1940).

Gardner (1936) has found that there is no sexual dimorphism in the mouse's pelvis before puberty. The masculine and feminine divergences in shape of this part of the skeleton begin with the onset of gonadal activity. From this fact, and from the other observations on birds and men referred to above, it would seem that testicular hormones are responsible for the masculine figure.

The Bodyweight

(a) *The effect of castration.* Most experimenters have reported that male animals which have been castrated when young are lighter when adult than non-castrated littermates. Ancel & Bouin (1906) castrated young male guinea-pigs and gave to some of them injections of testicular extract every alternate day. Other guinea-pigs were kept as controls, being neither castrated nor treated with testicular extract. At the end of 88 days the untreated animals were the heaviest, the castrated and injected came next, and the castrated and uninjected were the lightest (Table 127).

TABLE 127. The effect of castration and testicular extract on the
weight of guinea-pigs (Ancel & Bouin, 1906)

Treatment	Gain of weight during 88 days (g.)
None	320
Castrated and given testicular extract	306
Castrated	266

Stotsenburg (1909) castrated rats when they were 14 or 15 days old, and thereafter kept them with non-castrated littermates. At a later date it was found that the castrated rats weighed rather less than their non-castrated brothers though the difference was not great (Table 128, see also Table 131, p. 249).

TABLE 128. Weights of rats castrated during infancy compared with
weights of non-castrated littermates (Stotsenburg, 1909)

Average age in days	Average weight Castrates (g.)	Non-castrates (g.)
143	179·9	191·4
185	182·6	190·7

Donaldson & Hatai (1911) obtained comparable results in rats, finding that castration was followed by a reduction of the bodyweight and of the weight of the brain and spinal cord. Their figures are based on eight litters of which thirteen individuals were castrated and eleven were left intact. Their tabulated figures are placed in two groups according to whether the rats were losing or gaining weight at the time of death (Table 129).

TABLE 129. The effect of castration on the bodyweight and on the weights of
brain and spinal cord in the rat (Donaldson & Hatai, 1911)

Condition	Gaining or losing weight at time of death	Average age at death (days)	Body-weight (g.)	Body-length (mm.)	Brain weight (g.)	Spinal cord weight (g.)
Normal	Gaining	363	267	194	2·014	0·640
Castrate	Gaining	251	214	186	1·897	0·595
Normal	Losing	272	211	194	1·970	0·635
Castrate	Losing	271	189	192	1·890	0·618

Van Wagenen (1928), in a similar experiment with rats castrated at the time of weaning, found that the castrated animals did not in the end attain the same weight as non-castrated littermates, though the retarding influence of castration was not noticeable at first, becoming apparent only 100 or 150 days later. A delay in the effect of castration on the growth curve was noticed by Rubinstein, Abarbanel & Kurland (1939), who castrated rats when they were 22 days old and killed them 58 days later. At first their increase of weight was equal to that of non-castrated controls, but from the 40th day of life onward the castrated rats gained weight less rapidly than their intact fellows. Lawless (1936) castrated Wistar rats when they were 21 days old, and weighed them 67–84 days later, at which time they weighed less and were slightly shorter in the body than non-castrated controls.

It might be argued that the ultimate lowering of bodyweight, as recorded in the experiments just quoted, was perhaps caused by a retardation of nutrition resulting from the operation apart from gonadectomy. The question thus raised is answered to some extent by Commins (1932) who, using inbred rats kept on a stock diet, castrated some, operated on others as for castration but without removing the testicles, and left others intact. The weights of all were recorded when they were 165 days old and indicate that the reduction in weight in the castrated animals is attributable to absence of the testes and not to the operation.

The results so far refer to males only; in females gonadectomy in early life apparently leads to an increase of bodyweight (Evans & Simpson, 1937).

(b) *The effect of androgen on bodyweight.* To test the effect of androgen on the rate of growth, Rubinstein, Kurland & Goodwin (1939) gave 1 mg. of testosterone propionate daily to rats from the 26th to the 76th day of age and found that this treatment caused a reduction of their length and bodyweight. Kochakian (1940) reported a similar retardation of growth in young male mice of the *dba* strain which had been given 0·2 mg. of testosterone propionate daily; no such effect was caused in females of the same strain. Perhaps the check imposed on growth in these experiments was due to the magnitude of the doses, for Rubinstein & Solomon (1940) found that 50γ of testosterone propionate given daily to rats, beginning when they were 26 days old and continuing for periods varying from 26–80 days, caused a significant increase in both the length and bodyweight of the treated animals.

Zuckerman & Parkes (1939) castrated a young adult Hamadryas baboon weighing 17·5 kg.; after the operation its weight fell to 12·66 kg. At this stage weekly injections of 100 mg. of testosterone propionate were begun and continued for a year, at the end of which the baboon weighed 16·19 kg. Zuckerman & Parkes (1938) had already recorded an increase in the weight of a castrated rhesus monkey after giving him testosterone propionate. In man also an increase of weight has been noticed after the administration of androgens to eunuchs or eunuchoids. For example, Villaret, Justin-Besançon & Rubens-Duval (1938) gave testosterone propionate in daily doses varying between 10 and 30 mg. to a man of 21 with atrophic testicles and pronounced eunuchoidism. At the beginning of treatment his weight was 55 kg.; after he had received testosterone throughout a period of 6 months his weight was 63 kg. Others, including Spence (1940), have recorded an increase in the weight of men under treatment with androgen.

An observation by Steinach (1916) may be recalled in connection with this discussion. He discovered that a guinea-pig spayed during infancy, its ovaries being replaced by the testicle of a littermate, eventually attained a larger size than the intact male. This fact might be regarded as additional evidence that the female is more sensitive to androgen than the male; though it might also be considered in relation with the fact reported by Evans & Simpson (1931) that, after gonadectomy, the female rat responds more readily than the male to pituitary growth hormone.

From the experimental results just cited there seems little doubt that castration of the male may reduce the ultimate bodyweight, and that androgens given in moderate amount tend to increase the bodyweight. In the present discussion additional factors have to be considered. The appetite is one of these. Men appear to eat more in proportion to their bulk than women. This may be due partly to habit and opportunity and partly to the greater output of physical energy which is perhaps a feature of natural man. However this may be, Koren-chevsky & Dennison (1934) found that the daily administration of 1·5 c.u. of androgen to young castrated rats increased their daily food intake by 4 per cent as compared with that of castrated controls. Another action of androgen which will cause an increase in weight—the effect in this instance being almost im-mediate—is the retention of water under its influence. In this respect androgens to some extent resemble corticosterone, and puffiness of the hands and face, and even oedema of the legs enough to cause pitting on pressure, have been observed in eunuchoids when receiving 25 mg. of testosterone propionate each week (Kenyon, 1938; Kenyon, Sandiford, Bryan, Knowlton & Koch, 1938).

We have to remember that the weight of an individual depends on several factors, including the size of his bones, the degree of his muscular and ligamentous development, and the amount of fat deposited in his tissues. The effect of androgen on the bones has already been discussed; its effects on muscle and fat remain for consideration.

The Effect of Androgen on Muscle

(a) *The general muscular system.* To some degree the development of muscle will be governed by the size of the skeleton, larger bones requiring stronger muscles to move them. Oestrogens restrain the growth of bone and so will curtail the development of muscle also. Androgens counteract the inhibitory effect of oestrogen on the growth of bone, therefore they must, it seems, indirectly en-courage the development of muscle. The volume of the muscles will depend largely on the amount of use to which they have been put, and in this the gonadal hormones may play a part by causing restlessness and increased activity.

The bull and horse are no doubt more nervous, pugnacious and difficult to manage, and more inclined to physical activity, than are the ox and gelding, and this nervous disposition may be expected to influence their musculature. Apart from such hypothetical considerations there is experimental evidence that androgens favour muscular development. Papanicolaou & Falk (1938) noticed a measurable difference between the temporal muscles in the male and female guinea-pig, these structures being much larger in the male. The administration of gonadotrophin, they found, caused hypertrophy of these muscles in both male and female guinea-pigs if the gonads were intact; no such effect was produced after gonadectomy. If males were castrated before puberty their temporal muscles remained small and flat as in the normal adult female; and if testosterone pro-pionate were given repeatedly to castrated immature males or to adult females, whether spayed or not, hypertrophy of the temporal and other skeletal muscles ensued. No muscular hypertrophy was induced by oestrogen or progestin, and these hormones did not prevent testosterone from inducing an increased develop-ment of muscle.

Apart from these experiments by Papanicolaou & Falk, there seems to have been little attempt to study the effect of androgen on the general muscular system.

Clinical evidence, though not so convincing as controlled experiments done in the laboratory, suggests that androgens have a potent influence over the voluntary muscles. In boys with premature virilism due to an excessive production of androgen, the muscular development is often a pronounced feature, so that the patient has been described as being of the 'infant Hercules' type; and women who have been relieved of virilism are apt to notice a loss of muscular power accompanying the cure. In some patients with pathological degrees of muscular weakness androgens have been reported to give some relief. Hesser, Langworthy & Vest (1940) made tests on two male patients suffering from muscular weakness due to 'myotonia atrophica', which is described as a myopathy accompanied by gonadal atrophy and degeneration of the adrenals. Intramuscular injections of 25 mg. of testosterone propionate were made on alternate days for 2 months. The flexor functional capacity of the muscles of the fingers were tested by a dynamometer before and during treatment, attention being given to (1) the strength of a single contraction, (2) the total work capacity and (3) the maximal rate of flexion. Improvement in muscle function was noticed within a week after the first injection and in the course of treatment the grip increased from 6 to 16 kg. in one case and from 8 to 16 kg. in the other. Total work capacity was increased by 50 per cent, and no significant change occurred in the rate of contraction.

(b) *The specific action of androgen on the musculature of the penis* has been discussed earlier (p. 228).

(c) *The effect of androgen on the weight and dynamic capacity of the heart.* It will be convenient to mention here an investigation by Korenchevsky, Hall, Burbank & Cohen (1941) into the effects of androgen on the weight and muscular power of the heart. Their tests were made on the isolated hearts of untreated normal or castrated rats, and of castrated rats which had been given repeated injections of androsterone or testosterone propionate during the previous 2 months. They found that castration alone caused a reduction in both the weight of the heart and its muscular capacity. These losses were largely or entirely prevented by androgen. Treatment with the latter increased not only the volume of the heart muscle as a whole, but caused enlargement of the individual muscle fibres as seen in sections under the microscope. To test the heart's capacity for work weights were attached and its contractions were recorded on a kymograph. The amplitude of the contractions was measured from the tracings and the number of contractions was counted for 1 minute before and 1 minute during the suspension of the weights. From the figures obtained in this way the potential

TABLE 130. The effect of androgen on the weight and energy of the heart muscle in the castrated rat (Korenchevsky, Hall, Burbank & Cohen, 1941)

Condition of rat	Weight of heart (mg. per 200 g. of bodyweight)	Relative areas of muscle fibres as seen in sections	Potential energy of the Heart-muscle (Woodsworth's quotient)	
			Before attachment of weight	During attachment of weight
Normal	522	28	81·3	73·5
Castrated and untreated	496	22	71·9	60·0
Castrated and treated with androgen	561	24	78·2	81·7

energy of the heart muscle was calculated by means of Woodsworth's quotient. The results are shown in Table 130.

The Effect of Androgen on Blood Pressure, the Distribution of Blood and the Formation of Erythrocytes

(a) *Blood pressure.* Clinical experience and general observation might have led us to suspect that androgen would cause an increase of the systemic arterial blood pressure. Greene (1938), however, investigated the arterial tension in fourteen patients who were given testosterone for the treatment of enlarged prostate and he found that the blood pressure was unaffected by the treatment.

(b) *Distribution of blood.* Edwards, Hamilton & Duntley (1939) have observed, by means of a spectrophotometer, a lack of arterial blood in the skin of men who have been castrated. The condition was remedied by testosterone propionate. This observation might be considered in connection with the other recognized effects of androgen on the skin (p. 239).

(c) *Erythrocytes.* There seems little doubt that androgens have an influence on the formation of red blood cells, for in various different species it has been noticed that the erythrocyte count is higher in males than in females. Steinglass, Gordon & Charipper (1941) castrated rats and noted that after the operation the number of red blood cells diminished. By giving testosterone to these rats the erythrocyte count was raised from 7·4 to 9·1. In spayed females testosterone caused hyperplasia of the marrow with increased erythrogenic activity. Vollmer, Gordon, Levenstein & Charipper (1941) have reported that testosterone in daily doses of 0·5 or 1 mg. caused a rise of the erythrocyte count in male rats, whether normal, castrated, or hypophysectomized. This result, it seems, can hardly be attributed to an increased concentration of the blood due to loss of water, for androgens cause an increased retention of water in the body (pp. 143, 236).

The Effect of Androgen on the Deposition and the Distribution of Fat

(a) *The deposition of fat.* Reference will be made later to the fact that the female body contains a larger proportion of fat than the male (p. 375). Castration and spaying have been done since ancient days on the farm, and one of the recognized advantages, no doubt, was the increased accumulation of fat in the body. This effect has been tested by Korenchevsky (1930), who castrated rats, recorded their rate of growth and eventually measured the amount of retroperitoneal fat in their bodies. Castration slowed the gain in weight though it led to a greater deposit of fat in the retroperitoneal tissues. Cryptorchidism caused an increase of both bodyweight and retroperitoneal fat (Table 131, see also Table 121, p. 237).

It seems probable from what has been said in earlier pages that the increased deposit of fat after castration will be accompanied by a reduction in the weight of

TABLE 131. The effects of castration and cryptorchidism on the bodyweight and deposition of fat (Korenchevsky, 1930)

Condition of rat	Average gain in bodyweight per week (g.)	Average weight of retroperitoneal fat (expressed as g. per 200 g. of bodyweight)
Normal (controls)	12·2	5·48
Cryptorchid	13·1	6·43
Castrated	10·7	8·14

muscle and bone; this may account for the fact that in experiments quoted earlier the castrated animals have been lighter than the normal controls in spite of the increased fat.

Clinical observation of men and veterinary experience have both shown that castration does not always lead to diminished stature or to obesity. A eunuch may be tall and thin, and the same features may be noticed in the castrated cat. Such differences in the effects of castration may be caused in some instances by the environment. It has to be remembered, too, that the endocrine system is intricate and to disturb the biological conditions in one part of the mechanism may upset the balance in other parts; the end results of an experiment may represent the consequences not of a single disordered function but of a widespread loss of equilibrium. Castration affects the activity of the thyroid, which in its turn regulates to some extent the rate of growth and the accumulation of fat; and it appears that gonadectomy in the male may not always lead to a lowered basal metabolism rate but to one which is above the normal.

The effect of artificially administered testosterone propionate on the deposition of fat in the abdomen has been tested in castrated and spayed rats by Korenchevsky, Hall, Burbank & Cohen (1941, see Table 121, p. 237).

(b) *The distribution of fat*. Apparently the gonadal hormones affect, not only the general accumulation of fat, but its distribution in certain parts of the body.

The writer cannot quote any experimental work on this subject, but it has been generally observed that the arrangement of fat in the human body is not the same in the two sexes, and there is enough evidence from clinical sources to suggest strongly that the characteristic distribution in men and women respectively is largely regulated by the gonadal hormones.

The Production of Sarcoma by Androgen

The occasional appearance of sarcomata in mice at the site of injected testosterone has been reported by Lacassagne (1939a). So far as the writer is aware no other kinds of cancer have been attributed to androgen. It has to be borne in mind that androgen may be converted into oestrogen in the body, and there is no doubt about the capacity of oestrogen to produce malignant growths, whether this capacity is attributable to the oestrogen as such or to some other compound into which it may become converted in the body. Lacassagne gave subcutaneous injections of testosterone propionate, testosterone acetate or androstenediol dipropionate to mice of the R_{III} strain. Among sixteen of the mice so treated which survived for 300 days, sarcomata appeared at the site of injection in four males and five females.

PART IV. OESTROGENS

Chapter XI. *Oestrogens*

General considerations. Sources, metabolism and excretion of oestrogen. Gradients of responsiveness. Reversibility of effects.

General Considerations

Definition and Terminology. The term 'oestrogen' denotes any substance which will induce cornification in the vagina of the adult mouse like that of natural oestrus. The meaning of the word 'oestrin' is not so easy to define exactly, or to define at all without a background of biological history. Apart from knowing that the sexual functions of the female depend on the ovary and are not controlled through its nervous connections (Knauer, 1900), our recognition of the hormonal basis of ovarian activity began with the observation by Marshall & Jolly (1905) that oestrus could be induced in spayed dogs either by the injection of extracts of ovary removed from another dog during oestrus or by implanting oestral ovaries into the peritoneum. These workers recognized that the ovary produces two different hormones, and that the secretion which causes oestrus is different from that formed later by the corpora lutea. A few years later Adler (1912) reported that oestrus could be brought on in guinea-pigs by intravenous or subcutaneous injections of extracts or press-juices obtained from whole ovaries or from corpora lutea. The next great advance was that of Allen & Doisy (1923, 1924), who discovered that liquor folliculi from the sow's ovary caused oestrus-like changes in the rat's vagina. The hormone thought to induce these changes thereafter came to be known as theelin, oestrin or folliculin. The active principle was at that time a hypothetical substance which could be recognized only by its biological effects. Later, the isolation of the hormone in crystalline form by Doisy, Veler & Thayer (1930) and also by Butenandt (1929), working independently, narrowed the meaning of the term oestrin, as used by British workers, to the single well-defined chemical compound now known as oestrone. This would not have led to terminological difficulty if other naturally produced oestrogenic compounds, including oestradiol, oestriol, equilin and equilenin (Fig. 5, p. 104), had not been identified later.

After these discoveries we find the word oestrin used either as a synonym for oestrone or as a name for whatever oestrogenic substances might be contained in an extract. In the present work when experiments are quoted it has to be left to the discretion of the reader to judge, by the historical background just referred to, what precise meaning is attachable to the word 'oestrin'. In the earlier papers the term is non-committal and indefinite, and in later ones it usually but not always denotes oestrone. Altogether excluded from the term 'oestrin' are several artificially produced chemical compounds which, though producing biological effects like those of oestrone, have not been found in the living organism and in some instances are different in molecular conformation from the naturally occurring oestrogens (Fig. 6, p. 105). For reasons given earlier the writer thinks that the term 'female hormone' as denoting an oestrogen should be discarded (p. 102).

International units of oestrogen. Until a hormone is available in a pure form for biological experiment it is advantageous to have some means for comparing quantitatively the potencies of different preparations. For this purpose international standards of activity were introduced. These are mere approximations and are not needed to describe the dosage if pure hormones are used. Unfortunately the term 'international unit' is still occasionally employed to denote the amount of a pure compound used, instead of expressing it in terms of weight. The international unit of oestrone is the activity of $0.1\,\gamma$ of a standard preparation of pure oestrone, and the international unit of oestradiol benzoate is the activity of $0.1\,\gamma$ of a standard preparation of pure oestradiol benzoate.

General scope of oestrogenic action. Throughout the vertebrata oestrogens are concerned with the female part of sexual reproduction. They are largely responsible, not only for the more obvious physiological mechanisms connected with reproduction, but for the subsidiary traits in appearance and behaviour which distinguish the feminine from the masculine. Like androgens they affect the reproductive organs in three different ways: (1) they provide an automatic check upon their own elaboration by inhibiting the output of gonadotrophin from the pituitary; (2) they stimulate directly the growth and functions of structures, other than the ovaries, which are associated specifically with the female sex; and (3) they diminish, modify or enhance some of the actions of androgen and progestin.

In addition, oestrogens exert a psychological effect and influence conduct.

The chemical nature of oestrone. As shown in Fig. 5 (p. 104) oestrone is a compound with a steroid nucleus; it is resistant to heat and to acid, alkali and reducing agents, but is susceptible to oxidization (Laqueur & de Jongh, 1928).

Sources, Metabolism and Excretion of Oestrogen

(a) Sources

The ovary. In the earliest days of inquiry into the action of sex hormones the effects of spaying were known well enough to indicate the ovary as a source of oestrogen. A specific experiment was done by Goodale (1914), who castrated a brown leghorn cock when it was 24 days old, and then gave it intraperitoneal and subcutaneous implants of ovarian fragments obtained from two of the bird's sisters. This treatment caused the capon to resemble a hen.

Later observations, especially the discovery by Allen & Doisy of the oestrogenic action of follicular fluid and the isolation therefrom of oestrone itself, suggested that this hormone might be derived from the granulosa cells. Such an origin has been widely accepted, yet it may be doubted whether the granulosa cells are the chief source of oestrogen. Allen (1932) pointed out that much oestrogen remains in the residual ovarian tissue after the aspiration of liquor from the large follicles, and that the human corpus luteum contains oestrogen until just before menstruation begins.

The granulosa cells of the ovarian follicles are readily destroyed by X-rays and their destruction is not followed by a disappearance of all phenomena attributable to oestrogen. Steinach & Holtzknecht (1916) exposed the ovaries of a guinea-pig 6 weeks old to a dose of X-rays sufficient to cause atrophy of the follicles. When the animal was $4\frac{1}{2}$ months old the ovaries showed an absence of normal follicles, the presence of a few atrophied and atretic ones, and an extensive development of interstitial tissue which had undergone luteinization or some change which resembled it. The advanced development of the uterus and mammae suggested that the animal had continued to produce abundant oestrogen although the follicles were atrophic. Zondek & Aschheim (1927a) found that implants of pituitary tissue induced vaginal cornification in mice even though the ovarian follicles had been destroyed by exposure to X-rays. This did not occur when pituitary implants were made into spayed mice. Parkes (1926, 1927a, b; see also Brambell & Parkes, 1927) found that X-radiation of the ovaries of adult mice did not arrest the oestrous cycles, and in mice treated in the same way before puberty oestrous cycles subsequently ensued. Genther (1931) made a somewhat similar observation. She described the results of exposing the ovaries of guinea-pigs to sterilizing doses of X-ray as degeneration of the follicles and hypertrophy of the theca interna so that the ovary becomes permeated with interstitial gland. Primordial follicles remain, she says, and new follicles partially mature and then undergo atresia, producing pseudo-corpora lutea. Mandel & Grisewood (1934) submitted week-old female rats to X-rays, the doses varying from 540 to 1,240 r. After a week these doses were repeated. In all the twenty-three rats so treated the vagina opened prematurely at the average age of 22 days. Vaginal smears were those of oestrus. In untreated littermate controls vaginal opening occurred between the ages of 43 and 60 days. Moricard (1933, 1934), also, has made a particular study of this matter. He says that if placental gonadotrophin is given to immature or mature mice the follicular development which ensues is accompanied by a hypertrophy of the interstitial glandular tissue of the ovary. After exposing the ovaries to doses of X-rays varying from 200 to 500 R (Solomon) periodical oestrus occurs for several weeks and after 2 or 3 months there may be persistent cornification of the vagina. At this time the follicles have disappeared and the ovary consists of a mass of epithelioid cells resembling those of corpora lutea and comparable with the interstitial cells of the normal adult ovary. Moricard believes that this interstitial tissue, which undergoes hypertrophy in response to placental gonadotrophin and becomes hypertrophied also after exposure of the ovary to X-rays, is the source of oestrogen. The granulosa tissue, he says, is avascular and does not possess any of the characters of a secreting gland, whereas the interstitial cells, which he regards as derived from the theca interna, have a voluminous Golgi apparatus. The interstitial tissue of the ovary he calls the 'oestral gland'.

Schmidt (1936) exposed the ovaries of guinea-pigs to six daily doses of X-radiation till a total of 2,160 r had been given. As a consequence the ovaries were so changed as to consist entirely of 'interstitial gland', and the guinea-pigs showed irregular oestral cycles with prolonged periods of vaginal cornification.

All these experiences suggest that destruction of the granulosa cells of the ovarian follicles by X-rays does not suppress the output of oestrogen from the ovary, and that these cells therefore, if they produce oestrogen at all, are not the only source.

Experiments of a different kind lead almost certainly to the same conclusion. Aschheim (1926) implanted portions of human ovarian tissue into spayed mice and found that neither (*a*) pieces of cortex without follicles nor (*b*) granulosa tissue scraped from the wall of a large follicle caused oestrous-like changes in the mouse's vagina, whereas ovarian tissue containing atretic follicles or thecal tissue caused these changes. Aschheim found also that corpora lutea or ovarian cortex removed during pregnancy induced cornification of the mouse's vagina in one-third of the trials. He believed that the cells of the theca interna, and not the granulosa cells, are the source of ovarian oestrogen. The results of experiments carried out by Allen, Pratt & Doisy (1925) and Allen, Pratt, Newell & Bland (1930) do not agree with those of Aschheim. Allen and his colleagues assayed parts of human ovaries obtained at operations for their oestrogen content by implanting them into spayed rats and using vaginal smears as criteria of effect. They found that recent corpora lutea, namely those removed between the 13th and 17th days of the menstrual cycle, showed a high content of oestrogen; whereas corpora lutea excised between the 20th and 22nd day of the cycle yielded much less oestrogen. Corpora lutea removed toward the end of pregnancy gave no oestrogenic response. Liquor folliculi and pieces of follicular wall consisting mainly of granulosa cells were rich in oestrogen. Although at first sight the experiments just mentioned seem to indicate the granulosa cells as the source of oestrogen, they do not place such a conclusion beyond doubt. The presence of oestrogen in the liquor folliculi or granulosa tissue does not prove that the hormone has been formed in the latter, nor does it exclude the possibility that the interstitial cells may be a source of ovarian oestrogen, as most of the experimental work reported seems to indicate.

The belief that granulosa cells are at least not essential for the production of oestrogen by the ovary is further supported by some observations of Cole, Hart, Lyons & Catchpole (1933), who showed that the gonads of the foetal horse contain a high concentration of oestrogens. This, they say, is true of both testis and ovary. As the ovary at this stage consists almost entirely of interstitial cells, it seems probable that these are the source of the oestrogen.

Occasionally in women and girls, and in the lower animals also (Strong, Gardner & Hill, 1937), an ovarian tumour develops the cells of which produce oestrogen in abundance. This tumour is commonly described as a 'granulosa cell' tumour. Possibly the name was given to it because oestrogen was thought to be formed by granulosa cells, and since the tumour formed oestrogen it was believed to consist of granulosa cells. But the histology of these tumours, though often of granulosa-cell type, in some instances (thecoma) more nearly resembles that of lutein tissue (Novak, 1934*a*, 1941; Novak & Gray, 1936) and can hardly be said in any case to corroborate the origin of oestrogen from granulosa cells. Wallart (1929) states that in these tumours all gradations may be seen between thecal and granulosa cells.

So far there seems to be little proof that granulosa cells are a source of oestrogen beyond the fact that it is present in a high concentration in the follicular fluid.

During the process of luteinization there is not a rapid and complete disappearance of oestrogen from the ovary; it is still present in the corpus luteum during the first 3 or 4 months of gestation in women (Aschheim & Zondek, 1927; Allen, 1932). E. Allen (1941) believes that all the ovarian epithelial cells—granulosal, thecal, interstitial and luteal—may secrete oestrogen, but that the follicular epithelium is probably the normal primary source. He regards the theca as part of the follicle and says that during maturation it can be shown by Dustin's colchicine technique that nearly as much mitosis occurs in the thecal cells as in those of the granulosa.

Age at which oestrogens are produced by the ovary. Lipschütz (1925b), using guinea-pigs of various ages, grafted ovaries into castrated males. When the grafts were implanted into adult recipients they soon caused mammary development, showing that they produced oestrogen regardless of the age of the donor. With immature recipients, no such hormonal effects were induced, whether the grafted ovary had been taken from an immature or mature donor. Lipschütz concluded that the functional activity of the ovary depended not upon its own age but upon the age of the host. This does not necessarily mean that the ovary can produce oestrogen only in an adult. Zéphiroff, Drosdovsky & Dobrovolskaya-Zavadskaya (1940) observed that the ovary of the calf is as rich in oestrogen, relatively to its weight, as that of the cow. They estimated by tests on rats the amount of oestrogen in ovaries of 0·8 to 3·7 g. from calves and in ovaries weighing 12 g. from cows; the oestrogen concentration was the same in both, namely 1 r.u. per 3 g. of fresh tissue.

Although in normal conditions the ovary is the chief source of oestrogen, it is not the only one, for small quantities can be detected in the urine of women whose ovaries have been removed or who have passed the menopause (Robson, MacGregor, Illingworth & Steere, 1934; Frank, Goldberger & Salmon, 1936). Moreover, the condition of the endometrium in women who have passed the menopause often shows that oestrogen or some oestrogen-like hormone is still being produced in effective amount (Novak & Richardson, 1941).

The placenta. Considerable amounts of oestrogen probably are formed by the placenta. Herrmann (1915) found that placental extracts contained some factor which caused changes in the mammae of male and female rabbits. Allen, Pratt & Doisy (1925) demonstrated the presence of oestrogen in extracts of the human placenta and umbilical cord, and Allen & Doisy (1927) extracted from the placentas of several species an oestrogen identical with or very similar to that which they had obtained by the same methods from the ovary (see also Allan, Dickens, Dodds & Howitt, 1928, and Aschheim & Zondek, 1927).

In the sheep and cow the chorionic villi extend into pits of the uterine mucosa, from which they can be withdrawn so that the foetal and maternal components of the placenta can be separated. Parkes & Bellerby (1926) found an equal concentration of oestrin in these two components. Allen (1932) made the same observation. He says that both embryonic and maternal parts of the placenta contain much oestrin, which is abundant also in the amniotic fluid, though extracts of embryos contain very little. He states that oestrin appears in the human foetal membranes before the 3rd month of gestation, and that cattle, horses and sheep resemble man in this respect.

Hart & Cole (1934) removed the ovaries from a mare at about the 200th day of pregnancy and estimated the output of oestrin in the urine at frequent intervals during the next 150 days. Gestation was unaffected and parturition and lactation were normal. A drop in the output of oestrin followed removal of the ovaries, with a subsequent rise, and during the last 60 days of pregnancy the urinary content of oestrin was about the same as in three normal mares at the same stage of gestation. The conclusion from their experiment is that in pregnancy oestrin may be derived elsewhere than from the ovary, the placenta being the probable source.

Brindeau, Hinglais & Hinglais (1934) record the case of a woman who, having had one ovary removed 2 years previously, became pregnant. In the 2nd month of gestation the remaining ovary was found to be cystic and was removed. In spite of this the pregnancy continued and at the 8th month assays of her urine showed that she was excreting 1,200 r.u. of oestrogen per litre. After labour the placenta, which weighed 580 g., yielded 650 r.u. of oestrogen. During the next few days repeated assays of urine were made and these showed rather a slow reduction in the amount of oestrin excreted, which is not easy to explain, for we know that normally there is little storage of oestrogen in the body, and its disappearance is rapid. Twenty-four hours after delivery there were still 1,050 r.u. of oestrogen in a litre of urine, and 48 hours after delivery the assay showed 800 r.u.; a quick fall then occurred to 40 r.u. at the end of 72 hours and a mere trace at the end of 96 hours.

Lipschütz (1937c) castrated pregnant guinea-pigs and found that the progestational condition of the uterus persisted and the symphysis became separated. As a supply of both oestrin and progestin are required for these results he concluded that they must be derived from some other organ than the ovary, probably from the placenta. Allen, Diddle & Elder (1935) obtained oestrone and oestriol from the placentae of two chimpanzees at full term, and Westerfeld, MacCorquodale, Thayer & Doisy (1938) isolated oestrone in crystalline form from extracts of human placenta. These and other facts show almost with certainty that oestrogen is formed in the placenta.

The testis. The excretion of oestrogen is not confined to females; it occurs in males also, and in the latter its chief source appears to be the testicle. Laqueur Dingemanse, Hart & de Jongh (1927) assayed the pooled urine from a number of men and found oestrin present in amounts up to 50 m.u. per litre. They also demonstrated the presence of oestrin in the testis. Zondek, as already mentioned (p. 105), discovered very large quantities of oestrogen in the urine of stallions, the hormone being derived apparently from the testis, which gives a larger yield of oestrogen than any tissue hitherto examined. Deulofeu & Ferrari (1934) and Beall (1940b) have confirmed Zondek's observations. Deulofeu (1939) states that but little oestrogen is found in the urine of geldings. The oestrogen present in stallions' urine was identified as oestrone by Cartland, Meyer, Miller & Rutz (1935).

Brouha & Simonnet (1928) found that an extract of bulls' testes when injected into adult spayed rats caused vaginal cornification and changes in the uterus similar to those produced by oestrin.

Courrier (1934) ground testes from the pig, horse and bull, and extracted them with chloroform; all contained oestrogen; the horse's and pig's testes showed large amounts, there was much less in that of the bull. He found as much oestrogen in pigs' ectopic testes as in those which had completely descended, a fact which suggests that interstitial cells are the source.

The adrenal. The adrenal is another source of oestrogen. Engelhart (1930) and Callow & Parkes (1936) found that extracts made from the adrenals of oxen, horses and other animals contained oestrogen, and Beall (1939, 1940a) separated oestrone and another unidentified oestrogen from adrenal material obtained from oxen. The small amount of oestrogen found in the urine of patients whose ovaries have been removed (Frank, Goldberger & Salmon, 1936, and others) is probably derived from this gland. Carnes (1940) has extracted appreciable quantities of oestrogen from the adrenals of human foetuses and newborn babies.

Frank (1934) discovered large amounts of oestrogen in the urines of two women suffering from adrenal cancer. Burrows, Cooke & Warren (1936) likewise found a large excess of oestrogen in the urine of a man with the same disease.

An experiment, the outcome of which suggests that oestrogen had been produced by the adrenals, has been reported by Woolley, Fekete and Little (1941). They castrated male *dba* mice at birth. Six months later these mice showed hypertrophy of the adrenal cortex and, in many cases, development of the mammae. Cancer of the breast developed in one of these males.

Oestrogen in food. The possibility of ingesting effective quantities of oestrogen with our food appears to be regarded as negligible. This attitude does not seem quite warranted because appreciable amounts of oestrogen have been detected in several animal, vegetable and mineral substances. It may be present in milk (p. 267), and Riboulleau (1938) found in fertilized, but not in unfertilized hens' eggs, 1·5 to 2γ per gram of yolk and Chanton (1938) obtained between 70 and 80γ of oestrogen from 100 g. of hens' feathers. Donahue (1940) extracted oestrogen, as recognized by the vaginal reactions of spayed rats, from four species of marine invertebrates, namely a sea urchin, a reef urchin, a holothurian, and a lobster. The formation of oestrogen is not confined to the animal kingdom. Dohrn, Faure, Poll & Blotevogel (1926) induced vaginal cornification and enlargement of the uterus in spayed mice by extracts obtained from beetroot seeds, potato tubers, parsley roots and yeast. Dingemanse (1938) extracted from 1 kg. of honey amounts of oestrogen varying from 40 to 600 i.u. and from 1 kg. of dried sage Kroszczynski & Bychowska (1939) obtained 6,000 i.u. of oestrogen.

Aschheim (1933) found from 1,000 to 2,000 i.u. of oestrogen in a kilogram of mineral oil, and a kilogram of naphtha yielded 8,000 m.u. He states that 10,000 m.u. were obtained from a kilogram of natural asphalt, and that oestrogen was present also in coal.

With these facts in mind, and the knowledge that some oestrogens are highly effective when ingested, it will perhaps be wise to defer a final opinion as to the possibility of taking sufficient oestrogen in our food to produce biological reactions.

Artificially administered androgen. Experience has shown that the excretion of oestrogen in the urine may be increased after the artificial administration of androgen, and it seems certain that a transformation of androgen into oestrogen may occur in the body (p. 167).

Distribution of Oestrogen in the Body

(a) *Permeability of the placental blood vessels to oestrogen.* Several phenomena in the newborn human child suggest that it had been subjected before birth to gonadotrophin and oestrogen. Examples are mammary enlargement and the secretion of 'witches' milk', hypertrophy of the uterus and prostate, and epithelial metaplasia in the uterus masculinus. Possibly the effective hormones are produced by the foetal part of the placenta, or they may have been derived wholly or in part from the mother. It has been proved by experiment that oestrogen can pass from the maternal into the foetal circulation. Courrier (1930a) gave between 80 and 100 r.u. of an aqueous preparation of oestrin to guinea-pigs during the last 6 days of pregnancy. At birth the newborn females had cornification of the vagina and vulval swelling, their mammae secreted a little colostrum and the uterine horns were distended. On the 2nd postnatal day a leucocytic invasion of the vagina occurred. Zuckerman & Van Wagenen (1935) saw comparable effects in the newborn rhesus monkey following the injection of oestrone into the mother shortly before parturition.

Skowron & Skarzynski (1933) found that in the rabbit at the 20th day of gestation the concentration of oestrogen in the amniotic fluid corresponded approximately with that of the maternal blood, an equality of distribution which indicates a free exchange of oestrogen across the walls of the blood vessels in the placenta. Further, they gave to rabbits on the 20th day of pregnancy two injections of oestrin, each of 1,000 m.u., with an interval of 3 to 6 hours between the two doses. Six to 8 hours after the first dose estimations were made of the oestrin content of the amniotic fluid and foetal tissues. Similar estimations were made without previous injections of oestrin. The assays showed that the amount of oestrin in the foetal tissues and amniotic fluid of the oestrin-treated group was 2 or 3 times as much as that in the untreated group.

(b) *Fixation of oestrone by inflammation.* It is known that finely particulate or colloid matter present in the blood will become concentrated in inflamed tissues. Brunelli (1935) showed that oestrone becomes localized in this way after its introduction into the bloodstream. He shaved the abdomens of twelve male rabbits and treated each of them as follows. Gauze wet with chloroform was applied for some minutes to one part of the shaved area to cause local inflammation, and to another part of the shaved area gauze wet with normal saline solution was applied. A watery solution containing 1,000 m.u. of oestrone was then injected into an ear vein. One hour later each rabbit was killed, and the two treated areas of skin were separately excised with the underlying tissues, dried and extracted. When tested for oestrogenic activity the extracts prepared from the inflamed tissues gave positive results in every instance, whereas those prepared from the non-inflamed saline-treated skin were all negative. The author (Burrows, 1932) has summarized the conditions leading to the localization of electronegative colloids from the bloodstream; such a process occurs not only in the reticulo-endothelial organs and in regions of inflammation, but also in the neighbourhood of growing tissues and in the placenta. Brunelli's work suggests that oestrin is subject to a similar concentration and retention in these tissues.

(b) Inactivation, Metabolism and Excretion of Oestrogen

Inactivation. The ephemeral nature of the responses to gonadal hormones when given artificially, shows that these hormones are not stored in an active form in the body and that they must be continually produced in order to sustain a biological effect. The quick inactivation of an oestrogen after its introduction into the bloodstream was demonstrated experimentally by Fee, Marrian & Parkes (1929). Using a heart-lung-kidney preparation they found that if 100 r.u. of oestrin were added to the circulating blood only about 1 per cent was recovered from the urine (B.P. 90–120 mm. Hg., T. 36°–37° C.), and in a short while after its introduction none could be detected in the blood. The disappearance of oestrin from the blood was not attributable to its storage in the tissues of the heart or lung, for little if any could be detected in these organs. Further, they found that oestrin was not made inert by blood *in vitro* during 3 hours at 37° C. They thought that the inactivation of oestrin in the bloodstream might be caused by oxidation in the lungs.

Zondek (1934 c) demonstrated the rapid inactivation of oestrone in the living body by other methods. He found that after injecting as much as 20,000 m.u. into rats no detectable amounts appeared in the urine. This disappearance cannot be explained by storage of oestrone in the body, for he gave large subcutaneous doses of oestrone to infantile rats and killed and minced them 3 hours later; extracting the mince plus urine and faeces by organic solvents, he was able to recover only very small quantities of oestrogen. Not more than 1 per cent of active oestrogen was recovered 72 hours after the injection of 40,000 m.u., though by hydrolysis the yield could be increased. He thought that the inactivation probably took place in the liver.

Frank, Goldberger & Spielman (1934) injected 20,000 m.u. of oestrin into the ear vein of a castrated rabbit and found the blood free from detectable oestrogen within half an hour. By killing and extracting the animal 24 hours later barely a trace of oestrogen, they say, could be obtained from the body.

Israel, Meranze & Johnston (1937) have investigated the inactivation of oestrogen in dog's blood *in vitro* and in the circulation. They find that oestrogen is not inactivated in standing blood nor in heart-lung preparations, whereas 1,000 r.u. of oestrogen disappeared entirely from the blood circulating in a heart-lung-liver preparation within 15 minutes. They believe, like Zondek, that the liver is the site of inactivation.

Golden & Severinghaus (1938) made homotransplants of ovaries to the mesenteries of rats so that the ovarian hormones, including oestrogen, would mostly pass through the liver before entering the general circulation. In other rats their ovaries were transplanted into the axillae. The transplants became established at both sites. The rats with ovaries attached to the mesentery remained anoestrous throughout (40 days); those with ovaries in the axillae showed normal oestrous cycles. Furthermore, when the pituitaries were assayed for gonadotrophic potency, those of the rats with mesenteric ovaries showed an enhanced potency equal to that of spayed rats. In some of the rats with mesenteric ovaries these were removed from the mesentery and implanted into the axilla. In all but one of these animals oestrous cycles were resumed within the next 8 to 20 days. The results seem to indicate that ovarian oestrogens are inactivated in the liver.

Several later experiments have demonstrated the inactivation of oestrogen by the liver. To female rats between 21 and 25 days old Talbot (1939) gave 0·1 c.c. of an equal mixture of carbon tetrachloride and 95 per cent ethyl alcohol, so as to cause degenerative changes in the hepatic cells. The rats were killed 1, 2, 3 and 4 days after the injection of carbon tetrachloride, and examination revealed that during the first 3 days a progressive enlargement of the uterus occurred. No such enlargement took place in spayed rats, and the result therefore could not be attributed to any direct action on the uterus of the drug used. The enhanced oestrogenic effect observed appears to have been the consequence of an inability of the injured liver to inactivate at the normal rate the oestrogen produced by the animals themselves (Table 132).

TABLE 132. Injury of the liver as preventing the inactivation of naturally formed oestrogen (Talbot, 1939)

Days after the dose of CCl₄	Number of rats	Average weight of uterus (mg.)
0	11	20
1	9	24
2	15	37
3	11	48
4	5	35

Pincus and Martin (1940) performed an experiment on the same lines as that of Talbot. They spayed rats and gave to some of them carbon tetrachloride; 24 hours later all the rats received 1·5γ of oestrone. Vaginal smears at the end of 48 hours showed cornification in 62·2 per cent of the animals whose livers had been injured by carbon tetrachloride and in 19·5 per cent of those with healthy livers.

Other experiments have shown inactivation of oestrogen by the liver. Biskind (1941 b) implanted pellets of oestrone into the spleen in normal adult male rats and displaced the spleen with its contained pellet of oestrone into the subcutaneous tissues of the loin. In some of these rats the splenic vessels were tied 16 days later. The animals were killed 42 days after the insertion of the pellets. In the rats with intact splenic vessels which had allowed immediate access to the liver of dissolved hormone the testes were normal and there was no evidence of any reaction to oestrogen in other organs; whereas in the rats with ligated splenic vessels the testes and accessory generative organs were atrophic, showing that the dissolved oestrogen which had been prevented from passing directly through the liver had not been so quickly inactivated.

Segaloff and Nelson (1941) by a slightly different method have shown that oestrogens are more rapidly inactivated when introduced directly into the spleen than when given by subcutaneous injection. By this means they demonstrated also that esterification affords some protection against inactivation, for example the doses of oestradiol required to produce vaginal cornification in rats, when introduced into the spleen, were as follows: oestradiol 50γ, oestradiol benzoate 24γ, and oestradiol dipropionate 5γ.

From later experiments, Biskind & Biskind (1942) have concluded that vitamin-B complex is needed to enable the liver to inactivate oestrogen. They implanted pellets of oestrogen into the spleens of spayed rats. As long as the animals were kept on a normal diet they remained anoestrous. If, however, they were placed on a diet deficient in vitamin-B complex protracted oestrus ensued. The addition of the necessary vitamins to the food caused the rats to become anoestrous again.

In some of the earliest work with oestrogens it had been noticed that these hormones were relatively ineffective when given by intraperitoneal injection; probably this was because they had early encountered the influence of the liver.

The inactivation of oestrogens, as shown in the experiments just quoted, apparently occurs in man. Zondek (1934c) found that when 54 mg. of oestrone were given to a woman in the course of 15 days, only 1·6 mg. or 3 per cent, of oestrogen appeared in the urine, and after a dose of 40,000 m.u. given by sub-cutaneous injection to a man, less than 5 per cent of oestrogen was excreted during the next 10 days. Given as a benzoate, oestrone was not so readily inactivated.

Robson, MacGregor, Illingworth & Steere (1934) observed a similar rapid disappearance of oestrone from the blood after intramuscular injection in women who were past the menopause or who had been deprived of their ovaries; only a small proportion of the injected oestrone could be recovered from the urine. Kemp & Pedersen-Bjergaard (1937) have estimated that to maintain the urinary excretion of oestrogen at a normal level in a spayed woman, daily intramuscular injections of about 2,000 m.u. of oestrone are required.

The influence of the liver on the inactivation of oestrogen in man is illustrated by the fact that hepatic cirrhosis may be accompanied by an excess of free oestrogen in the urine. Glass, Edmondson & Soll (1940) examined fourteen men who were suffering from chronic disease of the liver. Signs of excessive oestrogenic action were noticed, atrophy of the testicles being present in all, and gynaecomastia in eight. Assays of the urine showed a diminished output of androgen and a raised excretion of oestrogen; moreover, the urinary oestrogen was in a free form, showing that the liver had failed to conjugate it. To discover in what part of the liver oestrogen becomes inactivated Zondek & Sklow (1941) blocked the reticulo-endothelial system of immature rats of 50 g. by the intracardiac injection of 0·4 c.c. of a colloidal solution of copper. Half an hour later 0·25 mg. of oestrone was given subcutaneously. Control rats, untreated with copper, were given similar injections of oestrone. Four hours later the animals were killed and mashed. Extraction of the mash showed that in both groups 98 per cent of the oestrone had been inactivated, from which the experimenters conclude that the hepatic cells and not the Kupffer cells contain the inactivating factor.

Heller (1940) found that slices of liver inactivate oestradiol *in vitro*, and that this inactivation is prevented by sodium cyanide. Oestrone, also, is rendered biologically inert by the liver, but all the other tissues examined enhanced the potency of oestrone, probably, Heller thinks, by enzymatic reduction to oestradiol. The endometrium showed this enhancing ability in the greatest degree among the tissues examined. The inactivation of oestrone by liver Heller ascribes to enzymatic oxidation. Oestriol was less affected by the liver than oestrone. Heller says that the kidney, though to a less degree than the liver, has some capacity for inactivating oestradiol.

Westerfeld (1940) has stated that tyrosinase, on incubation, inactivates oestrone, oestradiol and diethylstilboestrol. Zondek & Sklow (1942) believe that the inactivation is caused by an enzyme (oestrinase) which, though having properties closely resembling those of tyrosinase, is not identical with it. They say that cauliflower, which contains no tyrosinase, is yet able to inactivate oestrone. These workers extracted an oestrinase from potato and say that (a) it was not affected by standing in open air for 20 hours, (b) its activity was destroyed by heating at 70° C. for 45 minutes at pH 7·2, (c) its range of activity was pH 5 to 9, the optimum being between pH 6 and 7, (d) it was destroyed by H_2O_2 (2 per cent) and by a $M/500$ solution of NaCN.

Apparently the thyroid may take part in the inactivation of oestrogen. Van Horn (1933) gave large doses of desiccated thyroid daily to rats. This treatment caused a loss of bodyweight and persistent anoestrus. In 20 among 24 of these hyperthyroid rats 3 r.u. of oestrone were required to produce vaginal cornification.

Zondek has shown that the capacity to inactivate oestrone is not limited to mammals; it occurs in the frog and also in plants, including the hyacinth root and potato.

It has been found by Smith & Smith (1938) that the inactivation of oestrogen is interfered with by progesterone, under the influence of which a larger amount of active oestrogen will be available for use in the body and the increased quantity will be found in the urine.

The effect of pregnancy on the detoxication of stilboestrol. This may be a convenient place to mention an observation by Zondek & Bromberg (1942). To induce abortion three patients when between 7 and 10 weeks pregnant were given diethylstilboestrol, the total doses ranging from 270 to 445 mg. spread over a few days; the desired effect did not follow and in each case the uterus was emptied by operation. Another patient whose baby was stillborn was given 340 mg. in the first few days of the puerperium to stop lactation. In three of the four patients toxic symptoms appeared during the puerperium, although the same doses given before the termination of pregnancy had been well tolerated. One of the women tolerated when pregnant 170 times as much stilboestrol as she could take without nausea when she was no longer pregnant.

Detection of oestrogen in the circulating blood. In spite of its rapid inactivation in the circulation, small amounts of oestrogen may at times be detected in the blood (Frank, Frank, Gustavson & Weyerts, 1925; Loewe, 1925). Frank and his colleagues made extracts from the blood of five sows during oestrus and tested them on castrated rats and mice by means of vaginal smears. Positive oestral responses were obtained with four of the five extracts. Frank & Goldberger (1926) made periodical estimations of the oestrogen content of human blood during the menstrual cycle. Counting from the 1st day of menstruation they found an abrupt increase between the 10th and 15th days, and thought that ovulation might be recognized by the sudden increase of oestrogen in the blood.

Loewe & Voss (1926) demonstrated the presence of oestrogen in blood obtained from the umbilical cord of the human baby, the hormone having been derived probably from the placenta.

In view of the rapid disappearance of oestrogen from the circulation, as shown by the experiments previously quoted, these findings by Frank & Loewe and their colleagues suggest that when demanded by female requirements the production of oestrone in the body must be both rapid and abundant.

(c) Excretion of Oestrogens

Excretion of oestrogen by the kidney. The primary oestrogen elaborated in the ovaries appears to be oestradiol (p. 104). According to Callow (1938) oestradiol is converted by oxidation into oestriol and oestrone, in which forms they are excreted by the kidneys. Oestrone, Callow says, has been isolated only from the urine; it may be regarded therefore as a waste product. Oestradiol is not normally present in the urine except perhaps in very small quantity. Smith & Smith (1938) noted the presence in some samples of urine of an oestrogen which was neither oestrone nor oestriol; they suspected it to be oestradiol. Later Huffman, MacCorquodale, Thayer, Doisy, Smith & Smith (1940) identified oestradiol in urine collected from women during labour. They had chosen this urine for the investigation because it was known to contain the largest amount of an oestrogen which, until then, had not been identified in the urine.

In their combined forms oestrone and oestriol are relatively inert, and in the earlier estimations of oestrogen in the urine only the amounts of free oestrone and oestriol were recognized. In spite of this drawback useful information was obtained.

The site of the change from oestradiol to oestrone seems uncertain. Fish & Dorfman (1941) gave 50 mg. of α-oestradiol propionate by mouth on two successive days to each of four female guinea-pigs, the first dose being given a few hours after the beginning of oestrus. The urine was collected during the next 6 days, during the luteal phase of the cycle. They found that the administered oestrogen was excreted as oestrone. The same observation was made with spayed females and normal males, from which it appears that transformation of oestradiol into oestrone does not necessarily take place in the ovary or the uterus.

Excretion of oestrogen by the newborn. As mentioned already, oestrogen can traverse the placenta and so enter the foetal circulation; though it is rapidly inactivated in the foetus, enough passes into the urine to be detected there. Philipp (1929) found oestrin in relatively large quantities in the urine of newborn babies until the 3rd postnatal day, after which it rapidly decreased, the tests becoming negative by the 6th day or thereabout. Bruehl (1929), Neumann & Peter (1931) and Lyons (1937b) have reported similar findings. In this connection it should be borne in mind that some oestrogen may reach the infant through the mother's milk, especially during the first few days of lactation.

Excretion of oestrogen by adolescents. Both androgen and oestrogen are excreted in small amount by boys and girls. At puberty there is a rise in the amount of oestrogen excreted in both sexes. A rise occurs at this time also in the output of androgen, but this is less pronounced in girls (see Tables 77 and 80, pp. 169, 170).

Urinary excretion of oestrogens during varying states of the sexual organs. Loewe & Lange (1926) discovered that the amount of oestrin in the urine of normal women varies at different times, the highest yield being obtained about the middle of the menstrual cycle. Siebke (1930, 1934) made periodical assays of urine throughout the menstrual cycle in normal women, and found the highest concentration of oestrogen, namely about 200 i.u. per diem, 12 to 10 days before menstruation, and the lowest during menstruation. Gustavson & Green (1934) and Gustavson, Wood & Hays (1936) carried out similar investigations and noted a sudden rise in the output of oestrogen between the 9th and 12th days of the cycle, about the time of ovulation, followed by a rapid fall to zero and a second rise between the 14th and 21st days of the cycle followed by a gradual fall. Frank, Goldberger & Spielman (1934) observed an increased excretion of oestrin on about the 10th day of the cycle followed by a fall and a second rise about 3 days before menstrual bleeding. Smith & Smith (1935) found in normal women the highest output of oestrogen on the 15th day of the cycle and the minimum on the 1st day of menstruation. Palmer (1937) obtained corresponding results. He says that on about the 14th day of the cycle there is a sudden large output of free oestrone, that is to say oestrone not combined with glycuronic acid. Yerby (1937) found two high peaks of oestrin excretion during the menstrual cycle, namely at about the middle of the cycle and just before menstruation. In a woman with a 21-day cycle the first peak occurred on the 9th day and the second peak on the 17th. In another woman with a 30-day cycle the first peak was recorded on the 15th day and the second peak on the 28th.

Spurrell & Ucko (1938) tested the oestrogen excretion of two normal women throughout the menstrual cycle. The early morning specimens of urine during each successive period of 4 days was bulked and assayed. In one of the women the maximum output of oestrogen took place between the 11th and 15th days, and in the other between the 16th and 20th days. In both the lowest level occurred at menstruation.

Allen, Diddle, Burford & Elder (1936) made similar assays of urine during the menstrual cycle in the chimpanzee. Oestrin was present in the urine throughout the cycle, the highest yield being at about the time of greatest genital swelling, midway between two menstruations and corresponding approximately with the time of ovulation; the lowest yield was obtained during the menstrual flow. Palmer (1940) says that menstrual bleeding is accompanied by the excretion of uncombined oestrogen in the urine, and this is found also to accompany normal labour, abortion and menorrhagia.

Others have confirmed these observations (Smith, Smith & Pincus, 1938; Von Haam & Rothermich, 1940). It seems possible that the second high level of oestrogen excretion during the oestrous cycle may be attributed to the action of progesterone in preventing the inactivation of oestrogen or in assisting its conversion into oestriol, which is more readily excreted (Smith, Smith & Pincus, 1938).

Reference has been made already to the formation and excretion of oestrogen by males, and to the fact that the artificial administration of androsterone or testosterone to men is followed by an increased concentration of oestrone in the urine (pp. 103, 107). The results about to be quoted show that the output of gonadal hormones in the urine is not widely divergent in the two sexes in adult life; in normal conditions men, regarded collectively, excrete more androgen and less oestrogen than women, but the average amount, in each case, lies within the limits of individual variation in the other sex. This is shown by the results of assays carried out by Gallagher, Peterson, Dorfman, Kenyon & Koch (1937; see also Gallagher, Kenyon, Petersen, Dorfman & Koch, 1937), some of whose estimations are set out in Table 133.

TABLE 133. Urinary output of androgen and oestrogen in normal men and women (Gallagher, Peterson, Dorfman, Kenyon & Koch, 1937)

Number	Subjects	Ages	Androgen per diem (c.u.)			Oestrogen per diem (equivalent to γ of oestrone)		
			Max.	Min.	Average	Max.	Min.	Average
4	Normal men	26–35	79	13	40	29	2	10
4	Normal women	23–34	51	13	28	60	28	27

The influence of progesterone on the excretion of oestrogen. Qualitative as well as quantitative differences have been recognized in the excretion of oestrogens during the menstrual cycle. Smith & Smith (1931) noticed that in eight women who were being treated with chorionic gonadotrophin the concentration of oestrin in the blood and urine decreased after 4 or 5 days of treatment. Following up this observation they experimented on rabbits and found that by the administration of either a luteinizing gonadotrophic extract or progesterone to normal female rabbits the urinary output of oestrin could be increased tenfold. No increase occurred in spayed rabbits under the influence of gonadotrophin. If a dose between 600 and 700 r.u. of oestrin were given intravenously into spayed rabbits, about 30 r.u. were excreted in the urine during the next 4 days. The same dose of oestrin, if progesterone also was given, was followed by a recovery of 500 r.u. of oestrin from the urine. From these results Smith & Smith concluded that progesterone facilitates the excretion of oestrin. In a later paper (1938) they give the results of urine assays made in a woman of 27 at different stages of the menstrual cycle. Throughout the cycle the output of oestrone was greater than that of oestriol but the amount of oestriol excreted was increased during the luteal phase. They believe that progesterone facilitates the formation of oestriol, which is a less active oestrogen and is more readily excreted than oestradiol or oestrone. Their views are supported by observations, about to be quoted, on the output of oestrogens during pregnancy and pseudopregnancy.

Excretion of oestrogens in pregnancy and pseudopregnancy. Cohen, Marrian & Watson (1935; see also Marrian, Cohen & Watson, 1935) have investigated the urinary content of oestrogens during pregnancy in women. They find that during most of this period 99 per cent of the excreted oestrogens appear in a combined and relatively inactive form and that oestriol is the chief oestrogen excreted during pregnancy. After the 3rd month of gestation there is a large increase in the output of combined oestrogens, the average daily output 1 week before labour being 22 mg. of combined oestriol and 3 mg. of combined oestrone. Shortly before parturition the amount of combined hormone drops sharply and free oestriol with a little free oestrone appears. Schachter & Marrian (1936) showed that this is not a general zoological phenomenon, for in the mare parturition is not accompanied by pronounced changes in the ratio of free to combined oestrogen.

Patterson (1937) says that by a modification of Kohr's colour test oestriol may be detected in the urine at the same stage of early gestation as the positive Friedman reaction. He used *B. coli* to hydrolyse the oestriol glycuronate. By this method the colour test gave the same results as the Friedman test in all but one of sixty-five cases.

Smith & Smith (1938), like Marrian, Cohen & Watson, find that oestriol predominates over oestrone in the urine during pregnancy. At the time of labour they have observed a fall in the output of oestriol and a rise in that of oestrone.

Pincus & Zahl (1937) experimented with rabbits. After the injection of 300γ of oestriol into a pregnant or pseudopregnant rabbit, that is to say an animal with active corpora lutea, the oestrogen was largely excreted as oestriol. After the injection of 300γ of oestrone in similar conditions both oestriol and oestrone were excreted, whereas when the same dose was given to a rabbit whose ovaries had been removed 2 months previously, oestrone only appeared in the urine. The same dose of oestrone injected with 500γ of progesterone in a rabbit which had been spayed a long while previously was followed by the appearance of oestriol in the urine. This, however, did not occur in the absence of the uterus. Pincus & Zahl conclude that the excretion of oestrogen in the form of oestriol depends on the progestational or gestational uterus under the influence of progesterone.

The effects of varying sexual conditions on the excretion of oestrogen is shown in some experimental results obtained by Pincus (1937) in rabbits. To these were given 300γ of oestradiol, and the urine was thereafter assayed. In this way it was shown that in non-spayed rabbits the oestrogen was excreted almost entirely as oestrone, though in pseudopregnancy there was a considerable output of oestriol also. In spayed rabbits the oestrogen appeared in the urine as oestradiol (Table 134). The results suggest that oestradiol is converted into oestrone by the ovary and into oestriol by the pseudopregnant uterus. Oestriol injected in any of the conditions mentioned in the table was excreted as oestriol. If 0·5 mg. progesterone were given at the same time as 300γ of oestrone the latter was largely excreted as oestriol unless the uterus had been removed previously, in which event the urine contained oestrone only.

TABLE 134. Excretion of oestrogen by the rabbit after the administration of 300γ of oestradiol in varying conditions of the reproductive organs (Pincus, 1937)

Condition of rabbit	Total oestrogen excreted (γ)	Oestradiol (γ)	Oestrone (γ)	Oestriol (γ)
Oestrous	40	0	40	0
Pseudopregnant	115	0	80	35
Hysterectomized	30	0	30	0
Spayed	24	24	0	0

Excretion of oestrogen by the liver. Apparently the liver not only inactivates oestrogen but takes part also in its excretion. Cantarow, Rakoff, Paschkis & Hansen (1942) made biliary fistulae in dogs and examined the bile for oestrogen. In 24-hour specimens of bile from untreated dogs no oestrogen was detected.

After a single intravenous injection of 250,000 i.u. of oestrone, 120,000 i.u. of total oestrogen, of which 100,000 i.u. were free, were detected in the bile excreted in the 24 hours succeeding the injection, and 120,000 i.u., of which 88,000 i.u. were free, were excreted in the next 24 hours. These workers believe there may be an 'enterohepatic circulation of oestrogens' like that of bile. It is clear that further experiments will have to be done before any estimate can be made of how much active oestrogen is excreted by the kidneys and liver respectively, and how much is inactivated in the body.

Excretion of oestrogen by the mamma. Although the breast can hardly be regarded as an important factor in the elimination of oestrogen so far as quantities are concerned, the presence of oestrogen in milk, even in small amounts, cannot be ignored in view of the effects it may have on the young. Courrier (1930a) observed that if daily doses of 80 or 100 r.u. of oestrin are given to a guinea-pig which is suckling, after 3 or 4 days not only will the mother show a response but so also will the nurslings. The effects on the latter include opening of the vagina with keratinization, enlargement and muscular activity of the uterus, and excessive development of the mammae. Lacassagne & Nyka (1934) found oestrogen in colostrum 1·5 c.c. of which, given subcutaneously, caused vaginal cornification in spayed mice. They also detected oestrogenic activity in the fluid obtained from cysts in the non-cancerous part of a human breast in which a carcinoma was present. No oestrogen was discovered in the cancerous tissue.

Gradients of Responsiveness

Widely extended gradations of responsiveness to oestrogen exist, not only between one organ and another, but also between the different parts of a single organ. Loeb (1928) called attention to the graded intensity of epithelial reaction which can be seen in the different parts of the vagina and uterus of the guinea-pig both during natural oestrus and after the administration of an oestrogenic ovarian extract. The responses occurred most readily in the vagina, then in the cervix and later in the uterine horns. 'Espinasse (1934) has noted that, although they are both müllerian structures, the oviduct and uterus of the mouse do not show the same reactions to ovarian hormones.

Gradations of susceptibility to epithelial metaplasia in the generative organs of male mice under the influence of oestrogens have been described by the writer (Burrows, 1935a, b). The first appearance of a cornifying epithelial metaplasia in these animals is seen in a vestigial cyst, when this is present, which is regarded as an occasional surviving relic of the uterus masculinus; in this structure a cornifying metaplasia may be seen before it has taken place elsewhere. The next organ to be affected is the coagulating gland, and after this the seminal vesicle; still later the prostate will show metaplasia. Gradients of responsiveness can be distinguished not only between one organ and another but between different regions of the same organ. Thus in the coagulating gland of the mouse the earliest changes occur in the ducts, from which foci they spread in both directions, that is downward to the urethra and upward to the body of the gland and along the secretory epithelium, within the gland, the glandular cells nearest to the ducts being first affected. In consequence of this gradual spread one may see at the same time an advanced metaplasia in the ducts and adjacent part of the gland, an intermediate region in which secretion has ceased though metaplasia is absent, and a peripheral zone in which the cells are normal in appearance and are actively secreting. Similar gradients of reactivity have been noticed in the prostate and in Cowper's glands (Burrows, 1935c, 1937a).

Gradations of sensitivity to oestrogens have been demonstrated by Bachman, Collip & Selye (1935) in the sexual skin of rhesus monkeys. They administered at first 200γ of oestriol 3 times a day and later gave a single daily dose of 500γ. The first effect, seen within 48 hours, was an oedema of the anus and adjacent base of the tail. During the next 2 days the skin of the genital region reddened and began to swell. Gradually the oedema spread so that after 8 or 10 days the entire area of the pubes, inner sides of the thighs, perineum and proximal third of the tail had become oedematous.

By using Dustin's colchicine technique Allen, Smith & Gardner (1937) were able to define very neatly the gradation of responsiveness to oestrone between the vagina and uterus. Mice were spayed and 2 months later each was given a single injection of 50 i.u. of oestrone in oil. After intervals varying from $9\frac{1}{2}$ to 48 hours the mice were killed, having received $0\cdot1$ mg. of colchicine $9\frac{1}{2}$ hours before death. The mice which had been given oestrone and colchicine $9\frac{1}{2}$ hours before death showed abundant mitoses in the vaginal epithelium, but very few in the uterus; other mice killed $13\frac{1}{2}$ hours after the dose of oestrone showed numerous mitoses in the uterine glands and epithelium.

An interesting example of a graded response to oestrogen is seen in the general skeleton. Apart from the bones forming the symphysis pubis, oestrogen causes super-calcification throughout the osseous system, but this effect is most pronounced in the pelvis, spine and femora; in the tibiae, though less obvious, it is still rather pronounced; in the bones of the forelimb it is slight.

The distribution of osseous tumours in mice, as reported by Pybus & Miller (1940), corresponds with the gradient of responsiveness to oestrogen as outlined above. These tumours occur in females more frequently than in males and oestrin is thought to play a part in their development.

Reversibility of the Effects

A remarkable character of most, though not all, of the tissue changes induced by oestrogen in the adult animal is their reversibility. The fact itself is familiar in connection with the changes in the vagina and uterus which occur during the oestral cycle, and is of general application to non-malignant epithelial metaplasia induced by oestrogen. In mice the coagulating glands and seminal vesicles which have been converted under the influence of an oestrogen into mere cysts lined with a cornified epithelium and perhaps filled with cornified debris, on cessation of the abnormal supply of oestrogen will revert to their original form, become lined once again with a secreting epithelium and once more be able to perform their peculiar functions (Burrows, 1935 b, c).

The subject is one of importance in connection with the therapeutic use of gonadal hormones in man. Oestrogens are largely used for the treatment of gonorrhoeal vaginitis in children. This kind of therapy would hardly be possible if the changes produced by it in the organs of the body were permanent. Under the action of oestrogen the ovaries become atrophic, and this consequence could not be regarded with equanimity if it were to endure after the treatment had ceased. Allen & Diddle (1935) disposed of this question by giving oestrone to immature female monkeys for periods of 28 and 39 days. Immediately after the last injection one ovary was removed and examined. It showed atrophic changes, including a slight decrease in size with follicular atresia. After the monkeys had been without further oestrone injections for 30 days the second ovary was removed and was found to be normal, having quite recovered from the effects of the excessive supply of oestrone.

Gradients of recovery. Such a return to normal takes place in the reverse order of that in which the pathological changes first occurred. Not only are the organs which were affected last the first to recover, but within each organ the parts last to become altered show the earliest return to normal. In the mouse the peripheral alveoli are the first parts of the prostate, seminal vesicle and coagulating gland to regain their glandular histology and secretory functions, and they may arrive at this stage while the proximal alveoli, nearest the main ducts, still show a pronounced degree of abnormality. In the mouse the ducts and the part of the coagulating gland nearest the urethra are usually the last portions of the generative system to regain their normal state.

Irreversible changes. Some of the effects of oestrogen, especially if given during embryonic or early postnatal existence, are permanent. As examples may be mentioned the conformation of the skeleton, hypospadias, and the persistence of Müller's ducts in the male. Malignant neoplasia may be mentioned as another instance of an irreversible change.

The capacity to respond to oestrogen and the degree of responsiveness are innate cellular characters. In an earlier page (277) the epithelial metaplasia induced by oestrogen in the accessory generative organs has been described as 'spreading' from one part of an organ to another. This expression is not meant to convey the idea that some agent causing metaplasia passes from cell to cell while the process is actually going on. The capacity for individual cells to react to an oestrogen, though finely graded according to their anatomical situation, appears to be an innate property of the cells themselves.

Although in many regions of the body the boundary between reactive and relatively inactive tissue is ill defined, a region of responsiveness gradually passing into one of irresponsiveness, there are certain places where the transition is abrupt. Thus the proximal segment of the urethra from the bladder to the opening of the prostatic ducts resists the metaplastic influence of oestrogen, whereas the distal portion of the urethra is very readily affected; and, as Allen (1940) has noted, the mitoses induced by oestrone in the mouse's endometrium cease abruptly at the utero-tubal junction.

Champy (1922) made a similar observation with regard to the action of thyroxine on amphibian larvae. Under the influence of thyroxine the epidermal cells of the anterior limb-bud area proliferate while those of the adjacent gill area degenerate. It seems clear that the graded responsiveness to hormonic influence

is a property inherent in the individual cells and does not depend on the transmission of some agent along neighbouring structures.

The capacity to respond in a particular way to a given hormone is retained by
tissues when transferred to other parts of the body. Friedman (1929a) transplanted rabbits' ovaries into the rectus abdominis muscle and found that follicles
matured and ovulation followed mating, thus showing that despite their altered
surroundings the ovaries still reacted to gonadotrophin.

Stricker & Grüter (1929b) autografted mammae together with their nipples and
surrounding skin into the ears of immature rabbits. Between 4 and 5 months
later the transplants showed the structure of normal mammae in repose. If the
animals became pregnant the transplanted mammae grew, showing abundant
alveoli on the 13th day, and lactation occurred in them when the rabbits had
littered.

Danforth (1930a, b), transplanting the skin of fowls into birds of the opposite
sex, found that, while retaining their racial characters, the grafts responded to the
sex hormones of the host in the same manner as in their original situation.
A cock's skin transferred to a hen produced the female type of feathers. Raynaud
(1930) spayed adult guinea-pigs and transplanted small portions of their vaginae
into the pectoral region. Two or three weeks afterwards injections of oestrone
were given and the animals were killed 4 days later and autopsied. The grafts
were found to have reacted to oestrone in the ordinary way and with the same
intensity as the intact vagina.

Markee (1933) transferred pieces of endometrium from a rhesus monkey into
the anterior chamber of its eye and in this way was able to observe the endometrial changes accompanying menstruation; they were the same as those occurring in the normally situated endometrium.

Additional evidence that a tissue retains its responsiveness to oestrogen in
spite of its displacement to another part of the body is afforded by an experiment
done by Bachman, Collip & Selye (1936) on the female rhesus monkey. By means
of a pedicle flap they transplanted a piece of skin from the reactive area in the
thigh to the abdomen. After division of the pedicle the daily administration of
500γ of oestrone to the monkey caused the same reactions in the grafted skin as it
would have shown in its natural surroundings in the thigh. The same workers also
transplanted a piece of skin from a non-reactive area of the abdomen to the genital
region. During treatment of the monkey with oestrone the transplant remained
as a pale island of skin in brilliantly coloured surroundings. Zuckerman, Van
Wagenen & Gardiner (1938) have repeated the former of these two experiments
and obtained the same results as Bachman and his colleagues. For another
procedure of this kind see p. 101.

Variations in sensitivity to gonadal hormones, and the possibility of acquiring
resistance to their action, have been mentioned earlier (pp. 133, 138) and there
is no necessity to discuss the subject here.

Chapter XII.   *The Action of Oestrogen on the Embryonic Gonads and Müllerian and Wolffian Systems*

Embryonic gonads. Embryonic Müllerian and Wolffian systems.

The Embryonic Gonads

IN reviewing the actions of oestrogen in the living body we have first to consider the question: 'Can oestrogen influence the sex of the embryo; can it reverse the genetic trend towards a male or female development of the gonads?'

Experimental difficulties have prevented a study of the actions of oestrogen on the mammalian embryo in the earliest stages, and therefore the question proposed above must be left open as regards mammals. In amphibia and birds evidence of sex reversal has been brought forward, but even in these the reversal is incomplete (p. 184).

Burns (1925) joined in parabiosis larval salamanders (*Amblystoma punctatum*). Eighty such pairs survived until it was possible to identify the sexes by histology. In every instance the pairs were of the same sex, the twins were both male in forty-four instances and both female in thirty-six. No intersexual abnormalities were discovered, every pair consisting of perfect males or perfect females. Apparently in this experiment a reversal of sex must have occurred in approximately half the number of original individuals. The cause of sex reversal in this case cannot be stated; it may or may not have been due to gonadal hormones, but the experiment does seem to show that in these salamanders the genetic determination of sex is not necessarily final in every respect.

Ackart & Leavy (1939) experimented with larval axolotls (*Amblystoma tigrinum*). When the larvae were 30 mm. in length, at which stage sex differentiation was beginning, biweekly injections of 25γ of oestrone were started and continued until the larvae had a length of 65 mm., the total dose of oestrone being 350γ. At the end of this treatment there were no typical males, nine of the fifteen survivors being normal females, the remaining six having gonads which resembled ovaries and on microscopical examination were found to be ovotestes. From this it appears that the genetically determined sexes of these larvae had not been changed by the treatment with oestrone, though the male gonads had been modified.

Foote (1940) performed the same sort of experiment, using larvae of two species (*Amblystoma maculatum* and *A. tigrinum*). Treatment of the former was started when the larvae were 57 days old and continued until metamorphosis occurred between the 110th and 142nd days. With *Amblystoma tigrinum* treatment with oestrogen was maintained from the 50th to the 221st day, metamorphosis taking place meanwhile. Oestrone, oestradiol dipropionate or testosterone propionate was dissolved in alcohol and added daily to the aquarium water up to a concentration of 500γ to the litre (1 in 2,000,000). The results (Table 135) show that intersexual forms of males were common among the controls and testosterone-treated animals though the sex ratio was not altered, whereas in those treated with oestrogen sex reversal seems to have occurred in view of the final ratio of 53 females to 15 intersexual males. Oestrogen, in addition to its action on the testes, had stimulated the growth of the müllerian ducts.

TABLE 135.   The effect of oestrogen in determining the sex of larval salamanders (Foote, 1940)

	Females	Intersexual males	Normal males
Controls	27	14	12
Testosterone	34	27	7
Oestrogen	53	15	0

Much work of this nature has been performed on embryo fowls. Minoura (1921) grafted small portions of ovary on to the chorioallantoic membrane of embryo chicks between the 2nd and 16th days of incubation, with the result that in males the testes became modified and the individuals had the appearance of females.

Kozelka & Gallagher (1934) injected 100γ of oestrone or oestriol dissolved in ethylene glycol into the albumen of eggs of white and brown leghorns on the 3rd day of incubation and examined the chicks on the 20th day of incubation. They also injected eggs with 10 c.u. of androgen prepared from men's urine, plus 6·3 r.u. of oestrogen. Their results, which are summarized in Table 136, show that oestrogen can modify the testes so that they tend to resemble ovaries. In two instances the ovary had been altered under the influence of oestrone so as to have the appearance of a testis. In the third experiment, in which a high proportion of gonadal abnormality was induced by the combined action of oestrogen and androgen, the results are difficult to interpret because the effects attributable to each of the hormones used cannot be distinguished and separated.

TABLE 136.   The action of gonadal hormones on the gonads of embryo chicks (Kozelka & Gallagher, 1934)

Number	Hormones injected	Sex of chicks at end of incubation		Bisexual or abnormal gonads	Changes in müllerian system
		Female	Male		
37	Oestrone (1000γ)	21	16	4 (2 males, 2 females)	—
31	Oestriol (100γ)	17	14	4 (males)	—
56	Androgen (10 c.u.) *plus* oestrogen (6·3 r.u.)	21	17	18 (ovotestis or right testis and left ovary)	8

Willier, Gallagher & Koch (1935) injected from 56 to 2,250 r.u. of oestrone or oestriol into the albumen of eggs which had been incubated for 24 hours, incubation being continued till the 19th day. On examination it was found that the left testes were flattened and resembled ovaries. With the larger doses the right testes also were flattened. Dantchakoff (1936) injected fowls' eggs with oestrone on the 4th day of incubation. On hatching, the chicks all appeared to be females. The right gonads in the males were perhaps 2 cm. long and 1·5 cm. wide, and had the appearance and structure of ovaries. On the surface of an altered gonad of this kind small vesicles were present, visible to the naked eye, and there were follicles containing ova. But this ovary-like organ, Dantchakoff says, is not identical with the functional left ovary; it rather resembles the organ of Bidder of amphibia. The male feminized in this way, she says, is not a female though behaving as such; the male genetic tendencies are not abolished in the cells and tissues generally, and in time remasculinization occurs. Dantchakoff's conclusions are interesting in connection with the results of experiments reported by Gaarenstroom (1937). He injected oestrone dissolved in olive oil into fowls' eggs in the first few days of incubation, and among the embryos examined after this treatment there was a large predominance of females, together with a considerable number of birds described as intersexual, having a right gonad considerably smaller than the left (one-third to two-thirds the size). In those regarded as females the left gonad only was present. Controls were given injections of olive oil alone. The results are shown in Table 137.

TABLE 137. The results of injecting oestrone into fowls' eggs at an early stage of incubation (Gaarenstroom, 1937)

Number of eggs	Substance injected	Number of embryos	Female	Male	Intersexual
200	Oestrone	68	43	8	17
150	Olive oil	43	24	19	0
150	None	91	43	48	0

Microscopical examination of the left gonads in thirty-five of the chicks were made and the results are shown in Table 138.

TABLE 138. The effect of oestrone on the left embryonic gonads of the chick (Gaarenstroom, 1937)

13 Females			8 Males			14 Intersexuals		
Ovaries	Testes	Ovotestes	Ovaries	Testes	Ovotestes	Ovaries	Testes	Ovotestes
13	0	0	3	2	3	6	1	7

Gaarenstroom states that the influence of oestrone in sex determination in his experiment could be traced if it were given on or before the 6th day of incubation, and if given after the 7th day this effect on the gonads was not produced. These experiments leave little doubt as to the restraint exercised by oestrone on the development of the male gonad in the chick, if given at an early enough stage of embryonic development.

Riddle & Dunham (1942) studied the effects of oestrogen when given before ovulation and perhaps before fertilization. They used the ring dove (*Streptopelia decaocta*, syn. *risoria*). The bird lays only two eggs in each nesting cycle, the second egg being laid about 40 hours after the first. It is thus possible to anticipate ovulation of the second egg. Between 26 and 34 hours before the second egg was laid, each of seventeen doves was given an intramuscular injection of 0·5 or 1 mg. of oestradiol benzoate. After hatching the young genetic males can be distinguished from the females by the different colour of their plumes. The results of the experiment were as follows: Females were apparently unaffected by the injections of oestradiol; most of the seventeen genetic males on the other hand showed anatomical aberrations: (*a*) the left testis possessed an abnormally thick ovarian cortex and sometimes resembled an ovary in shape, and müllerian ducts were unusually large and persistent. The right testis was unaffected. The experimenters explain that in the embryo male of this species the left testis normally shows a slight temporary development of ovarian cortex, which is usually maintained until hatching and then completely disappears. It seems that in this experiment the amount of oestrogen transferred from the mother to the egg must have been very small.

In the preceding experiments, except those of Dantchakoff, the chicks were not allowed to grow up, so that no evidence was provided of the permanence or otherwise of the changes which oestrogens had caused. Later work has confirmed Dantchakoff's observation that the gonadal changes induced in chicks by introducing oestrogen into the eggs at an early stage of incubation are for the most part transitory. Domm (1939) injected oestradiol in doses ranging from 1,500 to 3,000 r.u., or oestrone in doses of 0·5 or 1 mg., into the eggs of brown leghorn fowls between the 3rd and 5th days of incubation. The number of eggs so treated was 410 and from these 69 chicks hatched and 51 survived to maturity, of which 24 were males and 27 were females. The sexes could be distinguished at the time of sexual differentiation, but after the first moult many of the males developed 'henny' plumage. Their behaviour was that of males somewhat lacking in vigour. On killing and dissecting them the left testis was in every case found smaller than normal, flattened and like an ovary in shape, and with an uneven surface. The right testis showed the same deformations in a less pronounced degree. In some instances oviducts were absent, in others they were present; in yet others only the cranial ends had persisted. The left oviduct when present was better developed than the right. There were vasa deferentia, though not so large or convoluted as in normal cocks. Microscopically the left gonad was found to be an ovotestis beneath the cortex of which were seminal tubules showing spermatogenesis. In a single instance follicles of varying size, containing normally developed ova with yolk granules and nucleus, were present in the cortex. The right gonads showed similar changes in a less pronounced degree and no follicles were present.

Clarification of the effect of oestrogen on the chick's gonads has followed later experiments by Gaarenstroom (1939). On the 2nd day of incubation 300γ of oestradiol benzoate or diethylstilboestrol were injected into the eggs of white leghorn fowls. The chicks were examined at the end of the incubation period, and among a total of 85 so treated there were 72 'hens', 13 'intersexuals' and no cock. In every case microscopical examination showed the left gonad to be an ovary. In another experiment (Gaarenstroom, 1940) a large number of white leghorn eggs were treated in the same way as before on the second day of incubation, and 159 of the chicks hatched. These were killed at three different periods after hatching and they revealed the fact that the feminizing effect of the treatment on the gonads largely disappeared in the course of time. (1) Of those killed 1 week after hatching 80 per cent had 'ovaries' and müllerian ducts. (2) Among the chicks killed 2 months after hatching 50 per cent only were normal hens; all the others had testicular tissue, although in some of them well-developed ovarian follicles were present also. (3) The remaining group of fowls, killed 9 months after hatching, showed a normal sex ratio. Most had normal testes, though müllerian ducts had persisted; in two a few small ovarian follicles were seen.

There appears to have been among the males in this series first a modification of the testis to the semblance of an ovary, followed later by a transition of the pseudo-ovary into an ovotestis and finally into a testicle.

Although, as appears from the observations just related, the feminizing effect of ovarian hormones on the testis is especially pronounced in the earlier stages of development, it is not necessarily confined to embryonic life. It will be remembered that in hens only the left gonad evolves into a functional ovary, the right gonad remaining rudimentary. Domm (1929) noticed that if the ovary were removed from a young chick the right rudimentary gonad was apt to develop into an organ resembling a testis to the naked eye, and in some instances it produced spermatozoa. The experiment shows that in the hen a normally developing ovary restrains the latent masculine potency of the right vestigial gonad.

Later experiments have shown that the influence of the ovary in suppressing the male elements of the gonads is dependent mainly or entirely on oestrogen.

In mammals the administration of oestrogen to the mother early in gestation is apt to prevent nidation of the ova, or, if given shortly after fixation of the blastocysts, to cause their deaths; for the most part we can observe the influence of oestrogen on the development of the testis in mammals only when the hormone has been administered in the later stages of foetal life or after birth. The two effects on the testis constantly seen in these circumstances are (1) its retarded development and (2) non-descent. Both of these results may be attributed to the inhibitory effect of oestrogen on the gonadotrophic potency of the pituitary. Greene, Burrill & Ivy (1938c) gave oestradiol dipropionate, in doses varying from 0·375 to 4 mg., to pregnant rats on the 13th, 14th and 15th days of gestation. In eight of the twenty-four males born from these mothers the testes were abnormally high in position and in three were in the position of ovaries, at the caudal end of the kidney.

For a note on the meaning of the term 'sex reversal' see p. 184.

The embryonic Müllerian and Wolffian systems

Experiments have shown that a specific function of oestrogen is to preserve and stimulate the development of the müllerian apparatus. On the wolffian system an inhibitory action has been occasionally noticed (Greene, Burrill & Ivy, 1938c; Raynaud, 1939a), but this usually has been slight and has not been observed as a constant phenomenon.

Lacassagne (1934b) gave weekly injections of 1,000 i.u. of oestrone benzoate dissolved in olive oil to a male rabbit from the day of its birth onward, the doses being later increased to 10,000 i.u. The rabbit was killed when 7½ months old and was found to have a large bicornuate uterus masculinus which was lined in parts by columnar cells and in other parts by a squamous keratinized epithelium. Close to the vas deferens and parallel with it was another small duct which Lacassagne thought might be a persistent relic of Müller's duct, that is to say an oviduct.

Burrows (1934 a, 1935 a, c) treated 373 male mice by applying various oestrogens twice a week to the skin of the back for periods of 50 days and upward. In seventeen of these there were present at death keratinized cysts dorsal to the prostatic urethra and lying between the vasa deferentia. There seems little doubt that these cysts, the largest of which was $10 \times 13 \times 13$ mm., were remnants of a uterus masculinus, preserved, enlarged and keratinized under the influence of oestrogen. Rauther (1904) and Raynaud (1938a) have both shown that a uterus masculinus is present in newborn mice, though it disappears in normal circumstances before adult life is reached.

Zuckerman & Parkes (1935) regard cornification of the uterus masculinus as an indication that the organ, or the part of it which undergoes cornification, is derived not from the müllerian ducts but from the urogenital sinus.

Raynaud (1939a) injected oestradiol into pregnant mice with the result that in the male offspring the müllerian system was preserved, a uterus, occasionally well developed, was present, and in some instances oviducts were found opening against the wall of the testis. The wolffian ducts were either normal or partially atrophied, that is to say, the epididymis was not well developed and the diameter of the vas deferens was reduced. Similar results were obtained in the offspring by giving 1 mg. of stilboestrol to mice on the 13th or 14th day of pregnancy (Raynaud, 1939b). In these experiments Raynaud noticed also some inhibition of the development of the prostate, seminal vesicles and other male accessory reproductive organs, and Greene, Burrill & Ivy (1938c) have noticed similar inhibitions in the rat under the same experimental conditions. They have also recorded the development of a vagina in some of the males.

Burns (1939a) gave injections of oestradiol to young male opossums from the 4th to the 18th day of pouch-life. This treatment caused preservation of the müllerian ducts and hindered the development of the prostate.

Persistence of müllerian structures in male birds after injection of oestrogen into the egg in the early stages of incubation has been recorded by several observers. Willier, Gallagher & Koch (1935) introduced oestrone or oestriol into the albumen of fowls' eggs on the 2nd day of incubation, the doses varying from 56 to 2,250 r.u. The chicks were examined on the 19th day. The müllerian ducts (oviducts) had persisted in the males and were much enlarged in both sexes. The wolffian apparatus showed no abnormality. Similar results have been obtained by Dantchakoff (1936) and Wolff (1936b).

Forbes (1938) gave injections of oestrone to twenty-two alligators, 15 months old. Each received a total dose of 46,500 i.u. during a period of 80 days. Twenty-two untreated alligators were kept as controls. The treatment caused preservation of vestiges of the müllerian duct in the males.

Chapter XIII.   *The Action of Oestrogen on the Anterior Lobe of the Pituitary, and on the Gonads after their Differentiation*

Pituitary. Ovary. Testis.

The Pituitary

THE action of oestrogen on the gonadotrophic function and histology of the pituitary have been dealt with (pp. 41, 88). Here we shall refer briefly to other effects.

Enlargement of the pituitary: pituitary tumours. Evans & Simpson (1929b) noticed that in the female rat the pituitary is larger than in the male. In ninety-two male rats the average weight of the pituitary was 8·8 mg., and in eighty-four females it was 11·6 mg. Hohlweg (1934) found that in the female rat repeated doses of oestrin cause enlargement of the anterior pituitary lobe with characteristic histological changes. Selye, Collip & Thomson (1935) noted that 800γ of oestrone given to lactating rats daily for 10 days caused enlargement of the pituitary to an average weight of 18·5 mg. Zondek (1936a) treated male rats by injecting large doses of oestrin, 5,000 or 10,000 m.u. being given twice a week. The effect of this treatment on the size of the pituitary is shown in Table 139, in which the action of oestrin in checking the general bodygrowth is also seen. Zondek says that whereas the pituitary of the male rat is always enlarged after prolonged treatment with oestrin, that of the female treated in the same way is usually macroscopically unaltered, though in one instance a pituitary tumour developed in a female rat at the end of 7 months during which weekly doses of 5,000 m.u. of oestrone had been given. The pituitary of the female rabbit also, Zondek says, fails to enlarge under the influence of oestrone. Other workers have found that the pituitary becomes enlarged in both sexes under the influence of oestrogen.

TABLE 139.   The effect of large doses of oestrin on the weight of the rat's pituitary (Zondek, 1936a)

Number of rats	Duration of treatment (weeks)	Treatment	Mean body-weight after treatment (g.)	Mean weight of pituitary after treatment (mg.)
5	16·8	Olive oil	141·8	9·7
5	16·8	Oestrin in olive oil	92	26·08

Noble (1938b), Deanesly (1939) and others have observed an increase in the size of the pituitary in rats and mice under the influence of oestrogen. Noble found that stilboestrol caused enlargement of the pituitary in male and female rats, the two sexes reacting alike in this respect. Brooksby (1938) has recorded enlargement of the pituitary in spayed rats under treatment with oestradiol. Kuzell & Cutting (1940) noted enlargement of the pituitary in normal and spayed rats under treatment with oestrogen, the increase in size being accompanied by a considerable increase of mitotic activity. Deanesly has noted that, like most of the effects of oestrogen in the adult individual, the enlargement of the pituitary is a reversible change, the organ returning to its normal size after cessation of treatment. Her experiments consisted of the subcutaneous implantation of tablets of oestrone (2·25 to 16 mg.) or oestradiol (11·8 to 16·5 mg.), which were removed after varying periods. Cramer & Horning (1936) reported that of twelve mice treated by them for prolonged periods with oestrin, eleven had enlarged pituitaries, eight of which were normal in shape while three were adenomatous. The writer has noticed that the mouse's pituitary under treatment with oestrogen may retain its natural shape even when much enlarged, though eventually it becomes deformed and nodular. Among his own oestrogen-treated mice he has seen a morulated pituitary tumour which occupied one-third of the total cranial space. When sufficiently advanced these pituitary tumours cause proptosis, slowness of movement, drooping of the head and other signs from which a diagnosis may sometimes be made before death. The tumours consist mainly of chromophobe cells, and have not in the writer's experience shown any sign of malignancy.

Lacassagne & Nyka (1937) drew attention to the fact that different strains of mice may show different pituitary reactions to equal doses of oestrogen. They used four different strains, namely R_{III}, 17 n.c., 30, and 39 and treated them with oestrone from birth onward. In the R_{III} mice so treated changes in the pituitary were seen in both males and females within 3 months; the capillaries were dilated, colloid secretion was more than normal, chromophiles were less conspicuous or were absent, chromophobes were increased in number, and the pituitary as a whole was enlarged. The pituitaries of the 17 n.c. mice reacted in much the same way as those of the R_{III} mice, though not quite so readily, so that after one year of treatment acidophile cells had not quite disappeared. The pituitary was much enlarged and contained adenomatous nodules of chromophobe cells. The pituitaries of the strain 30 mice still appeared normal after 6 months of treatment; at the end of a year there was some general enlargement. In strain 39 there was no enlargement of the pituitary even at the end of one and a half years of continued treatment. Gardner & Strong (1940) have confirmed the observation that the pituitaries of different strains of mice react differently to oestrogen. They say that among 700 oestrogen-treated mice of the following inbred strains, A, C_3H, CBA, $C_{12}I$, JK and N, no pituitary adenoma appeared; whereas among 106 mice of the C57 strain treated in the same way pituitary tumours developed in fifteen. Like Lacassagne & Nyka, Gardner & Strong failed to detect any correlation between the tendency to form pituitary tumours and a liability to mammary cancer, a finding which accords with the experimental results of Bonser & Robson (1940). These workers, also, have found that the response of the pituitary to oestrogen varies somewhat in different inbred strains.

Nelson (1941 b) made the interesting observation that enlargement of the pituitary in rats under the influence of oestrogen does not progress steadily with time, but shows a pronounced acceleration after the treatment has continued during a considerable period. He gave 50γ of diethylstilboestrol daily to twenty-eight normal male and female rats for 8 months or longer, and noticed that the enlargement of the pituitary increased rapidly after the 8th month. His results are as given below:

Period of treatment (months)	Average weight of pituitary (mg.)
8	75
10 and 11	150
12	210
13 and 14	325

The increased size of the pituitaries was due to the presence of chromophobic encapsuled adenomata.

Although, as Zondek remarked, the pituitary which has become enlarged under the influence of oestrin as a rule has a decreased gonadotrophic function, this perhaps may not be so always. The writer (Burrows, 1936 d) has examined a male mouse whose pituitary, after prolonged subjection to oestrone, was much enlarged, being 5×5 mm. in its two accessible diameters; it consisted mainly of chromophobes, though some of the cells were coloured faintly by haematoxylin and others by eosin. In this mouse features were present which suggest that the pituitary was still producing trophins in spite of the continued application of oestrone; among these features were a thyroid adenoma, extensive mammary development with secretion, and testes of normal size showing spermatogenesis.

There can be no doubt that the pituitary tumours occurring in oestrogen-treated mice are attributable to the oestrogen, for the spontaneous occurrence of such tumours in mice is a rarity. Slye, Holmes & Wells (1931) state that in 11,188 mice only one instance was seen. This was a very small infiltrating adenoma of the pituitary. Gardner, Strong & Smith (1936) have reported an example which is of particular interest. The mouse was an untreated breeder, 695 days old. The pituitary was $4 \times 3 \times 3$ mm. and the pars anterior consisted of chromophobe cells with a few scattered eosinophiles. Both ovaries were much enlarged, being $10 \times 8 \times 8$ mm. and $10 \times 6 \times 7$ mm. They contained no follicles or corpora lutea and consisted of tumour tissue resembling that of the so-called granulosa cell tumours of human ovaries. Cystic endometrial hyperplasia was present, and there were six mammary cancers. In this case it seems, as with similar ovarian tumours in man, there must have been an excessive output of oestrogen to which the other abnormalities, including the pituitary enlargement, may be attributed.

Oberling, Guérin & Guérin (1936) have made some curious observations on the occurrence of pituitary tumours in rats. They castrated immature male rats and grafted them with ovaries from littermates. Among seven of these which lived for more than a year after the operation pituitary tumours were present in four. In one of these the ovarian graft had disappeared, in the other three the grafted ovaries were present and functionally active. In another series of ten rats a crystal of $3 : 4$-benzpyrene was placed under the pia mater in contact with the brain. Three of these rats survived for more than 10 months and all of them had pituitary tumours.

An instance which may perhaps exemplify enlargement of the pituitary in man under the influence of oestrogen has been recorded by Zondek (1940 a). The patient was a woman of 26 whose breasts had been removed for cancer. Metastases appeared, and on this account she was given daily doses of 0·6 g. of oestradiol benzoate during the 60 remaining days of her life. After death her uterus showed advanced cystic endometrial hyperplasia and her pituitary weighed 710 mg. as compared with the normal of 595 mg. (the average for nullipara of her age, but see p. 92) and microscopical examination revealed an adenoma or localized hyperplasia of eosinophil cells.

THE ACTION OF OESTROGEN ON THE GONADS
AFTER THEIR DIFFERENTIATION

The Ovary

A. *The immature ovary.* In the experiments on chicks and amphibian larvae which have been quoted in the previous chapter oestrogens as a rule appear to have had little effect on the embryonic ovary, though occasional instances were seen (Kozelka & Gallagher, 1934) in which the ovary was altered so as to resemble macroscopically a testicle. Such a result has not been observed in experiments on mammals, possibly because in these the administration of oestrogen has begun, through necessity, much later in the course of embryonic development. In mammals the influence of oestrogen on the ovary after birth depends to some extent on the age of the recipient (Hohlweg, 1934); when the hormone is given to immature animals the development of the ovary is retarded or arrested; the organ remains small and the follicles do not ripen. This result has been recorded by Golding & Ramirez (1928), Kunde, D'Amour, Gustavson & Carlson (1930), Doisy, Curtis & Collier (1931), Leonard, Meyer & Hisaw (1931), Katzman (1932), Pincus & Werthessen (1933), Bialet-Laprida (1933), Hohlweg (1934), and others. Katzman (1932), using rats, showed that the arrested development is not permanent; if the supply of oestrin is stopped the ovary soon becomes normal, oestrous cycles appear and pregnancy with normal littering can ensue. The following experiments indicate the kind of dose required to arrest the development of the ovary. Golding & Ramirez (1928) injected 1 r.u. of oestrin twice a day into rats 12 and 18 days old, the doses being gradually increased. Even with these relatively small doses the ovaries were slightly smaller than those of uninjected controls at the end of 22 days. Selye, Collip & Thomson (1935) gave daily injections of 100γ of oestrone daily to 21-day-old female rats, with the result that the ovaries remained in an atrophic state.

It has been established that oestrogens inhibit the production of gonadotrophin by the pituitary and it seems that this is why oestrogens, when given to an immature animal, arrest the development of the ovaries and testes.

B. *The mature ovary.* (a) *Inactivity and atrophy.* The most obvious effect of oestrogen on the adult ovary is like that produced on the immature ovary as just described provided that the doses of oestrogen are large enough and that their administration is long enough continued. Noble (1938b) implanted crystals of diethylstilboestrol subcutaneously into adult rats, the amounts varying from 10 to 100 mg. The atrophy of the ovaries after this treatment in some ways resembled that which follows hypophysectomy. Assay of the pituitaries of these rats showed them to be deficient in gonadotrophin. Even with large doses of diethylstilboestrol the ovaries, Noble says, responded by increased weight to a gonadotrophic extract of pregnancy urine.

The arrest of ovarian activity caused by oestrogen is not permanent; the ovaries soon resume their normal appearance and function if the excessive supply of oestrogen is stopped.

(b) *Gonadotrophic action.* In spite of the atrophic effect of oestrogens on the ovary which is so obvious when they are supplied persistently and in large enough amount, several observers have noticed that oestrogens in certain circumstances exercise an effect comparable with that caused by gonadotrophin. As already mentioned androgens in suitable conditions exert a similar influence on the ovary and testicle (pp. 191, 198).

Williams (1940) implanted tablets of stilboestrol into immature hypophysectomized rats weighing between 40 and 50 g., with the result that ovarian atrophy was prevented or retarded and the response of the ovarian follicles to gonadotrophin from pregnant mare's serum was greatly increased. The rats were killed and examined 15 days after the operation and implantation. The same doses of stilboestrol did not cause any increased weight of the ovaries in normal rats of the same age and weight. Pencharz (1940) performed the same kind of experiment on rats with the same results. Hypophysectomy was done when the animals were between 21 and 23 days old and was followed at once in some instances, and 7 days later in others, by the implantation of tablets of diethylstilboestrol, oestradiol dipropionate, or testosterone propionate. To some of the rats chorionic gonadotrophin ('Antuitrin S') was given. Both of the oestrogens used caused enlargement of the ovaries, which contained numerous mediumsized, closely packed, follicles, with a reduction in volume of the interstitial tissue. The largest ovaries were found after the combined administration of oestrogen and gonadotrophin (Table 140). Testosterone propionate had no effect in preventing ovarian atrophy after hypophysectomy (cf. p. 191). It is curious that the addition of 'Antuitrin S' did not cause enlargement of the ovaries in the rats treated with testosterone.

TABLE 140. The gonadotrophic action of oestrogen on the ovaries of the hypophysectomized immature rat (Pencharz, 1940)

Treatment	Daily dose	Number of rats	Average weight of ovaries (mg.)
None	—	5	7
Diethylstilboestrol	130–170 γ	8	28
Oestradiol dipropionate	40–63 γ	5	13
Testosterone propionate	2·8 r.u.	3	8
'Antuitrin S'		9	14
Diethylstilboestrol *plus* 'Antuitrin S'		8	103
Oestradiol dipropionate *plus* 'Antuitrin S'		3	21
Testosterone propionate *plus* 'Antuitrin S'		3	8

The effect of oestrogen in stimulating the ovarian follicles of the immature hypophysectomized rat, and in augmenting the action of gonadotrophin under the same conditions, has been confirmed by Fluhmann (1941) and Simpson, Evans, Fraenkel-Conrat & Li (1941).

(c) *Ovogenesis.* New egg cells are produced in the ovarian cortex by mitosis in the germinal epithelium. Allen (1923) counted the cells undergoing mitosis in the germinal epithelium at various stages of the oestrous cycle in mice. He found that mitoses were least numerous during dioestrus and suddenly became abundant immediately after ovulation. Allen concluded that ovogenesis is a cyclical phenomenon, each oestrum being associated with the production of new ova. His observations have been confirmed and extended (Bullough & Gibbs, 1941, Bullough, 1942 a, b) and it seems proved that the development of new egg cells is dependent on oestrogen.

(d) *Multiplication of the granulosa cells.* Bullough also noticed that mitoses occur most abundantly in the granulosa cells which lie nearest to the oocyte and the follicular fluid so that a gradient of mitotic activity exists through the thickness of the wall of the follicle, being highest next to the follicular fluid and lowest in the theca externa. Furthermore, in the developing corpora lutea mitoses are commonest where, in the earliest stage of luteal development, the cells are still in close proximity to the remaining reservoir of follicular fluid. By inference, Bullough's observations suggest that oestrogens are responsible, not only for the progressive development of the ova, but for the growth of the granulosa cells also.

(e) *Maturation of follicles.* At all ages an excessive supply of oestrogen will prevent the maturation of follicles by checking the supply of gonadotrophin (FRH) from the pituitary. In consequence of this action no fresh corpora lutea will form, and if none are already present when the administration of oestrogen is begun *the ovaries will become diminutive* compared with those of untreated controls. Even if corpora lutea are present at first, when they eventually degenerate the ovary will become infantile in type and remain so as long as the treatment with oestrogen is continued (Bialet-Laprida, 1933).

(f) *Development and maintenance of corpora lutea; enlargement of the ovary.* Oestrogens, while preventing the output of FRH from the pituitary, enhance the output of LH; consequently they will cause any corpora lutea already present to hypertrophy and to remain in an active condition. Because of this abnormal development of the corpora lutea *the ovaries will be enlarged* as compared with those of untreated control animals (Hohlweg, 1934). This enlargement of the ovaries caused by the maintenance of luteinization led to some confusion as to the influence of oestrogens on the ovary in some of the earlier experimental work in which changes in ovarian weight alone were used as criteria of effect.

The luteinizing action of oestrogen on the ovary is not induced until puberty has been nearly or quite attained, ripe or nearly ripe follicles being essential for the reaction. Bachman (1936) gave single subcutaneous injections of 2,000γ of oestrin to rats which were 24 days old, without producing any evidence of ovarian luteinization. Merckel & Nelson (1940) noticed that the luteinizing effect of oestrin on the adult rat's ovary depended upon what stage of the oestral cycle had been reached at the time of the administration. They gave 40 r.u. daily to mature rats, beginning during oestrus. The result was a prolonged period of dioestrus lasting about 3 weeks, during which the corpora lutea were enlarged and apparently functioning as shown by mucification of the vaginal epithelium; whereas in an untreated rat the very short period of dioestrus is not accompanied by this vaginal response. Different results were obtained if the same doses of oestrin were given during dioestrus; the corpora lutea in these circumstances were neither enlarged nor maintained, and a continued vaginal cornification was induced. This occurred whether normal rats were used or rats which had been spayed or had undergone hypophysectomy. Wolfe (1935b) observed that, in a number of rats which had received 200 r.u. of oestradiol benzoate for about 12 days, the ovaries contained corpora lutea which were much larger than normal, being similar to those of the second half of pregnancy.

Several facts suggest that oestrogens co-operate with FRH to enhance the output of LH. Selye, Collip & Thomson (1935) found that if rats were given pituitary extract alone the heaviest ovaries obtained weighed 64 mg., whereas when oestrin in daily doses of 100γ was administered in conjunction with the same pituitary extract the ovaries weighed 165 mg., the increase being due entirely to the enhanced size of the corpora lutea. Other experiments bearing on the influence of oestrogen in maintaining the corpora lutea will be mentioned later (p. 401).

(g) *Ovarian tumours and cysts.* Champy (1937) states that in mammals an overgrowth of the ovarian rete is invariably induced by persistent doses of oestrone, and that small adenomata are apt to appear in this region. Nearly always, he says, with a prolonged period of oestrin injections invaginations of the germinal epithelium occur and cause the formation of mucoid cysts within the ovary. These cysts, according to Champy, follow two kinds of development: (1) they may remain as cysts lined by a mucous epithelium or (2) complicated papillomata may develop in their walls. Androgens, he says, produce similar abnormalities.

When considering the various effects of oestrogen on the ovary just mentioned it is to be remembered that the result is mainly indirect, being the consequence of oestrogen acting on the pituitary so as to alter the supply of gonadotrophin. In addition to these indirect actions, it seems that oestrogen may affect the ovary directly. Robson (1937b) induced pseudopregnancy in rabbits by injecting gonadotrophin intravenously. Later he ascertained the presence of corpora lutea by direct inspection. He then removed the pituitaries and thereafter gave daily doses of 10γ of oestrone or 5γ of oestradiol. As a result of these injections and in spite of the absence of the pituitary the corpora lutea were maintained in an active condition. Merckel & Nelson (1940), as mentioned above, found that the corpora lutea in the rat's ovary could be maintained by oestrogen after hypophysectomy.

The Testis

The Action of Oestrogen on the Immature and Mature Testis

(a) *Atrophy.* The main pathological effect of oestrogen on the differentiated male gonad is atrophy. Given to the infantile animal, oestrogens prevent maturation of the testes, which remain small and fail to descend into the scrotum; given to adults they cause arrest of function and diminution in size of the testicles. Both the atrophy and non-descent are probably the results of an inadequate supply of gonadotrophin from the pituitary. Laqueur, Hart & de Jongh (1926) gave daily injections of oestrin to rats, beginning the treatment when they were 3 weeks old. As a result their testes remained atrophic and spermatogenesis was absent.

Golding & Ramirez (1928) injected 1 r.u. of oestrin twice daily into young rats 12 or 18 days old, gradually increasing the dosage as the experiment proceeded and continuing the injections for 22 or 36 days altogether. At the end of these periods the testes were still infantile and had not descended from the abdomen.

Control rats of the same age had normal testes lying in the scrotum. No permanent damage was done to the atrophic testes, for they developed normally and descended into the scrotum within 15 days after the injections of oestrin had ceased. Wade & Doisy (1931) performed a similar experiment, giving daily doses of 5 or 20 r.u. of oestrin daily to immature rats, 21–30 days old. Littermate controls were kept. At the end of 19 days the rats were killed. The effect of the treatment on the weight of the testes is given in Table 141.

TABLE 141.   The effect of oestrin on the development of the testes in immature rats (Wade & Doisy, 1931)

Category	Number of rats	Daily dose of oestrin (r.u.)	Average gain of bodyweight (g.)	Weight of testes expressed as percentage of the bodyweight taking the normal relationship as 1
Controls	5	0	77	1·00
Oestrin treated	8	5	69	0·87
Controls	5	0	84	1·00
Oestrin treated	7	20	72	0·53

Spencer, Gustavson & D'Amour (1931) performed a similar experiment on rats with the same result, the testes of the oestrin-treated rats at the end of the experiment weighing only 29 per cent of the weight of those of control rats which had been given injections of oil alone.

Halpern & D'Amour (1934) treated twenty normal adult male rats with daily doses of oestrin, as follows:

Group 1 were given 5 r.u. for 3 weeks and 20 r.u. for 1 week.

Group 2 were given 5 r.u. for 3 weeks and 20 r.u. for 4 weeks.

The rats were then killed and their testes weighed; those of Group 1 equalled 65 per cent and those of Group 2 equalled 23 per cent of the weight of testes from untreated control rats. In Group 2 there were neither spermatozoa nor spermatids and but few primary spermatocytes: but mitosis was present in abundance in the spermatogonia. The interstitial tissue was considerably reduced.

In mice as in rats oestrogens cause a reduction in the size of the testes, an interference with spermatogenesis, and in some instances a return of the testes into the abdomen. In pigeons, also, oestrogen has been found to cause atrophy of the gonads (Riddle & Tange, 1928).

(b) Hyperplasia and neoplasia of the interstitial glandular tissue of the testicle. After prolonged dosage of mice with oestrogen, another kind of change than atrophy may be seen in the testicles, affecting the interstitial glandular tissue (Burrows, 1935b, 1936a, 1937a). It consists of a multiplication and swelling of the Leydig cells, which become loaded with lipoid material. In some instances the lipoid-laden cells coalesce in a manner which resembles that described as occurring in the adrenals under treatment with oestrin. Gradually the Leydig cells increase in number and size so as to form a broad framework surrounding and separating the seminal tubules, the epithelium of which begins to disappear. As a rule the whole of the testis is not equally affected in this way, nor is the change established at the same time or in equal degree in the two testes. In an advanced case a large part of the organ may consist of lipoid-laden interstitial tissue alone, while in other parts tubules may still be visible. In some places these may be represented merely by cylindrical spaces devoid of epithelium, in others, usually at the periphery of the testis, spermatogenesis may be in progress. Between spermatogenesis and complete disappearance of seminal epithelium all degrees of atrophy may be seen, perhaps in the same specimen. The testes affected in this way become enlarged, rounded and yellow in colour, and give the impression of tumour-formation, though an extended experimental test with various oestrogens carried out by the writer on mixed, non-pedigree mice did not result in the growth of a single testicular tumour with malignant characters (cf. p. 286).

Hyperplasia of the interstitial tissue of the testis brought about in this way may be associated with an absence of the usual atrophic effects of oestrogen on the accessory genital organs (Burrows, 1937a). This, the writer believes, is due to an increased production of androgen by the hyperplastic Leydig cells. If this explanation is correct it seems to represent an addition to the known defensive mechanisms of the body, namely the neutralization of one hormone which is in excessive supply through an increased production of an opposing hormone by tissue newly grown for that special purpose. Gardner (1937) has confirmed the occurrence of hyperplasia of the Leydig tissue under the influence of oestrogen and states that the changes described are more easily brought about in mice of some strains than in others. In his experiments large doses of oestrone benzoate (500 i.u.) or equilin benzoate (0·1 mg.) were given once a week to mice of the following strains: Strong A, C_3H, CBA, N, F. Only in the Strong A mice were hypertrophic changes seen in the interstitial glandular cells of the testis. Bonser & Robson (1940) also have found that mice of the Strong A strain are especially liable to this condition when treated with oestrogen. They gave 3 mg. of triphenylethylene subcutaneously once a week to mice of three inbred strains, R_{III}, CBA and Strong A. After 49 weeks of such treatment hyperplasia of the Leydig tissue was present in the Strong A mice but not in those of the other two strains except to a slight degree in some of the CBA mice. Atrophy of the testes with arrest of spermatogenesis ensued rapidly in mice of the R_{III} and CBA strains, whereas in the Strong A mice the testes were not reduced in size, the interstitial glandular tissue was hyperplastic and though some atrophy of the seminal tubules was present, it varied in degree, and in several of the testes which were affected in this way spermatogenesis was taking place in some of the tubules. Occasionally the changes suggested the presence of malignant neoplasia and in one instance it was thought that metastasis had occurred. Autografts and grafts into young mice of the same strain failed to become established. Bonser & Robson noted that hyperplasia of the Leydig tissue in the Strong A mice was accompanied by an absence of squamous metaplasia and keratinization in the accessory genital organs, and they agree with the writer that this probably is due to an increased supply of androgen. They suggest further that this may be the explanation of the maintenance of spermatogenesis in some of their Strong A mice in spite of continued treatment with oestrogen. In a later paper Bonser (1942) has reported that bilateral interstitial-cell tumours developed in the testicles of every one of eight Strong A mice which had received 3 mg. of triphenylethylene subcutaneously each week and had survived for 50 weeks. Hooker, Gardner & Pfeiffer (1940) have described a malignant testicular tumour with metastases occurring in a mouse of the Strong A strain which had been given weekly doses of 0·05 mg. of oestradiol benzoate. The condition of the accessory generative organs in this animal showed that androgen had been formed in effective quantity.

In a later paper Hooker and Pfeiffer (1942) have reported the artificial induction of many testicular tumours by oestradiol and stilboestrol given once a week for 8 months or longer to mice of the Strong A strain. Their results are as follows:

Oestrogen given	No. of mice surviving 8 months	No. with testicular tumours	No. with metastases	No. showing evidence of increased androgen production
Oestradiol	24	10	7	10
Stilboestrol	32	29	22	25

Shimkin, Grady & Andervont (1941) have recorded several instances in mice of hyperplasia of Leydig tissue under the influence of stilboestrol. From their illustrations and description the change appears to have been identical with that previously described by the author. Some of the swellings, they say, represented tumours, although mitoses were not conspicuous. In three cases there were metastases which were limited to the region of the adrenal except in one in which small tumour emboli were found in the lungs.

Hyperplasia of Leydig tissue, as described above, seemed to the author to have been caused by equilin more readily than by the other oestrogens which he had been using at that time. Attempts were made therefore to cause testicular tumours in rabbits by injecting equilin directly into the testes. The experiment was begun in November 1934; four rabbits were used, their testes (usually, but not invariably, the right) being injected once a week with 0·5 mg. of equilin dissolved in sesame oil, thirty-four injections being given to each animal. One of the rabbits died within a month and another at the end of 2¼ years. Of the survivors, one was killed 7 years after the first injection of equilin because of a tumour of the left testis. Metastases were present in the left orbit and left infraspinatus muscle. No remains of testicular substance could be identified in the primary tumour mass; the right testis showed no abnormal changes, other than those attributable to age, and some spermatogenesis was still in progress. The tumour was a seminoma, and neither it nor the metastases showed any structure comparable with the hyperplasia of the interstitial tissue as described above. It would be unsafe to attribute the neoplasia in this one instance to the action of equilin, though possibly this was the main agent. The fourth rabbit of this experiment died in the 8th year without any tumour in the testes or elsewhere.

The histological changes induced in the testes by an excess of oestrogen may be to some extent caused by retention of the gonads within the abdomen; such an explanation cannot be applied to hyperplasia of the Leydig cells as just described because in the most pronounced examples in which oestrogens induced this condition in mice the enlarged testes occupied the scrotum. Fell (1922) observed comparable testicular changes in a pig, including lipoid degeneration, coalescence of the lipoid-laden interstitial cells and almost complete atrophy of the seminal epithelium, which was represented by a single layer of atrophic cells. The animal was a pseudo-hermaphrodite with testes in the scrotum. The penis resembled a clitoris and both vagina and uterus were present. No trace of ovarian tissue was found. The same sort of hyperplasia of the interstitial glandular cells has been noticed in cryptorchid testes in several species.

It seems possible that the abnormal frequency of tumours in retained testes may be connected with a local hormonal disturbance. We know that the ovary if transplanted into the ear or elsewhere in the body where the temperature is less than that of the abdomen will produce an excess of androgen. Perhaps, though it remains to be proved, an abdominal testis will produce an excess of oestrogen; certainly it forms too little androgen. Facts rather strongly suggest that neoplasia in an undescended testis may depend on an excess of oestrogen, or a deficiency of androgen, and among these facts is the close similarity between the histological changes induced in the testes by artificially administered oestrogen and those which are commonly seen in testes which have been retained within the abdomen.

Greulich and Burford (1936) reported three examples of tumours growing in the cryptorchid testes of dogs. In every case there were changes in the mammae and prostate like those experimentally caused by oestrin, and two of the dogs were sexually attractive to male dogs.

Hamilton & Gilbert (1942), basing their report on 1,466 consecutive human cases of testicular tumour, say that the incidence of cancer in testes retained in the abdomen is 50 times that among testes which have been arrested in the groin. Another feature which suggests that testicular cancer may be caused by hormonal action is the frequency with which the disease affects both testes. Hamilton & Gilbert state that in man, if a malignant tumour appears in one testicle the probability that cancer will develop in the other testicle is 100 or 1,000 times greater than a chance incidence.

(c) *Non-descent of testes.* Lacassagne (1934b) gave oestrone benzoate dissolved in oil to a rabbit from the day of its birth. The doses were gradually increased from 1,000 i.u., reaching 10,000 i.u. at the end of the 2nd month. The rabbit was killed at the age of $7\frac{1}{2}$ months. At this time the testes were undescended and were only 4 mm. in diameter.

Burrows (1936 a, c) examined the activities of several oestrogens by applying them—dissolved in benzene, chloroform or alcohol—twice a week to the skin of mice in the interscapular region. The compounds used in this way included oestrone, oestrone methyl ether, oestradiol, oestriol, equilin, equilenin, the synthetic compound 9:10-dihydroxy-9:10-di-*n*-propyl-9:10-dihydro-1:2:5:6-dibenzanthracene (Cook, Dodds, Hewett & Lawson, 1934), and a similar compound with butyl substituted for propyl. Every one of these compounds when given to young mice arrested the development of the testes and prevented their descent into the scrotum. Raynaud (1939 a, b) gave oestradiol to pregnant mice with the result that the testes in the male offspring did not descend, remaining in some instances close to the kidney. Raynaud obtained similar results with stilboestrol given to pregnant mice on the 13th or 14th day of pregnancy, and the writer has found that triphenylethylene will cause retention of the testes in young mice. In rats, also, oestrogens cause retention of the testes in the abdomen or inguinal canals.

When these results are considered it will be noted that non-descent of the testes is accompanied by their backward development. We know that mere retention of the testes in the abdomen in most mammals will prevent spermatogenesis and cause the testes to be undersized. It might be suggested perhaps that oestrin prevents descent of the testes and that the atrophic condition is merely secondary to the retention. Against this explanation is the fact that oestrogens will cause atrophy of the testes even though they have already descended.

Atrophic testes do not descend into the scrotum, and oestrogens stop the pituitary from producing the gonadotrophin on which the testes depend for their full development. These two facts appear to explain why oestrogens prevent descent of the testes (Moore, 1930, 1932), even though the mechanisms of testicular descent still remain ill-defined or unknown. The writer has observed that a diet deficient in vitamin-E which causes destruction of the seminal epithelium will also prevent descent of the testes, although in this case, as already mentioned, 'castration' changes are found in the pituitary associated with an increased output of gonadotrophin (p. 68). It can be argued that the gonadotrophin in this case is not of the right quality to cause testicular descent. Speculation on such a matter is dangerous, and for a full knowledge of why the testes descend we must await the results of further investigation. The matter has been discussed earlier in connection with the influence of androgen on testicular descent (pp. 184, 202).

Chapter XIV. *The Action of Oestrogen on the Accessory Genital Organs after their Differentiation, with a special reference to inguinal hernia*

External Genitalia

Penis. Clitoris. Scrotum and Perineum.

(i) *Before complete development.* When considering this subject the period of our inquiry is not limited by the date of birth, for at this time the external generative organs in many species have not attained their permanent form. This is notably the case with mice, rats and rabbits; these animals are born in a very immature state.

Hypospadias. Oestrone when given to the mother shortly before parturition, or to the young soon after birth, may restrict in the male the growth of the scrotum and penis, so that these structures resemble those of the female in character (Lacassagne, 1934*b*). In the female *hypospadias* is readily produced. This result has been recorded by Hain (1935*a*), Wiesner (1935), Greene & Ivy (1937) and Turner & Burkhardt (1939). It should be explained that normal female rats and mice possess a penis somewhat like that of the male, the urethra in both sexes terminating at the free extremity of the organ. In the condition produced by oestrogens in female rats and mice, and here described as hypospadias, there is a median ventral cleft in the penis so that the urethra opens close to the vagina, and the external genitalia conform to the pattern of those organs in the female of the human and many other species. Hain (1935*a*) injected 1 mg. of oestrone into pregnant rats between the 17th and 20th days of gestation. All the female young which survived this treatment had ventral clefts of the urethra. A similar condition could be caused shortly after birth by giving oestrone to the lactating mother, the oestrogen being conveyed to the young in the milk. This result occurred after the administration of 2 mg. of oestrone to the mother during the first 4 days after parturition, or after 2 mg. were given as a single dose on the 3rd postnatal day, or after 3 mg. given during the first 2 days. Hain (1936) found that the same condition could be produced by injecting oestrone directly into the newborn young. Greene & Ivy (1937) obtained similar results in rats by injecting 2 or 3 mg. of oestradiol into the mother before parturition or 0·2 to 0·4 mg. into the newborn females.

The writer also has caused hypospadias in female rats and mice by the injection of oestrogen into the mother before parturition or into the young after birth, and has seen the same condition following injections of testosterone or progesterone (Burrows, 1939*c*). Hypospadias may occur spontaneously in female mice, and in them minor degrees of it are common. The writer has seen several examples among a stock of breeding mice of Murray's *dba* strain. The permanent hypospadias induced in mice by the artificial supply in early life of gonadal hormones in excess is not necessarily a bar to subsequent fertility. The writer has observed mice with hypospadias and others with a male-like penis possessing a free glans and a well-developed os penis, which have had repeated litters. The abnormalities in these instances had been caused by testosterone given to the mother before littering or to the newborn young. In one mouse persistent infertility was associated with spontaneously occurring hypospadias, this being the only abnormality detected.

Greene, Burrill & Ivy (1940) have explained the manner in which hypospadias occurs. They say that the caudal surface of the penis in the untreated newborn female rat is occupied by a median groove, the urinary meatus at this period being at the level of the perineum. Normally in early postnatal life this median groove, they say, becomes roofed in by fusion of the two lateral ridges between which it lies. By this fusion the groove is converted into a canal so that the urethra now opens at the tip of the penis instead of at the base. The effect of oestrogens is to prevent fusion of the lips of the groove—possibly by causing a premature differentiation of epithelium—so that hypospadias results.

The hypospadiac cleft may extend beyond the level of the perineum so that the urethra opens into the vagina. In a rat given 20γ of oestrone daily on the first few days after birth by the writer such a condition was present when the rat died a year and a half later. The vagina was patent and appeared normal; the urethra opened into the ventral wall of the vagina 3 mm. from its opening in the perineum.

The male rat at birth already has a completed penile urethra, but by giving high doses of oestrogen to the mother at an early enough stage of pregnancy, Greene, Burrill & Ivy have caused hypospadias in male rats.

Defective development of the perineal portion of the vagina. Another defect produced in female rats by giving large doses of oestrogen to the pregnant mother is atresia of the vagina. Greene, Burrill & Ivy (1940) attribute this deformity to a failure in the division of the urogenital sinus into urethra and vagina. The septum which divides the urogenital sinus into these two parallel channels develops, they say, from above downward. Oestrogens prevent this process from being completed, so that the lower ends of the vagina and urethra form a single channel representing the original urogenital sinus.

Hypertrophy of the labia majora and minora. When given in early stages of postnatal existence the chief effects of oestrogen on the external genitalia of the young female is hypertrophy of the labia majora and minora. This result has been noted in the human child (Lewis, 1933).

Apart from these changes oestrogens appear to have little influence on the development of the external genitalia in the female (Wiesner, 1935).

(ii) *External genitalia after puberty.* Apart from hypertrophy and oedema, which will be discussed elsewhere, the changes produced in the external genital organs of the adult *female* are not as a rule obvious. Hypertrophy of the vulva appears to be a direct action of oestrogens. Hill & Parkes (1933) report its occurrence under the influence of oestriol in ferrets which have been hypophysectomized or have had their ovaries removed.

The coloration of the sexual skin in some animals at oestrus may be mentioned as a curiosity; it is comparable perhaps with other manifold marital adornments to be seen throughout the animal world. Another change is an increased cornification and thickening of the skin extending for a variable distance around the vaginal orifice. This effect has been utilized in the treatment of senile vulvovaginitis in women, a condition in which much local irritation results from a deficient cornification of the skin.

In the external genital organs of the immature or adult *male* pronounced and measurable effects are induced by oestrogens.

(*a*) *Atrophy of the musculature of the penis.* This is an early and constant reaction, and is easily observed. The bulbocavernosus and erector penis muscles, which in the normal rat or mouse form a red, firm, rounded mass giving substance to the perineum, rapidly become much attenuated, flabby and pale under continued treatment with an oestrogen.

(*b*) *Squamous epithelial metaplasia in the urethra and urethral glands.* The earliest sites of epithelial metaplasia induced by oestrogen in the penis are (1) the neighbourhood of the verumontanum and (2) the ducts of the urethral glands. From the latter the metaplasia proceeds in both directions, that is to say, toward the lumen of the urethra and toward the acini of the glands. Eventually the glandular epithelium becomes entirely squamous and cornified and the whole penile urethra from verumontanum to meatus becomes lined by a cornified squamous epithelium which may become the site of leucocytic infiltration with shedding of the horny epithelium. The part of the urethra which intervenes between the bladder and the verumontanum remains free from this cellular metamorphosis; so also does the urethra of the female. These changes have been defined in the mouse by the writer (Burrows, 1935 c). Changes which are similar in character and distribution occur in the urethra of other species under the influence of oestrogen and have been observed in monkeys and apes (Courrier & Gros, 1934; Van Wagenen, 1935 a).

(*c*) *Paraphimosis and balanitis.* Another abnormality which arises occasionally in mice or rats under continued treatment with oestrogen is paraphimosis with balanitis. This has been recorded by Lacassagne (1934 a) in the rabbit and by Harsh, Overholser & Wells (1939) in mice under treatment with oestrin and has been seen by the writer in both rats and mice in the same circumstances. Eventually the exposed glans penis becomes ulcerated and scabbed, and death may result from obstruction to the outflow of urine. Gangrene of the glans has not been recorded in this condition in rodents, so far as the writer is aware.

Inguinal Hernia

A curious though inconstant result of giving oestrogens to the male mouse is the development of scrotal hernia. This was first described by the author (Burrows, 1934 b, 1936 a, c) and has since been amply confirmed. Unless perhaps as a rarity the lesion does not arise in mice except under the influence of oestrogen. The author examined over 4,000 mice, some of which had received no treatment, whilst others had been grafted with tumours or subjected to various tests in connection with the production of cancer; none of them had received applications of any oestrogen. In the whole of this series no instance was found in which abdominal viscera were present in the scrotum.

On the other hand every oestrogen tested by the author has been shown capable of causing hernias in mice. In this list are oestrone, oestradiol, oestriol, oestrone methyl ether, equilin, equilenin, 9:10:dihydroxy-9:10-di-*n*-propyl-

9:10-dihydro-1:2:5:6-dibenzanthracene, a similar compound with butyl substituted for propyl, diethylstilboestrol and triphenylethylene. No hernia has been induced by androgen or progestin.

Gardner (1935*b*) has recorded the occurrence of scrotal hernia in two among fifty-seven mice which had ovarian implants in their testes but otherwise had been untreated.

The influence of the testis on hernia formation. Among groups of mice submitted equally to the same oestrogen it was noticed that some got hernias and others did not. Hernia occurred most readily in those mice whose testes were fully developed and lying well down in the scrotum when the applications of oestrone were begun. Most of the experiments were carried out on young mice freshly arrived from the dealer, and said to be between 5 and 6 weeks old. Among 1,410 of these mice examined before treatment, the testes were undescended in 427. When such mice with undescended testes are submitted to regular biweekly applications of oestrone, some of the testes come down into the scrotum, but many remain undescended so long as the treatment with oestrone is maintained, and as long as the testes remain in the abdomen no hernia occurs. Two facts seem to have emerged: (i) the presence of a more or less mature testis is necessary for the formation of scrotal hernia under the influence of oestrone; (ii) oestrone may prevent maturity and descent of the testis, in which event no hernia will develop. It should be added that hernia does not invariably follow the administration of oestrogens even when the testes have descended into the scrotum.

Bonser & Robson (1940) have shown that mice of different inbred strains are not equally susceptible to the hernia-producing effects of oestrogen. They treated mice of the R_{III}, Strong A and CBA strains with weekly injections of 3 mg. of triphenylethylene. Scrotal hernias developed in 23 of 30 R_{III} mice so treated, in 11 of 23 Strong A mice, but no scrotal hernias appeared among the CBA mice.

The effects of artificial cryptorchidism. To study further the connection between hernia and descent of the testis 38 mice whose testes were fully descended were rendered cryptorchid by dividing the gubernaculum on each side and displacing the testes into the abdomen (Burrows, 1936*c*). In 18 of the mice the operation was performed by means of a laparotomy wound so as to avoid injury to the scrotum, and in 20 the exposure was made through the scrotum. Of the 38 mice treated in this way, 28 were given biweekly applications of oestrone to the skin (1 drop of a 0·1 per cent solution in benzene), and 10 were kept as controls. Hernias subsequently occurred in 15 of the mice treated with oestrone, and no hernia occurred among the controls (Table 142).

TABLE 142.   The effects of artificial cryptorchidism combined with applications of oestrone on the development of hernia (Burrows, 1936*c*)

Number of mice	Site of operation	Subsequent treatment	Mice with hernia
18	Abdomen	Oestrone	9
10	Scrotum	Oestrone	6
10	Scrotum	None	0

Sixty-eight days after the operation on these mice there were 34 survivors, of which 25 had been treated meanwhile with oestrone, and 9 had not been given oestrone. Among those treated with oestrone descent of the right testis had occurred in 19 and descent of the left in 18. Among the 9 which had received no oestrone the right testis had descended in 8 and the left in 6. In most and perhaps in all the instances in which retention occurred it was attributable to fixation of the testis in the abdomen by post-operative adhesions.

In some of the mice in this experiment a bilateral hernia was found, although one of the testes had been retained in the abdomen by post-operative adhesions. In one instance the left testis had failed to descend because of adhesions, and the right testis had descended into the left scrotum. In another mouse both testes had come down into the left scrotal pouch. In these mice bilateral hernias were present. It seems that a gubernaculum is not essential for the descent of the testis, though it is indispensable as a guide to a correct disposal of the testis in the scrotum. Further, although a well-developed testis is necessary for the production of a scrotal hernia by oestrone, that testis need not be in the scrotum while the hernia is forming.

The effect of unilateral castration. To obtain additional light on the subject the right testis was excised from 10 mice, and applications of oestrone were thereafter made to them and to 10 intact controls. Bilateral hernias appeared in 4 survivors with a single testis and in 6 among the 10 intact mice (Table 143). In each series no hernia occurred in the mouse with undescended testes.

TABLE 143. Bilateral hernias induced by oestrone in mice having one testis only (Burrows, 1936c)

Number of mice	Preliminary treatment	Surviving after 65 days	Number with retained testis	Mice with bilateral hernias
10	Unilateral castration	5	1	4
10	None	10	1 (both testes retained)	6

The effect of bilateral castration. The testes were removed from thirty mice, which were thereafter treated with applications of oestrone till their death. In none of these mice did any hernia appear.

The effect of vitamin deficiency. The healthy development of seminal epithelium is dependent upon adequate supplies of vitamins in the diet. Vitamin-E is of particular importance, a deficiency leading rapidly to degeneration of the seminal epithelium. Thirty mice were fed on an ordinary diet, another twenty were given a diet deficient in vitamin-E but otherwise adequate, and ten were kept on white bread only. All were given the same biweekly doses of oestrone by cutaneous application (0·01 per cent in benzene). While the experiment was in progress an epidemic of ectromelia occurred, and the murine mortality, especially among those on a vitamin-deficient diet, was high. The incidence of scrotal hernia among the three groups of mice is shown in Table 144.

TABLE 144. The effect of vitamin-E deficiency on the induction of hernias in mice by oestrone (Burrows, 1936c)

Number of mice	Diet	Surviving for 6 weeks or longer	Mice with hernia
30	Normal	25	12
20	Vitamin-E deficient	7	0
10	White bread only	5	0

The effects of long-continued applications of oestrone.   In some cases the hernias have progressively increased so as to reach very large dimensions, and there have been instances in which almost all the abdominal contents except parts of the liver and stomach have become transferred to the scrotum. In examples of this kind the entire spleen is sometimes found in the right scrotal pouch and some-times in the left. Even large hernias have diminished in size on cessation of the dosage with oestrone, though they have not disappeared. Small hernias which have not long existed may become cured spontaneously if no more oestrone is given.

The author has not seen hernia formation as the result of oestrogenic treatment in rats or other animals than mice, nor is he aware of any records in the literature of such an occurrence.

Chapter XV. The Action of Oestrogen on the Accessory Genital Organs (continued)

Coagulating gland. Seminal vesicle. Prostate. Uterus masculinus. Vas deferens and Epididymis. Ampullary glands. Cowper's glands. Preputial glands.

A. *Stimulating effects.* A small supply of oestrogen appears to be favourable if not essential for the well-being of the male accessory reproductive organs. The quantity required depends on the amount of androgen available at the same time. When both androgen and oestrogen are present in correct proportions they co-operate to produce healthy conditions; if androgen is lacking or oestrogen is in excess pathological changes will ensue. Co-operation between the different gonadal hormones has been discussed in detail earlier (p. 115) and all that need be added here is a reference to the good effects exerted by small quantities of oestrogen on the accessory reproductive organs of the castrated male even when no androgen is administered. Oestrogen in these circumstances, given in small amount soon after castration, will check to some extent post-castration atrophy of the accessory reproductive glands, or, if sufficient time has elapsed for them to have become atrophied, oestrogen will to a slight degree restore the epithelium and to a greater degree the stroma, and so will increase the volume of the atrophied structures; this beneficial action, however, will not enable the epithelium to resume secretion.

David, Freud & de Jongh (1934) castrated rats when they were between 20 and 30 days old. Injections of oestrogen were started 3 days later, ten injections being given in the course of 5 days. The restorative effects of this treatment on the weights of the accessory genital glands are shown in Table 145.

TABLE 145. The effect of oestrogens on the weight of accessory reproductive organs in the immature castrated rat (David, Freud & de Jongh, 1934)

| Oestrogen | Daily dose γ | Weight of organs expressed as mg. per 100 g. bodyweight | | |
		Prostate and coagulating glands	Seminal vesicles	Preputial glands
None	0	22 in one series	14	34
		17 in another series	13	34
Oestrone	2	44	25	30
	10	39	27	48
Oestriol	2	28	18	22
	10	48	19	38
Oestradiol	2	25	46	35
	6	74	45	45

B. *Restrictive effects.* Although the lesions caused in all the accessory reproductive organs of the male by excessive doses of oestrogen are much alike, there are some minor differences of response which demand a separate consideration of each organ. Unless otherwise stated the following descriptions refer to the mouse.

The Coagulating Gland

The fact that oestrogens may cause lesions in the accessory reproductive organs of the male was discovered independently by Lacassagne (1933 a), de Jongh (1933) and Burrows & Kennaway (1934), all of whom recorded the appearance of stratified, squamous keratinizing epithelium in these organs in mice after continued treatment with oestrin. The lesions observed by these workers affected the so-called anterior lobes of the prostate which are in fact the coagulating glands (Burrows, 1935 c).

Camus and Gley (1896, 1897) discovered that the guinea-pig's seminal fluid rapidly coagulates in the air, the vagina and even in the male urethra. This coagulation, they learned, is caused by an enzyme present in the secretion of adjacent glands which they regarded as prostatic. Later the effective enzyme was found by Walker (1910) to originate in a special pair of 'coagulating glands'; these differ from the prostate in function, in histological appearance and in the readiness with which they undergo functional and structural changes under the influence of oestrogen. The 'coagulating glands' are adjacent to the seminal vesicles and communicate with the urethra close to the openings of the seminal vesicular ducts. The copulation plug which occupies the vagina after mating in the mouse, rat and guinea-pig owes its presence to the action of the coagulation glands.

The first obvious effect of oestrogen on the coagulating gland of the mouse is a reduction or complete arrest of secretion. Both the cytoplasm and nuclei of the epithelial cells diminish in volume and the gland as a whole shrinks from this cause and through the loss of its normal contents. As the nuclei lessen in size they become more darkly stained in sections prepared for the microscope. At this stage an epithelial metaplasia begins in the ducts and spreads gradually and continuously down towards the urethra and up to the gland. In the gland itself the spread of the metaplasia is from the areas adjoining the duct to the rest of the gland, the parts farthest from the duct being the last to be affected. The first sign of this change is the appearance, between the basement membrane and the pre-existing atrophic epithelium, of large clear cells with vesicular nuclei. These cells multiply to form a stratified epithelium by which the atrophic secretory epithelium is replaced. As the layers of stratified epithelium increase the more superficial ones become squamous and eventually keratinized. With a continuation of this process the alveoli disappear and the glands become converted into smooth walled sacs lined with a thick keratinized epithelium and distended with shed keratinous debris. It is difficult to account for the disappearance of what in the normal gland look like alveoli unless we suppose that the structure is merely a bag with a much plicated wall. These changes are accompanied by some hypertrophy of the fibromuscular stroma of the gland. Sooner or later a leucocytic invasion of the keratinized tissue is apt to occur.

The Seminal Vesicle

The changes caused in the seminal vesicles of the mouse by oestrogen are sufficiently like those in the coagulating glands to need no separate consideration except to say that, in the mouse, the metaplasia occurs later and spreads in a less orderly manner. It begins in the duct near its entrance into the urethra and later appears in patches in the body of the gland. These patches eventually coalesce so as to form a complete lining of keratinized epithelium. While these changes are in progress the fibromuscular wall of the seminal vesicle undergoes a considerable hypertrophy (de Jongh, 1934 a, b; David, Freud & de Jongh, 1934; Korenchevsky & Dennison, 1935; Overholser & Warren, 1935; Van Wagenen, 1935 b; Wells, 1936 a). This effect of oestrogen upon the stroma of the seminal vesicles has been recorded in the mouse, rat, ground squirrel and monkey.

Van Wagenen (1935 b) gave daily injections of oestrone for 34 days to an immature male monkey weighing 2,450 g. At the end of this period the seminal vesicles weighed $5\frac{1}{2}$ times as much as those of an untreated control monkey weighing 3,000 g. The increased weight of the seminal vesicles, she says, was due entirely to hypertrophy of the fibromuscular stroma.

The Prostate

The abnormalities caused by oestrogens in the epithelium and stroma of the mouse's prostate are essentially the same as those caused in the coagulating gland and seminal vesicle. They occur later than in the coagulating gland, and the prostatic alveoli, though eventually they become lined by a stratified cornifying epithelium, do not lose their form or disappear as they do in the coagulating gland and seminal vesicle. A noteworthy feature of these transformations in the epithelium of the prostate is the sequence in which they happen in the different regions of the gland. The changes are seen first in the ducts and with time gradually spread from these foci toward the periphery, so that in some cases, while an advanced metaplasia may exist in the ducts and proximal alveoli, the peripheral regions may be little affected. A stratified epithelium may be present in the ducts and proximal alveoli, a pronounced epithelial atrophy in an intermediate zone, while the peripheral regions still show secretory activity (Burrows, 1935 b, c).

Although in the mouse, as in man, the greater part of the prostate arises by outgrowths from the dorsal wall of the urethra, so that the main ducts open into or near the prostatic grooves on each side of the verumontanum, there are in addition two or more lobules which discharge through the ventral wall of the prostatic urethra. These ventral lobules, like those which constitute the main body of the gland, exhibit epithelial metaplasia under the influence of oestrogens.

In a single instance the writer has seen a proliferation of the metaplastic prostatic ducts under the influence of oestrone histologically resembling a carcinoma but not accompanied by metastasis. This was found on the 155th day in a mouse which had been treated with one drop of a 0·1 per cent solution of oestrone in benzene applied twice a week to the nape (Burrows & Kennaway, 1934).

Hyperplasia of the fibromuscular stroma of the prostate accompanies the atrophic and metaplastic effects of oestrogen on the epithelial components, and seems to have occurred in all the species examined, including monkeys (Courrier & Gros, 1934, 1935; Van Wagenen, 1935 a, b; Parkes & Zuckerman, 1935). The gradation and character of the effects produced by oestrogen on the mouse's prostate suggested to de Jongh (1935 a) and to the writer (Burrows, 1935 c) that the pathological changes in the prostate of elderly men might possibly be caused by a relative excess of oestrogen, for the lesions in both cases have a similar anatomical distribution. A comparable gradient of metaplasia occurs in the enlarged prostate of the human foetus during the late stages of gestation (Aschoff, 1894, and Schlacta, 1904) during which it is subjected to large supplies of oestrogen. Surgeons well know that in performing the operation of so-called suprapubic prostatectomy they merely remove an enlarged fibro-adenomatous core leaving behind an envelope of more or less healthy glandular tissue (Wallace, 1904). Experiments on the monkey rather support this view concerning the etiology of prostatic enlargement in man. Zuckerman (1938) gave daily doses of oestrone to two young castrated rhesus monkeys for 1 year. In one case the doses were 100γ each, in the other $2 \cdot 5\gamma$. The results were qualitatively the same in both (Zuckerman & Sandys, 1939). In the monkey which had received the major dose the prostate as a whole was greatly enlarged, the area of a transverse sectional plane being 3 times that of the normal. The increase was due entirely to overgrowth of the musculo-fibrous stroma intervening between the urethra and the glandular part of the prostate.

Dogs suffer much from prostatic enlargement and it seemed desirable to investigate the condition in these animals, but it has been difficult to obtain the necessary material. On applying for permission to examine the corpses of dogs killed in a well-known dogs' home, where annually large numbers of pet animals are provided with euthanasia, the writer met with a refusal. He was told that the dogs in the home were not available, alive or dead, for experimental inquiry. A few specimens were kindly supplied from other sources—too few to establish a generalization—but they suggest that there may be two forms of enlarged prostate in the dog: (1) a pure hypertrophy with excessive secretory activity in relatively young dogs, and (2) a pathological enlargement occurring in old dogs and comparable with the condition which occurs in elderly men.

Kok (1936; see also de Jongh & Kok, 1935) gave daily doses of 4,000 i.u. of oestrone to two dogs for $5\frac{1}{2}$ weeks, at the end of which the prostates were 5 and 8 times as large as the prostates of untreated control dogs.

The action of oestrogen on the prostate of the human infant has been demonstrated by Sharpey-Schafer & Zuckerman (1941). They gave 5 mg. of oestradiol benzoate daily to two infants for periods of 46 and 50 days respectively. One of the children was suffering from congenital heart derangement and the other from hydrocephalus. One died when 7 weeks old and the other at the age of $7\frac{1}{2}$ months, the injections having been continued until death. In each case the prostate showed the changes typical of oestrogenic action, namely a squamous metaplasia of the crista urethralis, utriculus masculinus and the proximal prostatic ducts.

Wells (1936a) noticed a difference in the responses of the prostate to oestrin, depending on the presence or absence of functioning testes. He gave daily injections of oestrone to ground squirrels during a period of 30 days. Given to animals in the non-breeding season or after castration the treatment caused an increase in the size of the prostate owing to hypertrophy of the musculo-fibrous components. Squirrels treated in the same way during the breeding season showed a diminution in the size of the prostate owing to a reduction in the size of the epithelial cells and the amount of secretion produced by them. The writer has noticed that prostatic metaplasia in the mouse is more readily induced by oestrogen in the castrated than in the non-castrated animal.

Attempts to utilize the observations just quoted for human therapy have failed, and the administration of androgens to patients with enlarged prostate has not had any definite remedial effect. The condition does not appear to be caused, as was first thought possible, merely by an excess of oestrogen or a deficiency of testosterone. Owen & Cutler (1936) examined the urine of patients with malignant and benign enlargements of the prostate and found that no excessive excretion of oestrogen had occurred. Assays of urine excreted by men with enlarged prostate by Dingemanse & Laqueur (1940) showed that the output of both androgen and oestrogen was diminished, the oestrogen being reduced more than the androgen. The urine of normal men of the same.age was used as the standard for comparison.

Nevertheless, and in spite of these failures, there can be no doubt that the cause of benign prostatic enlargement in man lies in the testicles, though the nature of their influence remains obscure. The lesion does not appear after castration, and in those in whom it is already present removal of the testicles will relieve the trouble, unless malignancy has supervened. There was a time when castration was practised to a limited extent, though with good results, for the cure of enlarged prostate in man. The history of this method of treatment offers material for philosophic thought to those inclined toward such exercise, for the operation was suddenly and universally abandoned on the ill-substantiated fear that it might be quickly followed by mental decay.

The Uterus Masculinus

The uterus masculinus, which is present at birth in mice, usually disappears later. In some instances, apparently, it persists as a vestigial structure and under the influence of oestrogen becomes converted into a cyst lined by cornified epithelium. This is the writer's interpretation of cysts which he has found occasionally in mice treated for long periods with oestrogen (Burrows, 1934a, 1935c). These cysts are situated behind the bladder and between the lower ends of the vasa deferentia, in contact with the coagulating glands. In the only two instances in which complete serial sections were made, the cysts, though adjacent to the prostatic urethra, did not communicate with it. Squamous metaplasia in the uterus masculinus as a response to oestrogen has been observed by Courrier & Cohen-Solal (1936) and Laqueur (1936) in the guinea-pig, Van Wagenen (1935a), Courrier & Gros (1935) and Parkes & Zuckerman (1935) in monkeys, and Sharpey-Schafer & Zuckerman (1941) in man.

The uterus masculinus, however, does not react alike to oestrogens in all animals; in some it responds by a squamous metaplasia, in others apparently it does not. Zuckerman & Parkes (1935; see also Zuckerman, 1936) explain this difference on the supposition that in those species in which the uterus masculinus reacts to oestrogens by a squamous metaplasia, the organ represents the vagina and has been derived from the urogenital sinus, whereas in other species including man the utriculus is a müllerian derivative. Such an explanation seems to fail somewhat in its application to man, for the utriculus of the newborn baby is lined by a stratified, squamous epithelium; moreover a cornifying metaplasia is apt to occur in the uterine horns in rats and mice under the influence of oestrogen, and it is hard to believe that these structures have been derived from the urogenital sinus.

Guyon (1939) examined the utriculus of stallions and geldings and found that the lower end has a stratified epithelium and the upper part has a secretory epithelium with glands recalling those of the uterus. In spite of the large excretion of oestrogen by the stallion no cornification occurs. This observation agrees with the fact that little or no cornification occurs in the mare's vagina.

The Vas Deferens and Epididymis

Under the influence of oestrogen the epithelium of the *vas deferens*, which is normally columnar and ciliated, becomes flattened so as to become more or less cuboidal, and the cilia disappear. In more advanced stages, instead of a single layer of cells arranged in an orderly manner the epithelium consists of several layers of small rounded cells with darkly staining nuclei. The stroma of the vas is increased in thickness, the submucosa being most affected and giving the appearance of slight oedema. The submucosa of the lower ends of the vasa deferentia or ejaculatory ducts, close to their openings into the urethra, undergoes a striking change, becoming greatly thickened and assuming a myxomatous appearance like that seen in sections of the cock's comb (Heringa & de Jongh, 1934; Burrows, 1935a; Harsh, Overholser & Wells, 1939). This curious formation diminishes rapidly as the distance from the urethra increases.

When given for a period of 30 days to ground squirrels in the non-breeding season or after castration, oestrone caused a doubling of the size of the vas deferens, chiefly because of an increase in the volume of the musculo-fibrous stroma (Wells, 1936a).

The effect of oestrogen on the epididymis. This is like that on the other accessory generative organs, namely an atrophy of the epithelium with loss of secretory activity and an increase of the stroma. In advanced cases in the mouse a squamous, cornifying metaplasia may occur; but this is a late and uncommon sequel.

The Ampullary Glands

This small group of glands surrounding the urethral ends of the vasa deferentia are affected by oestrogen in the same manner as the other accessory glands of the male, though metaplasia is less readily induced in the ampullary glands and is likely to be observed only after prolonged administration. In a few instances only, among many investigations, has the writer seen small areas of squamous metaplasia in these glands in mice.

Cowper's Glands

In rats and mice these paired glands lie on the dorsal surface of the bulbar muscles and are applied to the lateral walls of the rectum. Each gland is connected with the urethra by a duct which traverses the muscles of the bulb. The body of the gland consists of one or more plicated sacs lined by secretory cells; the duct is lined by a single layer of flattened or cuboidal cells. The writer studied in mice the response of these glands to oestrone (Burrows, 1935b, 1937a). The first effect is a diminution in size of the glands owing to a reduction of the cyto-plasm and secretory activity of the epithelial cells. At a later stage, when the urethral epithelium has already become cornified, a similar metaplasia begins in the ducts of Cowper's glands close to their urethral openings and gradually ex-tends along them toward the body of the gland, on reaching which the metaplasia continues to spread along the glandular epithelium so that the secretory cells become replaced by a stratified epithelium. Accompanying this epithelial trans-formation a widespread atrophy and disappearance of acini in the periphery of the gland has been seen, the process recalling that which occurs in the mammary alveoli during natural involution. In most examples examined by the writer the metaplasia has not involved the whole gland, shrunken remains of the most peripherally placed portions of glandular epithelium being visible here and there even after prolonged treatment with oestrogens. The cytoplasm of these shrunken cells is coloured pink in sections stained with eosin and haematoxylin, whereas the cytoplasm of the large secreting cells of a healthy gland are coloured a pale blue.

In some instances cystic dilatation of the gland occurs under the influence of oestrogen, perhaps because of a gradually increasing blockage of the cornified duct at a time when the gland, or part of it, is still producing secretion. As in other situations where a cornifying metaplasia occurs leucocytic infiltration is apt to follow and the cavity of the gland may become filled with pus.

Harsh, Overholser & Wells (1939) gave 500 i.u. of oestrone daily for 30 days to rats, and at the end of this time found that the ducts of Cowper's glands had be-come lined with a stratified squamous epithelium. The effect of oestrone on Cowper's gland was investigated also by Wells (1936 a), using the ground squirrel for his observations. This mammal has a prolonged non-breeding season during which the testes and accessory reproductive organs remain atrophic and inactive. The average weight of Cowper's glands in the breeding season was found to be 341 mg. and in the non-breeding season 34 mg. Oestrone given to adult males, whether castrated or not, during the non-breeding season caused enlargement of Cowper's glands, the increase being due chiefly to hypertrophy of the fibro-muscular stroma. After 31 days of treatment with oestrone a squamous, corni-fying metaplasia was seen in the ducts. Given to males in the breeding season oestrone caused a reduction in the size of the glands owing to a great diminution of the secretion contained in the lumina. The same metaplastic changes were induced as in the non-breeding season. In immature males the secretory cells responded differently, showing an increased size such as might be induced by androgen.

The Preputial Glands

In mice these are well-developed active organs in both sexes. They respond to a continued excessive supply of oestrogen in the same way as Cowper's glands; their ducts become lined by stratified epithelium and eventually the glandular alveoli show a similar metaplasia so that the organ is converted into a cornified cyst. Suppuration is an occasional sequel.

Chapter XVI.   *The Action of Oestrogen on the Accessory Genital Organs* (continued)

Vagina. Uterus.

The Vagina

(a) *Cornification of the vaginal epithelium.*   In 1917 Stockard & Papanicolaou found that a periodic cornification of the vagina occurred in certain animals and could be easily recognized by the microscopical examination of vaginal smears. A simple test was thus provided by which the sexual cycle could be followed easily and with precision.   In 1923 Allen & Doisy obtained from the liquor folliculi of the sow's ovary a hormone which would induce cornification in the mouse's vagina. These two discoveries, followed as they were in 1929 by the preparation of oestrone in a pure crystalline form, have been the foundations of a great deal of subsequent research.

Oestrone given in an effective dose to a spayed mouse will produce a change in the vagina within 24 hours.   An early result is a rapid multiplication of the epithelial cells, which not only increase numerically but undergo a metaplasia, so that the vagina instead of being lined by two thin layers of atrophic cells becomes lined by a squamous, keratinizing epithelium many cells in thickness.   If no more oestrone be given a copious infiltration of the epithelium with leucocytes occurs, and the epithelium desquamates.   On the other hand, if the dosage with oestrone is continued at sufficiently short intervals, leucocytic infiltration may be absent, so that the vagina becomes filled by a keratinous mass (Burrows & Kennaway, 1934). These vaginal changes are not equally pronounced in all animals. They are relatively inconspicuous in the rabbit, and in the mare it is said that vaginal cornification does not take place in spite of a plentiful production of oestrogen. Papanicolaou (1933) has shown that vaginal cornification occurs in women between the 8th and 12th day of the menstrual period.   At this stage, he says, there is a complete or almost complete absence of leucocytes from the vaginal smears while the epithelial cells, though usually nucleated, are cornified.   Mucus at this stage is abundant. He states that the commonest date for a typical oestral smear in women is the 11th day of the cycle.   Papanicolaou & Shorr (1935) gave daily doses varying from 100 to 2,000 r.u. of oestrone subcutaneously to women who had undergone oophorectomy or were past the menopause. The results of this treatment were an increase of vaginal secretion, and a gradual lining of the vagina with a squamous stratified epithelium, the smears showing cornified epithelial cells with pycnotic nuclei together with mucus and relatively few leucocytes. These effects were cumulative and disappeared within a few days if the treatment ceased.

Vaginal cornification, sexual attraction and copulation.   Though vaginal cornification is commonly associated with oestrus it does not appear to control the tendency to copulate at this period.   As already mentioned, in some species the vaginal epithelium does not become cornified during the oestrous cycle, moreover cornification of the vagina induced by oestrogen in spayed rats does not lead to copulation. Some facts suggest that vaginal cornification, or the excretion of oestrogen in high concentration, may be associated with attraction of the male, and that progestin in addition is required for the acceptance of copulation (Dempsey, Hertz and Young, 1936). Arthus and Malan (1936) say that if rats are kept on a diet of powdered milk, vaginal cornification does not occur, and yet the rats will copulate. Beach (1942) gave 500 r.u. of oestradiol benzoate to each of 7 spayed rats and tested them for sexual reciprocity 67 hours later. The test showed that although the males were attracted and attempted to mate they were not accepted. If however 500 r.u. of oestrogen were given to spayed rats and 0·5 or 1 mg. of progesterone were given 48 hours later a copulatory response was displayed in every instance.

The effect of hyperaemia and trauma on vaginal cornification. On general physiological principles it might be thought that local hyperaemia would accentuate the effects of oestrogen on the vagina and other organs. There is evidence to support the suggestion; indeed it has been shown that gentle trauma if frequently repeated will induce cornification in the spayed rat's vagina even if no oestrogen be given. Wade & Doisy (1935 b) have found that in spayed rats repeated swabbing of the vagina for the examination of smears may alone cause full cornification. Normally the vagina of a spayed rat is lined by epithelium only two cells thick. If daily smears are taken, they say, this epithelium gradually becomes stratified until it is perhaps twelve cells thick, with desquamation of the surface layers. When examined 3 times a day 25 per cent of their rats gave full vaginal cornification on the 3rd or 4th day. Their method of taking vaginal smears was a gentle one; cotton wool was wound round the end of a toothpick, moistened with water, rotated a few times in the vagina and then smeared on a glass slide. Hu & Frazier (1935) have confirmed these observations. The cornification produced in this manner is, they say, transient and may disappear although the swabbing is continued; but the reaction is sufficiently pronounced to affect materially the results of biological assays.

Hechter, Lev & Soskin (1940) state that atropine prevents cornification of the vagina in the spayed mouse under the influence of oestrogen, whereas yohimbin causes the vagina to become cornified. From this they conclude that cornification is secondary to hyperaemia, this being caused by stimulation of parasympathetic nerves and the consequent release of acetylcholine.

Vaginal cornification caused by deficiency of vitamin A. One of the earliest changes recognized as a consequence of vitamin-A deficiency was persistent cornification of the vaginal epithelium. It is now recognized (Wolbach & Howe, 1925) that a squamous, keratinizing epithelial metaplasia resulting from this form of dietary deficiency is seen throughout the organs of the body and is in no way peculiar to the sex organs or especially connected with their reproductive functions; vaginal cornification in these conditions is merely part of a general epithelial response.

(*b*) *Stratification of the vaginal epithelium.* It has been convenient to refer in the first place to cornification of the vagina because this was the earliest effect of oestrogens on the vagina to be known and investigated. Stratification of the epithelium precedes the deposition of keratin and appears to be the essential response of the vaginal epithelium to oestrogen, and in the absence of inflammation or other complication, the squamous epithelium, whether keratinized or not, is maintained only so long as adequate supplies of oestrogen are available.

Freud (1938) has made the following summary of the various types of vaginal epithelium seen in rats:

Type I. Epithelium is only two or three cells thick, the long axis of the cells being perpendicular to the surface. This type is seen in infantile, spayed, hypophysectomized and lactating rats.

Type II. A stratified epithelium many cells thick but without cornification or mucification. It is seen during dioestrus and prooestrus.

Type III. A stratified epithelium with mucification of the superficial cells. This condition accompanies pregnancy and pseudopregnancy.

Type IV. A stratified epithelium undergoing cornification and showing in sections in addition to the layer of stratified epithelium seen in Type II three more strata, namely granulosum, lucidum and schollen, the latter being the desquamating cells from which the nuclei have almost or quite disappeared.

Types II, III and IV have in common a stratified epithelium which depends on oestrogen for its existence. For this reason Freud thinks the common phrase 'oestrus inhibition' is misleading because, as usually applied, it does not mean a reversion to Type I but to Type II. Freud's criticism is supported by the fact that vaginal cornification does not accompany oestrus in all animals. In some species, including the mare, it does not normally occur. Zondek & Sulman (1940) showed that cornification of the vagina is produced by oestrogen, if large enough doses are given, in an animal (*Microtus guentheri*) which does not naturally display the phenomenon. In this mouse 500 i.u. of oestrone or oestradiol benzoate were insufficient to cause cornification in the vagina, but cornification followed doses of 5,000 i.u.

Metaplasia of vaginal epithelium in response to oestrogens is not limited by age. The metaplastic response of the vaginal epithelium is not confined to the sexual life of an individual. It occurs in the foetus, the newborn and in the aged.

Courrier (1930a) found that if 80 to 100 r.u. of oestrone were given to a guinea-pig about to litter, the newborn young will have an open vulva and keratinized vagina. A similar effect on the young was produced by giving oestrone to a nursing mother guinea-pig.

Alexiu & Herrnberger (1938) examined vaginal smears from the newborn human baby. Between the 2nd and 5th postnatal days the vaginal smears, they say, show non-nucleated cornified cells only. On the 7th day leucocytes appear and by the 11th day the smears consist of leucocytes and nucleated epithelial cells. In the male child at birth and for a short time afterward a squamous metaplasia may be seen in the prostatic utricle (Aschoff, 1894; Schlacta, 1904).

Evidence of the efficiency of oestrogens in causing vaginal cornification in advanced age is not limited to experience with animals in the laboratory; oestrogens are used successfully by the clinician to cure the defective cornification which accompanies senile vulvitis in women.

The influence of neighbouring tissues on vaginal metaplasia in response to oestrogens. The capacity of vaginal epithelium to respond to oestrogens is not affected by its situation in the body. The property is innate in the cells, and is retained by them after transplantation into another part of the body or into another individual (p. 114). One example here will suffice to illustrate the text. Raynaud (1930) removed the ovaries from adult guinea-pigs. Three or 4 weeks afterward a piece of mucosa was removed from about the middle of the vagina and grafted subcutaneously in the pectoral region of the same animal. Two or 3 weeks later the guinea-pig was treated with oestrone. The grafts were found to react to this treatment with the same intensity as the normally situated vagina.

The action of oestrogen on the vaginal epithelium is direct. The hyperplastic and metaplastic changes induced in the vagina by oestrogens appear to be independent of the pituitary. Smith (1932) found no essential difference in the sensitivity of the vagina to oestrogen in the intact rat, and in rats which had undergone hypophysectomy or ovariotomy; Sammartino & Arenas (1940) have made a similar observation in dogs. Moreover, by direct application of oestrone to the vagina, cornification may be induced by doses which are too small to cause this reaction if administered elsewhere in the body (Berger, 1935; Lyons & Templeton, 1936).

(c) *Hyperaemia, hypertrophy and oedema of the vaginal stroma.* Oestrogens do not act on the vaginal epithelium alone. They cause a general hyperaemia and a moderate hypertrophy of the muscular and fibrous stroma. In an animal which has been spayed the resulting atrophy of the vaginal wall will largely disappear under treatment with oestrogen.

One of the effects of oestrogen on the uterus is a thickening of the submucosa, which assumes an oedematous or myxomatous appearance. This condition extends in a less pronounced degree to the vagina, especially in the part nearest to the uterus where, in sections prepared for the microscope, it may resemble the so-called myxomatous tissue of the cock's comb or the swollen sex-skin of the monkey (Gardner, Allen & Strong, 1936). A similar condition occurs in the distal ends of the vasa deferentia and around the ends of the ejaculatory ducts in the male (Burrows, 1935b), and is, almost certainly, to be explained by an increase of turgor, for the power to retain water in the tissues is an attribute of all the gonadal hormones, though the different types of hormone differ considerably in their potency to cause this retention and in the anatomical localities where they produce the effect. As a general rule the trophic effects of a gonadal hormone on any particular tissue are accompanied by a local retention of water in that tissue.

(d) *Leucocytosis.* This is a regular consequence of the changes occurring in the vagina and uterus during oestrus. In the vagina polymorphonuclear leucocytes penetrate the cornified epithelium and escape into the lumen even before desquamation has become general, and they continue to pass into the vagina for a while after desquamation has become completed so that they are a characteristic feature of metoestrus. The leucocytic invasion, like the uterine bleeding seen in some animals, may perhaps be the consequence of a more or less sudden diminution in the supply of oestrogen. As to the cause of this leucocytic infiltration of the vagina and uterus there has been some difference of opinion and further investigation is needed to settle the problem. For a discussion of the matter see p. 308.

(e) *Mucification of vaginal epithelium.* Under the combined influence of oestrogen and progestin, or of oestrogen and androgen, cornification may be replaced by mucification; that is to say, the vagina is lined by a stratified epithelium of which the superficial layers instead of being squamous and cornified are swollen, rounded or columnar, and contain abundant mucus-like material. As stated earlier (p. 304), the stratification is attributable to oestrogen. Stratification is not essential for mucification, for it is absent during lactation although the vaginal epithelium is of the mucus-producing type. Mucification is discussed in connection with the effects of progestin, androgen and corticosterone (pp. 214, 413, 451).

(f) *Vaginal acidity caused by oestrogen.* The discovery that oestrogens are of much value in the treatment of gonococcal vaginitis, especially in young children, was attributed at first to the metaplasia induced by the hormone in the vaginal epithelium, and this explanation is perhaps true in part. But it has been learned that the keratinizing metaplasia is accompanied by a pronounced alteration of the hydrogen ion concentration in the vaginal secretion. Lewis & Weinstein (1936) found that the mean pH of the vagina in seventeen normal children ranging in age from 4 months to 10 years was 7·2, and in ten children with gonococcal vaginitis the mean pH was about 7. It is known that gonococci grow best in an alkaline medium and that acidity checks their multiplication. Under treatment with oestrin, Lewis & Weinstein say, the vaginae in these children become acid. In no instance among twenty-four cases of gonococcal infection could gonococci be identified in smears when the vaginal secretions were more acid than pH 6. In one patient under treatment with oestrin the vaginal pH was 4.

Schockaert & Delrue (1936) state as a well-known fact that after removal of both ovaries, or after the menopause has become established, there is a reduction of acidity in the vagina. The vaginal pH in the normal woman is about 4·5 to 5, whereas after the menopause or removal of the ovaries it is 7 or more alkaline than 7. By giving oestrin to women after the menopause or after removal of the ovaries the vaginal acidity may be raised, they say, almost to a pH of 5.

Hall & Lewis (1936), using immature monkeys (*Macacus rhesus*) under amytal anaesthesia, estimated the pH of the vaginal fluid by glass electrodes placed against the vaginal wall. Before injections of oestrone the average vaginal pH was found to be 7·7. After treatment by daily subcutaneous injections of 100 r.u. of oestrone during 2 weeks the vaginal pH was 5·5. Four weeks after cessation of the injections the average vaginal pH was 8·7 (Table 146).

TABLE 146. The effect of oestrone on the pH of the vagina in immature macaques (Hall & Lewis, 1936)

Serial number of monkey	Total oestrone given in 2 weeks (r.u.)	Vaginal pH		
		Before treatment	At the end of treatment	4 weeks after cessation of treatment
1	None	7·5	—	—
2	None	7·5	—	—
3	None	8·1	—	—
4	1,700	8·0	5·4	8·8
5	1,400	7·6	5·5	—
6	1,400	7·7	5·7	8·6
Averages		7·7	5·5	8·7

Hall and Lewis say that in healthy children the pH of the vagina is between 7 and 7·2; in children with purulent vaginitis the pH was 6·8 or 6·9, and in the same children during treatment with oestrin it varied from 5 to 5·9. This increase in acidity they attribute to the formation of lactic acid from the glycogen which is present in the vaginal epithelium.

Weinstein, Wawro, Worthington & Allen (1938) treated six immature rhesus monkeys with repeated doses of oestrin. Before treatment the vaginal pH ranged from 6·8 to 7·2. The injections of oestrin caused a rise of acidity which became evident in about 20 days and reached a pH of 3·8. Eighteen days after the treatment with oestrin had been stopped the vaginal pH had returned to the normal, that is between 7 and 7·2. The increase of acidity was directly related, they say, to the concentration of glycogen in the vaginal mucosa, and to the thickness of the mucosa. During the acid stage the Döderlein bacillus became predominant, other bacteria being inhibited by the acidity; the numerous Döderlein bacilli being in their opinion the result and not the cause of the acidity. Comparable results were obtained with oestrogens in another macaque (*Macaca mulatta*) by Dow & Zuckerman (1939a), in women (Beilly, 1939, 1940), and in rats (Beilly, 1939).

(g) *Presence of glycogen in the cornified vaginal epithelium.* Robertson, Maddux & Allen (1930) showed that glycogen granules were present in the cornified epithelial cells of vaginal smears, and that the amount of glycogen increased as cornification progressed. Glycogen was present also in the epithelium of the endometrial glands. Van Dyke & Ch'en (1936) made observations on the vaginal pH in female macaques, and learned that during immaturity and after spaying the vaginal secretions are alkaline but become acid under the influence of oestrone. Normally, they say, the vaginal mucous membrane contains more glycogen than any other tissue in the body except the liver, and they suggest that the vaginal acidity following the administration of oestrogen may be due perhaps to lactic acid derived from the large store of glycogen. After removal of the ovaries the glycogen content of the vagina rapidly decreases.

(h) *Changes in electrical potentials in the vagina.* Burr, Hill & Allen (1935) observed that ovulation may be accompanied by a sudden rise of electrical potential in the vagina. Rabbits were mated and 6 hours later anaesthetized with sodium amytal. Hair was shaved from the abdomen and the electrodes of a potentiometer were applied to the vagina and the abdominal skin. The readings up to 8 hours after mating were uniform. At about the 9th hour the first ovulation occurs and is accompanied by a sudden rise of the vaginal potential followed by a fall. With each subsequent ovulation a similar disturbance of potential occurs, the number of ova shed corresponding with the number of the sudden changes. The potential differences noted were between 7 and 20 mv. The same phenomenon appears to accompany ovulation in women (Burr, Musselman, Barton & Kelly, 1937). In a human patient electrodes were kept in contact with the pubic skin and the interior of the vagina at the time of ovulation which, as proved by laparotomy, occurred on the 21st day of the cycle. Before ovulation the vaginal potential was 30 mv. negative to the pubic skin; at the time of ovulation the vaginal potential showed a sudden change to 50 mv. positive in relation to the pubic skin.

Rogers (1936, 1937, 1938a, b; see also Rogers & Allen, 1937) made investigations on rats. He clipped the pubic hair and, while they were under sodium amytal anaesthesia, placed one electrode of a potentiometer in the vagina and one on the skin over the symphysis and observed the potential relationships between these two spots at different stages of the oestrous cycle. At oestrus a great change in the potential differences took place, the P.D. rising to 20 mv. or more; in repeated tests the results were consistent. In spayed rats the changes of P.D. characteristic of the oestrous cycle did not occur spontaneously, but if enough oestrin were given to cause vaginal cornification the electrical disturbance was produced about 19 hours after the injection and by giving repeated doses of oestrin could be maintained. Obviously in this instance ovulation played no part in causing the electrical changes in the vagina. Rock, Reboul & Snodgrass (1938) also have made observations on the electrical conditions which accompany ovulation, and their observations rather suggest that ovulation, that is to say rupture of a follicle with release of the contained ovum, is not the essential factor in causing the change of electrical potential. They recorded the relative potentials of the vagina and abdominal skin in ten women before laparotomy. At the time of the abdominal operations the ovaries were examined and it was found that the maximum P.D. was associated with ovulation in seven cases and in three with maturation of follicles which had not ruptured. In these examinations the potentials of the vagina in relation to the skin of the abdomen ranged from − 10 to + 30 mv.

Rogers (1938a) noted that the characteristic changes in the electrical potential of the vagina caused by oestrogen were still brought about after hypophysectomy. He also noted (1938b) that during pseudopregnancy the relation between the potentials of the vagina and the symphysis resembled that seen in spayed rats.

Leucocytic Invasion as an Accompaniment of Epithelial Metaplasia

The presence of leucocytes in vaginal smears is a regular feature of the oestrous cycle (p. 305). The leucocytes appear when the cornified epithelium is being shed. The explanation of their appearance is at present speculative. It has become natural to associate the presence of leucocytes with bacterial infection and some people have ascribed oestral leucocytosis of the uterus and vagina to this cause. It is conceivable that the change of pH in the vagina which accompanies oestrus might be favourable to the rapid development and activity of symbiotic acidophilic bacteria and thus bring about leucocytosis. This explanation for such a constant and general phenomenon as we are now considering seems to the writer unacceptable in the lack of conclusive proof. There are other causes of leucocytic invasion than bacterial infection: aseptic inflammation for example. Moreover, an infiltration with polymorphonuclear leucocytes appears to be a frequent adjunct to cornifying metaplasia wherever it may occur and whether caused by the action of oestrogens, a deficiency of vitamin-A in the diet, or other agency (Burrows, 1935b). To the writer it seems likely that the leucocytosis is secondary to local differences of electrical potential. Those who have investigated the bacterial conditions accompanying oestral metaplasia of the uterus have found bacteria in company with the leucocytes. Gardner & Allen (1937), using ten mice 32 to 42 days old, excised the anterior three-quarters of a uterine horn and transplanted it into the inguinal region. The mice were then given 500 i.u. of oestrogen once a week for periods varying from 4 to 8 months. At the end of this treatment the grafts were distended with clear fluid containing very few or no leucocytes. The epithelium, it may be noted, was cuboidal or columnar. Leucocytosis and pyometra were found in the remnants of uterus which had been left in continuity with the vagina. In another paper Weinstein, Gardner & Allen (1937) give the results of a similar experiment together with the bacteriological findings which are summarized in a general manner in Table 147.

TABLE 147. The influence of oestrogen in favouring infection of the uterus in mice (Weinstein, Gardner & Allen, 1937)

	Number of mice	Results observed after treatment with Oestrone benzoate	
		With infected uteri	Without infected uteri
Treated with oestrone benzoate	58	42	16
Untreated	44	2	42

Gardner & Allen regard these results as indicating that oestrogenic hormones injected for long periods of time in large enough amounts induce changes favouring uterine infection or preventing the control or removal of infection, and that pyometra cannot be attributed to a direct action of oestrogenic hormones.

Variations in susceptibility to leucocytic infiltration. The liability to leucocytic invasion does not correspond with the readiness of an organ to display epithelial metaplasia. For example, cornification in the epididymis under the influence of oestrogen is a late and unusual phenomenon in mice and when it occurs it seems to be accompanied almost at once by suppuration. In the mouse the coagulation gland undergoes metaplasia in advance of the seminal vesicle, and yet suppuration in the latter commonly precedes suppuration in the coagulating gland. In the prostatic alveoli, as in the epididymis, leucocytic invasion may be almost coincident with cornification.

It is not to be inferred from these remarks that leucocytic infiltration is seen first in the structures that have been the last to undergo epithelial metaplasia. The interval between the onset of metaplasia and suppuration is shortest in these as a rule; but the onset of metaplasia may be so long delayed in them in comparison with, for example, the coagulating gland and seminal vesicle that the latter organs may be the seat of an abscess before any metaplasia has occurred in the former situations. Usually, in the mice observed by the writer, the seminal vesicles have been the first organs to show suppuration and the coagulating glands the next.

A squamous-cell metaplasia is not essential for local leucocytosis. An abundant discharge of leucocytes takes place in the uterus during the oestrous cycle in the absence of any squamous metaplasia.

The Uterus

Like the other accessory organs of reproduction, the uterus depends for its well-being and functional utility on a continued supply of hormones from the ovary; and a deficiency of these hormones, caused by removal of the pituitary, spaying or senile involution of ·the ovary, is followed by atrophy of all the component parts of the uterus. Knauer (1900) demonstrated by experiment the dependence of the uterus on the ovaries. He spayed rabbits and in some he replaced the ovaries either between the fascia and the abdominal muscles near the laparotomy wound or attached them to the mesometrium. After spaying, atrophy of the uterus followed, but if one of its ovaries were transplanted into another part of the animal and survived, atrophy of the uterus did not ensue. Uterine atrophy could be prevented also, he found, by the successful transplantation of an ovary from another rabbit, and this procedure restored to a normal condition the uterus which had been atrophied for many months as the consequence of spaying. Later work by numerous investigators has shown that by giving oestrogen, also, this 'castration atrophy' can be countered and the uterus restored almost to its normal state.

Between the years 1890 and 1900 the beneficial effects of retaining portions of the ovary during gynaecological operations on women were recorded by several surgeons, and it was found also that these beneficial effects could be secured by ovarian autografts.

(a) *Hyperaemia, oedema, hypertrophy*.   Courrier & Potvin (1926) were perhaps the first to describe in detail the uterine changes induced by the artificial administration of oestrogen in excess of normal requirements. They removed the ovaries from a virgin rabbit and 5 weeks later excised part of one uterine horn for inspection. The uterus had become involuted, being small, pale, and lined with a flattened epithelium; the muscle and submucosa were atrophied also and the latter was almost devoid of uterine glands. Three days after this operation the rabbit was given daily injections of 2·5 c.c. of follicular fluid obtained from cows' ovaries, and after a week of this treatment the animal was killed. The uterus was now congested and hypertrophied; its epithelial cells were no longer flat but swollen and rounded; the submucosa was thickened, hyperaemic and oedematous, and penetrated by the uterine glands. Hyperaemia and swelling of the uterus were noted by Loeb & Kountz (1928) in guinea-pigs after subcutaneous injections of follicular fluid; Reynolds (1939) has correlated the hyperaemia with a release of acetylcholine. Allen (1928c) gave oestrin twice daily for 3 weeks to a monkey (*Macacus rhesus*), the total dose during this period being over 1,000 r.u. At the end of this treatment the uterus was enlarged with hyperplasia of the muscle, glands and epithelium. Smith (1932) found in rats that these reactions of the uterus to oestrin took place with the same readiness after removal of the pituitary or ovaries as in intact animals.

The responses of the uterus to oestrin occur rapidly and, says Moricard (1934), come about in all animals, whether normal or spayed, which have been studied in connection with the matter. Immature mice weighing 7 g. were given one dose of 1γ oestrin. Slight enlargement of the uterus was present 24 hours later and after 48 hours was more pronounced. With continued administration of oestrin the uterus became distended with clear fluid.

In addition to an obvious hyperaemia oestrogens cause the uterus, especially its submucosal layer, to become oedematous. Fagin & Reynolds (1936), when examining the effects of oestrone on the uterus in spayed rabbits, noted that vaso-dilatation and oedema were early effects and preceded the onset of rhythmic contractions. Van Dyke & Ch'en (1936) estimated the water content of the uterus at different periods of the menstrual cycle in monkeys (*Macaca mulatta*) and found that it was highest during the stage of epithelial proliferation and that menstruation was associated with a local dehydration. Astwood (1938, 1939*a*) made the same sort of observation. He used normal adult rats and estimated the water content of the uterine horns during different stages of the oestrous cycle. The tissues were weighed, then dried and weighed again. The water content, he says, is highest at prooestrus and falls rapidly with the appearance of cornified cells in the vaginal smears. In pseudopregnancy, the water content of the uterus rises during the first 5 days and then falls to below the dioestrous level. Six hours after the injection of 2γ of oestradiol given during dioestrus the water content of the uterus increases, but this does not happen, Astwood says, if the oestrogen is given during prooestrus when the uterus is already oedematous. In immature rats between 21 and 23 days old a single subcutaneous injection of $0 \cdot 1\gamma$ of oestradiol caused an increase in the uterine weight which reached a maximum of 60 per cent more than the normal weight in 6 hours. The increased weight, as in the adult rat in similar circumstances, was due to an accumulation of water, which occurred mostly in the submucosa.

Allen, Smith & Gardner (1937; see also Allen, Smith & Reynolds, 1937), using Dustin's colchicum technique with rats and mice, studied the histological changes in the uterus which follow the administration of oestrogen. They used animals which had been spayed 2 months previously. The earliest effect of oestrin, they say, is a hyperaemia and oedema in the uterine cornua. This is followed by mitosis in the epithelium and glands. The more deeply lying parts of the glands are the first to show mitosis; with time, cell division ceases in the glands and they begin to secrete while mitosis continues in the epithelium lining the uterine cavity. In the muscle the most active mitosis comes about when the uterus is distended with the products of secretion; that is to say, the muscular hypertrophy is later in appearing than epithelial mitosis. Perhaps the muscular hypertrophy is caused by the distension alone and not by a direct action of oestrogen on the uterine muscle, for the mere insertion of a paraffin pellet into the uterine cavity of a spayed rabbit will cause hypertrophy of the uterus proportional, within limits, to the amount of distension produced by the pellets (Reynolds, 1937). Oestrin hinders this hypertrophy in response to distension. Reynolds found that the optimum effect in causing hypertrophy of the uterus was produced by pellets if their diameter was about equal to that of the uterus. Their effectiveness was less when the pellets were less than half or more than twice as thick as the uterus (p. 426).

Allen (1940), by the use of colchicum, observed mitosis in the uterine muscle in response to the presence of paraffin pellets.

The actions of oestrogens on the uterus which have been described above appear to be the result of a direct influence. According to Hill & Parkes (1933) they will occur in the absence of the pituitary or ovary. These investigators gave daily subcutaneous doses of 0·5 to 2 mg. of oestriol to (a) hypophysectomized and (b) ovariectomized ferrets for periods of 5 to 15 days. Judging by the weight of the uterus and the development of the endometrium at the end of the treatment the responses were the same as those obtained in normal ferrets. Sammartino & Arenas (1940), using dogs, also find that the effects of oestrogen on the endometrium and uterine stroma occur in the absence of the pituitary.

(b) *The condition of the human uterus after the menopause.*   It might have been supposed that the quiescence of the ovaries after the menopause would result in atrophy of the uterus. In most cases such a supposition would be correct, but not always. Novak & Richardson (1941) examined the endometrium in 137 women who had passed the menopause, and in 31 per cent of these the uterine epithelium was hyperplastic as though it had been recently submitted to the action of oestrogen. This hyperplasia probably can be attributed to adrenal activity.

(c) *Cholinergic action of oestrone on the uterus.*   Reynolds (1938) noted that when oestrin is injected into a spayed rabbit it causes a hyperaemia of the uterus, this effect reaching its maximum between 30 and 60 minutes after the administration. Extracts prepared from the uterus at this stage show its acetylcholine content to be significantly increased. Oestrogens have been found to have a cholinergic effect on the vagina also.

(d) *The effect of oestrogen on the respiration of the uterus.*   Khayyal & Scott (1931, 1934 and 1935) found by *in vitro* tests that the consumption of oxygen by the rat or mouse uterus shows cyclical variations, being almost constant during oestrus and the first part of dioestrus whereas it is nearly doubled during the latter part of dioestrus. After spaying, they say, no such variation takes place. Subcutaneous injections of fresh bovine follicular fluid caused a rise of oxygen consumption in the rat's uterus beginning within 1 hour and reaching a maximum between 3 and 6 hours. Similar tests by David (1931) did not give quite the same results. He estimated the oxygen consumption by the normal mouse's uterus to be about 1·7 or 1·8 c.c. of dry gas per gram of moist tissue per hour at N.T.P. No increased consumption of oxygen was observed in natural oestrus or when oestrus was induced in the spayed animal, though in immature mice oestrone caused an increase in the take up of oxygen to 2·3 c.c. per gram per hour. Kerly (1940) found that in the uteri of untreated rats anaerobic glycolysis showed variations. After the injection of oestrone it was increased, the rise beginning between 6 and 12 hours after the injection and reaching a maximum between the 24th and 36th hours. Though the oxygen consumption was increased this effect was less marked than the rise of anaerobic glycolysis. The latter result was not altered by hypophysectomy. Kerly did not detect any effect of oestrone on the respiration of the isolated uterus *in vitro*. Carroll (1942) says that injections of 50γ of oestradiol dipropionate into rats caused an increased uterine respiration. Both aerobic and anaerobic glycolysis were increased.

(e) *Epithelial hyperplasia and metaplasia: cystic glandular hyperplasia.*   With a continued administration of oestrogen the epithelial proliferation and other changes mentioned above progress, the uterine glands extend and penetrate more deeply into the submucosa and become distended with fluid. The cystic glandular hyperplasia thus produced is well recognized in human pathology.

Wolfe, Campbell & Burch (1932) spayed guinea-pigs and rats and thereafter submitted them to repeated daily injections of oestrin in doses ranging from 20 to 50 i.u. They found that cystic glandular hyperplasia of the uterus could be produced in guinea-pigs within 8 to 15 days by these moderate doses of oestrin. In rats the results were the same except that cystic dilation of the uterine glands was not always induced within this period. Lacassagne (1935 a) described the same condition in a rabbit which had been under treatment with oestrone during a period of 3 months. In mice he noted that, with equal doses, these effects of oestrone on the uterus varied in degree in different inbred strains.

(f) *Metaplasia of epithelium: adenoma and cancer.* At a more advanced stage of treatment with oestrin, squamous metaplasia appears in the endometrium. This occurs earliest in the cervix and especially in the part nearest the vagina, where metaplasia is more readily induced by oestrogen than in the rest of the uterus. Overholser & Allen (1933) treated rhesus monkeys with daily injections of oestrin, the doses being from 50 to 100 r.u. The duration of the experiment varied from 16 to 90 days with six spayed monkeys and lasted for longer periods in three others whose ovaries had not been removed. While the experiment was in progress the cervix was injured either by scissor cuts made at intervals of a few days or by the attachment of a metal clip. Much epidermization took place in the cervices of these monkeys and in one there was an appearance of infiltrating epidermoid carcinoma, and in another, according to James Ewing, a pronounced precancerous lesion. The other monkeys all showed a squamous keratinizing metaplasia. In further experiments Overholser & Allen (1936) gave 400 r.u. of oestrin 3 times a week or 300 r.u. daily to monkeys during periods varying from 6 months to 2 years. During the experiment the cervix was injured in some and left intact in others. In the three which had received the largest doses of oestrin adenomas of the cervical glands were present together with cystic changes and extensive epidermization. Overholser & Allen came to the conclusion that cervical trauma had no relation to the degree of the changes produced. Engle & Smith (1935) also gave daily injections of oestrin to monkeys for several months with the result that a squamous, keratinizing metaplasia occurred in the cervical glands. Hisaw & Lendrum (1936) performed similar experiments on monkeys with the same result. In one instance there was a development of squamous epithelium in the glands of the cervix which resembled the beginning of cancer in women; none of the lesions produced showed invasive properties. In the course of these experiments, they say, the cervical metaplasia induced by oestrin disappeared if the supplies of this hormone were discontinued. They observed also that the epithelial metaplasia was checked by progesterone; daily doses of 4 rb.u. of progesterone prevented daily doses of 100 r.u. of oestrone from causing metaplasia of the cervical glands. Nelson (1937 b) has made a similar observation.

The writer has treated many mice for long periods with oestrin, and among them cancer-like metaplasias have occurred in the uterus; but in no instance has a mouse so treated died of uterine cancer, nor has any case of indubitable squamous epithelial cancer of the uterus with metastasis been found among these animals. One of the mice died with a large uterine sarcoma which may or may not have been caused by the oestrin given. In the light of his own work the author is sceptical concerning the interpretation of some of the results just quoted. A squamous metaplasia of the uterine epithelium certainly is caused by oestrogen, but there seems still to be a doubt whether oestrogen, unaided by some additional factor, can induce squamous cell cancer of the uterus in mice. It may be, as with mammary cancer, that certain genic and nongenic transmissible factors are required in addition to oestrogen. Such a suggestion gains force from the fact that carcinomas of the uterus have developed in mice of the CBA strain, or in crosses of this strain with C57 under the influence of oestrogen. Allen & Gardner (1941) have reported that uterine cancer appeared in twenty-five (56·8 per cent) of forty-four such crossbred mice which had been treated with oestradiol benzoate and survived for more than 1 year. The dosage in this experiment usually had been 16·6γ per week. The tumours were squamous carcinomata. Miller & Pybus (1941), also, have observed uterine tumours in CBA mice which had been treated with oestrone. In some instances the tumours were fibromyomata or fibrosarcomata, but in others squamous epithelial cancer was present. Among twenty-six spayed CBA mice which had been given 300 i.u. of oestrone once a week, cervical tumours eventually appeared in twelve, and among twenty-eight non-spayed CBA mice treated in the same way uterine tumours developed in eighteen. In 130 untreated female CBA mice cervical tumours occurred in nine.

These cases in mice connected with the CBA strain seem to be almost the only undoubted examples of uterine squamous cell carcinomata caused experimentally in mice by oestrogen. A solitary instance of uterine cancer in a mouse of the C_3H strain treated with oestrogen has been reported by Gardner, Allen, Smith & Strong (1938).

In connection with the possibility of accessory factors taking a share in the etiology of uterine cancer in mice the following observations may be mentioned. Lacassagne (1936a) recorded the development of squamous metaplasia in the uterus of a mouse which died of pyometra after being treated for 5 months with oestrone benzoate and an alkaline extract of fresh pituitary gland (horse). In this case the epithelium at one place invaded the submucosa as far as the muscle into which it had slightly penetrated. Mitoses were numerous; no metastasis was discovered. Whatever the nature of the lesion may have been the administration of pituitary extract may perhaps have played a part in its production. Work carried out by Perry & Ginzton (1937; see also Perry, 1936) may also be mentioned here. They treated 150 spayed mice persistently with the carcinogen 1:2:5:6-dibenzanthracene, and to seventy-five of them gave oestrin in addition. Among those which had received oestrin cancer of the uterus appeared in three.

Domestic rabbits seem to be extraordinarily subject to a different form of uterine cancer. Greene & Saxton (1938) have reported eighty-three cases of adenoma or adenocarcinoma occurring among their own stock of rabbits. The writer (Burrows, 1940) also has drawn attention to the same subject; among twenty-five female rabbits which had lived for 900 days or longer under his observation fifteen died with adenocarcinoma of the uterus and in three of these primary cancer of the breast also was present. In every one of these rabbits the ovaries showed advanced changes, and in most of them cystic mastopathia and other conditions were present, which suggested that the rabbits had been subjected to abundant supplies of their own oestrogen. It is possible that progesterone co-operated with oestrogen in the causation of these tumours, for several and perhaps all of the rabbits had been repeatedly pseudopregnant, evidence of progestational changes in the endometrium not infrequently accompanied the neoplasia, and the ovaries in the tumour-bearing animals were always enlarged and appeared to be heavily luteinized. The fact that the primary tumours are commonly multiple and occupy successive intervals in the uterus so as to give it the configuration of pregnancy might also be taken to suggest a progestational influence. In a recent paper on these uterine tumours in the rabbit Greene (1941) says that there is abundant evidence that oestrin represents the exciting agent. His rabbits had suffered from hepatic disorder which, he suggests, may have assisted in the neoplasia by interfering with the inactivation of oestrogen.

In some of the writer's rabbits fibromyomata were present besides adenocarcinoma of the uterus, and this combination might be regarded as further evidence that excessive supplies of oestrone had been available.

(g) *Leucocytosis of the uterus and pyometra.* Loeb (1914) observed migration of leucocytes into the uterine cavity following oestrus, and discovered that this process was much increased in guinea-pigs by the act of copulation. In this animal, he says, leucocytes are seen migrating through the mucosa into the lumina of the uterine glands within 1 hour of copulation, and in 3 to 8 hours after copulation the leucocytes may enter the uterus in such quantities as to make its cavity resemble that of an abscess. Stockard & Papanicolaou (1917) correlated leucocytosis in the uterus and vagina of the guinea-pig with a particular phase of the oestral cycle, and Allen (1922) found that the same held good for the mouse, that is to say at the end of oestrus a leucocytic invasion takes place. As with uterine bleeding, the shedding of epithelium and local leucocytosis appear to depend upon a reduction of the supply of oestrogen.

The writer has shown that a local infiltration by leucocytes is a common accompaniment of epithelial metaplasia, whether induced by oestrogens or by other means, in whatever organ and whichever sex the metaplasia may have occurred (Burrows, 1935b). The matter has been dealt with (p. 308) in connection with the effect of oestrogen on the vagina, and no more will be said about it here except so far as concerns the uterus. Mice were treated by applying one drop of a 0·1 per cent of oestrone dissolved in benzene to the skin of the interscapular region twice a week. These applications caused an excessive degree of keratinization in the vagina, together with proliferative changes in the uterine epithelium. Accompanying the latter was a profuse infiltration of the endometrium with leucocytes. In several instances the uterine horns became greatly distended with pus and the mice died, often with purulent perimetritis. The pyometra may have been caused by a blockage of the vagina, which in some instances was distended by a mass of keratinized debris. The primary uterine leucocytosis was, however, not dependent on any obstruction in the vagina, and seems to have been caused by oestrogen (Burrows & Kennaway, 1934).

(*h*) *Hyperplasia of the uterine stroma.*   David, Freud & de Jongh (1934), having given daily injections of oestrin to spayed rabbits and rats during a period of several weeks, noted a pronounced hypertrophy of the fibromuscular stroma of the uterus. Among the oestrogens tested, oestradiol, oestradiol benzoate, oestradiol diacetate and reduced equilin all had a greater effect than oestrone on the fibromuscular components of the uterus. Lacassagne (1935 *a*), using mice, also found that oestrogens cause hyperplasia of the uterine stroma, noticing at the same time that these reactions occur more readily in some strains of mice than in others. He made observations on rabbits, to one of which he gave daily injections of oestrone from the day of its birth onward. When killed at the end of $2\frac{1}{3}$ years the uterine horns were greatly enlarged, knobbly, and contained subserous cysts. The lumen was narrowed and in places obliterated, and at these spots there were swellings, which resembled fibromyomata. The uterine muscle was hypertrophied throughout. McEuen, Selye & Collip (1936*a*) and del Castillo & Sammartino (1938) have found hypertrophy of the uterine stroma in rats after prolonged treatment with oestrone. The muscular hypertrophy caused in the uterus by oestrogen has been studied by Barks & Overholser (1938), who gave daily subcutaneous injections of 5 or 10 r.u. of oestrone to rats, most of which had been spayed. The injections were continued for periods varying from 10 to 30 days. The treatment caused an increase in the length and diameter of the uterus, the musculature was hypertrophied—the circular layer being more affected than the longitudinal—and the individual muscle cells were enlarged.

Loeb, Suntzeff & Burns (1939) say that in normal mice there is an increasing deposit of collagen in the stroma of the vagina and uterus with advancing age. The increase begins in infancy, does not advance much during sexual life and progresses later. The administration of oestrin over long periods enhances, they say, this deposit of collagen.

The following examples of hyperplasia of the uterine stroma induced by oestrogens may be correlated with general hypertrophic fibrosis and fibromyomata of the uterus in women.

(*i*) *Fibromyomata.*   Discrete fibromyomata were produced in none of the laboratory experiments quoted above except Lacassagne's. Differences in the reactions to oestrin by different species are probably accountable for this. Nelson (1937*b*, 1939) was able to induce fibromyomata of the uterus in guinea-pigs by means of oestrogen. He used thirty-two guinea-pigs and treated them with oestrogen for periods varying from 2 to 10 months, and in six of these animals discrete fibromyomata appeared in the uterus—four after injections of oestrone and two after oestradiol. Associated with the tumours was an adenomatous hyperplasia of the endometrium with metaplasia of the crypts. In guinea-pigs less intensively treated, cystic glandular hyperplasia and metaplasia were regularly present in the uterus. Moricard & Cauchoix (1938) gave weekly doses of oestradiol benzoate to guinea-pigs for periods up to 32 weeks, the total dose during this period being 160 mg. After 4 months of this treatment uterine fibromyomata were constantly present. Lipschütz & Iglesias (1938) obtained similar results in guinea-pigs with oestradiol; the uterine tumours so produced were fibromata or fibromyomata.

In further experiments (Vargas & Lipschütz, 1938) it was found that small fibroid or fibromyomatous tumours might develop elsewhere than in the uterus under the influence of oestrogens. Ten guinea-pigs were spayed and to each was given 3 times a week a subcutaneous injection of 80γ of oestradiol benzoate. The animals were killed at the end of 2 to 4 months. In four of them multiple, small discrete subserous fibroid or fibromyomatous tumours were present scattered over the stomach, spleen, mesentery and parietal peritoneum. In none of these guinea-pigs was any tumour present in the uterus. Similar tumours followed the administration of oestrone (Lipschütz & Vargas, 1939). They were also produced in the same way in male guinea-pigs, whether castrated or not, though the growths were smaller and took longer to appear in males than in females (Koref, Lipschütz & Vargas, 1939). Spaying accelerated their appearance. The dose of oestradiol or oestrone required to maintain the uterus when given 3 times a week in spayed guinea-pigs was only about one-thirtieth of that necessary for the production of fibromyomatous tumours (Lipschütz, Rodriguez & Vargas, 1939). In castrated males the tumours showed a special tendency to appear at the site of ligation of the spermatic cord. The fibromata and fibromyomata occurring in these various experiments were mostly subserous and were widely distributed over most of the abdominal organs, including the stomach, spleen, pancreas, liver, kidney, urinary bladder, omentum, mesentery, abdominal wall, diaphragm, uterus and oviducts. Occasionally similar tumours appeared within the stroma or in the submucosa of the uterus (Lipschütz, Iglesias & Vargas, 1940).

Perloff & Kurzrok (1941) implanted small pellets of oestradiol benzoate in the uterus, and by this means induced the formation of fibromyomata at the treated site in nine of twelve guinea-pigs in the course of periods ranging from 32 to 150 days.

It is of interest to note that the fibromyomatous tumours regressed when the administration of oestrogen ceased (Lipschütz, Iglesias & Vargas, 1939).

Stilboestrol appeared to be more potent than oestradiol or oestrone to produce fibromyomata. Lipschütz & Vargas (1940) say that 100γ given subcutaneously 3 times a week for 3 months will produce tumours in guinea-pigs.

The fact that oestradiol benzoate produced fibromyomata more readily in castrated than in non-castrated guinea-pigs led Lipschütz to inquire whether progestin or androgen might afford some protection against this tumour-producing effect of oestrogens. It was found (Lipschütz, Murillo & Vargas, 1940) that when given in a proportion by weight of 150 : 1 progesterone prevented the formation of fibromyomata under the influence of oestradiol. A similar preventive action against oestradiol monobenzoate was obtained (Lipschütz, Vargas & Ruz, 1939) by testosterone propionate when given in a proportion of 22 : 1, but not when given in a proportion of 13 : 1.

Lipschütz, Vargas & Palma (1941) state that although the fibroid tumours of the peritoneum induced in guinea-pigs by stilboestrol, hexoestrol or oestradiol are of similar character in males and females, larger doses are required to produce the same result, as measured by the size of the tumours, in males than in females. Furthermore, this difference in response is not attributable to the gonads because the same difference is seen between the two sexes after gonadectomy. Jedlicky,

Lipschütz & Vargas (1939), using oestradiol-caprylate, noted that to cause these fibrous tumours in male guinea-pigs the dosage required was about 8 times that which would produce a comparable effect in females.

Observations made on women seem, so far as they go, to be in accord with some of the experimental work on animals which has just been recounted. Among 466 cases of uterine hyperplasia in women, Tietze (1934) found in all in which histological control was possible, with one exception, large cystic, granulosa-bearing follicles in the ovaries. Their presence suggests that the patients were subject to excessive supplies of oestrogen. Witherspoon (1933, 1935 a, b) studied forty-four women with endometrial hyperplasia. In none of them was an uterine fibroid discovered. In twenty of the patients a laparotomy was performed and the absence of fibroid tumours was verified; in all the cases multiple follicular cysts were seen in the ovaries. After an average period of nearly 5 years every one of these patients returned for operation on account of multiple fibroid tumours of the uterus. Such an experience supports the view that oestrogen is the cause of endometriosis and fibroid tumours in women.

It might be possible to collect many clinical records which bear on this theme, but for the present one more must suffice. Novak (1934a) describes the case of a woman who suffered from a so-called granulosa-cell tumour of the ovary. In addition to this oestrogen-producing tumour she had an enlarged uterus containing a number of small fibromyomata together with endometrial hyperplasia.

(j) *Endometrial moles: deciduomata.*   Selye, Harlow & McKeown (1935) say that in rats deciduomata are not produced in response to uterine trauma after the 8th day of pregnancy, but another kind of tumour forms which they describe as an 'endometrial mole'. In twelve rats one uterine horn was slit on the 12th or 13th day of gestation. Five days later a gelatinous tumour of the mucosa was present at the site of injury, resembling in appearance the connective tissue of a hydatidiform mole. They repeated the operation on other pregnant rats and thereafter gave them daily subcutaneous injections of 30γ of oestrone. This caused larger moles than those produced in the first experiment. By the same means large endometrial moles, but not deciduomata, were induced in lactating rats; though the same procedure would cause deciduomata if no oestrogen were given. Selye, Harlow & McKeown conclude that endometrial moles result from uterine trauma under the combined influence of oestrone and progestin. Apparently the oestrogen must be in a larger proportion than is required for the production of deciduomata. In later experiments Selye & Friedman (1940) found that a similar condition could be caused in adult spayed rats by treating them with 100γ of oestradiol and 3 mg. of testosterone propionate or 3 mg. of ethinyltestosterone for 9 days and then injuring the uterus. They compare the appearance of the endometrial moles so produced with the so-called myxomatous condition of the turgid cock's comb, the mucoid stroma of the sex-skin of monkeys which have been treated with oestrin, the mucoid condition of the stroma around the ejaculatory ducts of mice which have been subjected to oestrogen and to the same condition of the stroma surrounding the cloaca in the breeding season in the lamprey and triton. A feature of all these conditions appears to be an excessive local retention of water.

A lesion which seems to resemble that just described is found occasionally in women. This has been named 'stromatous endometriosis' by Goodall (1939), who attributes the condition to the action of oestrogen. Of sixteen cases which have come under his observation with a diagnosis of uterine sarcoma, fourteen were examples of stromatous endometriosis. He describes this kind of tumour as a hyperplasia of the submucosal stroma which invades the surrounding tissues of the uterus and may spread along the broad ligaments, though it is not malignant.

Brooksby (1938) gave 500γ of testosterone propionate daily for 5 days to normal, non-pregnant rats and injured the uteri after 5 days of such treatment. The injections were continued for a further 5 to 7 days and then the rats were killed. No deciduoma or endometrial mole had occurred in any. The experiment was repeated on four rats which had littered, the injections beginning on the 1st day of lactation. The rats were killed on the 10th day of the experiment when typical deciduomata were discovered in three, and an endometrial mole in the other. When contemplating the result of this experiment readers may be reminded that parturition in the rat is followed almost immediately by oestrus, so that at this time there is a sudden copious production of oestrogen. From what has been already said it seems that the production of a deciduoma or of an endometrial mole in response to uterine trauma soon after parturition would depend upon the time the trauma was inflicted in relation to the postparturitional oestrus and the large output of oestrogen which accompanies it.

Differing effects of oestrogen on the uterus when given continuously or with intermissions. According to Lipschütz, Rodriguez & Vargas (1941) the effects on the uterus of a given total dose of oestrogen spread over a relatively long period of time are much greater if it is given 3 times a week continuously, than if it is given in larger doses for a week followed by an interval of 1 or 2 weeks when no hormone is given. This observation applied to the induction of general enlargement of the uterus, adenomatosis, formation of polyps and fibromyomata.

(*k*) *Inhibition of the oestrous cycle.* As already mentioned (p. 51), oestrogens inhibit the gonadotrophic activity of the pituitary. As the oestral changes depend upon the varying activity of the pituitary in forming gonadotrophin, they will be arrested so long as there is an abundant supply of oestrogen.

(*l*) *Uterine bleeding following the cessation or diminution of supplies of oestrogen.* Surgeons have long known that removal of both ovaries may be followed a few days later by a brief period of bleeding from the uterus. The result is attributable to the loss of something which the ovary supplied. The two chief hormones provided by the ovary are oestrogen and progestin, and experiments show that a more or less sudden deprivation or diminution of either of these hormones after they have been in ample supply will lead to uterine bleeding.

Another fact to be considered is that the periodical bleeding known as menstruation, which occurs only in the primates, can be prevented by the continual administration in sufficient quantity of either oestrogen or progestin. Experiments which illustrate these phenomena are as follows:

Allen (1928 *a*, *b*) spayed twelve rhesus monkeys. In nine of these the operation was followed 3 to 6 days later by uterine bleeding. Of the remaining three in which bleeding did not follow, one was old and the other two were immature.

That the uterine bleeding which follows removal of the ovaries is the consequence of stopping the supply of oestrogen was shown by Hartman (1940), who was able to prevent the uterine bleeding after spaying a rhesus monkey by the subcutaneous implantation of 3 mg. of oestrone.

Allen (1928a) observed that if spayed monkeys were treated continuously with oestrogen for a considerable time, bleeding from the uterus ensued soon after the treatment had been stopped. The oestrogenic extracts used in his experiments had been obtained from human placenta and sows' liquor folliculi. A similar reaction of the uterus after the cessation of treatment with oestrogen can be obtained, Allen states (1937), with infantile monkeys. Allen's observations on the monkey have been confirmed by others (Robertson, Maddux & Allen, 1930; Van Wagenen & Aberle, 1931; Corner, 1935; Hisaw, 1935; Zuckerman, 1936; and Engle, 1937). Engle & Crafts (1939) gave by the mouth a single tablet containing 1 or 2 mg. of ethinyl oestradiol to three spayed rhesus monkeys in all of which uterine bleeding occurred 18 to 20 days later. It was found also that single doses of diethylstilboestrol given orally or by intramuscular injection in amounts ranging from 1 to 200 mg. caused uterine bleeding in spayed rhesus monkeys; the interval between giving diethylstilboestrol and the onset of bleeding was about the same whether small or large doses had been given.

To cause uterine bleeding the complete withholding of oestrogen is not necessary; a mere reduction of the supply is sufficient. Zuckerman (1937c) gave to spayed rhesus monkeys daily doses of oestrone gradually increasing from 400 to 5,000 i.u. On the 15th day the doses were reduced to between 50 and 300 i.u., and continued at the new low level; under these conditions uterine bleeding followed the reduction of the dosage.

Zondek (1940b) states that in rabbits a single intravenous injection of 500–750 i.u. of oestrone will be followed after 4 or 5 days by uterine bleeding. Gonadotrophin given intravenously to immature rabbits, 100 r.u. being administered daily for 5 days, also will induce bleeding after the injections have been withheld. This result can be attributed to stimulation by gonadotrophin of the immature ovaries to produce oestrogen; for the response does not follow in spayed rabbits.

Uterine bleeding after the withdrawal of oestrogen occurs in women. Werner & Collier (1933) treated each of five spayed women, between 21 and 38 years of age, by intramuscular injections of oestrone. Daily doses of 200 r.u. were given during the first 4 weeks, 300 r.u. during the next similar period, and 400 r.u. during the last 4 weeks. Uterine bleeding occurred while the injections were being continued and also after their cessation. The effects on the uterus included enlargement and branching of the glands and a general increase in the thickness of the endometrium. Elden (1936) gave 10,000 r.u. of oestradiol benzoate twice a week to five spayed women until a total of 50,000 r.u. had been given; the treatment was then stopped. A week later uterine bleeding occurred in all. The treatment was repeated and 4 days after the last dose of oestrogen five daily doses of 10 r.u. of progestin were given. The latter procedure caused a delay in the appearance of uterine bleeding but did not prevent it. Bishop (1938) also observed in a girl of 20, both of whose ovaries had been removed, that when a course of treatment with oestrogen was stopped uterine bleeding ensued after an interval of days provided that the dosage had been sufficiently large.

Jones & MacGregor (1936) treated ten women who were past the menopause with oestradiol, six receiving 1,500,000 m.u. in 20 days. In seven of the ten women uterine bleeding followed cessation of the injections.

To refer to the loss of blood from the uterus after treatment with oestrogen has been stopped as 'menstruation' would be erroneous. As shown elsewhere the main fall in the natural output of oestrogen occurs near the middle of the oestrous cycle and not just before menstruation (p. 264). Uterine bleeding rarely follows this mid-menstrual fall in oestrogen production because it is prevented by progesterone. As the output of oestrogen by the ovary falls that of progesterone rises. It is the later drop in the supply of progesterone consequent on degeneration of the corpus luteum that causes the bleeding of menstruation. Removal of a corpus luteum is followed in women by uterine bleeding (Whitehouse, 1926).

Maintenance of cyclical uterine bleeding in the spayed monkey by oestrogen. In contrast with the results just quoted, Corner (1935) found that by continued daily injections of small amounts of oestrone periodical uterine bleeding resembling that of menstruation could be caused in the spayed monkey.

Menstrual bleeding. When supplied continuously in large enough amount oestrogen will prevent menstruation. Corner (1935) gave 15 to 90 r.u. of oestrone benzoate daily by subcutaneous injection to a number of female rhesus monkeys. This treatment did not prevent the first menstrual bleeding but the second bleeding was delayed for 2 to 8 weeks beyond the expected time. Zuckerman (1936) finds that daily doses of 250–1,000 r.u. of oestrone will inhibit the next menstruation in the rhesus monkey only if given before ovulation. If the injections are given after ovulation when a corpus luteum has already formed menstruation will occur at the expected time in spite of oestrone. After this by a long continued administration of oestrone menstruation may be perpetually inhibited.

The same effect can be brought about in women. Bowman & Bender (1932) were able to arrest completely the menses in women by the daily administration of between 20 and 80 "units" of oestrone given hypodermically or by vaginal pessary.

(m) *Influence of oestrogen on nidation of ova, gestation and parturition. Progestational changes: nidation of ova.* The role of oestrogens in preparing the uterus for nidation of fertilized ova will be discussed in a later chapter (pp. 415, 426). Briefly it may be said that the progestational changes in the uterus necessary for nidation are caused by progesterone and that this action only takes place in a uterus which is still under the influence of an earlier dosage with oestrogen. In other words nidation cannot occur unless the uterus has been prepared by oestrogen.

Prevention of pregnancy: abortion: death of foetuses. Although a preliminary subjection of the uterus to oestrogen is essential for the progestational changes which lead to nidation of ova, pregnancy will follow only if the supply of oestrogen ceases or is much reduced. Parkes & Bellerby (1926) noticed that, with mice, injections of oestrin early in gestation always terminated pregnancy. In the later stages, too, death of the foetuses occurred if large enough doses were given. More than twice as much oestrin, they say, is required in mice to produce abortion toward the end of pregnancy as at the beginning. Courrier & Kehl (1931) investigated the effects of oestrin on the progestational developments in the uterus which follow coitus in the rabbit. For this purpose 300 to 800 r.u. of oestrin were given during the period from the 4th to the 8th day after coitus. These doses completely arrested progestational changes in the endometrium; the uterus became thickened and violet in colour, its mucosa became oedematous and haemorrhagic, and bloodstained fluid collected in the lumen. Such effects of oestrin on the uterus were produced only when corpora lutea were present and progestational changes were already in progress. By later experiments (1932, 1933 a, b) on the rabbit Courrier & Kehl showed that 20 r.u. of oestrin given on each of the first 4 days after coitus merely postponed progestational changes, a normal dentelle formation taking place eventually. This dose was enough to prevent nidation of ova and conception. Larger doses—300 r.u. spread over the same period—entirely suppressed progestational developments in the uterus. If 20 r.u. of oestrin were given on the 4th, 5th, 6th and 7th days after coitus no interference with gestation followed; the same doses given on the 12th, 13th, 16th and 17th days caused abortion (see also W. M. Allen, 1937).

Kelly (1931) gave total doses ranging from 10 to 120 r.u. spread over 6 days to guinea-pigs, beginning the injections just after copulation. In every instance these doses prevented conception. In animals pregnant for 12 to 17 days, total doses of 75 to 150 r.u. given in six daily fractions caused abortion. Larger amounts at later stages of pregnancy invariably caused death of the foetuses, with degenerative changes in the placentas. The corpora lutea appeared to be unaffected. Kelly states that more oestrin is required to produce abortion toward the end than in the early stages of gestation.

D'Amour, D'Amour & Gustavson (1933) found that daily doses of 15 r.u. of oestrin given to rabbits after copulation prevented pregnancy. Larger doses—200 r.u. a day—given between the 15th and 25th days of gestation put an end to pregnancy. Hain (1935a), using rats, found, as Parkes & Bellerby had done with mice, that as gestation advances increasing doses of oestrin are required to terminate pregnancy. The administration is not followed immediately by abortion, which means that the termination of pregnancy is not caused by the action of oestrin on the muscular mechanism of parturition. The abortion appears to be secondary to the deaths of the foetuses and this is probably caused by injury to their placentae.

The prevention by oestrogen of pregnancy in women and its termination if gestation has been already established are well known to the modern clinician as practical measures. Indeed the induction of abortion in women by oestrogen appears to have been a recognized therapeutic resource in medieval times. Schmidt (1942) says that in *Ortus Sanitatis*, a pharmacopoeia by an unknown author published in 1494, it is stated that the residue from the evaporated urine of horses will cause pregnant women to abort. It will be recalled (p. 105) that the stallion's urine contains a very high concentration of oestrogen (Zondek, 1934a).

Further references to the effect of oestrin on progestational changes in the uterus and on the development of the placenta will be found in the pages which deal with the action of progestin (pp. 415, 426).

(*n*) *The influence of oestrogen on the muscular activity of the uterus.* In the experiments mentioned above the termination of pregnancy by oestrogen can be attributed largely, it seems, to an interference with the well-being of the placenta. There is another way in which oestrogens might affect the duration of gestation, namely by increasing the activity of the uterine muscle.

Brouha & Simonnet (1926, 1927), using rats and guinea-pigs, tested the effect of follicular fluid on the movements of the uterus *in vitro*. The uterus to be tested was excised and placed in Ringer solution at 37°. Follicular fluid, taken by puncture direct from the follicles of a mare's ovary, was added to the Ringer solution. If the uterus was motionless at the time, this addition caused it to undergo regular rhythmic contractions; if rhythmic contractions were present already their amplitude was increased. Bourne & Burn (1928) reported that oestrin increased the response of the guinea-pig's uterus to pituitrin *in vitro*.

Reynolds (1931) investigated the activity of the rabbit's uterus *in vivo* by means of a uterine fistula through the opening of which a rubber bag was inserted and connected with a recording apparatus. He noted that after spaying there was a complete loss of spontaneous uterine movement and that restoration of motility in these circumstances could be effected by giving oestrone, 20 r.u. per kg. of bodyweight being the minimal effective dose.

Jeffcoate (1932) reported that when doing the Aschheim-Zondek pregnancy test on mice, he noticed occasionally an abnormal result, the test mice showing responses which could be attributed to an excessive supply of oestrogen. The patients whose urine gave this reaction almost invariably aborted. He obtained similar responses when the A-Z test was done with urine from women in the first stage of labour. It seems possible from these experiences that an excess of oestrogen may have influenced the activity of the uterus and so caused abortion or parturition according to the stage of gestation at which the excess of oestrogen appeared.

Robson (1933*a*) demonstrated that oestrone amplifies the reaction of the uterus to oxytocin *in vitro* besides increasing its spontaneous activity *in vivo*.

It seems probable from the observations just quoted that parturition depends largely on a change in the balance between the available progesterone and oestrogen. Smith, Smith & Pincus (1938), having noted the output of oestrogen and progesterone throughout pregnancy, point out that the total urinary excretion of oestrogen rises until a few days before parturition. If, however, the excretion of oestrone and oestriol are separately estimated, it appears that the output of oestrone rises right up to delivery with a sharp rise immediately before the event, whereas the excretion of the less potent oestriol rises continuously until a few days before parturition and then shows a sudden fall. Thus the output of oestrogen, as measured by its potency, seems to be greatest at the time of delivery.

Oestrone sensitizes the uterus to preparations of ergot. Reisel (1936), using rabbits, observed the uterine responses through an abdominal window. In this way he found that one-eighth of the normal threshold dose of ergometrin chloride would cause a reaction in the uterus of a rabbit if oestrone had been given beforehand. The uterus was sensitized by oestrone also, though in a slighter degree, to ergotamin tartrate and the *extractum ergotae liquidum* of the British Pharmacopoeia.

Progestin inhibits the stimulating action of oestrogen on the uterine responsiveness to pituitrin and oxytocin.   The experiments by which Knaus and others demonstrated the inhibitory effect of progestin on the reaction of the uterus to pituitrin and oxytocin will be discussed later.

Paradoxical as it may seem (Robson, 1933 *b*), the presence of oestrogen has been found to be essential for the inhibitory effect of progestin on the uterine response to oxytocin.

Robson (1935 *a*) states that oestrin increases the reactivity of the uterus to oxytocin in the hypophysectomized as in the normal rabbit; he says also that the stimulating effect of oestrin on spontaneous uterine contractions is the same *in vitro* and *in vivo*, though the inhibitory action of progestin is seen only *in vivo*.

From its stimulating action on the uterine musculature it might be supposed that the administration of oestrone would accelerate the onset of parturition. Hain (1935 *a, b*) did not find such an effect in rats, though oestrone was apt to cause death of the foetuses. Sager & Leonard (1936) watched the movements of the uterus in spayed and non-spayed rabbits by means of uterine fistulae and found that a gonadotrophic extract ('Follutein') prepared from pregnancy urine inhibited uterine motility and that oestrin prevented this inhibitory action.

The effect of oestrogen on the response of the uterus to adrenaline.   Dale (1906) showed that the pregnant cat's uterus contracts in response to adrenaline, whereas the non-gravid uterus responds by relaxation. Handovsky & Daels (1937, 1938) found that this altered response does not occur in the earliest stage of gestation in the cat, being observed only when the embryos are more than 2 cm. long. It had been suggested that the inverted response during pregnancy was dependent on the presence of corpora lutea, but Handovsky & Daels found that it could be brought about by oestrogen provided that large enough doses were given and sufficient time was allowed. With the dose spread over 15 days, 30,100 i.u. of oestradiol caused an inverted response of the non-gravid cat's uterus to adrenaline.

Daels & Heymans (1938) state that in the dog and cat the action on the uterus of the drug piperidomethyl-α-benzodioxane is determined by the presence of oestrogen and progestin, as is the case with adrenaline. Their tests were performed *in vivo* and the results are shown in Table 148.

TABLE 148.   The effect of oestrogen and progestin on the uterine reaction to piperidomethyl-α-benzodioxane and adrenaline in the cat and dog (Daels & Heymans, 1938)

	Drug and reaction of uterus	
Condition of animal	Adrenaline	Piperidomethyl-α-benzodioxane
Normal, in follicular stage	Relaxation	Contraction
Normal, in luteal stage	Relaxation	Contraction
Prepared by large doses of oestrin	Contraction	Contraction
Prepared by large doses of progestin	Contraction	Relaxation

In performing tests of this kind it will perhaps be advisable to distinguish between the reactions of the cervix and those of the body or cornua of the uterus. Newton (1934) noticed that in the goat the responses of these two parts of the uterine musculature to oxytocin and adrenaline do not correspond.

DISTENSION OF THE BLADDER AND HYDRONEPHROSIS
CAUSED BY OESTROGEN

As a supplement to the effects of oestrogen on the neuromuscular apparatus of the uterus, a short reference may be made here to the influence which oestrogen may perhaps have on the vital mechanisms of urination.

Among the earliest occasional abnormalities reported in mice as a consequence of giving overdoses of oestrone were distension of the urinary bladder and hydronephrosis (Lacassagne, 1933a; Burrows & Kennaway, 1934). These results are seen both in female and male mice, and though various writers have attributed them to mechanical obstruction of the outflow of urine the writer has been unable to discover any such obstruction to account for the condition either in male or female mice. It has seemed to him that the failure to void urine must be attributed to some affection of the neuromuscular apparatus of urination. These observations on mice recall the ureteral dilatation which commonly occurs in women during the earlier stages of pregnancy, for it seems conceivable that the cause may be the same, or nearly so, in the two instances. Every attempt to demonstrate some mechanical obstacle to the outflow of urine in pregnant women has failed, and it seems to the writer that Kidd (1920) was probably correct when he suggested that the dilatation of the ureters in pregnant women may be caused by chemical substances present in the blood during gestation.

It may be perhaps that the ureteral dilatation in pregnancy, like the irresponsiveness of the uterine muscle to pituitrin during gestation, is caused by the combined action of oestrogen and progesterone. Recent work by Hundley, Diehl and Diggs (1942) encourages such a supposition. They found that 2 mg. of stilboestrol given twice daily for 10 weeks to a non-pregnant woman caused an increase of tone with more vigorous peristalsis in the ureters, whereas intramuscular injections of progesterone caused a reduction of the ureteral tone. In the absence of further data it would hardly be profitable to discuss the matter at length in the present essay, which is devoted to fact rather than theory. Some hypothetical aspects of the subject, correlated with objective findings in mice and in women have been briefly summarized elsewhere by the author (Burrows, 1936e) with special reference to pyelitis occurring in pregnancy.

Chapter XVII. *The Effects of Oestrogen on the Mamma*

Stroma. Nipple. Mammary Ducts and Acini. Lactation.

The Stroma

FEW observations have been reported concerning the action of gonadal hormones on the mammary stroma. Perhaps the local deposition of fat and the increased connective tissue framework of the breast characteristic of sexual activity in women is directly caused by the extending glandular structures, and is not an independent reaction to gonadal hormones. However, if the glandular tissue grows in response to stimulation by gonadal hormones in normal circumstances, the stroma also grows. Folley, Scott Watson & Bottomley (1941 a) noticed some enlargement of the udder in immature castrated male goats which had been submitted to a prolonged administration of diethylstilboestrol or its dipropionate, but there is no evidence to show whether this enlargement was merely secondary to the increased growth of mammary ducts or not.

Ingleby (1942) has noticed changes in the mammary stroma during the oestral cycle in women, the periductal connective tissue becoming more transparent and apparently oedematous during the second half of the cycle. Possibly this is an effect of progesterone alone or in co-operation with oestrogen.

The Nipple

Steinach & Holzknecht (1916) implanted ovaries taken from littermates into castrated male guinea-pigs and rats. One consequence was a pronounced enlargement of the nipples. Lipschütz & Tütso (1925) confirmed this effect in guinea-pigs and repeated the experiment on rats with the same result. In rats hypertrophy of the nipples was noticed 9 days after the grafting and they grew to as much as 6 mm. in length. Kunde, D'Amour, Gustavson & Carlson (1930) gave injections of oestrin in daily doses ranging from 25 to 800 r.u. to immature dogs of which one was a male and three were females. The injections caused enlargement of the nipples in each of the animals. The reaction was definite and pronounced. Moricard (1934) gave 1 mg. of oestrin divided into two doses to three immature guinea-pigs weighing 130, 140 and 145 g., and killed them at the 100th hour. Their nipples at this time were 4 times the length normal for their age. Nelson (1936c), in a review of the hormone control of the mamma, says that in males and spayed females of all species tested oestrogen induces growth of the nipples.

This effect is not confined to the natural oestrogens. Bottomley & Folley (1938a) gave daily doses of 10 mg. of 'anol', an artificially produced oestrogenic compound (Dodds & Lawson, 1937), to guinea-pigs, with the result that the nipples became more than 3 times the length of those in untreated control guinea-pigs. (Campbell, Dodds & Lawson (1938) have shown later that the active factor in the preparation used was not anol but hexoestrol.)

The amount of teat growth is not proportional to the amount of oestrogen supplied, except perhaps within narrow limits; a maximal growth of the teats follows a certain limited dosage and any increase of oestrogen beyond this amount will have no further effect in causing enlargement of the nipple. Lewis & Turner

(1942) gave 0·1 γ of diethylstilboestrol in oil daily to normal male guinea-pigs for 20 days, with the result that their nipples were $2\frac{1}{2}$ times as long as those of controls. No greater increase of nipple growth was induced by larger doses of diethylstilboestrol.

Folley & Bottomley (1941; see also Folley, Scott Watson & Bottomley, 1941 a) observed during experiments on the goat that the effect of oestrogen in stimulating growth of the nipples can be inhibited by progesterone (p. 430).

The *epidermis of the nipple* seems particularly sensitive to oestrogen. Uehlinger, Jadassohn & Fiertz (1941), using the colchicine technique, could discover only a very few mitoses in the epidermis of the nipples of untreated male guinea-pigs. If, however, male guinea-pigs had been treated with oestrone for some days and given an injection of colchicine $9\frac{1}{2}$ hours before death, abundant mitoses were seen in the epidermis of the nipples, which was 4 to 5 times as thick as that of control animals which had not received oestrone.

There is evidence which suggests that the existence of the nipples may depend upon a supply of oestrogen. Greene, Burrill & Ivy (1938 c, 1939 b) gave oestradiol dipropionate in doses ranging from 0·375 to 4 mg. to rats on the 13th, 14th and 15th days of pregnancy. Nineteen rats so treated carried to full term and among the litters were 24 males. In 14 of these, nipples were visible at birth, which is not the case with normal newborn male rats. Similar results followed the administration of diethylstilboestrol, in amounts varying from 12 to 42 mg., to pregnant rats. Nipples were present at birth in all the eighteen males born of the mothers so treated. Raynaud (1939 a), also, observed that the administration of equilin or oestradiol to pregnant mice was followed by the development and enlargement of the nipples in the newborn males. The last pair of inguinal nipples were the most enlarged and sometimes were the only ones present. The next in order of susceptibility were the pectoral nipples. No nipples can be detected in male mice in normal circumstances.

According to Lyons & Pencharz (1936) the effect of oestrone on the growth of the nipples occurs even after hypophysectomy. They removed the pituitaries from male guinea-pigs and then gave them repeated injections of oestrone ('Progynon') for a month, during which period each guinea-pig received a total of 1,800 r.u. of oestrone. At the end of this time the nipples of the hypophysectomized animals had grown equally with those of control guinea-pigs which had undergone the same treatment except that they had not been deprived of their pituitaries. Gomez & Turner (1936 a) observed the same effects. They removed the pituitaries from nine young guinea-pigs of which four were males and five were females. Immediately after the operation daily injections of oestrone or oestradiol were begun, the doses ranging from 25 to 1,000 i.u. At the end of periods varying from 20 to 65 days the animals were killed, and it was found that in all the nipples were well developed, whereas no growth of ducts had taken place in the mamma. A different result was obtained by Samuels, Reinecke & Petersen (1941) in rats, for in these 1,000 i.u. of oestradiol benzoate given every other day for a month failed to cause any growth of the nipples after removal of the pituitary.

When considering this subject it is of interest to recall that the nipples respond most readily to an oestrogen if it is applied directly to them.

It might be well to mention here that oestrogens are not the only hormones which cause the nipple to enlarge. The same effect has been brought about in male guinea-pigs by several androgens, and by adrenosterone and corticosterone also (Jadassohn, Uehlinger & Margot, 1938) (pp. 229, 452).

The Mammary Gland

Heape (1906) noticed a close association between the primary development of the mamma with prooestrum. This observation led him to attribute the early stages of mammary growth to some substance secreted by the ovary at this stage of the oestrous cycle. Marshall (1910) extended our knowledge by noting that the formation of corpora lutea is accompanied by events which differ from those occurring at prooestrum; he was thus able to infer that the ovary forms two different kinds of secretion each of which corresponds with a particular stage of follicular evolution.

It is now generally recognized that during each oestrous cycle changes occur in the mamma in response to the output of hormones from the ovaries. In women there is apt to be some fullness of the mammae, perhaps with discomfort or even pain, in the second half of the oestrous cycle. These happenings may be attributed largely to hyperaemia of the breast and distension of the mammary ducts by secretion, such effects being due to the combined actions of oestrogen and pro-gestin derived from a young corpus luteum. Astwood, Geschickter and Rausch (1937) observed in the rat that during prooestrus and oestrus the mammary ducts were narrow and empty whereas at metoestrus the ducts were distended with fluid. Doubtless progesterone shares in the production of these changes. The general effect of ovarian secretion on the breast was tested experimentally by Steinach and Holzknecht (1916), who grafted the ovaries of guinea-pigs into the kidneys of castrated male littermates; after this operation the nipples and mammae became enlarged, milk formed and the guinea-pigs, though male, were able to suckle young and showed an inclination to do so. Lipschütz & Tütso (1925) carried out similar experiments on guinea-pigs and rats and obtained results like those of Steinach & Holzknecht. Vintemberger (1925) gave repeated injections of fol-licular fluid obtained from cows into male and female rabbits and so induced an increased size of the mammae in both sexes. Hartman, Dupre & Allen (1926) observed a similar result in the opossum, and Loeb & Kountz (1928) in the guinea-pig. Allen (1927) reported that injections of oestrin caused extensive development of the mammary ducts in ovariectomized rhesus monkeys.

A detailed inquiry into the matter was carried out by Goormaghtigh & Amerlinck (1930), who treated female mice by daily subcutaneous injections of an extract prepared from the liquor folliculi of sows' ovaries. The injections were continued during several months, each dose containing 6 r.u. of oestrogen. This treatment caused a considerable growth of the mammary ducts, so that the area occupied by each mamma was much increased. Sometimes the ducts when actively proliferating appeared as solid cords; in other examples they were dilated, and occasionally they showed a condition comparable with the so-called chronic cystic mastitis of human pathology. It was noted that the changes were not equally distributed throughout the mammae; one part of a gland might appear unaffected or nearly so, while another part was markedly changed. Occasionally there were adenomatous formations visible to the naked eye, and in one instance mammary cancer developed.

Similar results in mice have been brought about by pure oestrogens (Burrows, 1935e). For most of these studies the male mouse has been used so as to eliminate interference with the results through ovarian activity.

The mamma of the normal male mouse consists only of a few short vestigial ducts, and the changes induced in this vestigial structure and in the mammae of females by oestrogen are as follows:

Mammary Ducts and Acini

(a) *Proliferation and ramification of ducts.* The chief early effect of oestrone on the mamma is an extension and ramification of the ducts with the occasional development of a few small lobules of alveolar tissue (Allen, 1928c; Turner, Frank, Gardner, Schultze & Gomez, 1932; Turner & Gomez, 1934; Gardner, Diddle, Allen & Strong, 1934, and others). The growing ends of the ducts are formed of solid cords, lumina appearing only at a later stage.

These changes produced by oestrogen in the mouse's mamma represent a general biological phenomenon, although variations of detail are noticeable in different species, and even among different individuals of the same species. The chief variation concerns the development of alveolar lobules, a subject that will be discussed presently.

Gardner, Smith & Strong (1935) reported that large doses of oestrone benzoate given to mice once a week caused only a stunted development of the mammary duct system as compared with that in mice which had received small amounts of oestrone daily; in this case the stunted ducts were distended with secretion and numerous alveoli were present.

(b) *Dilatation of ducts: cystic mastopathy.* An early effect of oestrogen is dilatation of the ducts. With continued treatment the dilatation may increase and cause cysts as in the so-called cystic mastitis in women, though as a rule there is no evidence of inflammation and so the term 'mastitis' would not be appropriate. The dilated ducts contain material which is stained pink by eosin and appears homogeneous. The cysts appear to be caused by a general increase in the length and diameter of the ducts causing tortuosity and kinking. The cystic cavities may be lined by a flattened, cuboidal or columnar epithelium, and it seems that a flattened epithelium is not necessarily the consequence of pressure; for cysts or dilated ducts may be seen which, judging by their outline, are flaccid and yet have a flattened epithelium, whereas other apparently tense cysts may have a cuboidal or columnar epithelium. Hyperplasia of the epithelium in the dilated ducts and cysts is not uncommon after prolonged treatment with oestrogen.

Similar changes have been induced by oestrin in the mammae of male monkeys by Geschickter, Lewis & Hartman (1934). They used two cebus monkeys which were treated by subcutaneous doses of 500 r.u. of oestrin ('Amniotin') for a period of 10 days, and five macacus monkeys which were treated in the same way for 6 weeks with smaller daily doses. Though the effects were most pronounced in the latter series they were alike in character in all the seven individuals; the mammae were enlarged, and showed, when examined microscopically, an increased length and dilatation of the ducts, an increase in the number of layers of epithelial cells lining them, and a proliferation of the periductal connective tissue.

In untreated unmated adult female mice, very wide differences are to be seen in the number of mammary ducts present and their calibre. Such differences depend partly on the stage of the oestrous cycle at which the mouse was killed. During oestrus the mammary ducts in mice become dilated, and, normally, return to their former thread-like dimensions during metoestrus (Cole, 1933). But this contraction is not always complete, and dilated ducts may be found in untreated female mice at all stages of the oestrous cycle. Among old unmated and untreated female mice great contrasts exist. In one the mammae may be represented by a few non-dilated ducts resembling in their form the vestigial ducts seen in the untreated male, while in another the ducts are numerous, dilated, and show here and there varying degrees of epithelial hyperplasia.

(c) *Increase of periductal fibrous tissue.* Occasionally in mice which have been given oestrogen one sees an increase of periductal fibrous tissue, chiefly after prolonged treatment, but it is not a constant or pronounced feature. In the mouse, as compared with some other species including man, there is, speaking generally, little tendency toward the formation of bulky fibrous tissue.

(d) *Leucocytic infiltration* around the ducts is sometimes seen, the majority of the invading cells being lymphocytes. The lymphocytic infiltration when present is not generalized, but appears in small localized regions only, and may involve only a small length of a single duct.

(e) *The development of acini* is an occasional accompaniment rather than a characteristic effect of oestrogenic action on the mouse's mamma, and when seen it is apt to be limited to one or more small clumps of ill-developed lobules. In exceptional instances there has been a copious development of acini throughout the whole breast, *gynaecomastia* being thus caused in males. It may be that the duration of treatment, the dosage used, the nature of the oestrogen employed, and the strain to which the treated mouse belongs have some influence in determining to what extent the acini are stimulated to grow. Gardner, Diddle, Allen & Strong (1934) compared the mammary response to oestrone in three different strains of mice (A, C₃H, CBA), the former two having a high and the other a low incidence of spontaneous mammary cancer among the females. Between these three strains no difference was observed in the rate or extent of development of the mammary ducts, but Gardner, Smith & Strong (1935) found that in one strain of mice (C₃H) oestrone benzoate in daily doses varying from 1 to 5 r.u. induced a localized excessive development of acini. A similar development was caused in other strains by oestrone if its administration was prolonged over many months. Bonser (1936) submitted spayed mice of two different strains (Bagg albino and Little's Black agouti) to prolonged treatment with oestrone and oestrone benzoate and observed that the formation of acini was frequent in the Black agoutis and infrequent in the Bagg albino mice. Spontaneous mammary cancer is frequent among the normal females of the Bagg albino strain and rare among the Black agoutis. The relative incidence of acinous formation in these two strains under treatment with oestrone is displayed in Table 149.

TABLE 149. Incidence of acinous development in mammae of spayed female mice of two different strains under the influence of oestrogen (Bonser, 1936)

Strain	Number of of mice	Number in which acini developed	Percentage
Bagg albino	31	4	13
Black agouti	33	22	66·6

The writer (Burrows, 1936a), using stock mice of no special strain, found that whereas a 0·01 per cent solution of oestrone caused an extension of the duct system with but little tendency to lobular formation, a 0·1 per cent solution of equilin and a 0·05 per cent solution of oestradiol caused a growth of lobules and acini with little extension of the mammary duct system; in these experiments it was uncertain whether the differences in the results were due to the different doses and potencies of the oestrogens employed or to an essential distinction in their activity. In each experiment one drop of the solution was applied twice a week to the skin of the interscapular region.

In the rabbit (Parkes, 1930; Gardner & Van Wagenen, 1938), the cat (Turner and De Moss, 1934) and the dog (Turner & Gomez, 1934) the main effect of oestrone, as in the mouse, is an extension of the duct system with but little tendency to the formation of acini. This is not the case with all species or with the same species on every occasion. Halpern & D'Amour (1934) gave daily injections of oestrin to normal adult male and female rats for a period of 7 weeks. The doses were 5 r.u. for the first 3 weeks and 20 r.u. for the last 4 weeks. The rats were killed 4 days after the last injection and all showed a development of acinous lobules with secretion. In the guinea-pig oestrone causes not only an extension of the duct system but a copious growth of acini too, and even lactation (Laqueur, Borchardt, Dingemanse & de Jongh, 1928; de Jongh, Freud & Laqueur, 1930; Turner & Gomez, 1934).

In male rhesus monkeys Folley, Guthkelch & Zuckerman (1939) obtained with oestrone only a growth of the mammary ducts, but Turner & Allen (1933) after giving injections of oestrone to adult male rhesus monkeys found a considerable growth of lobules in addition.

There are reasons for thinking that oestrogens need the co-operation of progestin, androgen, or some other accessory factor to produce the full development of alveolar lobules in the breast, and that discrepancies in the results obtained with oestrogen alone may depend upon the presence or absence of the accessory factor, whether derived from the gonad, adrenal or elsewhere.

Several workers have demonstrated that progesterone may be an important factor in the development of mammary acini, and it appears normally to co-operate with oestrogen in producing this effect. As in so many other activities of the sex hormones a normal outcome depends upon the relative amounts of oestrogen and progesterone which are available. Lyons & McGinty (1941) gave daily to four groups of immature male rabbits 0·25, 1, 4 and 8 i.u. of progesterone respectively and all the rabbits received in addition daily doses of 120 i.u. of oestrone. After 18 days of such treatment the best development of ducts and acini was found in the rabbits which had received 1 i.u. of progesterone daily. The larger doses of progesterone checked the growth of the mammary ducts.

Scharf & Lyons (1941) determined that the optimum daily dose for the production of mammary ducts in their immature male rabbits was about 120 i.u. of oestrone daily. To six different groups they gave 30, 60, 120, 240, 480 and 960 i.u. of oestrone respectively on 5 days a week for 5 weeks and to all the rabbits 1 i.u. of progesterone was also given with each injection of oestrone. In all the rabbits lobular acinous growth occurred, without cystic dilatation of the ducts. The best development of ducts and alveoli was obtained in the animals which had received daily injections of 240 i.u. or more of oestrone with 1 i.u. of progesterone. The mammae of these rabbits attained a condition almost equal to that seen in the 2nd or 3rd week of pregnancy.

Although, as shown in these experiments, the co-operation of oestrogen with progesterone appears to be the normal cause of full mammary development, some experiments suggest that progesterone alone, if supplied in large enough amount, may bring about a full development of acini. Selye (1940a) spayed rats and 9 days later began to inject them daily with 15 mg. of progesterone. After 15 days of treatment with these large doses the mammae were fully developed and the uteri showed progestational changes.

Mixner & Turner (1941b) observed that total doses of 12·5 to 20 mg. of pregneninolone (ethinyltestosterone) given during 10 days to female mice caused full development of mammary acini in every one of sixteen mice so treated; but they found that a similar result could be achieved by smaller doses of pregneninolone (5 to 10 mg.) if oestrone (133 i.u.) were given at the same time. Mixner & Turner (1942) estimate that six times as much progesterone is needed to secure the same degree of acinous development in the mouse as will be required if oestrone is given too, although when supplied in excessive amounts oestrone will inhibit the action of progesterone in causing the development of mammary acini. Reece & Bivins (1942) obtained extensive growth of acini in the mammae of adult spayed rats by giving them 15 mg. of progesterone and 33γ of oestradiol benzoate daily for 10 days.

(f) *Arrest of the development of mammary ducts.* Apparently the stimulating action of oestrone on the growth of mammary ducts depends upon the dosage, for it has been noticed that large doses inhibit the growth of ducts. Astwood, Geschickter & Rausch (1937), experimenting with rats, found that the optimal daily dose of oestrone for mammary development was about 10γ; with larger doses the growth of the ducts was impeded. Gardner (1941) has made the same kind of observation on male mice in which excessive doses of oestradiol benzoate or dipropionate arrested the growth of the mammae almost completely.

(g) *Benign adenomata.* No instance occurred in the author's experiments on mice of well-defined, encapsuled fibroadenomata comparable with those occurring naturally in rats, dogs and women; though isolated clumps of acini were sometimes seen and might conceivably be regarded as adenomata. A failure to cause non-malignant mammary adenomata in mice by giving oestrogen seems to have been the usual experience in experiments of this kind.

(h) *Epithelial hyperplasia and metaplasia.* In the writer's experiments dilatation and hyperplasia were never seen in the mammary ducts of an untreated male mouse, though present in 40 per cent of the males which had been submitted to continued treatment with oestrogen. Epithelial hyperplasia was of less frequent occurrence than dilatation of the ducts, and ensued later. The two conditions do not appear to advance at an equal rate as the experiment proceeds, and on looking through a number of microscope specimens the impression is acquired that to some extent the two conditions are independent; epithelial hyperplasia may be pronounced in those parts of a mamma which show relatively little dilatation of the ducts, while in the presence of extensive dilatation of ducts there may be but little hyperplasia of the epithelium.

After the applications of oestrogen have been continued for a year or so a change may occur in areas of hyperplastic epithelium. The change consists in the appearance at several foci of larger cells, containing faintly staining cytoplasm with large, pale, vesicular nuclei and large nucleoli. These anaplastic-looking epithelial cells multiply and form small masses which replace the pre-existing type of epithelial cell, and represent, the writer believes, the initial stage of mammary cancer; illustrations of the condition are shown in his original paper (Burrows, 1935*e*).

(*i*) *The development of macroscopic cancer of the breast* will be discussed in a separate chapter; meanwhile a few words may be said about (1) the distribution of the reactions in the mamma to oestrogen, (2) the significance of cystic mastopathia and (3) the share, if any, of the pituitary in causing the mammary reactions to oestrogen.

(*j*) *The distribution of the effects of oestrogen in the mamma.* As noticed in the case of other organs oestrogens do not affect the breast tissues in an equal degree throughout the gland, nor can one observe an orderly progession of the responses spreading from a single focus. The reactions are patchy in both the time and distribution of their appearance, as recorded by Goormaghtigh & Amerlinck (1930) and confirmed by subsequent investigators. Even where obvious changes are present throughout all the mammary tissue visible in a single section, the nature of the lesions often varies in character and degree in different microscopic fields, and even in different parts of the same field. Dilatation may be more pronounced in one group of ducts than in another; epithelial hyperplasia may be advanced in one duct, whether cystic or not, and absent from other ducts in the same section.

(*k*) *Multifocal origin of mammary cancer.* The foregoing facts are important in connection with the treatment of cancer. The surgeon is commonly taught that cancer in women is a unifocal disease, and operations for removal of the cancerous breast are designed on this notion. Is it correct? Experience seems to justify the idea in practice though there is evidence (Cheatle, 1921; Nicholson, 1921) that cancer may be of multifocal origin even in women. In animals (mouse, rabbit, dog), separate primary tumours often are present simultaneously in different mammae or in different parts of the same mamma.

Pybus & Miller (1936) examined the mammary glands of 300 mice which had spontaneous cancer of the breast and they were able to detect every phase of tumour development, innocent and malignant, in a single gland.

(*l*) *Cystic mastopathy.* Another fact elicited in animal experiments which bears on human pathology concerns the association between mammary cysts and cancer. 'Chronic cystic mastitis', as it used to be called, is undoubtedly a result of oestrogenic influence, for it can be brought about in animals, whether male or female, by giving them oestrogen. In women, therefore, it may be regarded as a sign that their breasts have reacted too readily or have been subjected to excessive amounts of oestrogen which in the breast is an important factor in carcinogenesis. To this extent the presence of cysts may be regarded as a danger signal. However, as already mentioned, epithelial hyperplasia and cancerous metaplasia are not limited to ducts which have become the seat of cystic dilatation. In mice they seem to appear more often in ducts which are not cystic. The writer has seen multifocal cancer growing in relatively large mammary cysts in the rabbit. It must be concluded that cystic mastopathia neither leads of itself to cancer nor renders it unlikely that cancer will occur. It is an indication of excessive oestrogenic stimulation of the breast and is therefore a warning of danger, the significance of which may be affected perhaps by a knowledge of the patient's family and personal history, and in any event cannot be ignored.

Logie (1942), from an analysis of 330 biopsies, concluded that the association of cancer with cystic mastopathy in man is not a chance occurrence. In the 149 specimens in which cystic mastopathy was present cancer was discovered in 67, or 45 per cent, and in the 118 cases in which cancer was found, cystic mastopathy also was seen in 67, or 56·7 per cent. In 130 cases neither cancer nor cystic mastopathy existed.

THE ROLE OF THE PITUITARY IN THE MAMMARY RESPONSE TO OESTROGEN

(a) *The effect of oestrogen on the mamma after hypophysectomy.* Most workers who have investigated the matter report the absence of response by the mamma to oestrogen after the pituitary has been removed. The administration of oestrogen after hypophysectomy failed to cause mammary development or to sustain the mammary glandular tissue in the rat (Selye, Collip & Thomson, 1935; Reece, Turner & Hill, 1936; Samuels, Reinecke & Petersen, 1941); in the mouse (Gardner, 1936; Gomez, Turner, Gardner & Hill, 1937; Lacassagne & Chamorro, 1939); and in the guinea-pig (Gomez & Turner, 1936a; Lyons & Pencharz, 1936; Desclin, 1939).

Lacassagne & Chamorro (1939) submitted mice after removal of the pituitary to the continued influence of oestrogen; no mammary growth occurred. In another experiment they gave weekly injections of 50γ of oestrone benzoate to mice, beginning the treatment a few days after birth. After 2 or 3 months, when the mammae had reached a stage of pronounced cystic hyperplasia, hypophysectomy was done. In spite of continued injections of oestrone benzoate the mammary glands underwent rapid regression.

Although these experiments appear conclusive, an absence of mammary response to oestrogens after hypophysectomy has not been a universal experience, for Nelson (1935b) reported that mammary proliferation occurred in hypophysectomized rats which had received 40 r.u. of oestrin daily, though the response was less than that produced by the same doses in rats with intact pituitaries.

When judging the effects of hypophysectomy two facts have to be remembered. One is that if only a very small part of the pituitary remains a mammary response to oestrone may be obtained (Gomez, Turner, Gardner & Hill, 1937); the other fact is that in many animals, including the mouse, the pars tuberalis of the pituitary is not always removed entirely in an operation regarded as a complete hypophysectomy (Newton & Richardson, 1941), and it may be that this surviving part of the anterior pituitary is able in some instances to supply what is needed to cause growth of the mammary gland in response to oestrogen.

(b) *Mammatrophin.* The influence of the pituitary on the growth and functional activity of the breast was demonstrated by Evans & Simpson (1929a), who caused full mammary development and lactation in adult virgin rats by implants of bovine or rat pituitary substance. In further experiments (1931) they found that extracts from the urine of pregnant women caused mammary development in normal virgin rats. No mammary growth occurred either in male or in spayed female rats in response to pituitary extracts; Evans & Simpson therefore attributed mammary development to the action of pituitary hormone on the ovary. Weichert, Boyd & Cohen (1934) repeated these experiments in rats with the same results. Evans & Simpson saw clear evidence that mammary growth began before any lutein tissue existed in the ovary, from which they concluded that the secretion of the corpora lutea could not be the initial stimulus to mammary growth. Moreover, no mammary hyperplasia was caused in male rats by the same pituitary extract which caused mammary development in non-spayed females. Bradbury (1932) gave chorionic gonadotrophin to female mice and noted that full development of the ducts and lobules took place in the mammae, but this happened only if the ovaries were present. A complete response of this kind was not obtained after hysterectomy. Bradbury thought that the extract used by him ('Antuitrin S') required the presence of both the uterus and corpora lutea in order to induce full lobular development in the mamma. Like Evans & Simpson, and Bradbury, Parkes (1929) concluded that the ovary is essential for the development of the mamma under the influence of pituitary hormones. Anterior pituitary extract injected into virgin non-spayed rabbits caused ovulation followed by a growth of the mammae which with continued injections equalled that of pregnancy; similar injections given to males or spayed females caused no mammary changes. Reference will be made presently to results obtained by other observers which differ from those just recorded.

It is now recognized that the initial factor in mammary development is oestrogen, and the results observed by the workers just quoted, when considered alone, might be attributed to the direct action on the mamma of ovarian hormones produced under the influence of gonadotrophin from the pituitary or placenta.

There are reasons for suspecting that the interactions are not so simple as this and it may be that oestrogens induce mammary development by causing the pituitary to produce a mammatrophin. Such an explanation would account for the absence of a mammary response to oestrogen after hypophysectomy. Corner (1930) reported that extracts of sheep's pituitary given to spayed virgin rabbits caused a full development of the mamma with lactation, and produced in 2 weeks a condition scarcely distinguishable from that of a breast at the full term of pregnancy. A previous course of treatment with luteal extract was not required for the production of this effect. Corner's results might be explained, on hypothetical grounds, by the assumption that the pituitary material used contained mammatrophin. Asdell & Salisbury (1933) obtained full development of the mamma in spayed virgin rabbits by giving alkaline extracts of pituitary or extracts prepared from pregnancy urine. In the same paper they state that if, in a rabbit made pseudopregnant by a sterile mating, the ovaries are removed on the 7th day further development of the mamma is arrested.

Direct and final proof of the existence of a pituitary or placental mammatrophin cannot yet perhaps be claimed. Gomez, Turner & Reece (1937), having shown that after hypophysectomy neither oestrin alone nor oestrin plus progestin will cause growth of the guinea-pig's mamma, performed the following experiment. They removed the pituitaries from four male guinea-pigs. Into three of these animals they implanted daily for 20 days one pituitary taken from a rat which had received injections of 100 i.u. of oestrin daily for a period of 10 to 20 days before death. The fourth guinea-pig received in the same way twenty daily implants of pituitaries derived from rats which had not been treated with oestrin. In the three guinea-pigs receiving pituitaries from oestrin-treated rats the mammae showed an extensive alveolar development comparable with that induced by twenty daily injections of oestrin given to normal guinea-pigs. In contrast with these results the mammary duct system of the fourth guinea-pig remained rudimentary. The outcome of these experiments seems compatible with the suggestion that the action of oestrin is to evoke the production by the pituitary of one or more mammatrophins. Gomez & Turner (1937b) say that mammatrophin does not exist in detectable amount in the pituitaries of animals which have not been subjected to ovarian hormones, whereas its presence can be demonstrated in the pituitaries of animals which are pregnant or, in the absence of pregnancy, have been subjected to oestrogen. Such a generalization can hardly be accepted yet without some reserve, for Reece & Leonard (1939), experimenting on rats, did not find any evidence that oestradiol stimulated the pituitary to produce mammatrophin. Implantation of rat's pituitaries into hypophysectomized rats caused the same amount of mammary growth whether the donors of the pituitaries had been treated with oestradiol or not.

Gomez & Turner (1938) think that there may be two mammatrophins, a duct-growth factor whose production is caused by oestrin and an alveolar-growth factor dependent on oestrin plus progestin.

Lyons (1937a) has described how a mammatrophin can be separated from the pituitary together with adrenocorticotrophin, 1 g. of these two mixed hormones being obtained from a kilogram of sheep's pituitaries. When tested on young rabbits the hormone given alone does not cause the development of alveoli or lactation, but if given after a course of oestrin injections it induces both development of alveoli and lactation. The secretion of milk, he says, can be caused in male rabbits by giving them daily doses of 10 r.u. of oestrin followed by mammatrophin.

Trentin, Mixner, Lewis & Turner (1941) found that fresh pituitary material obtained from pregnant cattle and free from progesterone caused the development of alveolar lobules in the mammae of spayed mice.

The various experimental results quoted above seem to indicate that for the full development of the mamma both ovarian and pituitary hormones are necessary. Whether the pituitary influence is conveyed by a specific hormone which we may call mammatrophin, or by a non-specific general growth hormone in co-operation with oestrogen, and perhaps with progestin, must for the present remain uncertain.

Astwood, Geschickter & Rausch (1937) think that a failure of the mamma to react to oestrogen after hypophysectomy may be in part due to the depressed nutritional activity which follows this operation. Nathanson, Shaw & Franseen (1939), too, seem to attribute the influence of the pituitary over mammary development to a general rather than a specific growth factor. They had noted that a pituitary growth extract, which by itself had no direct effect on mammary development, enabled oestradiol to cause proliferation of the mammary ducts if the two were given together to hypophysectomized rats.

A fact to be remembered when the subject of mammatrophin is under consideration is that the local application of oestrogen to a mamma causes development of that mamma only (Speert, 1940). Such a result suggests that if growth of the mamma depends on the joint influence of gonadal and pituitary hormones, the co-operation of these must take place in the mamma rather than in the pituitary.

It may be that adrenal and thyroid activity are essential to mammary development, as it is to lactation, and that the failure of the breast to respond to oestrogen after removal of the pituitary is, in part at least, the consequence of a diminished supply of adrenocorticotrophin and thyrotrophin.

Perhaps the uterus is an accessory factor in preparing the mamma for lactation. Asdell & Hammond (1933) removed the uterus and cervix from immature rabbits and after puberty mated them. Ovulation followed in every instance. The rabbits were killed on the 24th day after coitus. The corpora lutea had been maintained and the mammae were enlarged but not to the degree seen in non-hysterectomized pseudopregnant rabbits at the same date after coitus. The mammae of the hysterectomized rabbits at this date weighed 41 g. as compared with 61 g. in the intact pseudopregnant rabbits.

THE ROLE OF THE PLACENTA IN MAMMARY DEVELOPMENT

It seems that the placenta can maintain mammary development after hypophysectomy. Pencharz & Long (1933) removed the pituitaries from rats at various stages of pregnancy and found that if the operation were done after the 11th day gestation continued. In these cases the mammae underwent normal development, though lactation did not ensue after delivery.

Desclin (1939) removed the pituitaries of pregnant guinea-pigs after the 40th day of gestation. Pregnancy continued and the development of the mammae proceeded normally. Removal of the foetuses, provided that the placentae were left undisturbed, did not curtail mammary development. After delivery the breasts quickly atrophied in spite of injections of oestrogen. The conclusions drawn were (1) that the placenta can supply the hormones required for the development of the breast in pregnancy and (2) that oestrogens do not affect the mamma if both the placenta and the pituitary are absent. In further experiments Desclin removed the pituitaries of male guinea-pigs in which mammary development to the colostrogenic stage had been induced by daily injections of 100 i.u. of oestradiol benzoate, the administration of which was continued after hypophysectomy as before. Under this treatment the nipples continued to grow and the areolae became pigmented, whereas the mammary glands underwent complete involution. A different result was obtained by Robson (1936*b*) in the rabbit. He removed the pituitaries from sixteen rabbits between the 22nd and 24th days of pregnancy. To avoid the abortion which usually follows this procedure in rabbits he gave injections of a preparation of corpus luteum or, in one case, crystalline progesterone. After hypophysectomy in these animals no mammary development occurred.

Summarizing the causes of mammary development we may say that the ovary and the pituitary play the leading parts. The process seems to be as follows: the pituitary forms gonadotrophins which cause the ovary to produce oestrogen and progestin, which in turn co-operate with the pituitary in causing full development of the breast; the part played by the pituitary at this stage being perhaps attributable to the formation of a specific mammatrophin. During pregnancy the placenta apparently can in some species supply the place of both the pituitary and the ovaries.

The failure of the mamma to respond to oestrogen after hypophysectomy may be due to the absence of mammatrophin, a general growth hormone, adrenocorticotrophin, thyrotrophin or some other factor.

Lactation

The first step toward milk production is the anatomical development of the mamma, as already described. When this development has proceeded far enough the onset of lactation and its continuance seem to depend mainly on the pituitary. Heape (1906) showed that lactation does not depend on any influence derived from the foetus or placenta for it occurs in pseudopregnancy.

Before considering what part oestrogens play in this process, attention may be given to the influence on lactation of the pituitary, gonads and placenta regarded generally.

THE INFLUENCE OF THE PITUITARY ON LACTATION

The dependence of lactation on pituitary hormones is proved by the facts (1) that it can be induced by pituitary extracts, and (2) that it is prevented by hypophysectomy.

(*a*) *The induction of lactation by pituitary extracts.* Stricker & Grüter (1928) gave repeated injections of an aqueous extract of pituitary to immature female rabbits between 1½ and 2½ months old. Under this treatment the follicles matured and ovulation took place with follicular haemorrhage and luteinization, but no appreciable change occurred in the mammae. Nine mature rabbits were submitted to unfertile coitus and 8 to 10 days later pituitary extract was given; lactation followed in all. Four rabbits were spayed 10 days after sterile coitus and a course of injections of pituitary extract was begun on the next day. Two or 3 days after the first injection lactation ensued and became abundant. In addition to these experiments Stricker & Grüter gave pituitary extract to a rabbit 15 days after she had weaned her young, with the result that lactation was resumed; and the same result followed a single injection of pituitary extract into a dog 10 days after she had weaned her puppies. (Stricker & Grüter, 1929 a; Grüter, 1930) found also that, provided the mammae were fully developed, lactation could be induced by pituitary extracts though the gonads were absent. The test animals were the dog, rabbit, cow and pig. These observations have been confirmed by Corner (1930) and Lyons (1937 a) in rabbits, Nelson & Pfiffner (1930) in guinea-pigs, Allen, Gardner & Diddle (1935) in monkeys, and Houssay (1935) in dogs. Houssay injected an alkaline extract of bovine pituitaries into male and female dogs; these injections caused lactation in adult females whose mammary glands were already well developed, but did not have this effect in males or immature females. In the latter, after preliminary treatment with oestrone, lactation could be induced by pituitary extract.

(b) *The prevention of lactation by hypophysectomy.* Collip, Selye & Thomson (1933 b; see also Selye, Collip & Thomson, 1933 a) removed the pituitaries from pregnant rats. The rats littered, and lactation followed but quickly ceased. If hypophysectomy were done on rats which were already nursing young, lactation stopped and the mammae became atrophic. The same results have been recorded by Pencharz & Long (1933) in rats and by Gomez & Turner (1936 a) in guinea-pigs.

Further experiments have shown that milk formation cannot be induced in cats, rats, mice or guinea-pigs by gonadal hormones in the absence of the pituitary (Selye, Collip & Thomson, 1934 a; Lyons & Pencharz, 1936).

In some species, including the dog, cat and ferret, the secretion of milk can be induced by galactogenic preparations in the absence of the pituitary. Gomez & Turner (1936 a, b), however, failed to cause lactation in the hypophysectomized guinea-pig by galactogen. When judging experiments of this kind it must be remembered that the preparation of a pure pituitary hormone is hardly possible and the results of a test may be greatly affected by the presence or absence of other hormones than the one under examination. When galactogen is given to hypophysectomized animals the presence or absence of adrenocorticotrophin, thyrotrophin or perhaps other pituitary hormones may greatly confuse the outcome.

The Influence of the Gonads and Placenta on Lactation

Although the ovary and placenta play a large part in preparing the mamma for lactation, neither is essential for the production of milk once this process has begun. As the following remarks will show, this does not mean that the ovary and placenta are without influence on the production of milk.

(a) *The inhibition of lactation by the gonads and placenta.* Selye, Collip & Thomson (1933 a; see also Collip, Selye & Thomson, 1933 b) treated virgin rats with placental gonadotrophin until their mammae were fully developed; no lactation occurred. If now the ovaries were removed an abundant secretion of milk followed within 36 hours. Thus it appears that the gonads can inhibit lactation. The placenta, possibly by its production of oestrogen, also prevents the lactogenic action of the pituitary, and there seems little doubt that it is the elimination of the placenta at parturition which enables lactation to ensue.

Nelson (1934d) removed the ovaries and fertile horn from a guinea-pig which was pregnant in only one horn, and lactation followed. If, however, the embryos and ovaries alone were removed while the placentas were left intact, no lactation occurred. Selye, Collip & Thomson (1934a) confirmed this observation.

(b) *The presence of the gonads is not required for lactation.* Marshall & Jolly (1905) showed that removal of the ovaries during the latter part of pregnancy does not modify lactation. Kuramitsu (1921) made a similar observation. Moreover, it has been noted that the induction of lactation in non-pregnant animals by pituitary extracts is not prevented by gonadectomy once the mammae have become fully developed (Nelson & Pfiffner, 1930; Allen, Gardner & Diddle, 1935).

(c) *The prevention of lactation by oestrogen.* Parkes & Bellerby (1926) demonstrated the inhibitory effects of oestrin on lactation in mice. They used the rate of growth of the litters as a criterion and found that injections of oestrin into the mothers retarded the growth of the young. De Jongh (1933) also reported this inhibition of lactation by oestrogen in rats and mice. In nursing mice 50 i.u. of oestrone given twice a day arrested lactation. Robson (1935b) recorded confirmatory results and Smith & Smith (1933) observed that established lactation in the pregnant or pseudopregnant rabbit could be inhibited by oestrin; still more readily the onset of lactation could be prevented. Anselmino & Hoffmann (1936) reported that 1,000 i.u. of oestrone, given daily, stopped lactation in the rat, and Folley & Kon (1938) showed that 0·3 mg. of oestrone per 100 g. of bodyweight inhibited milk production when given to lactating rats. Nelson (1934d) found that daily doses of 100 r.u. of oestrone, started within a few hours of parturition, prevented lactation in the guinea-pig. Waterman, Freud & De Jongh (1936) made an experiment with two cows which had calved a year previously. One of the cows was treated by applying oestradiol benzoate in olive oil twice a day to the udder and teats during a period of 33 days. The other was treated in the same way with olive oil only, as a control. At the end of treatment milk production was unaffected in the control cow; in the one which had received applications of oestrogen the daily milk production had fallen from 5 litres to half a litre (see also Folley, 1936). Bacsich & Folley (1939) found that 1 mg. of oestradiol benzoate given daily prevents lactation in the rat without causing involution of the mamma.

In women, as in the lower animals, oestrogens, if given in sufficient dosage, inhibit lactation and are widely employed by clinicians for this purpose (Foss & Phillips, 1938; Kellar & Sutherland, 1939; Macpherson & Haultain, 1942).

Artificial oestrogens, especially stilboestrol, are in common clinical use for this purpose. Chassar Moir (1942) has administered the oestrogenic compound triphenylchlorethylene with success. In nearly fifty cases the production of milk was checked, he says, without inconvenience to the patient. Tablets of 0·5 g. were given twice a day for the first 4 days of the puerperium and once a day for the next 3 days. These doses did not cause nausea or sickness. Macpherson & Haultain (1942) inhibited lactation without inconvenience to the patients in all but three of eighty women by single injections of 0·25–3 g. of triphenylchlorethylene in oil given at the time of delivery.

Sardi (1935) carried out a series of experiments on male guinea-pigs which illustrate well the influence of oestrogen on lactation:

(1) Pituitary extracts were given daily for 16 to 35 days to three non-castrated, and for 73 days to seven castrated male guinea-pigs. No enlargement of the mammae or nipples followed nor was there any lactation.

(2) Repeated subcutaneous injections of oestrone given to male guinea-pigs caused enlargement of the mammae and elongation of the nipples without lactation.

(3) Injections of oestrone were given to three non-castrated and seven castrated guinea-pigs for 15 to 30 days, after which pituitary extract was given. This treatment led to abundant lactation. Young guinea-pigs were suckled by these males, but lost weight.

(4) Simultaneous injections of oestrone and pituitary extract into male guinea-pigs caused growth of the mamma unaccompanied by lactation.

These results suggest that a preparation of the mamma by oestrogen is necessary in order that pituitary extracts may induce the secretion of milk, which however cannot come about until the supply of oestrogen has been stopped or greatly reduced. Reece, Bartlett, Hathaway & Davis (1940) noted that daily doses of 100 r.u. of oestrogen, while causing some inhibition of lactation in rats, did not prevent the supply of lactogenic hormone; the pituitaries of the mother rats treated with these doses of oestrogen contained more galactogen than those of untreated lactating rats. From this experiment, as well as from those of Sardi which have been mentioned above, it seems that oestrogen checks lactation by counteracting the pituitary lactogenic hormone rather than by stopping its production.

(d) *Lactation following the cessation of a supply of oestrogen.* These experiments may be considered in connection with the fact that lactation may follow a course of treatment with oestrogen immediately the supply of hormone is withdrawn (Laqueur, Borchardt, Dingemanse & de Jongh, 1928; Steinach, Dohrn, Schoeller, Hohlweg & Faure, 1928). As already suggested, the onset of lactation after delivery can be rationally explained on the basis of these experiments, because with the expulsion of the placenta a copious fount of oestrogen is suddenly removed.

(e) *The induction of lactation by oestrogen.* When studying the actions of a particular hormone it is not uncommon to find that the kind of effect produced depends largely upon the dosage. Thus, the recognized effect of moderate amounts of oestrogen on the mamma is a development of the duct system; excessive doses, however, do not have this effect; they check the growth of the ducts. So too, although the capacity of oestrogen to prevent lactation is certain, it seems equally sure that in small amounts oestrogen will favour the onset of lactation and will increase the amount of solids excreted. Frazier & Mu (1935) gave from 20 to 60 r.u. of oestrogen daily to adult male rabbits during a period of 250 days. This treatment caused enlargement of the nipples to the size of those in lactating females, together with the formation of milk. In two instances the rabbits fostered and suckled young. Folley, Scott Watson & Bottomley (1940) find that virgin goats can be made to produce milk by applying 1 g. of a 1 per cent oily solution of diethylstilboestrol dipropionate to the udder 3 times a week accompanied by daily milking. In a virgin goat after 30 days of this treatment a yield of milk suddenly began and eventually amounted to 1,500 c.c. daily; the milk was normal in composition. The yield of milk by this goat was increased more than 40 per cent by giving an alkaline pituitary extract (Folley & Young, 1941 a). How much of the effect in this experiment was due to the oestrogen and how much to the stimulation of milking is hard to say, for milking combined with inunction of an ointment not containing oestrogen was also found capable of starting lactation.

Walker & Stanley (1941) treated a heifer which had been spayed 10 months before with diethylstilboestrol dipropionate; during 9 months a total of 1,560 mg. had been given. In the latter part of this period 530 mg. of testosterone had been administered also. Lactation ensued under this treatment and eventually the daily yield of milk reached 16 pounds. (See also Folley, Scott Watson & Bottomley, 1941.) Perhaps in experiments of this kind a fuller development of mammary alveoli and consequently a copious production of milk would be obtained by including progesterone in the preliminary preparation of the mamma.

(*f*) *The enrichment of milk by oestrogen.* Folley (1936, 1941) has found that oestrogens given in small amount cause an increased concentration in the milk of fatty and non-fatty solids. This enrichment apparently represents an enhanced secretion of solids and not a mere curtailment of the output of water.

GALACTOGEN AND PROLACTIN*

The production of milk appears to depend on one or more pituitary hormones, which for brevity we will call galactogen. It has been suggested that there are two galactogens. Gardner & Turner (1933) found that extracts of sheep and beef pituitaries which were able to induce lactation failed to sustain it, and according to Folley & Young (1941 b) the pituitary hormone which initiates the production of milk is probably distinct from that which maintains the process.

Like other pituitary hormones galactogen has the reactions of a protein and has not been prepared in a pure form. Apparently it, or prolactin, is a normal constituent of the male as well as the female pituitary (Holst & Turner, 1939), a fact which is of interest in connection with cases of gynaecomastia and lactation in men.

The effect of galactogen on the mamma appears to be confined to secretory activity, and is in contrast with that of gonadal hormones and perhaps of mammatrophin, which are concerned with anatomical development. Gardner, Gomez & Turner (1935) gave daily doses of 'galactin' for a week to three young virgin rabbits, beginning immediately after they had been spayed. At the end of this treatment the mammary ducts were distended with milk though no development of alveoli had occurred. In other experiments on immature rabbits they found that, after twenty daily injections, galactin caused secretion in the mammary ducts and in the alveoli too if any were present, but did not stimulate growth of the glandular structures.

* The pituitary hormone which causes the production of milk in mammals is here referred to as *galactogen*, while that which stimulates activity in the pigeon's crop-gland is called *prolactin*. The identity of these two hormones has not been established, though they are sometimes regarded as identical.

The influence of oestrogen on the production of galactogen.   Possibly oestrogen in suitable dosage takes part in the formation of galactogen; it certainly seems to favour the production of prolactin, as will be shown presently.

Lewis & Turner (1941) gave daily doses of stilboestrol to adult spayed rats, and at the end of 10 days found that the galactogenic hormone present in their pituitaries was more than doubled. Meites & Turner (1942), by tests on the pigeon's crop-gland, found that the prolactin content of the pituitary in the rat, guinea-pig and rabbit is not increased during pregnancy or pseudopregnancy but rises rapidly soon after parturition. They learned, moreover, that the prolactin content of the rabbit's pituitary and blood could be increased by the administration of oestrogen; the larger doses used by them were not so efficient in this respect as smaller doses.

The dependence of continued lactation on suckling.   The maintenance of lactation seems to depend largely on a limitation of the production of FRH and consequently of oestrogen. But there are other factors. The mamma is no exception to the rule that retention of its secretion checks the activity of a gland; unless the milk is regularly removed lactation will decline and eventually cease. This effect is a local one and is confined to the neglected gland. Hammond & Marshall (1925) occluded all five nipples on the left side of a rabbit, leaving the nipples on the right side intact. The rabbit was allowed to suckle her young until at the end of 16 days of lactation she was killed. The mammae of the right side weighed 170 g. and those of the left weighed 56 g. The result suggests that involution of the mamma at weaning is not caused by a lack of sensory stimuli, but is probably due to the non-removal of milk from the gland. Hammond & Marshall also point out that in sows with small litters the mammae used by the piglets are large and active, while the unused mammae are small and atrophic.

The importance of thoroughly emptying the cow's udders at each milking in order to maintain the yield is generally known.

The influence of afferent nervous impulses on lactation.   Local nervous stimuli are needed to maintain lactation. Selye & McKeown (1934 *b*, *c*) have shown that in rats the stimulus of suckling will cause some secretion of milk even though its escape has been prevented by obstruction of the ducts. Furthermore lactation is affected by psychical impulses. Experience has shown the ill-effects of psychical disturbance on the production of milk by the human mother as well as by domestic animals. It seems probable that not only the sensory stimuli of suckling but the maternal solicitude for the young also may lead to an efflux of galactogen from the pituitary.

ACCESSORY HORMONAL FACTORS IN LACTATION

Thyroid and adrenal secretion.   The production of milk does not depend on the pituitary and gonadal hormones alone. Apart from good nutrition and the absence of nervous disturbance, the secretions of the thyroid and adrenals also may have a favourable influence on lactation. By giving suitable doses of thyrotrophin (Folley & Young, 1938), thyroxine or dried thyroid gland (Graham, 1934) to a normal cow its yields of milk and of milk fat can be increased. It will be remembered that the thyroid plays a part in the metabolism of oestrogen and this capacity may help to explain its influence on lactation.

Regarding the adrenals it has been shown that their removal is followed by a cessation of milk secretion. Nelson (1941) found that lactation could not be induced in the hypophysectomized guinea-pig unless adrenocorticotrophin or adrenal cortical hormone were given also; in this experiment desoxycortico-sterone failed as a substitute for adrenal cortical extract. In mice it has been shown that desoxycorticosterone may suppress lactation.

THE EFFECT OF LACTATION ON THE GONADS

Arrest of oestrous cycles. The cessation of the menses in women during lacta-tion is a commonly observed though not inevitable occurrence. In animals, too, an arrest of the oestral cycle has been noted during suckling. Apparently animals differ in this respect. Hammond & Marshall (1925) say that in some species (rat, opossum) maturation of follicles ceases during lactation, in others (guinea-pig) it does not. When lactation arrests the oestral cycles it probably does so by inter-fering with the production of FRH by the pituitary. This is suggested by some measurements which were made by Hammond & Marshall on puerperal rabbits, some of which had suckled their young for a month or more while others had not been allowed to suckle (Table 150).

TABLE 150. Comparative weights of organs in puerperal rabbits which nursed or did not nurse their young (Hammond & Marshall, 1925)

Classification	Average weight of mammae (g.)	Average weight of ovaries (g.)	Average number of large follicles in ovaries	Average weight of uterus (g.)
Sucklers	86	0·34	0	2·5
Non-sucklers	29	0·61	11	9·0

Oestrus occurs in the mouse within 24 hours or thereabout after parturition. Provided that she suckles her young the next oestrus does not take place until between the 21st and 26th day after their birth. But this temporary arrest of the oestrous cycles happens, according to Crew & Mirskaia (1930), only if there are four or more nurslings. If only one or two are being nursed, oestrus is not in-hibited. If the litter is of normal size so that oestrus is prevented, removal of the young a week or a fortnight after parturition will be followed by oestrus in the mother within the next 4 days.

Bates, Riddle & Lahr (1937) believe that the pituitary hormone responsible for lactation inhibits the formation of FRH, and they put forward experimental

TABLE 151. The effect of prolactin and other hormones on the pigeon's testes (Bates, Riddle & Lahr, 1937)

Hormone	Daily dose	Days of treatment	Weight of testes (mg.)
None (muscle extract given for control)	5 mg.	7	1,310
Prolactin	40 bird units	13	104
Androsterone	—	10	230
Progesterone	0·25 mg.	10	279
Oestrone	40 r.u.	10	213

evidence in support of the view. They gave to adult ring doves different hormones, including a preparation of prolactin free from FRH, and noted the effects of the various hormones on the weight of the testes. The figures given in Table 151 show that a greater degree of testicular atrophy was caused by prolactin than by any other hormone used.

EXPERIMENTS ON THE PIGEON'S CROP-GLAND

Such fine studies have been made of the hormonal influences which regulate secretion from the pigeon's crop-gland, and the bearing which these studies have had on our knowledge of mammary activity are so important, that it seems essential to make some reference to them here. The crop-gland forms a milk-like fluid for the nourishment of the young, and its hormonal reactions resemble those of the mamma. The gland was discovered by John Hunter, who noted that it produces a secretion, 'pigeon's milk', on which the young are fed. He observed also that the gland is present in both sexes and that males as well as females feed the young.

Riddle & Braucher (1931) tested the reactions of the crop-glands to hormones and found that enlargement and secretory activity could be induced in them in immature and mature non-brooding birds of both sexes by suitable pituitary extracts. The actual hormone 'prolactin' appears to consist partly of a protein-like component, as do the gonadotrophins; like the latter prolactin has not yet been isolated in a pure form.

Riddle (1937) states that prolactin is present in the pituitaries of all the vertebrates examined. Rich extracts have been prepared and assayed on birds using the crop-gland as the test object. These extracts also stimulate the production of milk in mammals if the mammae are already well enough developed.

The action of prolactin on the crop-gland and mamma appears to be direct. According to Riddle the hormone injected intramuscularly is effective in causing secretion from the crop-glands in pigeons after hypophysectomy, gonadectomy, thyroidectomy or adrenalectomy, and when injected into single milk ducts in mammals it causes a localized secretory response. The hormone has a pronounced antigonadal action in pigeons of both sexes and in 10 days will reduce the weights of the testes by 90 per cent, an effect which is readily overcome by small doses of gonadotrophin (FRH), from which it seems that prolactin interferes with the supply of FRH.

Perhaps because it has not yet been completely isolated from other pituitary hormones, there has been some discrepancy in the results reported by different workers who have investigated the action of prolactin in mammals.

Experiments which have shown prolactin effective in causing lactation in non-pregnant mammals are as follows. Allen, Gardner & Diddle (1935) treated ten female rhesus monkeys with oestrone followed by prolactin. In all the five females with mature mammary glands the injections of prolactin caused lactation. In the other five monkeys, which were immature, lactation did not occur. De Fremery (1936) found that lactation could be induced in virgin goats independently of the season, by six daily injections of 5 pigeon units of prolactin.

In contrast with these results may be mentioned others in which prolactin failed to show galactogenic properties in mammals. Gomez & Turner (1936a, 1937a) found that prolactin did not maintain or restore lactation in cats, rats and guinea-pigs after hypophysectomy, though fresh non-extracted sheep's pituitary was effective in the maintenance of lactation in the same circumstances in two guinea-pigs. They discovered, however, that lactation could be maintained or restored after hypophysectomy in the guinea-pig by injections of prolactin combined with adrenocorticotrophin (Gomez & Turner, 1936b). Folley & Young (1938) experimented on cows in which the production of milk was declining. Very large doses of prolactin were required to cause an increase of milk production. A crude saline extract of anterior pituitary gland was more effective. They conclude that prolactin as defined by the pigeon crop-gland test is not the only lactogenic substance in the hypophysis. They say that a thyrotrophic preparation increases the secretion of milk in a lactating cow.

Presence of prolactin in urine during lactation. Lyons & Page (1935) examined the urine of eight lactating women and were able in every instance to extract prolactin which gave positive results with tests on the pigeon's crop-gland.

Inhibition by oestrogen of the action of prolactin on the development of the pigeon's crop-gland. It is interesting to note that the development of the crop-gland both in male and female pigeons can be inhibited by oestrogen. Folley & White (1937; see also Folley, 1939) used two groups of twenty-four pigeons. To one group were given daily doses of 5 mg. of oestradiol in sesame oil for six successive days; the other group received sesame oil only for the same period. On the 6th day daily injections of 5 mg. of prolactin were given to all the pigeons and were continued daily for 6 days. Twenty-four hours after the last injection the birds were killed and their crop-glands weighed. The results (Table 152) show that the crop-glands in the birds which had received oestradiol were smaller than those which had not. The restrictive action of large doses of oestrogen on the development of the crop-gland in this experiment appears to be like that which large doses of oestrogen exert on the mamma (p. 332).

TABLE 152.   Inhibition of the functional development of the pigeon's crop-gland by large doses of oestrogen (Folley & White, 1937)

Number	Sex	Preliminary treatment	Daily dose of oestrogen (mg.)	Mean weight of crop-gland (mg. per 100 g. of bodyweight)
9	Male	Oestradiol benzoate in sesame oil	5	680
7	Male	Sesame oil only	—	1,490
15	Female	Oestradiol benzoate in sesame oil	5	830
17	Female	Sesame oil only	—	1,080

The effect of oestrogen on the production and output of prolactin by the mammalian pituitary. Reece & Turner (1936, 1937) found that the pituitary glands from sexually mature ovariectomized rats which had been injected with 500γ of oestrone daily for a fortnight contained more prolactin, as assayed on pigeons, than the pituitaries from sexually mature ovariectomized rats which had not received oestrone. Removal of ovaries without further treatment decreased the prolactin content of the rat's pituitary. Reece (1938) with the same kind of experiment obtained the results set out in Table 153.

TABLE 153.   The effect of oestrin on the prolactin content
of the rat's pituitary (Reece, 1938)

Number of rats	Hormone treatment	Daily dose (i.u.)	Duration of treatment (days)	Other treatment	Mean bodyweight at death (g.)	Mean weight of pituitaries (mg.)	Mean bird units of prolactin in pituitary
10	None	—	—	None	160	8·9	7·93
10	None	—	—	None	167	7·9	12·6
10	Oestradiol	200	15	None	161	15·7	11·10
10	Oestradiol	1,000	15	None	159	13·3	16·1
10	Oestradiol	1,000	15	Spayed	160	21·4	14·3

Lewis & Turner (1941) found that by treating adult spayed rats for 10 days with stilboestrol the prolactin content of their pituitaries was increased by 226 per cent.

Meites & Turner (1942) treated rabbits with oestrone for 10 days, and tested their blood for prolactin on the pigeon's crop-gland. Three different doses of oestrone were used, namely 500, 1000, and 5000 i.u. The tests showed that oestrone had increased the prolactin content of the blood, and that the lowest doses had caused the greatest increase. Meites & Turner say that the prolactin content of the pituitary in rats, guinea-pigs and rabbits is as low during pregnancy and pseudopregnancy as in normal, non-pregnant, non-lactating animals and that it rises rapidly after parturition.

Chapter XVIII. *Factors in the Causation of Mammary Cancer*

Oestrogen. Hereditary factors. Subsidiary agencies.

The subsequent discussion may be easier to follow if it is preceded by a statement of what appear, from experiments on animals, to be the principal causes of mammary cancer. They are as follows:

I. Oestrogen.
II. Hereditary Factors: (a) Genic; (b) Nongenic.
III. Subsidiary Factors.

The Role of Oestrogen in Mammary Cancer

(a) *The influence of the ovary.* Long before the identification of oestrone the ovary was thought to be an agent in the development of mammary cancer. In 1896 Sir George Beatson stated that 'we must look in the female to the ovaries as the seat of the exciting cause of carcinoma, certainly in the mamma, in all probability of the female organs generally'. Beatson's views on the causative influence of the ovaries were confirmed by experiments in the laboratory. Lathrop & Leo Loeb (1916), using female mice of several different strains, found that removal of the ovaries, if carried out before the age of 6 months, led to a pronounced decrease in the incidence of cancer of the breast in these animals, although it did not entirely prevent it. In the few cases in which cancer arose in spayed female mice, it appeared later in life than in non-spayed mice of the same strain. Spaying mice above the age of 6 months did not have any pronounced effect in preventing mammary cancer. In a second paper (1919) Loeb showed, with greater precision than before, a contrast between the results of early and late spaying. If the ovaries were removed from mice between the ages of 3 and 5 months, cancer of the mamma was almost entirely prevented. If the spaying was done between the 5th and the 7th months, the cancer rate was diminished and when cancer did appear it was at a later age on the average than in non-spayed mice. Removal of the ovaries after the 8th month of life had no appreciable effect in reducing the incidence of mammary cancer.

Cori (1927) followed up Loeb's investigations. He removed the ovaries from mice belonging to a strain in which 78 per cent of the females suffered from spontaneous mammary carcinoma. In forty-nine of these mice which had been spayed at ages varying from 2 to 6 months, and which lived to 19 months or more, mammary cancer developed in five only. By removal of the ovaries in the first 6 months of life, the incidence of cancer in these mice was thus reduced from 78 to 10 per cent. In a second series Cori spayed 100 mice belonging to the same strain when they were between 15 and 22 days old. At the end of 10 months eighty-six were still alive, and sixty reached the age of 20 months. Not one of them suffered from cancer of the breast; at 20 months of age, mammary tumours had appeared in 74 per cent of the untreated control mice belonging to the same strain.

Woolley, Fekete & Little (1939), using the *dba* strain of mice, had less striking results. They spayed eighty-two of these mice 24 hours after birth, and among these mammary cancer developed in 26·8 per cent; the incidence in normal un-mated females of the same strain was 50·4 per cent.

In a later paper the same workers (Fekete, Woolley & Little, 1941) describe some of the ultimate effects of spaying *dba* mice on the day after birth. They find that in mice so treated the vagina, uterus and mammae gradually recover from their early castration atrophy. This recovery begins when the mice are about 7 months old and is associated with irregular cycles of oestrus. Among seventy-five mice deprived of their ovaries on the day after birth, mammary tumours eventually appeared in thirty-seven between the ages of 14 and 27 months. From the condition of the accessory genital organs in these mice it appeared that, in spite of the absence of ovaries, considerable amounts of oestrogen were being produced.

Murray (1928), besides confirming the effect of early oophorectomy in retarding the development of mammary cancer in female mice, carried out further experiments which might be regarded as the reverse of those of Loeb & Cori. He used a strain of mice among which 80 per cent of the untreated breeding females eventually suffered from mammary carcinoma. Spontaneous tumours of the mamma were unknown in the untreated breeding males of this stock. Murray removed the gonads from 210 mice of each sex belonging to this strain when they were between 4 and 6 weeks old, and transplanted subcutaneously in the abdominal region of each male an entire ovary taken from a sister. His results are shown in Table 154.

TABLE 154.   The effect of gonadectomy and implantation of ovaries on the incidence of mammary cancer in male mice (W. S. Murray, 1928)

	Number observed	Number in which cancer of breast occurred	Per-centage	Mean age at which the tumours appeared (months)
Untreated breeding females	479	367	80	10·5
Spayed females	210	36	17·1	17·5
Castrated males with ovarian transplants	210	15	7	14·4
Castrated males without ovarian transplants	241	0	0	—

De Jongh & Korteweg (1935) repeated this experiment. Using a pure strain of mice they implanted ovaries into castrated males. Among sixteen males so treated mammary cancer occurred in nine and sarcoma in two. The result is significant because spontaneous mammary cancer very rarely, if ever, has been seen in untreated male mice.

Another kind of experiment has demonstrated the dependence of mammary cancer on the ovaries and the oestrogen which they produce. Furth & Butterworth (1936) submitted mice to a moderate dose of X-rays (200 to 450 r.); among the mice so treated several granulosa tumours of the ovary arose and were associated with cystic glandular hyperplasia of the uterus and sometimes with mammary cancer. This type of ovarian tumour is known in man to produce an abundance of oestrogen, and the presence of cystic glandular hyperplasia in the treated mice strongly suggests that the ovarian tumours in them also had caused an excessive supply of oestrogen, to which the mammary cancers might be attributed.

(b) *The influence of extracts of liquor folliculi.* Goormaghtigh & Amerlinck (1930) gave daily subcutaneous injections of an oestrogenic extract prepared from the liquor folliculi of sows to female mice during a period of several months. Each dose contained 6 r.u. of oestrogen. Adenomatous formations appeared in the mammae of some of these mice and in one a mammary cancer developed.

(c) *The influence of pure oestrogen.* The results of the foregoing experiments accord with those obtained later by using pure oestrogens. Lacassagne's (1932) work calls for particular recognition in this field. In an early experiment he gave weekly injections of 0·03 mg. of crystalline oestrone benzoate dissolved in sesame oil to five young mice commencing when they were between 10 and 18 days old. Three of the mice were males and two were females. They belonged to a strain (R_{III}) in which about 72 per cent of the untreated females eventually suffer from spontaneous mammary carcinoma, though males are not thus affected. In every one of the three males subjected to oestrone in this experiment cancer of the breast developed within 6 months, one of the mice having two separate tumours. Mammary cancer appeared also in one of the two females, at the early age of 4 months. Since Lacassagne's pioneer work a number of investigators have induced cancer of the breast in male mice, whether castrated or not, with oestrone (Burrows, 1935 d, e; Bonser, 1935 a; Gardner, Smith, Allen & Strong, 1936; and many others). Usually in these experiments the oestrogen was dissolved in oil and given by subcutaneous injection. This method of administration is not essential for success. Burrows used a solution of oestrone in benzene applied to the skin in the interscapular region, and Twombly (1939) injected dry crystals.

It is now known that mammary cancer may be induced in male mice by various oestrogens, both natural and artificial, including Girard's equilin and equilenin (Lacassagne, 1936b; Gardner, Smith, Strong & Allen, 1936b), oestradiol benzoate (Gardner, Smith, Allen & Strong, 1936), triphenylethylene (Robson & Bonser, 1938), diethylstilboestrol (Lacassagne, 1938). Geschickter (1939) induced mammary cancer in twenty-five of eighty-six rats by means of oestrone, and in others with diethylstilboestrol, the dosage in the latter case being 200γ daily for 100 days.

The artificial induction by oestrogen of mammary cancer in rats has a special interest, for spontaneous cancer of the breast in rats is rare, though non-malignant fibroadenomata are common.

In most of the experiments concerned with the induction of mammary cancer by oestrogen, male mice have been used because they do not naturally get tumours of the breast. The same effect is caused in female mice by the artificial administration of oestrogen; this treatment increases the incidence of mammary cancer in female mice and lowers the age at which it occurs (Suntzeff, Burns, Moskop & Loeb, 1936).

The carcinogenic effect of oestrogen on the mamma appears to depend largely on the magnitude of the dosage and the period of life during which the hormone is given, cancer being induced more readily if the administration is started when the mice are young. Moreover, whether a mouse will or will not get mammary cancer in response to oestrogen applied artificially will be decided largely by other factors to be considered presently.

The quantitative relationship between oestrogen and the induction of mammary cancer. It will be recalled that the growth of the mamma in response to oestrogen does not progressively increase with increasing supplies of the hormone. There is an optimum dosage by which a maximum development of the breast is attained; with larger amounts of oestrogen growth of the mammary ducts is diminished. Gardner (1941) concluded, from experiments on mice of the C_3H strain, that the highest incidence of mammary cancer was induced by those doses of oestrogen which were most favourable for normal mammary growth. If this observation is correct it would be erroneous without some qualification to attribute mammary cancer to an excessive quantity of oestrogen.

The extent of time during which the mammae have been subjected to effective supplies of oestrogen seems to be an important factor in the etiology of cancer of the breast. The subject was investigated by Burns & Schenken (1940). They used male mice of the C_3H strain, and treated them for varying periods. Once a week each mouse received 100 r.u. of oestradiol benzoate subcutaneously, beginning when the mice were 2 weeks old. The incidence of mammary cancer in these animals depended on the time during which this treatment was continued. A maximum effect was obtained with treatment lasting 16 weeks, after which the continued administration of oestrogen did not cause any further increase in the incidence of mammary cancer (Table 155).

TABLE 155. The relation between the incidence of mammary cancer in male mice and the duration of their subjection to oestrogen (Burns & Schenken, 1940)

Group	Number of mice	Weekly dose (r.u.)	Duration of treatment (weeks)	Number in which tumours developed	Incidence of mammary tumours (%)
I	38	0	—	0	0
II	11	100	4	0	0
III	20	100	8	3	15
IV	14	100	12	1	7·1
V	19	100	16	11	57·9
VI	19	100	20	9	47·4
VII	19	100	For duration of life	10	52·6

The role of the pituitary in mammary cancer. By its control of the ovaries and their production of oestrogen the pituitary must be closely concerned with the development of tumours of the breast. At present we are not aware of any work enabling us to apportion with confidence a specifically malign influence to mammatrophin, gonadotrophin, general growth hormone or other production of the pituitary. Loeb, Burns, Suntzeff & Moskop (1937) found that the repeated implantation of mouse pituitaries into mice, in conditions favourable to survival of the grafts, led to a definite increase in the incidence of mammary tumours. Gomez & Turner (1937b) say that mammatrophin is not present in detectable amount in the pituitaries of animals which have not been subjected to ovarian hormones, whereas it is demonstrable in the pituitaries of animals which have been treated with oestrogen, and they suggest the possibility that mammatrophin may play a part in the development of cancer of the breast in mice.

Loeb & Kirtz (1939) implanted pituitaries from littermates into mice of various strains, with the result that the incidence of cancer was increased in the high-cancer strains (A, D, C_3H) but not in the low-cancer strains ($C57$ and CBA).

Nongenic and Genic Hereditary Factors in the Etiology of Mammary Cancer

The existence of separate inbred strains of mice in the females of which the liability to spontaneous mammary cancer shows great contrasts—the incidence of this disease ranging in the different strains from almost nil to nearly 100 per cent—proves the existence of inherited agency. Some of the strains have been inbred through so many generations that the animals may be regarded now as homogenic; Murray's *dba* strain, of which the writer has made much use, has been inbred since 1909.

For convenience of recital and at the cost of verbal precision, different strains of mice will be distinguished by the terms 'high-cancer' and 'low-cancer', and the reader will understand that these epithets relate to the incidence of mammary cancer alone. We cannot refer to 'cancer-immune strains' and 'cancer-susceptible strains', because the various inbred mice differ only in the degree of their propensity to the disease. By controlled breeding a cancer incidence of nil or of 100 per cent has not yet been quite attained. In the accounts which follow it is hoped that this explanation will be borne in mind.

Experiments have proved that there are two different heritable agencies which favour the development of mammary cancer; these are nongenic and genic. It will be convenient to consider the nongenic agents first.

(i) THE NONGENIC HEREDITARY FACTOR IN MAMMARY CANCER

Inquiries have proved that a familial tendency to the development of mammary cancer in mice is largely dependent on an extrachromosomal factor. Lathrop & Loeb (1918) observed that if high-cancer and low-cancer strain mice are crossed, the offspring are apt to follow the mother as regards susceptibility to cancer of the breast. The father's influence in transmitting the tendency to mammary cancer is genic and, though definite, is relatively small.

Using the symbol HC to denote high-cancer strain, and LC to denote low-cancer strain, the results of crosses may be crudely expressed as follows:

Female	Male	Cancer incidence in offspring
HC × HC		Very high
HC × LC		High
LC × HC		Low
LC × LC		Absent

Little (1933), writing on behalf of his colleagues and himself, mentions four independently conducted experiments which were continued over three years, and the results all agree. In these experiments the incidence of tumours in the offspring of reciprocal crosses between mice of high-cancer and low-cancer strains were compared. In such crosses the chromosomal constitution of all the F_1 females is similar, therefore any significant difference in tumour incidence among them cannot be attributed to chromosomal influence. The results of these crosses in the first generation are summarized in Table 156.

TABLE 156. The effect on the incidence of mammary cancer in mice of crossing high-cancer (HC) and low-cancer (LC) strains (Little, 1933)

Experiment	Type of cross Female	Male	Percentage of mammary tumours in 1st genera-tion of females
I	HC ×	LC	36·06
	LC ×	HC	5·53
II	HC ×	LC	68·11
	LC ×	HC	7·41
III	HC ×	LC	86·3
	LC ×	HC	0·0
IV	HC ×	LC	90·0
	LC ×	HC	0·0

Korteweg (1936) obtained similar results. He crossed mice of Murray's *dba* high-cancer strain with mice of a low-cancer strain (C57), and the results, like those of Little, prove that the susceptibility to mammary cancer is conveyed mainly, though not entirely, through the mother. His experiments show further that the degree of susceptibility in the offspring of a low-cancer strain mother with a high-cancer strain father is passed on to the second (F_2) generation. Some of his results are quoted in Table 157.

TABLE 157. The effect on the incidence of mammary cancer in mice of crossing high-cancer and low-cancer strains (Korteweg, 1936)

	Mother	Father	Number of offspring	Number getting mammary tumours
(1)	*dba*	C57	7	7
(2)	C57	*dba*	67	2
(3)	*dba*	F_1 (C57 ♀ × *dba* ♂)	63	52
(4)	F_1 (C57 ♀ × *dba* ♂)	*dba*	30	3
(5)	C57	F_1 (C57 ♀ × *dba* ♂)	56	1
(6)	C57	F_1 (*dba* ♀ × C57 ♂)	2	0

The relatively small part played by the father and the predominant influence of the mother in transmitting susceptibility to mammary cancer has been demonstrated also by Murray & Little (1936) and by Bittner & Little (1937), whose results are set out in Tables 158 and 159, which should be studied together.

TABLE 158. Showing the predominant influence of the mother in transmitting susceptibility to mammary cancer in mice (Bittner & Little, 1937)

High-cancer strain mother		Low-cancer strain father	Number of progeny	Percentage incidence of mammary cancer in progeny
dba	×	C57	113	39·8
dba	×	M. bact.	69	68·1
A	×	CBA	44	90·9
C₃H	×	I	36	91·7
dba	×	C57	7	100·0
C₃H	×	N	46	97·9

TABLE 159. Showing the relatively slight influence of the father in transmitting susceptibility to mammary cancer in mice (Bittner & Little, 1937)

Low-cancer strain mother	High-cancer strain father	Number of progeny	Percentage incidence of mammary cancer in progeny
C57	*dba*	379	6·1
M. bact.	*dba*	27	7·4
CBA	A	16	31·3
I	C₃H	10	0·0
C57	*dba*	67	3·0
N	C₃H	18	27·8

Murray & Little (1939) have clearly established that the reduced liability to cancer which is effected by cross-breeding and persists in the second generation is not attributable to genic agency. They crossed *dba* (HC) with C57 (LC) mice (Table 160). In the table it will be seen that although the chromosomal constitutions of the dBF_1 and BdF_1 females are identical, as also are those of the dBF_2 and BdF_2, their liability to cancer differs considerably.

TABLE 160. Showing transmission to the second generation of the non-chromosomal influence in the liability of mice to mammary cancer (Murray & Little, 1939)

Number of mice	Crossings		Designation of offspring	Percentage incidence of mammary cancer among females
	Female	Male		
113	*dba* ×	C57	dBF_1	39·82
379	C57 ×	*dba*	BdF_1	6·06
664	dBF_1 ×	dBF_1	dBF_2	35·54
687	BdF_1 ×	BdF_1	BdF_2	5·96

Apparently the predominance of other than chromosomal factors in transmitting a liability to mammary cancer is not confined to mice; mankind has been found to present a similar familial tendency to mammary cancer (Wassink & Wassink van Raamsdonk, 1923; Waaler, 1932; Wassink, 1935), and this tendency, according to Greenwood in his abstract of Waaler's paper, is not amenable to any simple Mendelian interpretation. The phenomena represent a type of inheritance which is nongenic (see also Pybus & Miller, 1935 and Little, 1936).

Recently Passey (1942) has reported that an investigation carried out in Leeds by Miss Wainman has failed to reveal such a familial tendency to cancer of the breast in women as had been observed by the Norwegian and Dutch inquirers.

The manner in which the nongenic factor of mammary cancer is transmitted from mother to offspring. Having shown that the liability of mice to mammary cancer is transmitted mainly by non-chromosomal agency, we may now consider how and when the transmission can take place. Korteweg (1936), when calling attention to the nongenic nature of the transmission, gave reasons for thinking that the cytoplasm of the ovum might be the carrier; further work has suggested other explanations. The methods of conveyance may be considered in the light of three kinds of experiment, as quoted below.

(a) *Transference of fertilized ova.* Bittner & Little (1937) transferred fertilized ova from high-cancer strain (*dba*) to the uteri of low-cancer strain (C57) mice. Among sixty-one *dba* females obtained in this way no tumours had appeared at the time the paper was written, though forty-one of the mice at that time had reached an average age of 14·5 months, that is to say an age at which tumours would have been expected in *dba* mice reared normally. The experiment seems to show that either the foster-mother had suppressed the tendency to mammary cancer which otherwise would have shown itself, or she had failed to supply some carcinogenic factor which the normal *dba* mother would have supplied. As will appear, the latter alternative seems to be the correct one.

Fekete & Little (1942) transferred fertilized *dba* ova into C57 black females, and C57 black ova into the uteri of *dba* females, and noted the incidence of cancer in the subsequent generations of mice derived from these ova. In this way they found that the modified incidence of cancer was continued to the 3rd generation (Table 161).

TABLE 161. The incidence of mammary cancer in the descendants of mice which had been transferred as ova into the uteri of foster-mothers (Fekete & Little, 1942)

	1st generation		2nd generation		3rd generation	
	Number of breeding females	Percentage with mammary tumours	Number of breeding females	Percentage with mammary tumours	Number of breeding females	Percentage with mammary tumours
dba ova transferred to C57 uteri	97	11·3	115	8·7	87	16·0
C57 ova transferred to dba uteri	35	62·8	34	82·3	24	75·0

(b) *The transmission of the nongenic agent of mammary cancer by the mother's milk.* A chance observation in America showed that susceptibility to cancer of the breast might be transferred from mother to young after birth by the act of suckling. An orphaned litter of mice belonging to a high-cancer strain was nursed by a foster-mother of a low-cancer strain and it was observed that tumours did not develop in these fosterlings according to expectancy.

Bittner & Little (1937) transferred mice of a high-cancer strain (A) mother to the care of low-cancer strain foster-mothers with the result that the eventual incidence of cancer in the sucklings was much reduced. Some of their results are given in Table 162.

TABLE 162. The influence of a foster-mother on the incidence of mammary cancer in mice (Bittner & Little, 1937)

Strain of mother	Mammary cancer expectancy (%)	Strain of foster-mother	Mammary cancer expectancy in foster-mother's strain (%)	Incidence of mammary cancer in fostered A mice (%)
Strong A	83·2	CBA	2·8	23·1
Strong A	83·2	C57	1·7	4·9

In a later paper Bittner (1939a) points out that the effect of transferring mice born of a high-cancer strain mother to a foster-mother of a low-cancer strain on the susceptibility to mammary cancer is only seen if the transference is made soon after birth. If the young are left with their high-cancer strain mothers for 2 or 3 days before being handed over to the foster-mother their susceptibility to cancer still remains high. This fact is shown also by some experimental results obtained by Andervont & McEleney (1939). They used two strains of mice, namely C_3H, in which spontaneous mammary cancer occurs in nearly 100 per cent of the females, and C57, in which the incidence of mammary cancer is less than 1 per cent. Newly born young of C_3H parents were given C57 foster-mothers, and C57 young were fostered by C_3H mice. As Table 163 shows, this fostering, if started within 17 hours of birth, reduced the incidence of cancer of the C_3H mice from near 100 per cent to 25 per cent, and increased the cancer incidence in the

C 57 mice from less than 1 per cent to 5·3 per cent. The table shows also that these effects decreased with the length of time during which the baby mice had been left with their own mothers. Andervont & McEleney (1941) removed C_3H mice at full term from the uterus and placed them with C 57 black (low-cancer strain) foster-mothers. Among thirteen of such mice which lived to 15·3 months, no mammary cancer developed. Andervont, Shimkin & Bryan (1943) have found that the susceptibility of mice to the transmissible agent of mammary cancer begins to decrease soon after birth.

TABLE 163.   The influence of foster-mothers in transmitting susceptibility to mammary cancer in mice (Andervont & McEleney, 1939)

Strain	Foster-mother	Number of mice	Longest possible time with own mothers (hours)	Number getting mammary tumours	Percentage incidence	Average tumour age (months)
C_3H	C 57	12	7	3	25·0	13·1
C_3H	C 57	16	17	4	25·0	15·9
C_3H	C 57	19	24	12	63·1	13·2
C_3H	C 57	4	48	4	100·0	11·2
C 57	C_3H	19	17	1	5·3	20·5
C 57	C_3H	15	24	2	13·3	17·7
C 57	C_3H	10	48	1	10·0	20·0

Bittner (1937a) showed that the reduced liability to cancer caused by transferring newborn mice of a high-cancer strain to a foster-mother of a low-cancer strain was carried to the 4th generation. Mice of a high-cancer strain (A) were fostered by mothers of a low-cancer strain (CBA), and the incidence of cancer in these and in subsequent generations was recorded. The mice of the 4th generation still showed a reduced liability to cancer of the breast (Table 164). It appeared also that if in the course of time any of these fosterlings developed mammary cancer their offspring were more than usually liable to the same kind of neoplasia.

TABLE 164.   Persistence of foster-mother's influence in lowering the incidence of mammary cancer in subsequent generations (Bittner, 1937a)

	Generation	Number	Percentage in which mammary tumours appeared
Female descendants of A mice which had been nursed by CBA foster-mothers. The 2nd, 3rd and 4th generations were nursed by their own mothers	1	9	33·3
	2	40	35·0
	3	29	6·9
	4	13	15·4
	Total	91	23·1

(c) Transmission of the nongenic factor of mammary cancer by living tissues. Further experiments have revealed that the mother's milk is not the only medium by which the tendency to acquire mammary cancer later in life may be transmitted to the newborn young. Bittner (1939b, 1940b) took portions of liver, spleen or thymus from young high-tumour strain (A) females or portions of lactating mamma from older females of strain A and grafted these bits of tissue into baby mice of two groups, namely:

(1) Low-cancer strain (C 57 blacks) nursed by C 57 females.
(2) Low-cancer strain (C 57 blacks) nursed by high-cancer strain (A) mothers.

Inocula of spleen, thymus or mamma, but not those of liver, transmitted susceptibility to cancer of the breast (Table 165).

TABLE 165.   Showing the capacity of tissue grafts to transmit susceptibility to mammary cancer in mice (Bittner, 1939c)

Low-cancer strain (C 57) mice (total numbers used)	Grafted with tissue from high-cancer strain (A) mice	Percentage of test mice in which cancer of breast developed
48	+	2·1
37	+	13·5
46	+	21·7
11	+	18·2
586	−	0·5
104	−	10·6
108	−	1·9
112	−	0

Bittner (1940b) showed that the tendency to mammary cancer transmitted by the mother's milk to suckling mice is not limited to the first milk secreted but is present apparently during the whole period of lactation. The active influence conveying susceptibility to mammary cancer can be transferred, he says, to females at the age of 4 weeks by the oral administration of milk from mothers of a high-cancer strain.

Woolley, Law & Little (1941), using C_3H mice whose liability to mammary tumours had been lowered by foster nursing, injected each of them when between 1 and 3 months old with 0·5 c.c. of whole blood from non-fostered male and female C_3H mice aged from 2 to 4 months. These injections led to an increase of mammary cancer in the recipients, the incidence being 19·16 per cent as compared with 6·06 per cent in the uninjected fostered controls.

The various experiments mentioned above seem to show that whatever the nongenic transmissible factor may be which controls the susceptibility to cancer of the breast in mice of either sex it can be conveyed to the young by the mother's or the foster-mother's milk or by implanted tissues, and that it is transmitted to subsequent generations.

Varying resistance to the transmissible factor.   Andervont (1940) found that some mice which normally had a low liability to spontaneous mammary cancer were yet very susceptible to the transmissible agent derived from foster-mothers of a high-cancer strain. For example, high-cancer strain (C_3H) mice were fostered by low-cancer strain (C) mothers, and the young of C mice were fostered by mice of the C_3H strain. The outcome showed that the C mice did not themselves possess the transferable carcinogenic factor conveyed in milk but were very susceptible to it when it was conveyed to them by a foster-mother, whereas another non-cancer strain (C 57 black) showed a relatively low response to the transmissible agent (Table 166).

Murray (1941), also, has carried out experiments which suggest that some mice may be relatively resistant to the nongenic transmissible factor. By crossing C 57 and *dba* mice, and noting the effect of this factor on the incidence of mammary cancer in subsequent generations, Murray showed that the C 57 mice are less influenced by the factor than are mice of the *dba* strain.

TABLE 166. Showing the variable susceptibility of mice of different strains to the nongenic transmissible agent of mammary cancer (Andervont, 1940)

Nurslings	Foster-mothers	Percentage of nurslings ultimately getting mammary cancer
C₃H	C₃H	100
C₃H	C	50
C	C	0
C	C₃H	64
C 57	C₃H	14
C 57	C 57	0

Co-operation between oestrogen and the nongenic transmissible factor in the induction of mammary cancer.   The nature and biological action of the nongenic transmissible factor are unknown, but experiments have demonstrated that its presence greatly facilitates and is perhaps essential for the development of mammary cancer under the influence of oestrogen.

Lacassagne (1933 *b*) observed that the facility with which mammary cancer can be induced in male mice by oestrogen depends on the strain to which they belong. Cancer of the breast, he found, is more readily induced in male mice belonging to a strain in which 72 per cent of the untreated females suffer from mammary cancer, than in males of a strain in which the incidence of spontaneous cancer in the females is only 2 per cent. In a later paper Lacassagne (1935 *b*) mentions the following experience. Oestrone benzoate, in weekly injections of 0·3 mg., was given from the day of birth or soon afterward to male mice of two strains, one of which showed in the untreated females a high incidence of spontaneous mammary cancer and the other a low incidence. Among twelve males belonging to the high-cancer strain mammary cancer developed in eleven, the other one dying free from cancer in the 7th month. The first cancers appeared at 3½ and 4½ months respectively and the last at 10 months. In contrast with these results were those obtained in six male mice of the low-cancer strain treated in the same way. At the end of 8 months all the six males were alive without tumours, during the 9th month a mammary tumour developed in one and eventually between the 12th and 18th months mammary cancer developed in all these mice. Clearly, individuals of the low-cancer strain were not quite refractory to the carcinogenic influence of oestrone, though they responded less readily than mice of the other strain. Bonser, Stickland & Connal (1937), by means of oestrone benzoate, induced mammary cancer in five among thirty-two female mice belonging to Little's CBA strain in which spontaneous mammary cancer is rare.

Lacassagne (1939 *a*) has shown, moreover, that the liability of male mice to mammary cancer in response to oestrogenic action is transmitted in the same way as the liability of the female of the same strain to spontaneous mammary cancer. He crossed mice of a high-cancer strain (R$_{III}$) with mice (Agouti 39) among which no mammary cancer had occurred during 7 years. Soon after birth the young of these cross matings were given 50γ of oestrone benzoate by injection once a week. Brothers and sisters of the F_1 generation were mated, and their F_2 offspring were treated with oestrone benzoate in the same way. The aptitude of oestrone benzoate to cause mammary tumours in these mice is shown, roughly summarized by the writer, in Table 167.

TABLE 167. The influence of parentage on the susceptibility of male mice to the induction of mammary tumours by oestrone benzoate (Lacassagne, 1939*b*)

| Generation | Susceptibility of parent's strain to spontaneous mammary cancer | | Mammary tumours induced in males of F_1 and F_2 genera-tions by oestrone benzoate |
	Female	Male	
F_1	Low	× High	—
F_1	High	× Low	+
F_2	High	× Low	+

Similar results were obtained by crossing R_{III} mice with those of another strain (17 n.c.) in which no mammary cancer had occurred during 11 years.

Lacassagne (1939*b*) has further reported that the mere transfer of a newly born male of the low-cancer strain Agouti 39 to a foster-mother of strain R_{III} is enough to abolish the young mouse's relative immunity from mammary cancer under the influence of oestrone. In a male of strain Agouti 39 suckled by an R_{III} foster-mother and treated with oestrone benzoate from the 13th day of life onward, mammary cancer appeared 182 days after the first administration (Lacassagne & Danysz, 1939). Twombly (1940) also has found that the liability to mammary cancer in response to oestrone can be controlled to some extent in males by suckling. The C 57 strain is highly resistant to the induction of mammary cancer by oestrogen (Suntzeff, Burns, Moskop & Loeb, 1936). The R_{III} strain is highly susceptible to this form of cancer induction. Twombly transferred newly born male C 57 mice to R_{III} foster-mothers, and when they were 10 days old gave them implants of oestrone (0·07 to 0·18 mg.). In twenty-seven males of the C 57 strain so treated mammary tumours had already appeared in nine, fourteen of the other mice being still alive when the paper was written. Bittner (1941) allowed nurs-lings of high-cancer strain mice (C_3H and A) to be fostered by mothers of low-cancer strains (C 57 black and CBA). When the mice so fostered were between 4 and 6 weeks old they were given subcutaneous implants of 2 mg. of oestrone or 1·5 mg. of oestradiol benzoate. None of the mice nursed by the low-cancer strain foster-mothers eventually had mammary tumours, in spite of the fact that they belonged genetically to high-cancer strains, were subjected to a large dose of oestrogen and were known to be susceptible to its carcinogenic action in the presence of the unidentified transmissible factor.

Comparable results have since been obtained by many workers in different parts of the world and the contrasting degrees of susceptibility among the various strains of inbred mice to the induction of mammary cancer by oestrogen have been completely established, and mice of the various strains are now available to any investigator who intends to work at the problem.

The nature of the nongenic inherited factor of mammary cancer. From the ex-periments quoted it seems that whatever the nongenic factor may be, its presence is usually required for the production of mammary cancer by oestrogen. Perhaps the agent may be a virus, using this term in the sense of a self-multiplying protein particle. To assume that it multiplies after introduction into the living body seems justified, for its continued influence through successive generations is difficult otherwise to understand. Moreover, experiments have shown that its dimensions are compatible with its being a virus. Visscher, Green & Bittner (1942) extracted lactating mammary tissue from high-cancer C_3H mice with distilled water and centrifuged the material, first at 15,000 r.p.m. and then at 40,000 r.p.m. for 1 hour. The various fractions were tested by administering them to C_3H mice which had been suckled by foster-mothers of a low-cancer strain, and noting the eventual incidence of mammary cancer. The sediment obtained by the second spinning was the most potent in leading to cancer of the breast, and the results indicate, they say, that the agent is a colloid of high molecular weight and may be a virus.

Bryan, Kahler, Shimkin & Andervont (1942) found that the transmissible factor could be largely, though not completely, sedimented from mouse's milk by centrifuging it for 1 hour at 60,000 times the force of gravity.

As Lacassagne (1935 b) has pointed out we cannot, when discussing the induction of mammary cancer by oestrogen, distinguish between 'susceptible' and 'insusceptible' strains of mice, because the difference between the different strains is one of degree only.

Another point to which attention may be redirected is that a high incidence of mammary cancer in the females is accompanied by a high susceptibility among the males of the same strain to the carcinogenic action of oestrogen. In strains in which the females are but little subject to spontaneous mammary cancer, the males are resistant to the carcinogenic action of oestrogen on the mammae.

The ovary is not the source of the nongenic hereditary agent of mammary cancer. Reference may be made to an interesting experiment by Little (1936), in which he found that the ovary of a low-cancer strain mouse may be effective in causing mammary cancer in a male of a high-cancer strain. From this it appears that the susceptibility to mammary cancer, as distinct from the oestrogen by which it is caused, is not referable to the ovary.

There seems to be no evidence that the nongenic transmissible agent of mammary cancer takes any part in the causation of uterine cancer under the influence of oestrogen (Allen & Gardner, 1941).

(ii) THE GENIC FACTOR IN SUSCEPTIBILITY TO MAMMARY CANCER

Although the greater part of the liability to cancer of the breast in mice appears to pass from one generation to the next otherwise than by genes, there is clear evidence of some genic agency. The paternal influence in the transmission of susceptibility to mammary cancer, which was noticed by Lathrop & Loeb (1918) to be small though not negligible, is genic. Proof of this was adduced by Cloudman & Little (1936), who crossed male mice of two different low-cancer strains (C 57 and Zavadskaya's brachyuric strain) with the same high-cancer strain (*dba*) females and found that the incidence of mammary cancer in the virgin F_1 females varied according to the strain to which the father belonged. Korteweg (1936) came to the same conclusion. He crossed high-cancer strain (*dba*) mice with mice of a low-cancer strain (C 57) and noted that the offspring of a *dba* ♀ × C 57 ♂ cross showed a high-cancer incidence, whereas those of a C 57 ♀ × *dba* ♂ cross had a low incidence, but not so low as that shown by the offspring of a C 57 ♀ × C 57 ♂ pair.

Bittner (1940 a) and Andervont (1940) have both concluded from breeding experiments that the genic factor is inherited as a mendelian dominant.

In what way this chromosomal factor favours mammary cancer has not been completely elucidated. The effect which it produces may be defined loosely as a predisposition to mammary cancer. Andervont & McEleney (1941), by back-crossing F_1 mice, the female progeny of $C_3H \times I$ or Y mice, with I or Y males, obtained results which seemed to show that the genic factor largely controls the susceptibility of mice to the nongenic inherited factor of mammary carcinogenesis.

Whatever the nature of the genic influence may be, it appears to regulate the degree of susceptibility of the organism to the carcinogenic action of one or other or both of the two main factors. Its share in carcinogenesis may vary in nature and degree in different strains (Heston & Andervont, 1944).

Summary. From the experimental work quoted it seems that mammary cancer is caused by the conjoint action of three factors: (i) oestrogen, (ii) a nongenic agent, perhaps a virus, transmissible to the offspring through the mother's milk, and (iii) a genic, mendelian dominant factor which controls to some extent the susceptibility of the individual to the other two essential agencies. All three factors are required for the induction of mammary cancer. If oestrogen is given to males which already have factors (ii) and (iii) they probably will become subject to mammary cancer, but if factor (ii) is missing they will probably remain free. Thus oestrogens usually fail to evoke cancer in male mice of the C 57 black strain. If, however, newborn mice of this strain are fostered by R_{III} mothers they become amenable to the production of breast tumours by oestrogen, and moreover show a higher incidence of the disease in response to oestrogen than do R_{III} males fostered by C 57 blacks. These results suggest that the C 57 blacks, when nursed by their own mothers, remain free from mammary cancer, in spite of being more susceptible to the carcinogenic action of oestrogen than the R_{III} mice, because the nongenic transmissible factor is absent.

The development of mammary cancer depends not only on the mere presence of the three factors named. An individual may have all these etiological factors and yet remain free from cancer of the breast. This has been demonstrated by experiments, which show that the effectiveness of oestrogen as a cause of mammary cancer depends not alone upon its presence but also upon its quantity and on the environmental conditions.

THE SEARCH FOR RECOGNIZABLE DIFFERENCES BETWEEN MICE OF HIGH AND LOW SUSCEPTIBILITY TO CANCER OF THE BREAST

Mammary cancer in mice being largely confined to certain strains, it is natural to seek some characteristic by which to distinguish mice liable to mammary cancer from those that are not. Much work has been done for this purpose with disappointing results. At present there are no criteria, apart from a knowledge of its ancestry and nurture, by which the liability of any particular mouse to cancer of the breast can be foretold. Some of the attempts to elucidate the problem will now be considered.

(a) *The oestrous cycle.* Various workers have attempted to find some demonstrable difference between the oestrous cycles of mice belonging to strains with a high- and a low-cancer incidence respectively. Harde (1934) noted that mice of a low-cancer strain had a normal oestrous cycle, whereas the mice of two high-cancer strains had a more prolonged oestrum. Brunschwig & Bissel (1936) compared the cycles of the C57 black (low-cancer strain) with those of *dba* (high-cancer strain) mice and found that more oestrous cycles occurred within a given time in the former than in the latter. It appears, however, that the differences so recorded are casual and not associated with any special liability to, or immunity from, cancer of the breast; for the collected observations of several workers carried out on many different strains do not indicate any peculiarity of the oestrous cycles generally associated with a liability to mammary cancer. Loeb & Genther (1928) came to this conclusion and their findings have since received firm support (Bonser, 1935 *a*, *b*; Lacassagne, 1936 *c*; Burns, Moskop, Suntzeff & Loeb, 1936; Simpson, 1936). Bonser (1935 *a*, *b*) compared the oestral behaviour of two strains of mice, one (Bagg albinos) showing a high incidence of cancer of the breast and the other (Black agoutis) a low incidence, and she found but little difference between the two. The average duration of the cycles in the high-cancer strain was 5·7 days and in the low-cancer strain 5·8 days. A slight difference was observed in the vaginal reactions to oestrone, the low-cancer strain showing a slightly greater reactivity than the other. The mice used in this test had been spayed. Her findings are set out in Table 168.

TABLE 168. The absence of relationship between the duration of oestrous cycles or of oestrum and mammary cancer in mice (Bonser, 1935 *b*)

Age	Number of mice	Strain	Average duration of oestrous cycle (days)	Average duration of vaginal cornification (days)
Young	33	High-cancer strain	5·7	2·2
Young	26	Low-cancer strain	5·8	2·4
Old	10	High-cancer strain	8·3	1·8
Old	5	Low-cancer strain	6·4	2·3

Burns, Moskop, Suntzeff & Loeb (1936) noted the total period of vaginal keratinization during 32 days in ten different strains of mice, and found that, although the length of the sexual cycle differs much in different strains, no relationship can be seen between the character of the sexual cycle and liability to mammary cancer. Wolfe, Burack & Wright (1940) investigated the matter in rats, using a strain in which spontaneous mammary tumours are frequent and comparing their oestrous cycles with those of non-tumour-bearing rats. Daily vaginal smears were taken from fifty rats of each group and no correlation was detected between oestral characters and the tendency to have mammary tumours.

(b) *The duration of sexual activity as shown by oestrous cycles.* Concerning mice the writer does not possess any data referring to this matter. In women it has been thought that a prolongation of sexual life favours the development of

TABLE 169. The age of the menopause in women over 50 years old with mammary cancer (Olch, 1937)

Menopausal age	Number
Under 40	10
40–44	41
45–49	104
50 and over	187
Total	342

mammary cancer. Olch (1937) collected 342 instances of women who first sought advice for cancer of the breast when they were over 50 years old. Half of these women (54·7 per cent) were still menstruating or had continued to do so till the age of 50. This fact Olch regards as significant. His figures are given in Table 169.

(c)' *The vaginal response to oestrogens.* The varying sensitivity of the vagina to oestrogen among mice of inbred strains has also been investigated to discover whether animals prone to mammary cancer respond differently from mice not prone to this form of neoplasia. Although the various strains do not all react exactly alike it has not been possible to associate the differences with any susceptibility or insusceptibility to cancer. Lacassagne (1934a) noticed that smaller doses of oestrone caused vaginal cornification in mice of the high-cancer strain R_{III} than were required to cause the same change in the low-cancer xvii mice. In contrast with this result Bonser (1935a, b), using other strains than those on which Lacassagne's inquiry was made, found that the vaginal response to oestrone was more readily induced in mice of a low-cancer strain (Black agoutis) than in those of a high-cancer strain (Bagg albinos), though the difference was small. Van Gulik & Korteweg (1940) reported similar results. They compared the vaginal reactions to oestrone of *dba* (high-cancer strain) and C57 (low-cancer strain) mice, and found that the vaginas of the C57 mice reacted to oestrone more readily than those of the *dba* strain.

(d) *The appearance of oestrogenic phenomena in spayed mice of high-cancer strains.* Woolley, Fekete & Little (1940) have reported a remarkable difference between the mice of certain distinct strains. Mice of two high-cancer strains (*dba*, C_3H) and one low-cancer strain (C57 black) were spayed on the day after birth. Vaginal opening usually occurred in these mice only a few days later than in littermate controls, though in some of the C57 mice the vagina never opened, and in others the aperture remained very small. After they were a year old examination of the mice revealed a notable difference between those of the low-cancer and those of the high-cancer strains. In the latter the uteri had become enlarged so as to exceed those of normal females, and the uterine glands were highly developed and sometimes cystic. In C57 mice little or no mammary development had taken place, whereas in mice of the high-cancer strains the breasts were well developed. These and other differences, which are crudely summarized by the writer in Table 170, suggest that, although spayed, the mice of the two high-cancer strains had been continuously subjected to oestrogen, possibly derived from the adrenals.

TABLE 170. The relative conditions of the accessory reproductive organs in spayed mice of high-cancer and low-cancer strains one year after spaying (Woolley, Fekete & Little, 1940)

Strain of mouse	Susceptibility to mammary cancer	Vagina open	Cyclical vaginal cornification	Endometrial hyperplasia	Breasts developed	Occasional cancer of breast
dba	High	+	+	+	+	+
C_3H	High	+	+	+	+	+
C57	Low	+ −	−	−	+ −	−

(e) *The metabolism and excretion of oestrogen.* Pincus & Graubard (1940) gave single intramuscular injections of 2·1 mg. of oestrone to seven patients who had cancer of the cervix, and then assayed their urines for oestrone and oestriol. No oestriol was excreted; the same result was obtained if the injections of oestrone were accompanied by 10 mg. of progesterone. Normal women showed an increased excretion of oestriol after similar injections. Ross & Dorfman (1941) assayed the urine of four patients with mammary cancer, and found that there was no excess of oestrogen nor deficiency of androgen. Aub, Karnofsky & Towne (1941) failed to discover any significant difference in the urinary output of oestrogen or 17-ketosteroids by mice of a high-cancer strain (C_3H) as compared with mice of a low-cancer strain (C_{57}).

It is conceivable that a defective activity of the liver or kidney by delaying the metabolism or excretion of oestrogen might favour the induction of cancer of the breast.

(*f*) *Differences in the development of the mamma.* Gibson (1930) compared the development of the mammae in mice of high- and low-cancer strains. The latter displayed a more rapid and abundant branching of the lactiferous ducts and a better development of alveoli; and in advanced age the gland involuted normally. In the high-cancer strain mice the alveolar development was not so good and as age advanced the duct epithelium tended to metaplasia rather than involutional atrophy. Gardner & Strong (1935) compared the mammary development in ten different strains of mice, five of which had a high and five a low incidence of mammary cancer. They were unable to detect any structural feature in the mamma which could be associated with the predisposition to cancer. Bonser (1936) thought that a difference could be recognized between the responses of the mamma to oestrogens in male mice belonging to a high- and a low-cancer strain respectively. In the latter there was a greater tendency to the development of acini, and she thought that the mice in which mammary cancer was most likely to develop were those in which acinous growth was deficient. Her observations accord with those of Gibson.

(*g*) *Are mice of high-cancer strains more susceptible generally to carcinogenic agents than mice of low-cancer strains?* The reaction to tar of mice belonging to high-cancer and low-cancer strains was compared by Lynch (1925), who found that tar produced tumours with equal readiness in the two strains; mice in which mammary cancer had already developed spontaneously reacted to the neoplastic influence of tar on the skin in the same way as other mice. Kreyberg (1935) also inquired into this matter. He applied tar to the skin of a large number of mice belonging to high-cancer and low-cancer strains. Tar tumours appeared later and in a lower percentage in both males and females of the high-cancer strain than in those of the low-cancer strain. We may conclude therefore that a high incidence of mammary cancer is not correlated with an enhanced general responsiveness to carcinogenic agents.

(*h*) *Adrenal degeneration.* Cramer & Horning (1937 *a*, 1939 *a*, *b*, *c*) believe that adrenal degeneration is a visible sign of susceptibility to mammary cancer and that in mice of high-cancer strains the degeneration affects mainly the medulla. Neither of these opinions has received general support. Degeneration of the adrenal cortex is not a pronounced feature in mice of the C_3H and *dba* strains in spite of a high susceptibility to cancer of the breast. The advanced condition of adrenal degeneration seen in R_{III} mice seems to be a peculiarity of that strain. References bearing on the matter are Dobrovolskaya-Zavadskaya & Zéphiroff (1938), Dobrovolskaya-Zavadskaya & Pezzini (1939), Kreyberg & Eker (1939), Kreyberg (1939), Bonser (1939), Bonser & Robson (1940).

Subsidiary Factors in Mammary Carcinogenesis

(a) *Reproduction and Lactation.* Human statistics have shown that cancer of the breast has a higher incidence in unmarried and childless women than in those who have had babies (Lane-Claypon, 1926; Dublin & Lotka, 1935). Observers have differed as to whether, in mice, mammary cancer is more frequent in breeders or non-breeders. Lathrop & Loeb (1913, 1916), Cori (1927), Marsh (1929), Murray (1937), Suntzeff, Burns, Moskop & Loeb (1936) have all stated that mammary cancer occurred in their mice more frequently and earlier in life among breeders than among non-breeders. Slye (1916) records that in her stock cancer of the breast was about equally common in mated and unmated females, but appeared at a slightly younger age in the virgins. Pybus & Miller (1936), whose mice were derived from the Lathrop & Loeb stock, found no significant difference in the age at which mammary cancer developed in breeders and non-breeders. Bonser & Connal (1939) and Andervont & McEleney (1939) have recorded a greater incidence of mammary cancer in non-breeding mice. These discrepancies cannot all be attributed to racial peculiarities in the mice (Table 171) and one must conclude that they may have been the result of differences in the circumstances in which the mice were kept.

TABLE 171. Varying incidence of mammary tumours in breeding and non-breeding mice of the C_3H strain

	Breeders		Non-breeders	
Authors	Average age when tumour appeared (months)	Percentage in which mammary tumours appeared	Average age when tumour appeared (months)	Percentage in which mammary tumours appeared
Suntzeff, Burns, Moskop & Loeb (1936)	10·9	60·8	9	2·5
Andervont & McEleney (1938)	8·51–10·50	78–100	11·5–13·33	92·8–100
Andervont & McEleney (1939)	9·5	51	11·5	71

It has been suggested that the age at which pregnancy first occurred, or the number of pregnancies, might affect the incidence of mammary cancer, but Murray (1934), studying the *dba* strain of mice, found no support for these suggestions. The writer, too, in the same strain observed no correlation between the incidence of cancer and the number of litters produced (Table 172).

Bagg (1936) and Bagg & Hagopian (1939) reported that rapid breeding, suckling being prevented, caused an increased incidence of mammary tumours in mice and rats, but Little & Pearsons (1940) with the same strain of mice (C57) as that used by Bagg obtained a different result. Bagg (1926) thought that the

TABLE 172. The incidence of mammary tumours in *dba* mice in
relation to littering (Burrows, unpublished work)

Number of litters	Number of mothers	Number of mothers in which cancer of the breast developed
0	3	3
1	6	6
2	11	11
3	17	11
4	18	15
5	16	15
6	7	7
7	5	4
8	2	2
9	1	1
10	1	1

In two of these mice sterility was due to the early death of their mates, and in the third a genital malformation (hypospadias) was present.

prevention of suckling increased the liability to cancer in breeding mice; Bittner (1937) was unable to confirm this. The retention of secretion in the mammary ducts has been thought to favour the development of cancer (Adair & Bagg, 1925). Others have failed to confirm this opinion (Suntzeff, Burns, Moskop & Loeb, 1936; Fekete & Green, 1936). Fekete & Green suggested that blockage of the ducts might perhaps determine the site of a cancer though it would not cause it. In this connection it may be noted that oestrogen has been detected in colostrum and in some mammary cysts (Lacassagne & Nyka, 1934). Lewis & Geschickter (1934) found 6 r.u. of oestrogen per gram in a rat fibroadenoma. Mohs (1937) failed to find any such concentration of oestrogen in mammary fibroadenomata.

The foregoing summary seems to show that lactational activity has little, if any, part in the etiology of mammary cancer.

(b) *Diet and nutrition.*   Several kinds of food deficiency will check the oestrous cycles, the growth of the mamma in response to oestrogen, and the development of mammary cancer. For brevity these inhibitory effects of malnutrition will be considered together.

It has been noticed that if rats or mice are kept on a diet which is adequate in kind but deficient in amount the oestrous cycles will be checked (Evans & Bishop, 1922a; Asdell & Crowell, 1935; Tannenbaum, 1942), the mammae will fail to develop under the influence of oestrogen (Astwood, Geschickter & Rausch, 1937; Trentin & Turner, 1941), and the incidence of mammary cancer will be reduced (Rous, 1914; Tannenbaum, 1940a, b). In Tannenbaum's (1942) experiment the incidence of mammary cancer was increased by the addition of 1·4 g. of starch daily to the restricted food supply of each mouse.

The oestrous cycles and appearance of mammary cancer may be checked also by a lack of certain aminoacids in the food, namely lysine (Evans & Bishop, 1922a; Courrier & Raynaud, 1932; Bittner, 1935; Pearson, 1937; Voegtlin & Thompson, 1936; Voegtlin & Maver, 1936), cystine and methionone (Osborne & Mendel, 1916; Voegtlin, Johnson & Thompson, 1936; Morris & Voegtlin, 1940) and pantothenic acid (Morris & Lippincott, 1941).

Unfortunately the nutritional disturbances caused by these dietetic restrictions prevent their use for prophylactic purposes in man.

Attempts to Prevent Mammary Cancer

We may briefly consider the possibility of applying preventive measures against human mammary cancer in the light of knowledge gained by studying the disease in animals.

Heredity. The control of human mating so as to reduce the chance of mammary cancer in the offspring would, of course, be impracticable even if it were desirable. Some thought, however, might be given to the nongenic inheritable agent and it seems justifiable to propose that any mother whose family history suggests the presence of the transmissible agent of mammary cancer ought not to nurse her own female child. This is not an entirely new idea, for Haddow (1942) quotes the following from a French thesis by Rouzet (1818): 'L'hérédité du cancer une fois reconnue, on voit de quelle importance il est de remettre l'enfant nouveau né à une nourrice étrangère, lorsque la mère a presenté, soit avant, soit pendant la grossesse, quelques symptômes cancereux; puisqu'il n'est guère douteux qu'elle ne communiquât à son nourrison en l'allaitant, cette disposition au cancer à laquelle il avait peut-être échappé pendant l'époque de la gestation.'

Curtailment of oestrogen. It seems likely that if we could reduce the formation of oestrogen in the body or hasten its inactivation or excretion we might retard the development of mammary cancer. Limitation of the production of oestrogen by hypophysectomy, spaying, or severe restriction of diet, are not practical measures. There may be, however, some prospect for chemotherapy. Lacassagne (1939 *a*, *b*) and Nathanson & Andervont (1939) have shown that mammary cancer in mice may be largely prevented by the continued administration of testosterone propionate (Table 173). It seems conceivable that some prophylactic measure, applicable to man, may one day evolve from this observation.

TABLE 173. The incidence of mammary cancer in C_3H mice receiving large doses of testosterone propionate (Nathanson & Andervont, 1939)

Treatment	Number	Number with mammary cancer	Percentage incidence of mammary cancer
1·5 mg. testosterone propionate weekly	20	6	30
None	20	20	100

Chapter XIX. *The Effects of Oestrogen on Connective Tissues and Skin*

General Growth and Bodyweight. Skeleton. Adipose Tissue. Muscle. Skin. Induction of Sarcoma.

General Growth and Bodyweight

IN most mammals the male is larger than the female. Stotsenburg (1909, 1913) and others have shown that this difference is caused by ovarian action. Stotsenburg found that castration of male rats early in life had little or no effect on the subsequent growth curve, whereas the female spayed at the same period grows to a larger size than her intact sister. These observations have been supplemented by Steinach & Holzknecht (1916), who interchanged the gonads of young male and female littermate guinea-pigs, implanting testes into spayed females and ovaries into castrated males. The males bearing ovaries failed to attain the general dimensions or weight of normal males or females and the females grafted with testes grew to an unusual size—larger in fact than normal untreated males (Table 174).

TABLE 174.   The effect of interchanging the gonads between the two sexes on the ultimate bodyweight of guinea-pigs (Steinach & Holzknecht, 1916)

Subject	Bodyweight (g.)
Normal female	845
Normal male	1,002
Spayed female with grafted testis	1200
Normal male	980
Normal female	808
Castrated male with grafted ovary	516

Moore (1919) repeated the experiments and confirmed the observations of Steinach & Holzknecht on the guinea-pig. Moore (1922) also noticed that rats which had been spayed during infancy grew larger than normal females, whereas castration of males had the opposite effect.

Bugbee & Simond (1926) showed that repeated injections of a follicular extract retarded growth in male and female rats, whether their gonads had been removed previously or not. Riddle & Tange (1928) observed a similar effect of oestrogen prepared from sows' follicular fluid on the growth of pigeons. Wade & Doisy (1931), Spencer, D'Amour & Gustavson (1932*a, b*) and Spencer, Gustavson & D'Amour (1931) obtained the same results with oestrin in both male and female rats. The last named used twenty-four rats derived from three litters and divided them into two groups each containing six males and six females. As soon as they were weaned treatment was begun, Group I receiving 40 r.u. of oestrin in olive oil subcutaneously every other day, Group II being given olive oil only on the same occasions. The effect of this treatment on the general bodyweight is shown in Table 175.

Korenchevsky & Dennison (1934) gave to three groups of immature male rats 20, 60 or 100 i.u. of oestrin daily for a period of 43 days. At the end of this time the ratios of weight increase in the oestrin-treated rats compared with controls

TABLE 175. The effect of oestrone on the bodyweight of rats
(Spencer, Gustavson & D'Amour, 1931)

	Treatment	Total weight of group at end of treatment (g.)	Percentage gain in weight
Group I	Oestrin in olive oil	1,051	211·0
Group II	Olive oil only	1,348	262·3

which had received no oestrin were 88 : 127, 83 : 114, and 86 : 127 in the three groups respectively.

In a small experiment by the writer ten male mice of an inbred (Strong A) strain were used. They were between 5 and 6 weeks old and were divided into two batches of five. Twice a week one batch received cutaneous applications of testosterone (0·1 per cent) and oestrone (0·1 per cent) dissolved in benzene, and the other batch received applications of testosterone (0·1 per cent) only. At the end of 63 days the mice which had received oestrone in addition to testosterone were found to weigh less than those which had received testosterone alone (Table 176).

TABLE 176. The effects of oestrone on bodyweight in male mice (Burrows)

Treatment	Average bodyweight before treatment (g.)	Average bodyweight at end of treatment (g.)
Testosterone *plus* Oestrone	23·5	23·1
Testosterone only	22·4	26·6

An inhibitory influence on bodyweight is displayed by synthetic oestrogens, for example diethylstilboestrol (Gaarenstroom & de Jongh, 1939; Gaarenstroom & Levie, 1939; Noble, 1939). Noble implanted into rats various synthetic oestrogens in a dry form and noticed that they checked the rate of bodygrowth to a greater degree in females than in males and that this effect was not pronounced until a weight of between 100 and 120 g. had been reached, after which the bodygrowth became much retarded.

It seems that the growth-checking action of oestrogen is to some extent at least the result of an inhibition of hormone production by the pituitary. Reece & Leonard (1939) removed the ovaries and pituitary from immature rats and 5 days later gave them eleven daily pituitary implants as follows:

Three received pituitaries from normal adult females and two were given pituitaries from normal adult males.

Five received pituitaries from oestrogen treated females and one was given pituitaries from oestrogen treated males.

The oestrogen treatment of the donors consisted of 10 daily injections of 20 r.u. of oestradiol. The results, set out in Table 177, show a reduction of the growth-promoting influence of the pituitaries of rats under treatment with oestradiol.

Gaarenstroom & Levie (1939) found that the general growth of immature rats could be arrested almost completely by daily doses of 200 or 250γ of stilboestrol. This inhibition of growth was overcome by giving pituitary growth hormone at the same time as the stilboestrol. Gaarenstroom & Levie do not think, however, that these results must mean that the inhibition of bodygrowth by oestrogen is the direct consequence of a suppression of the supply of growth hormone from the

TABLE 177. The effects of oestradiol on the growth-promoting influence of the pituitary as shown by implantation in hypophysectomized rats (Reece & Leonard, 1939)

| | | Mean bodyweight of recipients | | |
Source of implanted pituitaries	Number of recipients	At beginning of experiment (g.)	At end of experiment (g.)	Weight of implanted pituitaries (mg.)
None (controls)	5	65	63	—
Normal rats	5	67	84	119
Oestrogen-treated rats	6	68	71	173

(The larger weight of the pituitaries derived from the oestrogen-treated rats will be noted.)

pituitary; the inhibition of growth might be due, they suggest, to an interference by oestrogen with the growth of bone at the epiphyses. Freud, Levie & Kroon (1939) point out that after hypophysectomy the membrane bones grow normally whereas long bones do not, and they believe that the terms 'growth hormone' and 'chondrotrophic hormone' are synonymous.

Griffiths & Young (1942) checked the growth of rats by hypophysectomy or by the subcutaneous implantation of 15 mg. of diethylstilboestrol. Thereafter pituitary extracts were given and records were kept of the bodyweight and rate of growth of the tail. It was found that whereas after hypophysectomy and the administration of pituitary extracts the increase in bodyweight and tail-growth were in constant proportion, this was not the case in rats which were treated by diethylstilboestrol and given the same pituitary extracts, a result which suggests that the arrest of growth cannot be explained by the lack of pituitary hormone only, nor attributed entirely to a deficiency of chondrotrophin.

Whether the result is brought about by an arrest of the supply of growth hormone or of chondrotrophin, or by some other cause, the fact remains that oestrogens restrain the general growth of the body.

Skeleton

(a) *Length of bones.* There is little doubt that the effect of oestrogen on the bodyweight as just described is largely the result of bony changes. Students of human anatomy have noticed a precocity of skeletal development in the female compared with the male. This precocity appears in two forms, namely (1) an earlier establishment of some of the centres of ossification, and (2) an earlier union of epiphysis and metaphysis. Frazer (1920) in his textbook of human anatomy states that the centres of ossification in the carpus appear from a few months to a year earlier in the female than in the male, and as examples of female precocity in the junction of the epiphyses to the bodies of the bones he mentions that epiphyseal union occurs at the lower end of the radius 6 months, at the lower end of the ulna 1 year, in the phalanges 1 year, and in the metatarsus 2 to 3 years earlier in the female than in the male. The secondary centres of ossification in the human vertebrae also appear earlier and reach full development sooner in girls than in boys. According to Beadle (1931) these centres of ossification have been seen as early as the 9th year in girls, whereas in males they have not been seen till the 14th year. The youngest age for completion of the bony epiphyseal rings in girls is 13 years and in boys 16 years, and their complete union with the vertebral bodies occurs also soonest by some years in the female sex. As the growth of a bone in length ceases when the epiphysis unites with the shaft, the fact that such union occurs earlier in the female than in the male accounts largely and perhaps entirely for the greater stature of the latter.

Steinach & Holzknecht (1916) demonstrated that the difference between the two sexes in skeletal growth is due to ovarian action. They grafted ovaries obtained from female guinea-pigs into castrated male littermates and after the animals had attained full growth various elements of the skeleton were measured. Some of the measurements are shown in Table 178.

TABLE 178.   The effect of ovarian implants on the length of bones in the castrated male guinea-pig (Steinach & Holzknecht, 1916)

Part of skeleton	Measurements of bones		
	Normal male (mm.)	Normal sister (mm.)	Castrated brother grafted with ovary (mm.)
Tibia	44	40	38
Femur	36	33	31
Ulna	33	31	29
Humerus	31	28	26
Vertebral column	160	148	136
Pelvis	47	43	38

It will be noticed that the effect produced on the bones by the ovary growing in the castrated male is greater than that of the ovary growing naturally in the female. The counterpart of this effect was noticed, inasmuch as the spayed females carrying grafted testes grew to greater dimensions than normal males. Spencer, D'Amour & Gustavson (1932 b) obtained the same effects on the length of bones in rats by injections of oestrin. They used nine litters comprising fifty-one individuals and divided them into two groups of males and two of females; one group of each sex received daily 20 r.u. of oestrin in olive oil. The other groups were given olive oil only. The average lengths of the femora in the four groups at the end of the experiment are shown in Table 179.

TABLE 179.   The effect of oestrin on the length of the rat's femur (Spencer, D'Amour & Gustavson, 1932)

	Normal male (mm.)	Oestrin-treated male (mm.)	Normal female (mm.)	Oestrin-treated female (mm.)
Average length of femur	30·4	28·7	28·7	26·5

Zondek (1936a) has reported similar results in rats treated with oestrin, their long bones being shorter than those of untreated controls. The length of other long bones and of the skull showed similar changes.

That the reduced length of bones under the influence of oestrogens is due to a premature union of the epiphyses with the shafts has been demonstrated in dogs by Tausk & de Fremery (1935) and by Gardner & Pfeiffer (1938 a) in mice. This effect was observed also in the vertebrae of the rat's tail by Levie (1938), who treated two groups of castrated male rats, 42 days old and weighing 80 g. at the start, with daily doses as follows: Group I, 500γ of oestradiol; Group II, 500γ of testosterone. Other rats of the same age were used as untreated controls. At the end of 14 days the tails of the rats treated with oestradiol had grown less than those of the control or the testosterone-treated animals (Table 180). Premature union of the epiphyses was found to be the cause of this diminished tail-growth.

TABLE 180.   The effect of oestrogen on the growth of the
rat's tail (Levie, 1938)

Rats	Treatment	Daily dosage γ	Average increase in length of tail mm.
Normal	None	—	30
Castrated	None	—	29
Castrated	Testosterone	500	31
Castrated	Oestrone	500	11

Oestrogenic effects of antirachitic agents.   When considering the effects of oestrogen on the epiphyses and general texture of bone it is interesting to recall that neoergosterol, calciferol and ergosterol have some oestrogenic capacity. Given in large doses to rats they cause vaginal cornification (Cook, Dodds, Hewitt & Lawson, 1934).

(b) *Texture and structure of bones.*   Oestrin has other effects on bones than the epiphyseal changes mentioned above. Zondek (1937) observed in rats and cocks which had been under continued treatment with oestrogen a supercalcification of the skeleton with partial obliteration of the marrow cavities. These changes were readily detected by X-rays, and on splitting a femur longitudinally it could be seen that the medulla was largely occupied by finely porous, easily crumbling, osseous tissue. A gradient of responsiveness to oestrogen could be detected in the skeleton, the effects being most pronounced in the femur and less advanced in the tibia. Changes in the texture of the bones were visible also in the spine, but the humerus and ulna were nearly normal. In addition to the change of texture, the bones were shorter than normal because the epiphyses had united earlier than usual with the shafts. Gardner & Pfeiffer (1938 a, b) also reported alterations in the structure of bones brought about by oestrogens. They treated thirty-four male and female mice with oestrone benzoate or equilin benzoate for periods extending to 348 days. After this treatment the femurs were white, opaque and very hard and brittle, and the marrow cavities were almost completely replaced by compact bone. These changes, which it seems are reversible, were shown by radiographs to be present throughout the skeleton except in the bones of the symphysis pubis. Testosterone propionate prevented the osseous changes induced by oestrogen, weekly doses of 1·25 mg. inhibiting the action of weekly doses of 1,000 i.u. of oestradiol benzoate. Sutro (1940) reports similar osseous changes in mice following weekly doses of oestradiol benzoate ranging from 150 to 1,000 r.u. He says that proliferation of new bone in the medullary cavities is especially pronounced in the femur and tibia, and that the calvarium is affected in the same way. Miller, Orr & Pybus (1943) have made a detailed study of the osseous changes induced by oestrogen in mice. They have observed, like Gardner & Pfeiffer, that the changes are reversible.

One effect of the increased formation of compact osseous tissue under the long-continued influence of oestrogen is an increased strength of the long bones. Gardner (1940) compared the resistance of the femur to a breaking strain in oestrogen-treated and normal mice. Weekly subcutaneous injections of 16·6 or 50γ of oestradiol benzoate had been continued from the time of weaning until the mice were killed at an average age of 478 days. The femurs of the oestrogen-treated mice were considerably more resistant to fracture than those of the controls, the weights required to break the bones in the same mechanical conditions being 2,499 and 1,655 g. respectively.

Like Zondek, Pfeiffer & Gardner (1938) noticed a gradient of susceptibility in the bones to the influence of oestradiol, the femur being the first long bone to show the characteristic effects, followed by the tibia, radius and ulna in that order. Further reference will be made to the effects of oestrogen on bone marrow in connection with changes brought about in the blood.

(c) *Tumours of bone caused by oestrogen.* From three females and one male belonging to Simpson's A strain of mice, all having tumours of bone, Pybus & Miller (1938 *a, b,* 1940) have raised an inbred stock in which osseous neoplasia occurs with considerable frequency (53·3 per cent). In the liability to this form of neoplasia there is a pronounced sex difference; the incidence being 77·3 per cent in the females and 29·6 per cent in the males. The mean age of onset is 15·3 months in females and 17·7 months in the males. The primary tumours are often multiple, vary greatly in character and undergo metastasis. To discover whether oestrone might take part in the osseous neoplasia, Pybus & Miller made implants of 5 mg. of oestrone subcutaneously in males when they were between 3 and 4 weeks old. In less than 3 months bone tumours had appeared in three of these mice. Since that paper was published Pybus & Miller have investigated further the influence of oestrogen on the production of the tumours and they say (1941) that several lesions, comprising bone cysts, osteoclastomata, sarcomas and changes like those of Paget's disease, can be produced in mice by oestrone: the degree of reaction depending on the dose, duration and method of application and on the strain of the mice.

The distribution of bone tumours in their mice (Table 181) appears to correspond with the graded effects of oestrogen on the skeleton as noted by Zondek and by Pfeiffer & Gardner; among the long bones the femur was the most frequent site of neoplasia.

TABLE 181. The distribution of spontaneous tumours of bone in mice
(Pybus & Miller, 1940)

	Number
Skull and jaws	55
Sternum and ribs	64
Spine including tail	135
Upper extremity including shoulder girdle	17
Lower extremity including pelvis	324
Femur	174
Tibia	129
Humerus	15

When considering the effects of oestrogen on bone it may be well to recall the disease known to clinicians as hyperostosis frontalis interna. This disease, which affects the bones of the skull and especially the frontal bones and the anterior and middle fossae of the skull, is usually first noticed at about the menopause and 98 per cent of all recorded examples are said to have occurred in women (Andrews, 1942).

THE SPECIAL EFFECTS OF OESTROGEN ON THE BONY PELVIS

(d) *General conformation of the pelvis.* To study the influence of gonads on the plumage and various other external manifestations of sex in birds Yarrell (1827) removed the ovary from young chicks. Among other results of this procedure he noticed a change in the skeleton. The spayed fowls when fully grown showed a peculiar shape of the lower part of the back from the want of the enlargement of the bones, observed in all normal females, by which a breadth of pelvis sufficient to allow a safe passage to the egg is attained.

The difference in shape of the male and female pelvis in man and the great importance of the subject in connection with childbearing had long been known, and in retrospect it seems curious that Yarrell's experiments on birds were not followed by recorded observations on mammals. The writer is not aware of any such observations having been made until the last few years.

Gardner (1936) observed in adult mice a difference of form between the male and the female pelvis. Until the age of about 30 days the pelves in the two sexes appear alike, but after this date the pelvic bones of the female become altered, so that in adult life there is a pronounced sexual dimorphism. This transformation to the adult female form, Gardner says, can be induced in males by grafting ovaries into them or by administering oestrogens (oestrone, oestriol, equilin).

In mankind, as in the mouse, the shape of the feminine pelvis is determined by the ovaries. Morton (1942) in an extensive inquiry found no sexual dimorphism of the human pelvis during foetal life nor after birth until the approach of puberty.

(e) *Separation of the pubic bones during gestation.* In view of its consolidating effect on the rest of the skeleton the opposite effect which oestrin exerts on the bones, cartilage and ligamentous tissue of the pubic portion of the pelvis seems remarkable. It has long been known that parturition in the normal guinea-pig is facilitated by a disappearance of the interosseous cartilage of the symphysis and a wide separation of the pubic bones, which at the time of delivery are united merely by a thin ligamentous band. Hisaw (1925) made a saline extract of desiccated sow's ovary and gave the material once a week by intraperitoneal injection to male and female pocket gophers (*Geomys bursarius*). It may be explained that immature pocket gophers, both male and female, have pubic bones united to form a symphysis. In the female at the breeding season, before copulation, resorption of the bones forming the symphysis occurs and continues during pregnancy. Such resorption does not naturally happen in males; but in both males and females treated with ovarian extract by Hisaw resorption of the symphysis took place. This curious change, which normally accompanies adolescence in the untreated female, could be prevented, he found, by means of testicular grafts. Courrier, Kehl & Raynaud (1929) removed the ovaries from guinea-pigs during the second half of pregnancy—a procedure which does not prevent the continuance of gestation to full-time. In spite of the absence of ovaries the pubic bones separated as in normal pregnancy, a phenomenon which can be explained perhaps by a supply of oestrogen from the placenta. Courrier (1930a) showed that the separation of the pubic bones in the female guinea-pig could be maintained after pregnancy by injections of oestrin, and Brouha (1933), Tausk (1933) and De Fremery, Lucks & Tausk (1932) all found that separation of the symphysis could be caused in the non-pregnant female guinea-pig by injections of oestrin. Pencharz & Lyons (1934) made the interesting observation that if hypophysectomy were done on a guinea-pig at the 40th day of pregnancy, gestation continued to full term, and in spite of the ovarian atrophy consequent on the operation, separation of the pubic symphysis took place. Burrows (1935 f) made applications of oestrone (0·01 per cent) or equilin (0·1 per cent) twice weekly to the skin of castrated and non-castrated male and normal female mice. In both sexes the interosseous cartilage disappeared and was replaced by a thin fibrous band. A gradually increasing separation of the pubic bones took place if the treatment with oestrogen was continued. A thinning of the pubic and ischial rami accompanied these changes; in advanced cases the symphysial ends of these bones disappeared. Normally when mice litter for the first time there is a mere loosening of the symphysis without much separation, the changes being apparently irreversible and increasing in extent with each successive pregnancy. The condition may be relied upon, apparently, to determine whether a mouse has or has not previously littered. Gardner (1935 a), also, has recorded separation of the symphysis with absorption of the adjoining portions of the pubic bones in male mice after the administration of oestrin in subcutaneous doses of 500 i.u. given weekly from the age of 28 days. These changes in the symphysis under the influence of oestrogen are not prevented by hypophysectomy (Gomez, Turner, Gardner & Hill, 1937).

The wide separation of the pubic bones which precedes delivery in the guinea-pig and some other animals is represented in other species, including man, merely by a softening of the ligaments of the symphysis pubis and a lessening of its rigidity.

The effect of oestrogen on the concentration of calcium in the blood and the occurrence of renal calculus will be discussed in the next chapter.

Adipose Tissue

There seems little doubt that the characteristic *distribution* of fat in the normal female is a result of oestrogenic activity. The effect, if any, of oestrogen on the *total quantity* of fat in the body does not seem to have been exactly defined. We know that androgens tend to reduce the deposit of lipoid in the tissues. Perhaps the excess of fat in the female is the consequence of an inhibition of this action of androgens, or it may be that oestrogen in normal amount encourages the storage of fat. However, there is no doubt that the normal female has a larger proportion of fat in her composition than has the normal male (see Table 121). Wilson & Morris (1932) determined the total fat content in the bodies of male and female angora rabbits which had been kept under identical conditions. In rabbits 11 months old the percentage of fat in the males was 1·73 and in the females 5·78. In rabbits 24 months old the percentages were 3·30 in males and 9·12 in females.

Voluntary Muscle

Apparently oestrogens are not responsible for the lesser development of voluntary muscle in the female sex, for Papanicolaou & Falk (1938), in experiments which are referred to in the chapter on androgens, found that oestrogens did not prevent testosterone from inducing an increased development of muscle in castrated guinea-pigs. It appears therefore that the relative weakness of the female sex is due to a lack of androgen rather than to a weakening effect of oestrogen.

The Skin

Regional effects. There are areas of the body surface where the specific effects of oestrogen are well recognized. These areas are not the same for all species; the reactions include those cutaneous peculiarities by which the two sexes can be distinguished, for example the hen-feathering of birds (pp. 101, 114, 270). Apart from these widespread, though regional, responses to oestrogen there are certain local responses which occur in most, and perhaps in all, mammals, namely a thickening of the epidermis of the nipples (p. 327) and of the vagina (p. 302).

General effects. Apart from the special effects to which reference has just been made, oestrogens seem to exert a general inhibitory action on the growth of epidermis. It seems a fact that, in many species at least, the male possesses a thicker hide than the female. Wilson & Morris (1932) compared the pelts of male and female angora rabbits which had been kept on the same food and were of the same age when killed. When 11 months old the average weight in ounces of the pelt in males and females was 9·08 : 7·58; in rabbits at 24 months the ratio was 8·92 : 8·25.

Gardner & De Vita (1940) gave oestradiol benzoate in weekly doses ranging from 133·3γ to 333·3γ to five male and five female fox terriers. Under this treatment the hair failed to regrow after shaving or clipping. The arrest of hair growth continued if testosterone propionate in weekly doses of 5 or 10 mg. were given also. Mulligan (1943) performed a similar experiment on dogs, using stilboestrol, with the same results, i.e. the hair did not become restored after clipping; moreover it became thinned over the abdomen, perineum and around the base of the tail. Von Wattenwyl (1941) used spayed guinea-pigs, into which he implanted tablets of oestradiol. The daily absorption from these varied from 26γ to 185γ. The treatment caused alopecia 100–150 days after the implantation.

Emmens (1942) studied the growth of hair in rats and concluded that in normal males the hair grows faster than in females, whereas in castrated males it grows at the same pace; he noticed further that by giving oestrogen the regrowth of hair in epilated regions of the skin was retarded. Hooker & Pfeiffer (1943) found that 83γ of oestradiol benzoate given subcutaneously twice a week to rats caused some loss of hair, retarded the regrowth of hair after shaving and caused a diminution in the size of the sebaceous glands. These effects could be prevented by giving testosterone.

Sarcomata induced by Oestrogen

Tumours of bone have been discussed (p. 373), but they are not the only kind of sarcoma caused by oestrogen. Gardner, Smith, Strong & Allen (1936 *a, b*) reported the occurrence of spindle-celled sarcomata at the site of repeated injections of oestrogen, in seven among 126 mice so treated; and Suntzeff, Burns, Moskop & Loeb (1936) recorded a similar experience. The writer has had several instances of this reaction in rats which had been given subcutaneous injections of oestrone dissolved in lard or sesame oil.

Chapter XX. *The Actions of Oestrogen on Organs other than those Considered in earlier Chapters*

Liver. Pancreas. Blood and vascular system. Adrenals. Excretion of water and electrolytes. Thymus.

The Liver

(a) *Weight*. Korenchevsky and his colleagues (Korenchevsky & Dennison, 1934; Korenchevsky, Hall & Ross, 1939) reported a decrease in the weight of the castrated rat's liver under the influence of oestrogen; this result was less obvious in non-castrated rats. Selye (1940b) noticed a reduction in the weight of the liver in rats which had been given oestradiol. Such an effect of oestrogen does not appear to be a pronounced or constant reaction. Griffiths, Marks & Young (1941) noted an increase in the weight of the liver in rats which had been treated with oestrogen.

(b) *Histological changes in the bile ducts* have been observed by Gardner, Allen & Smith (1941) in mice after the administration of oestrogen. The compounds used were oestradiol dipropionate or benzoate (16·6 to 50γ weekly), oestrone (250γ weekly) and stilboestrol (250γ weekly). This treatment caused the bile ducts to become thickened, rigid, white and somewhat nodular. The main duct was less affected towards the duodenal end. The cystic duct was thickened up to the neck of the gall-bladder. Microscopically the enlarged ducts showed an increase of the epithelial folds and of the glands, which sometimes reached as far as the serosa. The epithelium was hyperplastic and the ducts increased in extent.

Battaceano & Vasiliu (1936) say that when given to dogs oestrone causes at first a diminution in the volume of the liver which is followed by an increase. In a dog with a biliary fistula doses ranging from 1,000 to 5,000 r.u. arrested the flow of bile.

(c) *Hepatoma and haemangioma*. Miller & Pybus (1941) have noted a high incidence of hepatoma and haemangioma among mice of the CBA strain after their continued subjection to oestrogen. These tumours appeared in 44·4 per cent of the normal males in a group of mice which had received subcutaneous injections of 300 i.u. of oestrone in olive oil once a week. In castrated males treated in the same way hepatomas appeared in 6·5 per cent, but none of these tumours developed in normal females under similar conditions. Gorer (1940) states that among his CBA mice which lived 11 months or longer hepatomata appeared in 18 of 35 males (51 per cent) and in 10 of 43 females (23 per cent).

(d) *Liver glycogen*. Conflicting observations have been made concerning the effect of oestrogens on the glycogen content of the liver; a discussion of the subject is therefore hardly worth while. References to the subject are as follows: Brunelli (1935), Gilder & Phillips (1939), Janes & Nelson (1940), Griffiths, Marks & Young (1941), MacBryde, Castrodale, Helwig & Bierbaum (1942).

(e) *Blood sugar*. Several workers have recorded an increase of the blood sugar as a result of giving oestrogens. Riddle & Honeywell (1923) observed an increase of about 20 per cent in the blood sugar of pigeons at each ovulation period. The increase, they say, begins about 108 hours before ovulation and is maintained until the second ovulation 44 hours later and reaches the normal about 108 hours afterward. (It will be remembered that the pigeon lays only two eggs at each nesting.) Zunz & La Barre (1939) found that 300 or 400 i.u. of oestradiol benzoate given intramuscularly to dogs caused a rise in the blood sugar, but this effect was not produced if the adrenal blood vessels had been tied previously. Janes & Nelson (1940) injected 0·05 mg. of stilboestrol twice a day into rats weighing between 150 and 200 g. and estimated the blood sugar after 5, 10 and 20 days of this treatment. An increasing concentration of sugar in the blood was noted during this period of 3 weeks (Table 182).

TABLE 182.   The increase of liver glycogen and blood sugar in the rat under treatment with stilboestrol (Janes & Nelson, 1940)

Days of treatment	Glycogen in liver (mg. per 100 g.)	Sugar in blood (mg. per 100 g.)
No treatment	75	68
5	231	73
10	331	74
20	409	84

The Pancreas

Increased production of insulin

Goetsch, Cushing & Jacobson (1911) noticed that dogs with pituitary insufficiency withstand the effects of partial excision of the pancreas better than normal dogs. Some time later Houssay & Biasotti (1931) discovered that hypophysectomy reduces the glycosuria and prolongs the life of dogs after pancreatectomy, and they say that when deprived of the pituitary dogs become very susceptible to insulin coma, from which they recover if given injections of sugar. Barnes, Regan & Nelson (1933) found that, so far as the reactions just mentioned are concerned, effects comparable with those of hypophysectomy can be produced by giving oestrogen. Daily subcutaneous injections of 200 r.u. of oestrogen ('Amniotin') were given to four pancreatectomized female dogs, three of which received this treatment during 3 weeks before pancreatectomy; the other dog's pancreas was removed so as to cause copious glycosuria, and not till then were the daily injections of oestrogen begun. Glycosuria was checked by the injections and, though losing some weight, the dogs remained lively. At the end of 3 weeks the treatment with oestrogen was stopped and within 3 days the dogs suffered from severe glycosuria. Nelson & Overholser (1934) made the same sort of experiment on rhesus monkeys. Daily doses of 100 r.u. of oestrin were given for a fortnight, at the end of which the pancreas was excised, the injections of oestrin being continued for 6 days afterward; during this time no appreciable glycosuria occurred. Cessation of the injections was followed by glycosuria, which disappeared under the daily administration of 200 r.u. of oestrin.

Vasquez-Lopez (1940) noticed, in mice which had been treated with oestrogen for several months, in every cell of all the islets of Langerhans a hypertrophic Golgi apparatus larger than ever seen by him in the normal mouse. He suggests that this histological change indicates a functional hyperactivity of the cells resulting from the influence of oestrogen on the pituitary. Fraenkel-Conrat,

Herring, Simpson & Evans (1941) support this suggestion. They implanted
10 mg. of oestradiol into normal rats at weekly intervals. The rats were killed at
the end of 18 days and their pituitaries were implanted into hypophysectomized
rats, each recipient receiving nine pituitaries in 19 days. These rats were killed on
the 21st day. Control rats were hypophysectomized and treated in the same way,
receiving pituitaries of untreated rats. The insulin content of the pancreas was
then estimated for the donors and recipients of the pituitaries. In the pituitary
donors the administration of oestradiol dipropionate had caused a rise of the
pancreatic content of insulin by 54 per cent in one batch and 90 per cent in
another. The implantation of pituitaries from these oestrogen-treated rats into
hypophysectomized rats caused a rise in the pancreatic insulin of the recipients
of 39 per cent as compared with hypophysectomized controls which had received
pituitaries from untreated rats. Griffiths & Young (1940) implanted tablets of
oestrone or stilboestrol subcutaneously in Wistar rats and found that a rise of the
insulin content of the pancreas was thereby caused. One month after the implantation of 12 mg. of oestrone the pancreatic content of insulin amounted to
1·12 units per 100 g. of bodyweight, and a comparable result was obtained with
stilboestrol.

Blood and Vascular System

(a) *Lipaemia induced by oestrogen.* Lorenz, Chaikoff & Entenman (1938)
found a great rise in the fat content of the blood of fowls within 48 hours of an
intramuscular injection of 3,000 r.u. of oestrone, as shown below.

	Blood lipids 48 hours after the injection of oestrone (mg. per 100 c.c. of blood)
Untreated fowls	356–466
Oestrone-treated cocks	650–1087
Oestrone-treated hens	575–1961

Entenman, Lorenz & Chaikoff (1938) obtained the same kind of effect by injecting a pituitary gonadotrophin into immature hens. They say that all the blood
lipids were increased, including cholesterol, phospholipid and free and combined
fatty acids. Even more striking results were obtained by Zondek & Marx (1939)
by giving a cock 4 mg. of diethylstilboestrol on each of six consecutive days (see
Table 183). H. G. Loeb (1942) has noticed a similar, but less pronounced, effect
of oestrogen in rats. He kept male rats on a diet which was rich in fat and treated
them with oestradiol for a period of 24 days, at the end of which their blood
showed between 531 and 566 mg. of lipoid per 100 c.c. as compared with 351 mg.
per 100 c.c. in the controls fed in the same way but not given oestrogen.

(b) *The effect of oestrogen on the concentration of calcium in the blood.* In connection with the changes already mentioned as occurring in bone under the influence of oestrogen, it may be noted that oestrogens cause a rise of the blood
calcium. This was first observed in birds, and though in them it occurs in a pronounced form and probably assists the formation of egg-shell, it happens also in
mammals.

Riddle & Reinhart (1926) noticed that each ovulation in the pigeon is accompanied by a large rise in the blood calcium. The rise begins about 108 hours before ovulation and 123 hours before the beginning of the formation of the shell.

At the time of ovulation the concentration of calcium in the blood may rise to 19 mg. per 100 c.c. of serum. During sexual quiescence no difference was found in the blood content of calcium in the two sexes. As the result of further studies Riddle & Dotti (1936) state that gonadotrophin causes an increase of serum calcium in normal or hypophysectomized pigeons but not after gonadectomy. The reaction is obtained more quickly in males than in females, and is not caused by corticosterone or testosterone. Oestrone, they say, caused an increase of blood calcium in normal, castrated or hypophysectomized pigeons and rats, and in normal dogs, fowls and doves, but the reaction was not obtained in rabbits. Progesterone, they found, had the same effect in a less degree.

Pfeiffer & Gardner (1938) gave daily doses of 1,000 i.u. of oestradiol benzoate to normal male and non-laying female pigeons. Under this treatment the concentration of calcium in the birds' sera rose from between 8 and 9 mg. per 100 c.c. to between 13·12 and 21·65 mg.

Zondek & Marx (1939) found that the administration of oestrogen to fowls caused an increase of both calcium and lipoid in the blood. A cock was given 4 mg. of diethylstilboestrol on six consecutive days beginning on 8 November. The effects of this treatment on the concentration of calcium and lipoid in the blood is shown in Table 183.

TABLE 183.   The effect of oestrogen on the concentration in the cock's blood of calcium and lipoid (Zondek & Marx, 1939)

Date	Total dose of diethylstilboestrol (mg.)	Blood calcium (mg. per 100 c.c.)	Blood fat (mg. per 100 c.c.)
28 Nov.	0	11·2	125
1 Dec.	12	30·0	3,280
4 Dec.	24	41·4	5,438

Future experiments, no doubt, will decide what part, if any, the parathyroids take in the mobilization of calcium under the influence of oestrogen. Grafflin (1942) examined the parathyroids of the Virginian deer (*Odocoileus virginianus borealis*) throughout the year and did not detect any changes in the parathyroids corresponding with the shedding or renewal of antlers, nor was any seasonal variation observed in the concentration of calcium, phosphate or phosphatase in the blood or cerebrospinal fluid.

(c) *Urinary calculi.* A mobilization of calcium by oestrogen in some instances perhaps may cause urinary calculi. Burns & Schenken (1939) gave oestradiol dibenzoate in weekly doses which ranged from 100 to 1,500 r.u. to mice of the C₃H strain. The mice were fed on Purina fox chow with lettuce once a week. Among the mice so treated calculi were found with considerable frequency in the bladders, ureters and kidneys (Table 184).

TABLE 184.   The formation of urinary calculi in mice under treatment with oestrogen (Burns & Schenken, 1939)

	Number of mice	Number of mice with calculi	Percentage
Untreated controls	96	4	4·1
Treated	151	50	33·1

In a later paper Schenken, Burns & McCord (1942) reported a difference between the two sexes in their liability to renal calculus under the influence of oestrogen. They submitted a number of male and female mice of the C₃H strain to continued treatment with oestradiol benzoate, and the results showed a striking predominance of males among those animals which eventually had renal calculi. The relative numbers are given below.

	Number	Percentage with renal calculi
Males	169	43·7
Females	30	3·4

(d) *Effect of oestrogen on the excretion of citric acid.* Shorr and his collaborators (Shorr, Bernheim and Taussky, 1942; Shorr, Almy, Sloan, Taussky and Toscani, 1942) say that citric acid is a regular constituent of human urine in which its concentration is increased by alkalosis however caused, by the injection of citric acid or its precursors, or by giving oestrogen. In daily assays made of the urine of five healthy young women during six complete menstrual cycles the lowest concentration always coincided with menstruation and the highest levels occurred at the middle of the cycle. The administration of oestradiol benzoate (two cases) was followed by significant rises in the excretion of citrate, whereas testosterone given to a male (one case) was followed by a reduction. Testosterone propionate given to a girl with amenorrhoea also led to a diminished excretion of citric acid. The output of citric acid they say is regulated by the concentration of calcium and enhances its solubility. In two cases of recurrent renal calculi they found a persistently low concentration of citric acid in the urine.

Dickens (1941) compared the citric acid content of various tissues and found a relatively high concentration in bone. As much as 70 per cent of the total citric acid in the body may be, he says, in the compact tissue of bone.

(e) *Blood pressure.* Clinical experience suggests that oestrogens might tend to lower the blood pressure. A raised blood pressure is apt to accompany virilism, and severe hyperpiesis is more frequent in men than in women, and in the latter is thought to be associated with the menopause. Liebhart (1934) made tests on more than 300 individuals, including men, girls, and normal, spayed and postmenopausal women, giving doses of oestrone ranging from 100 to 5,000 m.u. and determining the blood pressure by a Riva Rocci apparatus. He concluded that oestrone has a capacity for lowering the blood pressure, and that this action is more pronounced in women than in men, and in adult women than in girls before puberty. In some instances a fall of the systolic pressure equivalent to 30 mm. Hg. was obtained within half an hour of giving oestrone. Guirdham (1941) has found that oestradiol lowers the blood pressure in cases of menopausal hyperpiesis, arterial sclerosis and renal disease, as well as in normal conditions.

Reynolds (1941) examined the effects of oestrogen on the blood vessels of the ear in the spayed rabbit, and also, by means of a plethysmograph, on the blood vessels of the human finger. The usual result of both tests was a dilatation of the small blood vessels which followed the injection of oestradiol within a few minutes and persisted for at least 2 hours. The human tests were performed on twenty men and on twenty-five women who had passed the menopause. Other experiments bearing on this matter have been done by McGrath (1935) and Thomas (1940).

A vascular phenomenon which clinical experience has shown to be controlled by oestrogen is the recurrent flushing experienced by women at the beginning of the menopause. This symptom is thought to be the result of an increased supply of FRH consequent on the reduced ovarian activity at this period. As shown in earlier pages, oestrogens inhibit the supply of FRH from the pituitary.

(*f*) *An effect of oestrogens on the permeability of the capillary blood vessels* in certain regions of the body has been reported by Hechter, Krohn & Harris (1942), who treated rats with oestradiol, oestrone or oestriol, and after an interval gave each of them an intravenous injection of 1 c.c. of a 1 per cent solution of trypan blue for 100 g. of bodyweight. In every instance there followed an abnormally large concentration of dye in the uterus and vagina. This was not the consequence of increased affinity of the genital tissues for the dye, because such tissues when minced and tested *in vitro* took up no more dye than was accumulated by tissues from other parts of the body (see also Goldmann, 1909).

The writer believes, for reasons which he has given elsewhere (Burrows, 1932), that it would be safer to describe a result like that just mentioned as an increased 'permeation' of the capillaries rather than an increased 'permeability'. The fact of an increased passage of dye through the capillary walls does not necessarily indicate any change in their permeability, for it is conceivable that the enhanced transit of dye is entirely due to a change in the forces, electrical or otherwise, which cause its transportation.

(*g*) *Action of oestrogen on the cellular constituents of the blood.* It seems probable, if the formation of *erythrocytes* is to some extent regulated by a pituitary haemapoietic hormone as Flaks, Himmel & Zotnik (1938) believe, that oestrogens might, in view of their depressing activity on the pituitary, hinder to some extent the supply of erythrocytes. Several workers have reported that red cells are normally more numerous in the male than in the female. Such observations have been made on many different species. Steinglass, Gordon & Charipper (1941) have investigated the matter, using an inbred strain of rats. They found that after gonadectomy both the red cells and the haemoglobin increased in the females and fell in the males. In spayed females oestrogen reduced the erythrocyte count from 8·9 to 6·7 millions per cmm. and in castrated males the same treatment caused hypoplasia of the marrow.

MacBryde, Castrodale, Helwig & Bierbaum (1942) found that in dogs daily doses of 10 mg. of oestrone, 1·66 mg. of oestradiol benzoate or 5 mg. of diethylstilboestrol caused changes in the marrow with a diminution of erythrocytes and haemoglobin in the circulating blood, and Tyslowitz & Dingemanse (1941) obtained similar results in dogs with daily doses of 1 to 5 mg. of oestrone.

Plum (1942) found that in women both the reticulocyte count and the amount of erythrocyte-ripening substances in the blood are at their maximum during menstruation and minimum in the mid-menstrual period. The rise in both is rapid and begins about 1 week before menstruation, falling rapidly after this phase. The observation accords with the supposition that oestrogen checks the formation of the red blood cells.

There is little doubt that oestrogens affect the production of *white cells*. Tyslowitz ((1939); see also Tyslowitz & Dingemanse, 1941) found that daily injections of 5 mg. of diethylstilboestrol to dogs brought about an agranulocytic anaemia within the next 25 to 50 days. In untreated mice various abnormalities of the white blood cells are frequently seen, and they comprise an unbroken series from a general leucaemia to lymphosarcoma. These conditions are more frequent in female than in male mice. Mercier (1938), basing his observations on 465 mice of a single strain, recorded lymphosarcoma in 64·9 per cent among the 254 females, and in 34·1 per cent among the 211 males. Others have called attention to the special proclivity of female mice to leucaemia and lymphosarcoma (Gorer, 1940; Pybus & Miller, 1941; Miller & Pybus, 1942). Cole & Furth (1941) point out that leucaemia occurs not only more frequently in female mice than in males but appears earlier in the females. It seems probable that this tendency to the development of leucocytic neoplasia among females is in part at least dependent on the action of oestrogen. Lacassagne (1936d) noticed the occurrence of lymphosarcoma and leucaemia among mice which had been under treatment with oestrogen.

Gardner, Kirschbaum & Strong (1940) noted that among 149 C$_3$H mice which had received oestrogen from the ages of 3 to 140 days until death, lymphoid tumours, some accompanied by leucaemia, occurred in 22 (15·4 per cent), whereas no such tumours appeared among 117 mice of the same strain which had not been given oestrogen.

The occurrence of these lymphoid tumours appears to be largely influenced by the gonads. Marine & Rosen (1940) observed that lymphoid tumours were much more frequent in castrated than non-castrated white leghorn fowls, and Miller & Pybus (1942) found that gonadectomy, whether in the male or female, was followed by a greatly increased incidence of lymphoid tumours in mice of the Edinburgh strain which had been submitted from an early age to weekly doses of 300 i.u. of oestrone.

At present no connection has been revealed between oestrogens and leucaemia in man. Cooke (1942) states that among 1,500 cases occurring in children 59·3 per cent of the total occurred in boys, and in all cases among adults males were affected in between 60 and 79 per cent.

Ledingham (1940) has observed a special effect of oestrogen on the Foà-Kurloff cells in the guinea-pig's blood. These cells, which represent a type of leucocyte hitherto found only in the guinea-pig, are seen, Ledingham says, very scantily in the blood and spleen at birth and during infancy, but rise in number between the 30th and 40th day of postnatal life. The cells are more numerous in pregnancy and also in non-pregnant guinea-pigs which are under treatment with oestrogen. A single subcutaneous injection of 60γ of oestradiol dipropionate on the 52nd day of life was followed by a rise of the Foà-Kurloff blood count from 20 per 1,000 leucocytes to 210 per 1,000, 13 days later. A similar dose given to males or females in the first few postnatal days caused a sudden and great increase in the Foà-Kurloff count, and this effect was produced also in foetuses by treating the mother with oestrogen. Testosterone did not produce these effects.

(h) *The formation of antibodies.* A few experiments have been done to learn whether oestrogens have any effect on the resistance of an individual to infection. Von Haam & Rosenfeld (1942 a, b, c) immunized rabbits against Type I Pneumococcus by giving increasing doses of this organism intravenously on four consecutive days, and repeating such a course in each of the next 4 weeks. They found that if throughout this process of immunization injections of oestrone (totals of 22 and 38 mg.) were given the concentration of specific agglutinins and antibodies was much greater than in rabbits treated without oestrone but otherwise in the same way. They obtained a similar result in mice.

The Adrenals

(a) *Increase in weight of the adrenals.* Laqueur (1927) noted that the adrenals of young rats which had been given injections of oestrin were larger than those of control, untreated rats of the same age. Andersen & Kennedy (1932), basing their observations on 127 unmated female rats, found that during oestrus the adrenals were increased in weight owing to enlargement of the cortex (Table 185). The cells of the fasciculate zone, and the cell nests near the medulla, they say, are enlarged during oestrus and contain an increase of lipoid; the glomerular zone is narrowed.

TABLE 185. Changes in weight of the rat's adrenal during the oestral cycle (Andersen & Kennedy, 1932)

Stage of cycle	Number of rats	Mean weight of adrenals (mg. per 100 g. of bodyweight)
Prooestrus	8	25·0
Oestrus	32	26·3
Postoestrus	29	24·0
Dioestrus	58	23·7

Leiby (1933) gave daily doses of 4 mg. of oestriol to adult spayed mice for 6 days. At the end of this period the adrenals were found to be considerably enlarged. Andersen (1935) spayed rats at about 80 days of age and 6 weeks later gave them 5 r.u. of oestrin 3 times a day for 2 days. She killed the rats at intervals of 24 to 120 hours after the first injection, and found an increased weight of the adrenals with an increase of lipoid in the cortical cells accompanied by congestion of the reticular zone. These changes were present even in the rats killed 24 hours after the first injection of oestrin.

According to Bourne & Zuckerman (1941) the increased volume of the adrenal found in normal rats during oestrus and in spayed rats which have been given injections of oestrone is chiefly due to changes in the fasciculate zone, and is the consequence of an increase in size without an increase in the number of the cells.

Korenchevsky & Dennison (1934, 1935) have shown that injections of oestrone cause enlargement of the adrenals in rats, and Ellison & Burch (1936) obtained similar results by giving oestrone, oestradiol or oestriol to rats whether castrated or not. The effects on the adrenal depended on the presence of the pituitary. Selye, Collip & Thomson (1935) and Selye & Collip (1936) found that daily doses of 100γ of oestrin did not cause enlargement of the adrenals in immature rats, nor did the continued giving of 200γ of oestrin daily prevent involution of the adrenal after hypophysectomy. Further, they state that with adult rats the adrenals enlarged by the influence of oestrin contain but little lipoid, whereas abundant lipoid is present in the adrenal cortex within 48 hours of hypophysectomy even if the doses of oestrin are continued. They believe that enlargement of the adrenal in female rats treated with oestrin is an indirect action mediated by the pituitary, and they suggest that the presence of corpora lutea may be necessary for the result. Some of their findings are shown in Table 186.

TABLE 186. The effect of oestrin on the female rat's adrenal as influenced by the presence or absence of the pituitary (Selye & Collip, 1936)

Number of rats	Hypophy- sectomized	Hormones given	Mean weight of adrenals (mg.)	Mean weight of ovaries (mg.)
6	+	Pituitary extract	69	41
6	+	Pituitary extract *plus* Oestrin	71	91
6	−	Pituitary extract	80	78
6	−	Pituitary extract *plus* Oestrin	100	155

Noble (1939) reports that the synthetic oestrogen, triphenylethylene, like the natural oestrogens, does not cause enlargement of the adrenal in hypophysecto-mized rats.

(b) *The effect of oestrogen on the x zone of the adrenal.* The x zone was first described by Elliott & Armour (1911) and by Thomas (1911) as occurring in man, and since then has been investigated in mice by many inquirers. The x zone is present in the adrenal cortex of infantile mice of both sexes and is composed apparently of the cells of the zona reticularis and the innermost parts of the zona fasciculata. These cells in early life are especially numerous and cause a relative increase in thickness of the cortex. The cells forming the x zone are smaller and stain more densely and evenly with eosin than those of the outer part of the zona fasciculata, the cells of which contain a varying amount of material which is not stained by eosin (p. 233).

There seems little doubt that the portion of adrenal cortex adjacent to the medulla is concerned in some way with sexual functions, and a reference may be made here to the observation of Broster & Vines (1933) that the inner part of the adrenal cortex displays a special staining reaction in cases of virilism in women.

Howard-Miller (1927) states that in male mice the x zone·ceases to grow in the 4th week after birth and has disappeared by the 39th day. The cells do not become vacuolated, she says, during this disappearance. Whitehead (1935 b) has made a similar observation. In female mice the x zone disappears rapidly if pregnancy supervenes, otherwise it may persist for a considerable period and degeneration when it occurs may be accompanied by vacuolation and coalescence of cells (Howard-Miller, 1927; Whitehead, 1932–33, 1935 b). Altenburger (1924) observed in male mice castrated at 3 to 8 weeks of age, and killed 3 months later, a pronounced increase of the adrenal cortex with qualitative alterations involving the zona reticularis, and Deanesly (1928) has found that in male mice castration before maturity may cause the x zone to reappear.

Martin (1930) studied in mice the effects of prolonged treatment with oestrone and observed that a total degeneration of the x zone is caused in im-mature castrated males and in normal and spayed females. In immature non-castrated males oestrone caused a persistence of the x zone and in non-castrated adult males a reappearance of the x zone. It seems that castration and the administration of oestrone have a similar influence on the presence of an x zone in the male mouse. However, the effects of oestrone on the x zone depend largely upon the size of the doses, the length of time during which they are given and perhaps on the age of the animal. Waring (1942) gave 0·01 mg. of oestrone during 14 days to spayed and non-spayed female mice which were 40 days old when the experiment began. These doses had no effect on the x zone, which remained intact.

(c) *Degeneration of the adrenal cortex.*   In old untreated female mice there is often to be seen a peculiar degenerative change in the adrenals (Burrows, 1936b). The chief feature is an accumulation of lipoid-like material in the cells of the zona reticularis and the innermost cells of the fasciculate zone, that is to say in the same situation as the x zone. As the condition advances the accumulation of lipoid increases and eventually adjacent cells coalesce to form rounded masses lying in the vascular stroma between the cortex and medulla. These masses are composed of (1) droplets of lipoid-like material which are not stained by the usual dyes, (2) a small but variable amount of cytoplasm stained pink by eosin and (3) scattered, often pyknotic, nuclei stained with haematoxylin. The vascular stroma occasionally appears to be increased in amount and a pronounced hyperaemia may be present. An identical change can be induced in the adrenals of male mice, whether castrated or not, by giving continued doses of oestrogen. The peculiar lipoid degeneration described calls to mind that which has been described by the author as occurring in the interstitial glandular cells of the testis under the influence of oestrogen. This form of degeneration in the adrenal cortex may often be seen in untreated old female mice and occasionally to a slight extent in untreated old males too. It is not a feature peculiar to mice or animals bearing tumours; the change has been recorded in old rabbits by Whitehead (1935b), Greene & Saxton (1938) and Burrows (1940), who has observed it also in old untreated female guinea-pigs and rabbits.

Cramer & Horning (1936, 1937a, 1939a, b) have recorded a similar condition in the adrenals of mice under the name of 'brown degeneration'. They regard the change as arising in the medullary cells (1936) or in the medulla and cortex (1937a, 1939a, b), and as having an association with the development of mammary cancer.

The writer feels bound to say that he has not seen any instance in which the medulla has shared in the degeneration now under consideration, and in this respect his observations differ from those of Cramer & Horning.

Excretion of Water and Electrolytes

(a) *Local and general cutaneous oedema caused by oestrogen.*   Vulval oedema resulting from the action of oestrogen has been mentioned (p. 305). In some species an oedema, commencing in the perineum and spreading from this region to the surrounding skin, has been noticed in both sexes as a consequence of administering oestrogen. Courrier & Gros (1935) injected variable amounts of oestrone into three immature Algerian macaques (*Macacus ecaudatus*) weighing between 1,800 and 3,000 g. After 10 days of this treatment there was a considerable perianal swelling which at the end of 25 days had attained the size of a large mandarin orange. The skin was red and oedematous, and the swelling involved the scrotum and penis in addition to the perianal region. Bachman, Collip & Selye (1935) made a more extended observation. They gave 200γ of oestriol thrice daily at first, and later 500γ once daily for periods extending from 12 to 110 days to monkeys (*Macacus rhesus*). Within the first 24 hours this treatment caused oedema of the anus and adjacent part of the base of the tail. In the next day or two the skin of the genital region began to redden and swell. At the end of 8 or 10 days the swelling had extended over a wide area, so that the skin over the pubes, inner sides of the thighs, perineum and proximal third of the tail was oedematous. In the area so affected pitting on pressure was elicited. With continued treatment the oedema subsided in the course of the next 10 to 40 days, but did not always quite disappear. Irrespectively of the foregoing results, when the treatment had continued for 20 to 25 days, a secondary colourless, cutaneous oedema of a different type and distribution was seen. Beginning along the course of the sartorius muscles in the thighs it appeared next in the lumbar region and gradually spread over the back and over all the hairy portion of the trunk until eventually the face and lower part of the abdomen were involved. This oedema persisted as long as the injections of oestriol were continued, and slowly regressed after the treatment had ceased. The oedema was limited to the skin and did not affect the subcutaneous tissues. Zuckerman & van Wagenen (1935) have made somewhat similar observations. They injected oestrone into a young rhesus monkey 6 days after birth and continued the injections for 43 days, during which the monkey received a total of 8,250 r.u. Very little change occurred during this time in the scrotum, but the skin over the buttocks and lower part of the back became thickened and swollen so that the hair follicles became unduly prominent (apparently a 'peau d'orange' effect). While the treatment was in progress the monkey was castrated, after which 'the scrotal response was much better than it had been during the first course of injections, when the testes were present'.

Ogston, Philpot & Zuckerman (1939) gave daily injections of 100 mg. of oestrone to female monkeys (*Macacus nemestrina* and *M. mulatta*). After varying periods of treatment the swollen sexual skin was removed; the exudate obtained from it was examined and found to contain an abnormal concentration of free chloride, and the suggestion is made that the oedema may have been caused by an increase of osmotic pressure caused by the excess of free chloride. In the exudate they detected material giving the reactions of mucoprotein. Aykroyd & Zuckerman (1938) say that in the monkey the swelling of the sexual skin is accompanied by enlargement of the connective tissue cells through their having taken up water.

Engle & Crafts (1939) found that 200γ of diethylstilboestrol given orally or by intramuscular injection to adult spayed rhesus monkeys caused oedema of the skin over the whole body.

When these effects of oestrogen are being considered it will be remembered that progestins also have an influence on the excretion of water (p. 437).

(b) *The effect of oestrogen on the excretion of electrolytes.* While considering the production of cutaneous oedema by oestrogen attention may be given to the fact that to some extent the gonadal hormones resemble the adrenal cortical hormones in the effect they have on the excretion of electrolytes. Thorn & Harrop (1937) found that a single dose of 5 mg. of oestradiol caused a pronounced and rather prolonged retention of sodium in a normal dog. Repeated injections in both male and female dogs did not prevent an ultimate return of the sodium excretion to normal. All the sex hormones investigated, they say, including oestradiol, oestrone, testosterone, testosterone propionate, progesterone and pregnanediol, have shown some sodium-retaining effect. They add that a single injection of 17 mg. of oestradiol, given to a patient with incompetent adrenals (Addison's disease) and maintained on a diet constant in fluid and mineral content, led to a retention of chloride, sodium and water with a gain in bodyweight and blood pressure. This effect lasted 72 hours. In normal dogs Thorn & Engel (1938) observed that oestrone, oestradiol and testosterone propionate all caused a decreased renal excretion of inorganic phosphorus and total nitrogen, and an increased output of potassium.

As a supplement to these findings it may be noted that, unlike progestin, oestrogens have little if any power to maintain life after adrenalectomy. Gaunt & Hays (1938) found that 0·1 or 0·2 mg. of oestrone given daily to two adrenalectomized ferrets, instead of prolonging life, appeared to be toxic.

The Thymus

It has long been known that the thymus involutes before adult life is reached. In the rat the gland, having attained its highest development, begins to shrink at about the 85th day of life, according to Hatai (1914). Pregnancy and lactation also, he says, are accompanied by a diminution in the size of the thymus. Chiodi (1940) observed that the rat's thymus becomes increased in size after gonadectomy, whether in the male or female, and that oestrogens (oestrone, oestradiol benzoate) cause atrophy of the thymic parenchyma of normal and gonadectomized rats of either sex. Ross & Korenchevsky (1941) reported similar results in the rat from gonadectomy. The effect of oestrone or oestradiol (0·014–0·024 mg. in a week) they describe as an atrophy of the lobules with a specific hyperplasia of the epithelial cystic structures which show a stratified epithelium.

Several investigators have noticed that involution of the thymus is readily caused by a variety of noxious agents; this fact does not seem to account for the effects attributed to oestrogen, for these effects follow the normal physiological production of oestrogen in the healthy body.

PART V. PROGESTINS

Chapter XXI. *Progestins*

Definition and Functions. Sources. Metabolism and Excretion. Development and Maintenance of Corpora Lutea.

THE progestins are active principles of the corpora lutea of the ovary. They are chemical compounds whose biological actions are chiefly concerned with conception, gestation, parturition and perhaps lactation. More than one substance

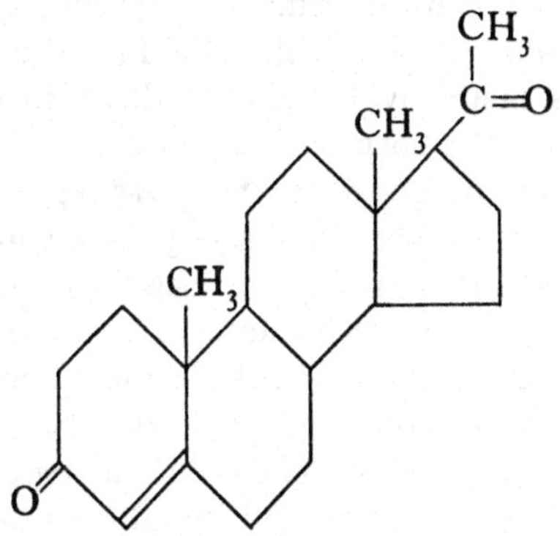

I. Progesterone

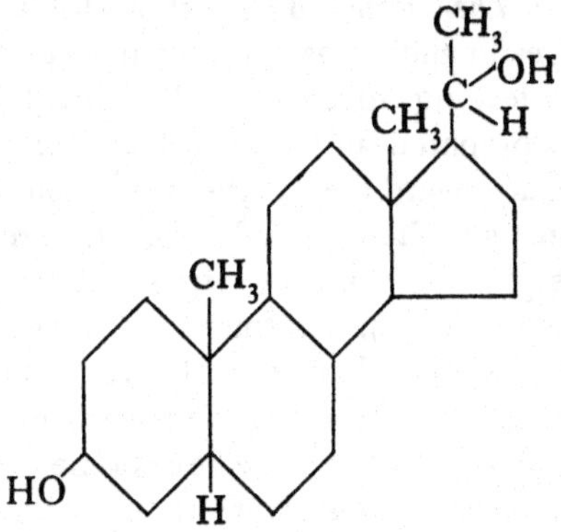

II. Pregnanediol

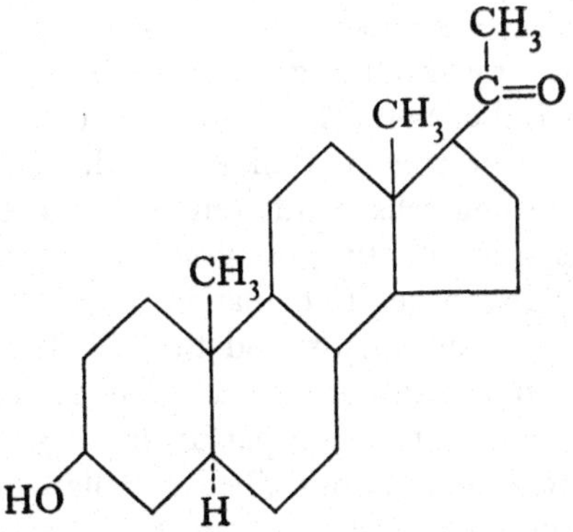

III. *epiallo* Pregnane-3-ol-20-one

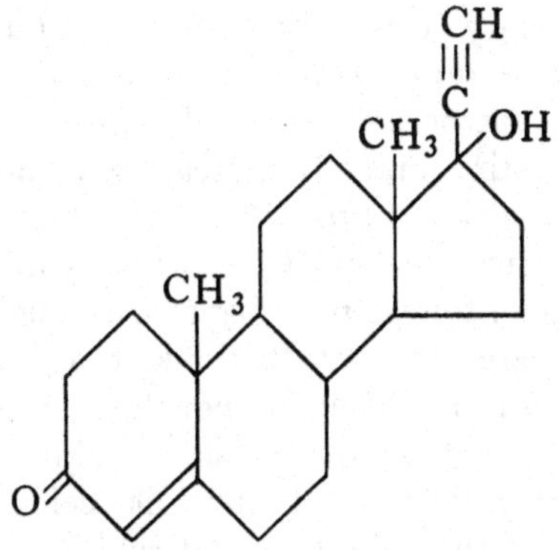

IV. Pregneninolone (Ethinyltesto-sterone, Ethisterone)

Fig. 7.

having the same kind of action as progesterone is known, and a generic name is therefore convenient. The author has used the term 'progestin' to denote chemical compounds having a biological action comparable with that of progesterone.* The structural formulae of progesterone and some allied hormones are shown in Fig. 7.

* 'Progestogen' has been suggested as an alternative generic name.

The molecular structure of progesterone closely resembles that of hormones formed by the testis and adrenal cortex; this resemblance extends to many of their biological actions (Fig. 1, pp. 98, 99).

The international unit. As with other types of gonadal hormone it has been found convenient to adopt an international standard unit of progestational potency as ascertained by biological assay, so that the progestational activity of materials whose hormonal content has not yet been identified can be compared and recorded. The international unit of progestational potency equals that of 1 mg. of progesterone when assayed in the immature rabbit.

Sources of Progestin

(*a*) *The ovaries.* The corpora lutea of the ovary are the chief natural source of progestin and it was from them that progesterone was first isolated and identified. The leading advance in this direction was made by Corner & Allen (1929), who prepared an effective luteal extract from the corpora lutea of sows. Not long afterward the isolation and identification of progesterone itself was reported almost simultaneously by four different groups of workers, namely Butenandt, Westphal & Hohlweg (1934), Allen & Wintersteiner (1934), Fels, Slotta & Ruschig (1934), Hartmann & Wettstein (1934). It seems uncertain whether any cells other than those derived from the granulosal portion of the follicle may participate in forming the corpus luteum and in elaborating progestin. Sandes (1903), who studied the development and fate of graafian follicles in a marsupial cat (*Dasyurus viverrinus*), thought that both the granulosal and thecal cells took part in forming the corpus luteum.

As to the hormonal functions of a newly developed corpus luteum, it is hardly necessary to remark that the output of oestrogen does not stop suddenly and become replaced by that of progestin. The transition is gradual and for a considerable period after ovulation both kinds of hormone are formed, the output of progestin gradually increasing while that of oestrogen declines. Perhaps both oestrogen and progestin are formed by the same cells at different stages of their existence. Allen (1941) says evidence is accumulating to show that a large graafian follicle may begin to change its secretion before ovulation.

There is reason to think, too, that the corpus luteum and the follicle from which it has arisen are not the only sources of ovarian progestin. Goodall (1920) found that the ovaries of a rabbit which had died during parturition consisted entirely of a mass of lutein-like cells, and the writer has described a similar condition of the ovaries in old rabbits. Although the ovaries in some of the latter instances consisted throughout of lutein-like cells and no corpora lutea could be seen, progestational conditions were present in the uterus, showing that progestin had been secreted (Burrows, 1940).

(*b*) *The placenta* is another source. Courrier, Kehl & Raynaud (1929) noticed that if a pregnant guinea-pig were deprived of her ovaries and then given 50 r.u. of oestrin in the course of 3 days, vaginal cornification did not occur as it would have done in a non-pregnant spayed guinea-pig so treated. They were aware that active corpora lutea inhibit this action of oestrin on the vagina, and therefore concluded that in the spayed pregnant guinea-pig the inhibitory agent must be produced elsewhere than the ovary, possibly in the placenta.

Lipschütz (1937c) made the same sort of observation; he removed the ovaries from pregnant guinea-pigs and found that vaginal mucification persisted for 4 weeks afterward. Mucification of the vagina being largely dependent on progestin it seemed that this hormone, during pregnancy, must have arisen in some organ other than the ovary.

Using the method devised by Allen & Meyer (1933), Adler, De Fremery & Tausk (1934) separated from placentae material having the activities of progesterone. They obtained 1 rb.u. from 600 g. of fresh full-term human placenta and 1 rb.u. from 500 g. of cow's placenta. Fels, Slotta & Ruschig (1934) and McGinty, McCullough & Wolter (1936) also found progestin in fresh human placentae.

Haterius (1935, 1936) discovered indications of the formation of progestin in the placenta by a different method. In a number of rats the oviduct on one side was tied so as to limit pregnancy to a single uterine horn. The rats were then mated. On the 12th or 13th day of pregnancy one ovary and all but one of the foetuses were removed, but all the placentae were left intact. Two days later the remaining ovary was removed. The rats treated in this way carried to full term but failed to deliver the young, which were recovered alive by operation 2 days after the normal date of delivery. At this time the placentae which had been deprived of their foetuses were viable and healthy. In another ten rats the same procedure was used except that all but one of the placentae were removed together with the foetuses, so that only one foetus with its placenta remained. After removal of the second ovary none of these rats carried to full term, at which date their uteri had undergone involution and no sign of embryonic tissue was present. A third set of rats treated in the same way, except that the second ovary was not taken away, carried to full term. The conclusion was that the placentae supplied the progestin needed after the ovaries had been removed.

Browne, Henry & Venning (1939) studied the excretion of pregnanediol in thirty-five women who were the subjects of threatened or habitual abortion, and found that a change in the output takes place between the 70th and 90th days of gestation. This is the time, they suggest, when the placenta takes over from the corpora lutea the function of producing progesterone. The exact date, they say, varies in different individuals and in the same individual during different pregnancies, but it represents a critical stage of gestation, the average date of abortion in 500 cases being the 81st day.

(c) *The adrenals* are yet another source of progestin. This was first suggested by Engelhart (1930). Using young unmated rabbits weighing between 1·5 and 2 kg. he excised one uterine horn for subsequent comparison and control. A week later he gave at intervals of 12 hours subcutaneous injections of a lipoid extract of adrenal cortex. On the 8th day of this treatment the remaining horn of the uterus was examined and found to have undergone progestational changes. Callow & Parkes (1936) prepared from the adrenals of cattle, horses, sheep and pigs extracts which caused full progestational changes in the rabbit's uterus. The horse's adrenal had the highest potency of those tested.

After these results had been recorded it was learned that other products of the adrenal than progesterone, e.g. deoxycorticosterone, can induce progestational changes in the endometrium. Nevertheless, the actual production of progesterone by the adrenal has been demonstrated by Beall & Reichstein (1938), who, from the adrenals of oxen, extracted progesterone and a kindred compound, *allo*pregnane-ol-3-20-one; this, though possessing little or no biological activity, was present in larger amount. The adrenals seem to be the probable source of the progestin found by Frank, Goldberger & Salmon (1936) in the urine of spayed, non-pregnant women, and by Hain (1942) in the urine of postmenopausal women.

In pathological conditions of the adrenal excessive amounts of progestin may be excreted in the urine. Malley & Bradshaw (1941) found an average daily output of 142 mg. of sodium pregnanediol glycuronate in a girl of 14 with adrenal virilism; the average daily output of 17-ketosteroids in this case was 58 mg. Two women with adrenal carcinoma and symptoms of virilism excreted between 9 and 20 mg. of pregnanediol daily (Salmon, Geist & Salmon, 1941); and Butler & Marrian (1937) isolated pregnane-3:20-triol from the urine of patients with adrenal hyperplasia. (See also Venning, Weil & Browne, 1939.)

(*d*) *The testes* were found by Callow & Parkes (1936) to show appreciable progestational activity. Bulls' testes contained the biological equivalent of 1 mg. of progesterone in 1,400 g. of fresh tissue, and boars' testes contained 1 mg. in 1,700 g. of fresh tissue. Possibly the progestational effects in this experiment were caused by androgen (see p. 208).

Absence of storage. Like oestrogen and androgen, progestin is not stored to any considerable extent in the body. Being rapidly inactivated and excreted, it must be elaborated almost or quite continuously to maintain its biological effects. Zondek (1939) injected progesterone into infantile rats and subsequently killed, minced and extracted them; 24 hours after the injection he was unable to find any progesterone in their bodies. Bloch (1936) failed to detect 1 mg. of progesterone in 500 c.c. of the blood of pregnant women.

Metabolism and excretion. Progesterone is largely eliminated from the body as pregnanediol, a relatively inert saturated alcohol which was first isolated from the urine of pregnant women by Marrian (1929) and given the name of pregnanediol by Butenandt (1930). Venning, Henry & Browne (1937) showed that injections of progesterone in women are followed by the excretion of pregnanediol in the urine. They estimated the output of pregnanediol during pregnancy and found that it increased until the 8th or 9th month of gestation. Within 24 hours after parturition it disappeared from the urine or was present only in very small amount. Venning & Browne (1937 *a*, *b*) estimated the urinary output of pregnanediol from ten women during the different stages of the menstrual cycle. They state that pregnanediol appears in the urine within 24 to 48 hours of ovulation. In further experiments Browne & Venning (1938) gave 24 mg. of progesterone by injection to each of two women who had undergone hysterectomy, but were unable to detect any pregnanediol subsequently in the urine. The result suggests that the uterus, at least when in a progestational condition, takes part in the conversion of progesterone into pregnanediol.

Hain & Robertson (1939 *a*, *b*) also have estimated the pregnanediol content of the urine in women, and they find that it is absent during the first half of the menstrual cycle though relatively large amounts are present in the urine between the 17th and last day of the menstrual cycle. The urinary assays were correlated with biopsies, as a result of which it was concluded that no pregnanediol is excreted while the endometrium is proliferating, whereas it is present in the urine in every case when the endometrium is in the secretory, progestational condition.

Hain (1940) states that during the last 3 weeks of gestation in women the excretion of pregnanediol, like that of combined oestrogen, rises, the largest output occurring 8 or 10 days before parturition.

Stover & Pratt (1939), investigating the urinary output of pregnanediol in non-pregnant women, found that the maximum excretion, amounting perhaps to 8 mg. in 24 hours, takes place between the 21st and 26th days of the menstrual cycle. Pregnanediol appears in the urine, they say, within 48 hours of ovulation, and ceases to be excreted in recognizable amount 1 to 4 days, usually 2 days, before the menstrual flow. During pregnancy pregnanediol is excreted in ever greater quantity until full term, there being no demonstrable drop in excretion before labour. In a case in which pregnancy was interrupted by hysterectomy, the excretion of pregnanediol continued during the next 5 days.

Cope (1940a, b) says that pregnanediol is excreted during the luteal phase of the oestral cycle, disappears from the urine just before uterine bleeding begins, and reappears about the 14th day of the next cycle. He suggests that the continued excretion of pregnanediol may be used as a very early indication of pregnancy. He has found that during the first few weeks of gestation the amount excreted is small, and then gradually increases to a maximum, which is reached in 8 or 9 months. The amount excreted in any particular stage of menstruation or pregnancy varies widely, he says, as between one individual and another, maxima varying from 30 to 120 mg. per diem being noted during pregnancy. To judge whether a given sample of urine submitted for assay is in fact a 24-hour sample, Cope suggests that the creatinine content, which ought to be more than 1,000 mg. in the 24 hours, may be estimated. Chronic nephritis and toxaemia of pregnancy, he points out, interfere with the excretion of pregnanediol.

According to Hamblen, Ashley & Baptist (1939) there are three main stages in the natural metabolism and elimination of progesterone. (1) In the endometrium, they say, progesterone is converted into pregnanediol; (2) in the liver the pregnanediol is combined with glycuronic acid; (3) as sodium pregnanediol glycuronate the hormone is excreted by the kidney.

Venning & Browne (1940) found that in a normal menstruating woman with a cycle of between 25 and 27 days, excretion of pregnanediol in the urine occurred from the 14th to the 25th day of the cycle, the maximum being on the 18th, 19th or 20th day. After hysterectomy—with or without removal of the ovaries—neither the injection of progesterone nor of pregnanediol, they say, leads to the appearance of combined pregnanediol in the urine, whereas the injection of 20 mg. of sodium pregnanediol glycuronate is followed by the identification of about 50 per cent of the injected compound in the urine. In the presence of active corpora lutea, when the endometrium is in a secretory state, intramuscular injections of sodium pregnanediol glycuronate or of progesterone lead equally to the excretion of sodium pregnanediol glycuronate in the urine. The foregoing observations support the view that the endometrium when in a secretory, progestational phase is largely responsible for the conversion of progesterone into pregnanediol. The fact that oestradiol benzoate given to women in doses of 20,000 to 80,000 i.u. reduces the urinary output of pregnanediol, as reported by Pattee, Venning & Browne (1940), seems to accord with this view, for these doses of oestrogen would suppress the progestational stage of the menstrual cycle.

Whether the endometrium is or is not the main site for the conversion of progesterone into pregnanediol, it can hardly be regarded as the only one, for Buxton & Westphal (1939) gave 30 mg. of progesterone daily to two men with Addison's disease and to one normal man aged 34, and in all three cases sodium pregnanediol glycuronate was recovered from the urine; and, as mentioned above, Stover & Pratt have reported the continued excretion of pregnanediol for 5 days after hysterectomy.

Pregnanediol is not the only excretory product of progesterone, for Westphal & Buxton (1939) failed to recover pregnanediol from the urine of monkeys, rabbits and cats which were pregnant or had received injections of progesterone (see also Astwood & Weil, 1940; Marker & Hartman, 1940; Hamblen, Cuyler & Hirst, 1940; and Elder, 1941).

Other compounds allied to, though not necessarily derived from, progesterone have been detected in the urine. Marker, Kamm & McGrew (1937) isolated an isomer of pregnanediol (*allo*pregnanediol) and another progestin, namely *epiallo*-pregnane-ol-3-one-20, from the urine of pregnant women; and Butler & Marrian (1937) identified pregnane-3:20-triol in the urine of patients with adrenal hyperplasia.

The excretion of pregnanediol is not invariably confined to the female of a species. Marker, Wittle & Lawson (1938) found higher concentrations of pregnanediol in the urine of sixteen bulls than occur normally in the urine of pregnant cows or in that of pregnant women. It is interesting in this connection to recall the large amounts of oestrogen excreted by stallions (p. 105).

The Corpus Luteum

After a ripe ovarian follicle has ruptured and discharged its ovum it becomes replaced under normal conditions by a corpus luteum. More than a century ago it was noticed by anatomists that these corpora lutea are largely developed and persistent during pregnancy and lactation. Such a constant association eventually led inquiring minds to define by experiment the connection of well-developed corpora lutea with anoestrus, pregnancy and lactation. Research proceeded in two ways; first, by observing the actions of the corpus luteum as a whole, and secondly, by the use of luteal extracts and the determination of their biological capacities.

Progress by the second of these methods was for long hampered by the fact that the extracts often contained a mixture of oestrogen and progesterone, and as the two hormones interfere with each other in some of their biological actions, an advance of knowledge by the use of extracts was hampered. Parkes & Bellerby (1928) showed that in cows the quantity of oestrogen in the corpora lutea depends largely on their state of development. If they contain free fluid considerable amounts of oestrogen are present, but later, when they have become solid, the corpora lutea yield but little oestrogen.

The fact that the ovary produces two different hormones was first made clear by Marshall & Jolly (1905). They determined by experiments on dogs that (1) bitches could be brought into heat by extracts or grafts of oestral ovaries, (2) corpora lutea did not have this effect, (3) corpora lutea were essential for gestation during the early stages of pregnancy (see also Marshall, 1910).

As soon as a practical method of separating the two hormones had been devised (Allen & Meyer, 1933) progress became rapid, and was further assisted when, in the following year, a number of chemists succeeded in isolating progesterone in a pure crystalline state (Allen & Wintersteiner, 1934; Butenandt, Westphal & Hohlweg, 1934; Slotta, Ruschig & Fels, 1934; Hartmann & Wettstein, 1934). The two isomorphous and interchangeable forms of the pure hormone so obtained were named α-progesterone and β-progesterone by Allen, Butenandt, Corner & Slotta (1935).

With the pure hormone available accurate experiments became possible, and observations made on the corpora lutea themselves became of less consequence, though, as will be seen presently, they have provided a great deal of useful information.

Before discussing in detail the actions of progesterone on the female reproductive organs it may be useful to consider some of the factors controlling the existence of the corpora lutea.

THE DEVELOPMENT AND MAINTENANCE OF CORPORA LUTEA

A. *The Development of Corpora Lutea*

The first event in the normal formation of a corpus luteum is ovulation, that is to say, the rupture of a graafian follicle with release of the contained ovum. Bouin & Ancel (1909) showed that in the rabbit the development of corpora lutea could be induced merely by opening ripe follicles with a needle or fine scissors. In natural circumstances in the rabbit, as in some other animals, the ovarian follicles rupture only in response to mating or to an equivalent stimulation of the pituitary which responds by supplying the amount and kind of gonadotrophin (LH) required for ovulation.

Evans & Long (1922 a, b) discovered that intraperitoneal injections of an extract made from the pituitaries of oxen caused the arrest of oestrus in the rat with the formation of great numbers of substantial corpora lutea and an absence of ripe graafian follicles. They found also that ovulation was not a necessary forerunner of luteinization, for this change took place in the unruptured 'atretic' follicles of rats under treatment with pituitary extract.

Hill & Parkes (1930) showed that a single intravenous injection of placental gonadotrophin, given to non-mated rabbits during oestrus, brings about ovulation followed by the development of corpora lutea and pseudopregnancy. Zondek (1934 b) obtained comparable effects in the bat. This animal copulates before hibernation and, so it is said, the sperm remains viable in the uterus until the winter sleep is over, when it fertilizes the ovum discharged from the ripe follicle. By giving prolan to bats Zondek caused ovulation followed by the formation of corpora lutea even during hibernation.

The rapidity with which the pituitary produces an effective supply of gonado-trophin in response to a suitable excitation is remarkable. Fee & Parkes (1929) found that ovulation and the formation of corpora lutea after mating could be prevented in the rabbit by removal of the brain and pituitary, but only if the pituitary were removed within less than 1 hour after copulation. If hypophy-sectomy were done when the interval since coitus had been long enough to permit ovulation, the development of corpora lutea was not prevented (Deanesly, Fee & Parkes, 1930). Smith & White (1931) allowed rabbits to mate, and removed their pituitaries slightly more than 1 hour later. Nevertheless ovulation occurred and was followed by luteinization, which progressed normally during the next 2 days.

Luteinization is not an inevitable consequence of the ripening of ova. In some species, as in man and dog, ovulation with the subsequent formation of functional corpora lutea usually follows the arrival of ova at maturity. In other species, among which are the rabbit and ferret, ovulation does not succeed maturation in the absence of mating or some other effective stimulus; consequently in these animals corpora lutea are not developed in ordinary circumstances in the absence of sexual excitation. In yet other species, including the rat and mouse, ovulation accompanies every oestral cycle, and corpora lutea form but rapidly degenerate without exercising much if any secretory function.

Whatever circumstances may cause the production of corpora lutea one essential factor appears to be a supply of LH, and there is some evidence that the co-operation of oestrogen is required for this supply.

Varieties of corpora lutea. The remarks made above suggest that different kinds or grades of corpora lutea exist. Six can be distinguished, as follows.

(i) *Atretic follicles.* Luteinization, or a change resembling it, occurs in atretic follicles, which are follicles whose development has been arrested at a stage before maturity and which have failed to rupture though submitted to the influence of LH. The condition may be seen in animals persistently treated with LH, or else with oestrogen and so deprived of FRH. These corpora lutea, if they can properly be given this name, are embedded in the ovary, remain small, and do not seem to exercise any progestational action.

(ii) *Non-functional, oestral corpora lutea.* As mentioned above, in some species, including the rat and mouse, corpora lutea develop with every recurrent oestrous cycle; but whatever their function may be, if any, it does not appear to be progestational in character and the corpora lutea rapidly undergo involution.

(iii) *Functional oestral corpora lutea.* In yet other species, including man, in which spontaneous ovulation as a rule accompanies every period of oestrus when adult age has been reached, a corpus luteum develops after the rupture of each follicle and exercises a progestational function.

(iv) *Corpora lutea of pseudopregnancy.* In some animals, the dog and rabbit for example, ovulation in the absence of conception, whether mating has or has not occurred, is naturally followed by pseudopregnancy. Between these two species there is this difference that in the dog maturation of the follicles proceeds normally to their rupture with the consequent results of luteinization and pseudo-pregnancy. In the rabbit mating or some other accessory stimulus is required to cause rupture of the mature follicles, but after the action of this stimulus and the consequent ovulation, pseudopregnancy will ensue as it does in the dog. The corpora lutea which accompany pseudopregnancy are distinguished from those of normal oestrus by their greater size, their more pronounced functional activity, and their longer persistence.

(v) *The corpora lutea of pregnancy* are not essentially different from those of pseudopregnancy, but usually they are larger and perhaps have a longer life, for in most animals in which the condition has been studied pseudopregnancy is of shorter duration than normal pregnancy.

(vi) *Corpora lutea of lactation.* After the brief period of oestrus which follows parturition corpora lutea form and, if lactation is in progress, become functionally active and persistent. These corpora lutea are larger than those of oestrus, but not so large as those of pregnancy (p. 399).

Sensory stimuli which lead to the formation of corpora lutea. It will be convenient to consider here what sort of stimuli will induce ovulation and the consequent formation of active corpora lutea in animals which, like the rabbit, do not ovulate spontaneously, or in animals which, like the mouse and rat, ovulate spontaneously but do not in ordinary circumstances produce functional corpora lutea. In these species an accentuated impulse is required, apparently, to stimulate the production by the pituitary of enough LH to effect efficient luteinization.

(i) *Mating.* As mentioned earlier, the effect of copulation in the rabbit is to cause an almost immediate release of LH from the pituitary, with the result that the ripe follicles rupture and give place to functional corpora lutea. These results follow whether the mating takes place with a normal male or with one which has been rendered sterile by a previous ligation of the vasa deferentia. The excitatory impulses appear to reach the pituitary through afferent paths of the nervous system, and to some extent may be regarded as involving psychic mechanisms (p. 39).

(ii) *Suckling and lactation.* Selye & McKeown (1934 a) made use of some non-pregnant rats which had previously borne and nursed young but had already weaned them and had been in oestrus since. To these rats were given young litters which, in some instances, they were able to suckle. As a consequence their mammary glands developed, they displayed constant dioestrus, their uteri underwent progestational changes, the ovaries contained large corpora lutea, and maturing follicles were absent.

(iii) *Stimulation of the cervix uteri.* Long & Evans (1922) found that the development of corpora lutea and pseudopregnancy could be induced in rats by inserting glass rods into the cervix uteri; this effect followed even if the ovaries had been removed and transplanted into another part of the body. The result therefore is not the consequence of any direct nervous excitation of the ovary.

The influence of afferent nerve impulses derived from the uterus or vagina on the output of gonadotrophin from the pituitary has been shown in several different ways. Meyer, Leonard & Hisaw (1929) repeated and confirmed the experiments of Long & Evans on the production of corpora lutea and pseudopregnancy by stimulation of the uterine cervix, and they noted that if the cervical stimulation were performed under anaesthesia these results followed less often. Haterius (1933) found that abdominal sympathectomy or removal of the cervical ganglia prevented the induction of pseudopregnancy in rats by mechanical or electrical stimulation of the cervix uteri, but did not interfere with its induction by sterile mating (Table 187). The induction of ovulation and pseudopregnancy by electrical stimulation of the cervix uteri has been recorded by Shelesnyak (1931), Marshall & Verney (1936), Greep & Hisaw (1938) and Astwood (1939 b).

TABLE 187. The effect of dividing nervous pathways on the induction of pseudopregnancy in the rat (Haterius, 1933)

Method used to induce pseudopregnancy	Operation	Number of rats	Number becoming pseudopregnant
Sterile coitus	Abdominal sympathectomy	17	17
	Cervical ganglionectomy	10	10
	None	24	23
Mechanical stimulation of uterine cervix	Abdominal sympathectomy	16	0
	Cervical ganglionectomy	10	0
	None	24	11
Faradic stimulation of uterine cervix	Abdominal sympathectomy	16	1
	Cervical ganglionectomy	10	0
	None	24	22

(iv) *Psychic stimuli.* Ovulation, followed by the formation of corpora lutea and pseudopregnancy, may be induced in the rabbit merely by placing her in a cage adjoining one containing an adult male or perhaps a female. It will be remembered that a virgin pigeon may be induced to ovulate, not only by permitting the proximity of a male, but also by the proximity of another female, or even by putting a looking-glass in the pigeon's cage by which she can see her own reflection (p. 39). The experiments of Haterius just referred to may also be thought to indicate the effectiveness of a psychic stimulus in causing ovulation and luteinization.

The absence of corpora lutea in oviparous animals. In the experiment with the pigeon just mentioned the formation of corpora lutea does not follow ovulation. These bodies do not occur in the bird's ovary, although according to Leonard (1937) the bird's pituitary produces LH. Fraenkel & Martins (1938) and Fraenkel, Martins & Mello (1940) examined pregnant viviparous snakes (*Crotalus terrificus* and *Bothrops jararaca*) and found in their ovaries corpora lutea resembling those of mammals and corresponding in number with the embryos present in the uterus. If in the course of pregnancy the corpora lutea were removed, the embryos died. Such corpora lutea, they say, are not found in oviparous snakes, and in viviparous snakes are present only during pregnancy.

B. *The Maintenance of Corpora Lutea*

To exert properly their functions in the living body the corpora lutea are controlled in accordance with temporary physiological needs. Such regulation involves not only their mere existence as functional organs, but includes also the degree of their activity; not so much progestin is required during lactation as during pregnancy and this difference in demand is reflected in the structure and to some extent in the size of the corpora lutea. Long & Evans (1922) measured the corpora lutea of rats and found the following maximum diameters:

		mm.
Corpora lutea of ovulation		1·2
,, ,, ,,	pseudopregnancy	1·4
,, ,, ,,	lactation	1·4
,, ,, ,,	pregnancy	2·0

(i) *The role of the pituitary in the maintenance of corpora lutea.* Before proceeding to discuss this matter an explanation already given will be repeated here to avoid a possible misunderstanding. The writer will refer to 'follicle ripening hormone' or FRH, and 'luteinizing hormone' or LH, as though they were established as distinct entities. Whether they are two different compounds or whether there is only one hormone whose actions vary with different circumstances is a problem yet unsolved. Meanwhile to discuss in a lucid manner the phenomena of maturation of ova, ovulation and luteinization without using a terminology which distinguishes between the causes of the two gonadotrophic effects of maturation and luteinization is difficult, and so the writer, for convenience, has taken the easier course of writing as though two distinct pituitary gonadotrophins were known to exist.

It might be thought that, having developed in response to LH from the pituitary, corpora lutea would depend for their maintenance on a continued supply of this hormone, and some of the earlier experimental work seemed to support such a proposition. Evans & Long (1922 *a, b*) found that the repeated administration of fresh anterior pituitary substance to female rats suppressed oestrus and caused abundant and persistent luteinization of the ovaries. Evans & Simpson (1929 *a*) implanted fresh pituitaries into pregnant rats and found that gestation was prolonged thereby for several days beyond the normal and that the ovaries were heavily luteinized, and Katzman, Levin & Doisy (1931) gave extracts from pregnancy urine, i.e. LH, to rats in the later stages of gestation with the result that delivery was delayed, and the ovaries were enlarged and showed excessive luteinization. Later work proved, however, that corpora lutea do not depend directly for their continued existence on pituitary LH. Smith (1932) reported the presence in the rat's ovary of well-formed corpora lutea, 0·5 mm. in diameter, 12 months after hypophysectomy. Bunde & Greep (1936), having induced heavy luteinization of the ovaries in young rats by injections of an anterior pituitary extract, then removed the rats' pituitaries and noted that this procedure was not followed by an immediate regression of the corpora lutea. Pencharz & Long (1933) removed the pituitaries of rats at various stages of pregnancy and observed that the operation if done after the 12th day of gestation had no detectable effect on the corpora lutea. In this case it might be possible to attribute the persistence of the corpora lutea to placental activity (see also Robson, 1938).

From these facts and others which will be mentioned later it appears that *a supply of LH is essential for the production of corpora lutea though not for their maintenance once they have developed. Having formed they tend to persist provided that FRH is absent.* It does not follow, however, that they remain functional in these circumstances. Evans, Simpson & Lyons (1941) say the corpora lutea induced in immature or hypophysectomized rats by gonadotrophin are usually non-functional as tested by their capacity to assist the formation of placentomas.

(ii) *Maintenance of corpora lutea by the placenta.* Many experiments, some of which will be cited presently, have shown that the corpora lutea of advanced pregnancy depend in some species upon the placenta, and here again there is a temptation to regard the LH produced by the placenta as the essential agent. The following experimental results may be quoted. Loeb (1914) caused extra-uterine pregnancy in the guinea-pig and found that it was accompanied by degeneration of the corpus luteum and a renewal of ovulation in spite of the presence of a living embryo in the abdomen. He concluded that the placenta was responsible for maintaining the corpora lutea and preventing ovulation during normal pregnancy. Pencharz & Long (1933) observed that hypophysectomy performed on rats after the 12th day of gestation had no detectable effect on the corpora lutea, whereas if done at an earlier stage it led to a cessation of luteal function and terminated pregnancy. Their results are summarized in Table 188.

TABLE 188.   Results of hypophysectomy at different stages of gestation
in the rat (Pencharz & Long, 1933)

Days after coitus	Result
4	Failure of nidation
7 to 10	Failure to maintain pregnancy
11 to 20	Persistence of corpora lutea with prolongation of pregnancy
21	Normal parturition at term

Removal of the pregnant uterus in the rabbit is followed by degeneration of the corpora lutea (Klein, 1933a), and it may be of interest to consider this fact in the light of other experiments. Hammond (1917) removed the foetuses from pregnant rabbits, leaving the placentae, which survived in some instances until the end of the normal term of gestation; but the corpora lutea degenerated and he therefore believed that they were dependent on the presence of the ovum. Klein (1933b, 1935) obtained a different result. From the unilaterally gravid rabbit he removed all the embryos on the 15th day of gestation, leaving the placentae intact. In spite of the absence of embryos the corpora lutea persisted in a functional state so that mucification of the vagina was maintained. Unless the placentae are intact, he says, the corpora lutea degenerate. If when the embryos are removed the placentae are detached but left in place, the corpora lutea degenerate as though the placentae were absent. Astwood & Greep (1938) showed the dependence of active corpora lutea on placental influence in a different way. They caused pseudo-pregnancy in rats by faradism of the cervix uteri during oestrus and 4 days later injured the endometrium. In all of thirty-five rats so treated deciduomata developed, showing that active corpora lutea were present. In other rats treated in the same way but hypophysectomized on the day of endometrial injury no deciduomata appeared. If, however, to these hypophysectomized rats suspensions of rat placenta were injected twice a day deciduomata developed, but this only happened if the ovaries were present and active corpora lutea had formed. Astwood & Greep concluded from these results that during normal pregnancy in the rat the corpora lutea depend on the placentae for their maintained activity, presumably by a luteotrophic hormone.

Deanesly & Newton (1941) found that hypophysectomy of mice, combined with destruction of the embryos, on the 12th day of pregnancy was not followed by atrophy or degeneration of the corpora lutea provided that the placentae were left *in situ*. If, however, the placentae were eliminated as well as the embryos on the 12th day of pregnancy, degeneration of the corpora lutea followed.

In these experiments on pregnant animals it seems certain that the maintenance of the corpora lutea depended upon the vital activity of the placenta. This would not necessarily mean that the LH, which we know to be produced by the placenta, is the only active agent. The placenta is not only a source of LH, but also of gonadal hormones including oestrogen, androgen and progestin, and it seems possible that these may be the chief factors in the persistence of corpora lutea in an active state. The gonadal hormones check the formation of FRH, on an absence of which, as mentioned earlier, the maintenance of corpora lutea seems largely to depend. (But see p. 408 for reference to an experiment by Hill which hardly seems to support this view in an unmodified form.)

(iii) *The influence of gonadal hormones on corpora lutea.*   Several workers have reported that the gonadal hormones may help to sustain the corpora lutea in a condition of functional activity. Selye, Collip & Thomson (1935) noticed that in rats the administration of *oestrogen* leads to the formation of many large corpora lutea. Eight nursing rats were given daily injections of 500γ of oestrone from the 4th to the 14th day of lactation. They remained throughout this time in the dioestral state with mucified vaginae. Post-mortem examination at the end of this period revealed a considerable enlargement of the ovaries due to an increased size of the corpora lutea. Wolfe (1935 b) gave 200 r.u. of oestradiol benzoate daily for 12 days to twenty-five adult female rats. Vaginal oestral smears were obtained in these rats during the first 3 to 5 days of the treatment, after which mucification occurred. The ovaries of these rats contained a normal number of corpora lutea but these were much larger than the corpora lutea of normal oestrus, being equal in size to those of the latter half of pregnancy. Allen & Heckel (1936) gave injections of oestrin to rabbits from the 11th day of pseudopregnancy onward, with the result that regression of the corpora lutea was prevented. Robson (1937 b, 1938) induced pseudopregnancy in rabbits by intravenous injections of gonadotrophin. Later the presence of corpora lutea was verified by laparotomy. Hypophysectomy was then done, after which the corpora lutea atrophied. This atrophy could be prevented either by (a) gonadotrophin or (b) oestrogen, provided that the administrations were begun at once. If withheld until the end of 3 days after hypophysectomy, whether oestrogen had been given meanwhile or not, gonadotrophin alone no longer maintained the corpora lutea. It is known, Robson says, that during such an interval of 3 days after removal of the pituitary the ability of the ovary to respond to gonadotrophin rapidly becomes less. The experiments suggest that gonadotrophin maintains the luteal function not by a direct action on the corpora lutea but by stimulating the ovaries to produce oestrogen. The daily doses of oestrogen used in these experiments were 10γ of oestrone or 5γ of oestradiol. Heckel & Allen (1939) observed that regular doses of oestrin given to rabbits in the later stage of pregnancy delay parturition. The presence of the ovaries is required, they say, for this effect, which they attribute to a maintenance of the corpora lutea in a functional state through the influence of oestrogen.

Greep (1940) states that injections of oestrin in hypophysectomized animals prevent involution of the corpora lutea. He says further that, whereas removal of the rabbit's pregnant uterus on the 15th day of gestation is followed by a rapid involution of the corpora lutea, this involution can be prevented by injections of oestrogen. Westman (1940), experimenting with rabbits, made the following observations: (1) If a mature follicle is allowed to remain after coitus (the surrounding ovarian cortex being destroyed by electrocoagulation), a normal corpus luteum does not develop; (2) if the ovarian cortex is destroyed around a corpus luteum which has already matured its growth and function are interrupted; (3) if oestrogen is given in either of the above circumstances the normal development and functional activity of the corpus luteum are preserved. Westman noted also that by the administration of oestrogen the active life of the corpora lutea in a pseudopregnant rabbit could be at least doubled. He found also that if hypophysectomy be performed on a pseudopregnant rabbit the corpora lutea degenerate between the 3rd and 5th days after; they can be maintained, however, for at least 14 days by the administration of oestrogen. He states that oestrone, oestradiol and diethylstilboestrol all have this preservative effect, which was not shown by progesterone, androsterone, testosterone or adrenocortical hormone. Further remarks on the influence of oestrogen in maintaining the corpora lutea in a functional state will be found on p. 282.

Androgens however, if given in adequate amount, appear to have a similar effect in maintaining the corpora lutea (p. 192).

Progesterone will cause the corpora lutea to persist according to McKeown & Zuckerman (1937), who gave daily injections of 1 mg. of progesterone to seven rats which were known to have normal oestral cycles. During the 9 to 11 days of this treatment the rats with one exception remained in dioestrus and corpora lutea persisted in all. The effect of progesterone in maintaining the corpora lutea has not been always observed. Neither Westman, in the experiments quoted above, nor Courrier & Kehl (1938b) confirm such an effect. The latter found that 10 mg. of progesterone given daily to a pregnant rabbit, while not interfering with gestation, caused involution of the corpora lutea. We have to remember that the actions of the sexual hormones are so complicated and depend so much on relative dosages and other circumstances that the observations of Westman and of Courrier & Kehl do not necessarily disprove the statement that progesterone, supplied in appropriate doses, may tend to preserve the corpora lutea in a functional state.

(iv) *Suckling and lactation.* Selye & McKeown (1934a), using multiparous rats which were shown at the time by vaginal smears to have regular oestral cycles, induced them to suckle the young of other rats. Although these foster-mothers produced little or no milk, they remained while suckling in continued dioestrus, their uteri underwent progestational changes, their ovaries contained corpora lutea which were much enlarged, and in some instances placentomata could be induced. Maturing ovarian follicles were not present. Selye & McKeown point out that the mechanism by which suckling causes the formation of persistently active corpora lutea is different from that which produces this result in pseudopregnancy. In the latter a single event will bring about a continued effect, whereas with lactation the corpora lutea persist only so long as the suckling is continued, and removal of the infants is soon followed by degeneration of the corpora lutea and the reappearance of oestrus.

Among the agents which inhibit ovulation and help to maintain the corpora lutea in a functional state are, it seems, prolactin and adrenocorticotrophin (Lyons, 1937 a). Active corpora lutea are not essential for the continuance of lactation, for McKeown & Zuckerman (1937) destroyed the corpora lutea in seven suckling rats which nevertheless continued to lactate and remained in dioestrus until 5 days after the young had been weaned. Clinical experience had already proved that removal of the ovaries in women does not prevent lactation (see also Lyons, 1942, and Lyons, Simpson & Evans, 1943).

(v) *Influence of the uterus on the survival of corpora lutea.* Selye (1934) showed that the placenta is not required to prevent the corpora lutea from involuting; he removed the foetuses and placentae from pregnant rats by laparotomy and introduced paraffin so as to cause *distension of the uterus.* This procedure led to prolonged dioestrus and persistence of the corpora lutea. E. Allen (1937) showed by Dustin's colchicine method that distension of the uterine cavity by pellets of paraffin is followed by proliferation of the uterine muscle, both circular and longitudinal, and much mitotic activity in the endometrium.

Another fact which suggests that the uterus has an influence on the corpora lutea was reported by Loeb (1927, 1932), who observed that *hysterectomy* in the guinea-pig, if performed shortly after oestrus, caused the life of the corpora lutea to be prolonged even for as long as 100 days, a condition similar to that of pseudopregnancy becoming established with mucification of the vagina and an arrest of ovulation. If in these circumstances the corpora lutea were removed ovulation was quickly resumed. These consequences of hysterectomy in the guinea-pig did not occur in the rat. In this animal active corpora lutea do not follow ovulation apart from pregnancy or pseudopregnancy. Asdell & Hammond (1933) obtained results comparable with those of Loeb. They performed hysterectomy on immature rabbits. At puberty these were mated. Corpora lutea then developed and retained their size and inhibited ovulation for periods varying from 24 to 29 days, that is to say about 10 days longer than the active life of corpora lutea in ordinary circumstances in the pseudopregnant rabbit. Hechter, Fraenkel, Lev & Soskin (1940) found that hysterectomy of pregnant animals is followed by persistence of the corpora lutea for almost the length of their normal period of activity during gestation (Table 189).

TABLE 189. The persistence of corpora lutea after hysterectomy during pregnancy (Hechter, Fraenkel, Lev & Soskin, 1940)

	Duration of normal pregnancy (days)	Length of survival of corpora lutea in animals hysterectomized during pregnancy (days)
Guinea-pig	65	60–100
Rabbit	32	25–29
Rat	21	21

The extended survival of active corpora lutea after hysterectomy may perhaps depend on the capacity of the endometrium to inactivate progesterone by converting it into pregnanediol, as mentioned a few pages earlier (p. 393). If this be so, hysterectomy during the non-progestational phase would have little or no effect upon any non-functional corpora lutea which happened to be present.

(vi) *Persistence of corpora lutea in the absence of FRH.* From the experiments quoted above it is obvious that the biological mechanisms on which the corpora lutea depend for their maintenance have not yet been exactly defined. In a general review of experimental results it is seen that the maintenance of corpora lutea is usually associated with an absence of FRH, whether this is the consequence of hypophysectomy, or of the inhibitory action of gonadal hormones. The reason why distension of the uterus seems to inhibit the output of FRH by the pituitary remains a subject for future experiment.

Concerning the effect of FRH on the formation of corpora lutea it may be recalled that the pituitary of the castrated male, unlike that of the normal female, produces a continuous supply of FRH. This is shown by placing a castrated male and a normal female in parabiotic union, in which state the formation of effective corpora lutea is prevented and continuous oestrus occurs. Lipschütz (1938) demonstrated the same phenomenon in another way. He found that an ovary grafted successfully into a castrated male guinea-pig remains in a constant follicular phase, luteinization being absent. A vagina transplanted into one of these ovary-bearing castrated male rats showed an oestrous condition lasting for several weeks.

As a rider to the above argument it might be said that the corpora lutea by producing progesterone are themselves able to check the supply of FRH by the pituitary and are to this extent self-supporting.

The influence of the thyroid on the corpus luteum. In some instances the thyroid seems to play a part in the development and maintenance of functional corpora lutea. Weichert and Boyd (1933) gave ¼ or ½ a gram of desiccated thyroid daily to female rats. This treatment caused dioestrus lasting for 13 to 24 days and recurring at intervals so long as the thyroid feeding was continued, one or two normal oestral cycles intervening between each prolonged dioestrum. When the thyroid feeding ceased normal oestral cycles became re-established. Apparently the prolonged dioestrous periods represented pseudopregnancy; their duration, the character of the corpora lutea, the presence of vaginal mucification and the formation of deciduomata in response to uterine trauma seemed to warrant such a conclusion.

The sensitivity of the corpus luteum to X-rays. By a series of experiments on mice Parkes (1927 a, b; see also Brambell, Fielding & Parkes, 1928) showed that corpora lutea are more resistant than ovarian follicles to X-radiation. By doses of X-radiation enough to cause complete degeneration of the ova as well as the granulosa cells, the corpora lutea were apparently unaffected, and after such treatment lactational dioestrus continued for its normal period of about 3 weeks. The fact that oestrus then supervened although follicular maturation had been arrested is another observation made by Parkes in the course of his experiments and is mentioned elsewhere (p. 253). Lacassagne, Fehr & Nyka (1936) applied varying doses of X-rays to the ovaries of rabbits a few hours after copulation, and on the 6th day transfixed the uterine horns with threads, the animals being killed on the 12th day. By this method they, also, found that the resistance of corpora lutea to X-radiation is greater than that of the ova. A dose of X-radiation sufficient to kill the ova did not prevent the activity of corpora lutea, which still produced enough secretion to permit progestational changes and the formation of deciduomata in the uterus.

Chapter XXII. *Progestins* (continued)

THE BIOLOGICAL ACTIONS OF PROGESTIN

Effect on pituitary and on embryological development: on testis and ovary: on accessory generative organs of male and female. Co-operation and antagonism between progestin and other gonadal hormones. Similarities in action between progesterone, androgen, oestrogen and adrenal cortical hormones.

THE natural history of the corpora lutea or luteal glands having been discussed in a general way, the biological activities of their special secretions may now be considered. To avoid giving the impression that progesterone is the only effective luteal hormone, the term progestin will be used in a generic sense to include cognate compounds.

The Pituitary

Progestin, like the other gonadal hormones, acts either directly on individual organs, or indirectly affects them through the changes which it induces in the anterior lobe of the pituitary gland. These pituitary changes have been dealt with already (pp. 57, 91) and need no further reference here except perhaps to recall the fact that progestin arrests the output of FRH and so interferes with the maturation of ovarian follicles and the initiation of ovarian cycles. For this purpose relatively large amounts are required.

Embryological Development

It will be remembered that both oestrogen and androgen have a pronounced influence over the development of the reproductive organs. Unfortunately we are not in a position to say much about the effects of progestin on the immature organs, for little experimental inquiry has been made into the subject. Burrows (1939c) found that progesterone, like androgen and oestrogen, when injected into newborn female rats and mice, causes hypospadias, the ventral wall of the clitoris becoming deficient so that the opening of the urethra instead of being at its apex is at its base, level perhaps with the margin of the vaginal orifice. Burrows found also that, when persistently given to young male rats, progesterone checks the development of the testes and prevents their descent into the scrotum. Much has been done to determine what influence progesterone exerts on the development of the mamma (p. 430). Beyond these observations we know little about the effects which progesterone may have upon the growth of the reproductive organs in early life.

Male Reproductive Organs

(a) *The testis*. Besides preventing their descent in juveniles progestin has other effects on the testes. Nelson (1937a) found that progesterone and pregnaneolone, like androgen, will maintain spermatogenesis for awhile in hypophysectomized rats if the administration follows the operation at once and is continued without intermission. In this action pregnaneolone appeared to be more effective than progesterone. In maintaining spermatogenesis after hypophysectomy 2·5 mg. of progesterone were about equivalent to 0·5 mg. of testosterone, these doses being given daily. When considering how this effect is brought about, it may be useful to recall the similar result induced by androgen. Cutuly, McCullagh & Cutuly (1937 a) showed that scrotal atrophy and return of the testes into the abdomen, which follow removal of the pituitary in rats and mice, may be prevented by giving androsterone or testosterone, and they suggest that the continuance of spermatogenesis in these circumstances may be attributed to the testes remaining in the scrotum. Against this explanation Nelson points out that although testosterone is the most effective stimulator of scrotal development, 0·25 mg. daily causing large pendent scrota even in hypophysectomized rats, yet the testes are thereby but poorly maintained; moreover, androsterone is more effective than testosterone in maintaining spermatogenesis, though it has less influence on the scrotum. Dehydroandrosterone or androstanedione in daily doses of 0·5 mg., he says, will maintain spermatogenesis after hypophysectomy, though they have only a slight influence on the accessory sexual structures; under their influence the scrota are fairly well maintained but are smaller than those of rats which have been given testosterone.

There is no doubt that the presence of the testes in the scrotum is favourable to spermatogenesis and the longer they can be kept there the longer is spermatogenesis likely to continue. Information appears to be lacking as to the effect of progestin on the position of the testes after hypophysectomy, and knowledge on this point is essential for a sound discussion of the matter.

The writer (Burrows, 1940) has observed additional evidence of a gonadotrophic activity of progesterone, for when given to adult rats during a short period it caused enlargement of the testes. He injected 5 mg. of progesterone on 4 successive days into young male adult rats which were killed on the day following the last injection. In every instance the testes of these rats were larger than those of controls, although by implanting their pituitaries into immature female mice it was found that the gonadotrophic potency of the progesterone-treated rats was considerably reduced. In mice also the testes became enlarged during a short course of progesterone given subcutaneously.

(b) *The male accessory reproductive organs.* The action of progesterone on the accessory genital organs of the male resembles to some extent that of androgens in character. Burrows (1936 f) found that progesterone had some capacity to protect the prostate, coagulating glands and seminal vesicles of non-castrated mice from the atrophic action of oestrogen. Van der Woerd (1938) states that progesterone has a slight though definite restorative effect on the preputial glands of castrated rats. Greene, Burrill & Ivy (1939 b) castrated rats when 19 days old and on the same day began a course of subcutaneous injections of progesterone. Some of the rats received 3 mg. daily for 20 days, others were given 6 mg. daily

TABLE 190. Androgenic action of progesterone on the accessory reproductive organs of the immature castrated rat (Greene, Burrill & Ivy, 1939 b)

	Daily dose of progesterone (mg.)	Days of treatment	Mean weight of ventral prostate (mg.)	Mean weight of seminal vesicles (mg.)
Treated rats	3	20	69·1	13·2
Controls	0	—	11·8	5·0

for 10 days. Under this treatment the prostate and seminal vesicles were largely maintained and displayed secretion. Untreated littermates were used for controls. Some of the results are shown in Table 190.

The clitoris. In the female guinea-pig, and perhaps in other animals, progesterone manifests androgenic potency, for it will cause the clitoris to develop into an organ resembling a penis. Such effects have already been described (p. 110).

The Ovary and the Oestrous Cycle

(a) *Oestrus and ovulation.* The influence of progesterone on the pituitary has been discussed (pp. 57, 91), and the experiments cited are appropriate to the present discussion, because the action of progesterone on the ovary seems to be mainly indirect and to depend on its capacity to check the supply of FRH from the pituitary and so to inhibit maturation of follicles and oestrus. Avoiding excessive repetition, some experiments which bear on the subject may receive attention here to clarify the argument.

Loeb (1914) noticed that extirpation of the corpora lutea from the guinea-pig's ovary at an early stage of the oestrous cycle led to ovulation before the date at which it would normally occur. Even during pregnancy removal of the corpora lutea was followed by ovulation. From these facts Loeb concluded that ovulation is prevented by the secretion of the corpora lutea. This conclusion has been well established by subsequent work.

Papanicolaou (1920) repeated and confirmed Loeb's experiments on guinea-pigs. Removal of the corpora lutea within 24 hours after ovulation caused the next oestrum to appear at the end of 11 days instead of 16 or 17 days which is the length of the normal oestrous cycle in the guinea-pig. Evans & Long (1922a, see Evans, 1923–24) gave repeated intraperitoneal injections of fresh bovine pituitary extracts to female rats and noted that maturation of follicles and oestrus were inhibited. These results they attributed to excessive and persistent luteinization of the ovaries, which were greatly enlarged as compared with those of untreated littermates, were extensively luteinized, and showed an absence of ripe follicles. Teel (1926) also observed that an alkaline extract of bovine pituitaries caused arrest of ovulation with follicular atresia in rats. The ovaries were greatly enlarged and were composed almost entirely of luteinized tissue. As many as twenty corpora lutea were counted in one ovary.

Some experiments by Hill (1933) may be mentioned here. He united twenty-four female rats in parabiosis with castrated male or female partners, and noticed that if corpora lutea were absent at the time of the operation, as was the case with immature rats, the female subjected to the gonadotrophin of the castrated partner soon showed continual oestrus. If, on the other hand, at the time of the operation corpora lutea were already present they persisted and the onset of oestrus was greatly delayed.

Experiments have proved that the inhibitory influence of the corpus luteum on ovulation is caused by progestin. Kennedy (1925) gave intravenous injections of an extract of cows' corpora lutea to female rabbits shortly before coitus and observed that the ovulation, which normally would have followed, did not occur. Papanicolaou (1926), with a similar extract, obtained the same result in the guinea-pig. Parkes & Bellerby (1928) prepared an extract from sows' corpora lutea which inhibited oestrus in the mouse. In the course of this work they noted that in young corpora lutea with fluid contents oestrogen was present, whereas it was absent in demonstrable amount from the solid corpora lutea of a later stage. Hisaw, Meyer & Weichert (1928) and Gley (1928) reported that an extract similar to that used by Parkes & Bellerby prevented oestrus in the rat, as shown by vaginal smears.

Selye, Browne & Collip (1936) gave daily subcutaneous injections of 4 mg. of progesterone to mature female rats for a period of 12 days. The consequences were a cessation of oestrus and atrophy of the ovaries, from which ripe follicles and corpora lutea were absent. The vagina was neither mucified nor cornified. The mean weight of the pituitaries of these rats was increased and that of the ovaries diminished (Table 191).

TABLE 191.   The effect of progesterone on the ovary and pituitary of the rat (Selye, Browne & Collip, 1936)

Treatment	Average weight of pituitaries (mg.)	Average weight of ovaries (mg.)
None (controls)	10·5	39
Daily doses of 4 mg. of progesterone	13·5	30

Makepeace, Weinstein & Friedman (1936, 1937) found that the ovulation which normally follows copulation in the rabbit can be prevented by five previous daily injections of 2 mg. of progesterone; if LH be injected at the same time ovulation follows as usual. The experiment seems to prove that the inhibition of ovulation caused by progesterone in this instance is attributable to its action in checking the output of gonadotrophin from the pituitary.

Phillips (1937) injected progesterone into rats weighing between 150 and 200 g. Daily doses of 1·5 mg. given each day inhibited oestrus; with smaller doses ranging from 0·25 to 1 mg. the oestrous cycle was prolonged though oestrus was not entirely prevented. In a rat which had been treated in this way and had remained anoestrous for 21 days the ovaries contained large corpora lutea and a few large graafian follicles; the uterine horns were in a progestational condition and were 1 cm. in diameter, the normal diameter being about 3 mm.

Dempsey (1937) gave 0·05 i.u. of progesterone daily for 20 days to female guinea-pigs, beginning on the 10th day of the oestrous cycle. Oestrus was inhibited and at the end of the period of injections corpora lutea were absent and large unruptured follicles were present. Oestrus and ovulation ensued within 24 hours of cessation of the injections and were followed by the formation of corpora lutea. The experiment seems to show that while inhibiting oestrus progesterone in relatively small quantities does not entirely and immediately prevent the growth of follicles.

Into seven rats which had normal oestral cycles McKeown & Zuckerman (1937) injected 1 mg. of progesterone daily. The treatment was continued for periods of 9 to 11 days, at the end of which period the rats were killed; with one exception the rats remained in dioestrus throughout and large corpora lutea were present in their ovaries. In another series of experiments McKeown & Zuckerman (1938) destroyed with an electrocautery the corpora lutea in seven rats between the 3rd and 7th days of pseudopregnancy, with the result that oestrus reappeared within 4 days, whereas the anoestrus in control pseudopregnant rats lasted for 12 to 15 days. They found also that oestrus occurred within 3 to 7 days after the destruction of corpora lutea in pregnant rats provided that the operation was done on the 6th or 7th day of pregnancy. In twelve rats the corpora lutea were destroyed between the 11th and 16th days of pregnancy. Oestrus did not follow in these instances.

It might be added that veterinary surgeons have long recognized a connection between sterility in cattle and the persistence of corpora lutea, and from the experiments just quoted it appears that the absence of conception in these cases is probably due to the influence of progestin in checking oestrus and ovulation.

The inhibition of ovulation by progesterone does not appear to be a universal phenomenon in the animal world, for Zwarenstein (1937) found that in *Xenopus*, during the reproductive season, progesterone induced ovulation, a result which can be caused also in this oviparous amphibian by oestrogen or by placental or pituitary gonadotrophin.

(b) *The oestrous cycle.* In view of what has been just said we can now consider the chemical mechanisms of the oestrous cycle and the part played by progestin in that phenomenon. The FRH of the anterior pituitary, which causes maturation of the ovarian follicles, is the initial factor; as the follicles ripen under its influence the ovary elaborates increasing quantities of oestrogen which produces in the generative and other organs those characteristic changes associated with oestrus. When follicular maturation is complete ovulation may or may not follow immediately. In some animals, including the rabbit and ferret, an additional stimulus such as mating is required for ovulation, and these animals do not show a regular sequence of oestrous cycles apart from the recurrent annual breeding season. In other animals, among which are the mouse, rat, dog and man, ovulation usually occurs when follicles have matured, and the follicles thereon become converted into corpora lutea which may or may not be functionally active (p. 396).

The act of ovulation appears to depend on a supply of LH, and the absence of spontaneous ovulation in the rabbit and ferret may be attributed to an insufficient supply of this gonadotrophin. The evidence in favour of such an explanation is twofold. First, ovulation can be caused in the adult oestral rabbit within 24 hours by giving an adequate dose of LH intravenously. Secondly, ovulation in women is accompanied by an increased excretion of gonadotrophin in the urine.

Mating is not the only natural stimulus which will cause enough outpouring of luteinizing gonadotrophin from the pituitary for the induction of ovulation in the rabbit; the mere proximity of a male, and perhaps of a female, in an adjacent cage may be sufficient (p. 39).

Ovulation is not necessarily followed by the development of functional corpora lutea. According to Allen (1940) functional corpora lutea do not form normally in the rat or mouse unless the animal is mated, from which it may be concluded that in these animals progestin is not an essential factor in the oestrous cycle. Further, Allen says that in the mouse and monkey oestrous cycles may occur without ovulation, and the same seems to be true occasionally of women.

In its simplest form, therefore, as in the unmated rat or mouse, the oestrous cycle may be regarded as follows: Under the influence of FRH from the pituitary the ovaries produce enough oestrogen to cause oestrus. But oestrogen, while causing this condition, inhibits the further supply of FRH from the pituitary; so, in turn, the formation of oestrogen by the ovary is curtailed and the phenomena of oestrus subside (p. 53).

In many and perhaps in most animals which have regular, recurrent oestrous cycles a luteal phase intervenes. Towards the end of oestrus, just before ovulation, luteinization begins. As the corpora lutea develop they produce increasing amounts of progesterone which, like oestrogen, checks the supply of FRH and so prevents further maturation of follicles. After a while the corpora lutea degenerate, their output of progesterone diminishes until the pituitary, relieved of its inhibitory action, once more supplies enough FRH to cause maturation of ovarian follicles, and a fresh cycle begins (p. 41).

(c) *Oestral bleeding and menstruation.*   In the course of the oestral cycle there are two periods at which bleeding from the uterus may occur, namely (a) at the end of the oestrogenic phase and (b) at the end of the luteal phase.

Oestral bleeding is the result of a more or less abrupt decline in the supply of oestrogen following ovulation. The phenomenon is familiar in dogs, and is of rare but occasional occurrence in women. In ordinary circumstances a sudden reduction in the supply of oestrogen, whether caused by removing the ovaries or by stopping a course of oestrogen injections in a spayed animal, is followed after a short interval by bleeding from the uterus. This form of uterine bleeding can be prevented by giving suitable doses of progesterone, and it seems that the occurrence of oestral bleeding in some animals and not in others must depend on the relative rapidity of development of the corpus luteum and the consequent rise in the supply of progesterone.

Menstrual bleeding, which occurs only in primates, is the result of degeneration of the corpus luteum and the consequent fall in the supply of progestin. Stockard & Papanicolaou (1917), in the course of experiments on animals, observed that as long as active corpora lutea are present the endometrium remains intact. When the corpora lutea degenerate the endometrium also degenerates. Fraenkel (1903) mentions several instances in which, during laparotomy in women, a corpus luteum was destroyed with a Paquelin cautery; this was followed by uterine bleeding 4 days later. During a series of forty laparotomies on women Halban & Köhler (1914) excised the corpus luteum. In thirty-seven of these cases uterine bleeding ensued between 2 and 4 days later and lasted for periods of 3 to 8 days. Pratt (1927) in the course of thirty laparotomies in women removed the corpora lutea; in twenty-eight of these cases bleeding from the uterus ensued within 48 hours. The same result followed in eleven instances in which both the corpus luteum and its ovary were removed.

Smith & Engle (1932) gave progesterone daily for 28 days to a rhesus monkey. No menstrual bleeding took place during this period, but on the 6th day after the injections had ceased uterine bleeding occurred. They say that if rhesus monkeys are subjected to daily doses of 100 r.u. of oestrin, uterine bleeding will occur within 6 days of ceasing the injections. They also say that the uterine bleeding which follows removal of the ovaries in these animals is prevented by daily doses of 3·7 rb.u. of progesterone. Corner (1935) found that 1 rb.u. of progesterone given daily prevented or delayed menstruation in the rhesus monkey, and that uterine bleeding occurred within 5 to 8 days of withholding the injections. Engle, Smith & Shelesnyak (1935) also have found that in monkeys the uterine bleeding which follows cessation of a course of oestrin injections may be prevented by giving progesterone. In contrast with this they find that if progestational changes have been already produced in the uterus by administering oestrogen followed by progesterone the bleeding which results from stopping the supply of progesterone is not prevented by giving 200 m.u. of oestrogen daily.

The sequence of histological changes in the uterus which culminate in menstruation can be brought about in spayed monkeys by giving them oestrogen and progesterone in succession. Hisaw, Meyer & Fevold (1930) spayed mature rhesus monkeys and a fortnight later treated them for 10 days with oestrogen, and for another 7 days with an extract of corpora lutea. The monkeys were then killed and their uteri were found to be in a premenstrual condition.

An approximate estimate of the amount of progesterone formed in the body at any time may be made by ascertaining the quantity of pregnanediol in the urine, for there is little or no storage of the hormone. Venning & Browne (1937 a, b) made periodical estimations of pregnanediol excretion in ten normal women throughout the menstrual cycle. Pregnanediol first appeared in the urine within 48 hours of ovulation, that is to say between the 13th and 20th days of the cycle, and disappeared between the 22nd and 29th days. Menstruation began between 1 and 3 days later.

Menstruation is thus shown to depend on degeneration of a corpus luteum after a period during which it has been functionally active. When corpora lutea persist in an active state, as they do during pregnancy and lactation, menstruation will not occur and pregnanediol will be excreted persistently in the urine.

Hain & Robertson (1939a) estimated the pregnanediol excreted during two successive menstrual cycles in a married woman whose average cycle was 25 days. Between the 10th and 24th days of the first cycle a total of 44 mg. of pregnanediol were excreted; during the same time in the next cycle 114 mg. were obtained. The largest quantity excreted in a single cycle in the investigations of Venning & Browne was 60 mg. Hain & Robertson thought that the relatively large output of pregnanediol during the second cycle in their patient indicated the persistence of a functional corpus luteum, that is to say pregnancy; events proved this to be correct. It would appear possible therefore to diagnose pregnancy by pregnanediol estimations at a very early stage.

As mentioned earlier the occurrence of ovulation depends upon the output of pituitary LH, and this may sometimes be detected in the urine of women when ovulation takes place, but the amount is small and may pass unrecognized in some cases. Venning & Browne (1937b) say that its excretion in recognizable quantity may be limited to part only of a single day, the amount being perhaps too small to be noticed in a 24-hour specimen of urine.

Accessory Reproductive Organs of the Adult Female

A. *The vagina: mucification of epithelium.* In many animals pseudopregnancy and pregnancy are accompanied by characteristic changes in the vaginal epithelium. This is stratified, and the superficial cells instead of becoming squamous and cornified, as they do in oestrus and in response to doses of oestrogen, assume a rounded or cylindrical form and become filled with mucus-like material. The completed condition is the result of two distinct processes, namely stratification of the vaginal epithelium, and mucification of its superficial cells. The stratification depends on the action of oestrogen; the mucification in the superficial cells is due to progesterone (p. 304). In other words vaginal mucification as seen in pregnancy or pseudopregnancy is the result of a co-operation between oestrogen and progesterone, these hormones being available in suitable amounts. During lactation the supply of oestrogen is deficient and only the action of progestin becomes manifest, so that mucification is present while stratification is absent.

It will be remembered that both the production of oestrogen by the ovary and ovulation are inhibited indirectly by progesterone through its action on the pituitary (p. 57); the prevention by progesterone of vaginal keratinization and its replacement by mucification appears to be a direct action. Parkes & Bellerby (1927, 1928) and Gley (1928) not only verified that ovulation could be inhibited in the mouse and rat by extracts made from corpora lutea of cows and pigs, but they ascertained also that functionally active corpora lutea interfered with the cornifying action of oestrin on the vaginal epithelium. Between 2 and 5 times as much oestrin, they estimated, was required to cause vaginal cornification in a mouse during lactation as in a non-lactating castrated mouse. Further, by giving single injections of oestrin they perceived a correlation between the amount of oestrin required to produce vaginal cornification in the lactating mouse and the size of the litter, that is to say the number of corpora lutea present; the larger the litter the larger was the dose of oestrin needed to cause cornification of the vagina. In mice spayed during lactation smaller amounts (2 m.u.) of oestrin were enough to cause vaginal cornification whatever the size of the litter.

Courrier (1930b) found that in the lactating guinea-pig oestrin in doses of 100 r.u. failed to cause vaginal cornification, and Levin, Katzman & Doisy (1931) noticed that large doses of oestrin or oestriol given to pregnant rats did not cause cornification of the vagina. Selye, Harlow & McKeown (1935) say that 2 mg. of oestrone given daily fail to elicit vaginal cornification in the rat if corpora lutea of pregnancy or lactation are present.

Later work has proved that the inhibitory action of corpora lutea against the cornifying influence of oestrogen is attributable to progesterone. Before this conclusion was established a hitch occurred in the sequence of experimental proof, for it was discovered that vaginal mucification could be induced in spayed adult mice and guinea-pigs, and in non-spayed immature guinea-pigs by giving doses of oestrin which were too small to produce cornification (Robson, 1930; Courrier, 1930a; Robson & Wiesner, 1931; Meyer & Allen, 1933). Thus for a while it seemed that vaginal mucification might be attributed merely to the action of sub-threshold supplies of oestrogen.

It has since been learned that the ovary is not the only source of progesterone, and that progesterone is not the only hormone which will induce vaginal mucification; androgens and adrenal steroids also have this capacity. In spite of gonadectomy, therefore, sufficient quantities of these hormones may be available to induce mucification, so long as there is not enough oestrogen present to inhibit this reaction.

Desclin (1934) noted that daily intraperitoneal doses of between 2 and 10 r.u. of oestrone given to immature guinea-pigs weighing about 200 g. caused vaginal opening with cornification on the 5th day. If one-twentieth to three-twentieths of a rabbit unit of a luteal extract were given at the same time mucification occurred in the vagina instead of cornification. Allen & Meyer (1935), using progestin extracted from corpora lutea and freed from oestrogen, found that large amounts when given to spayed mice had no visible effect on the vaginal epithelium. If to these doses of progestin oestrin were added in an amount which, if given alone, would cause cornification of the vaginal epithelium, mucification was the result. One rb.u. of progestin, they estimated, could prevent the cornifying effect of 10 r.u. of oestrin on the rat's vagina. Selye, Browne & Collip (1936), also, investigated the quantitative relationships required for the prevention by progesterone of the cornifying effects of oestrone on the vagina of the spayed rat. Daily injections of progesterone and oestrone were given for a period of 20 days. It was found that 400γ of progesterone prevented 30γ of oestrone from causing cornification of the vaginal epithelium. The results of these doses were mucification in the vagina and progestational changes in the uterus. Korenchevsky & Hall (1937), also using spayed rats, found that daily doses of 30γ of oestrone or oestradiol plus 400γ of progesterone caused only a slight and partial mucification of the vagina; full mucification followed daily doses of 1γ of oestrone plus 1500γ of progesterone.

Courrier & Cohen-Solal (1937), performing quantitative tests on spayed rats, showed that 8 mg. of progesterone would prevent cornification of the vagina in response to 0·02 mg. of oestradiol. Working with adult female golden hamsters (*Cricetus auratus*) which had been spayed 2 months previously Klein (1937) found that 5γ of oestrone given simultaneously with 0·5 to 1 mg. of progesterone caused a mucification of the vaginal epithelium similar in degree to that which accompanies pregnancy. Further Klein (1938b) showed that to produce this effect the simultaneous action of oestrone and progesterone were essential. In the doses used oestrone alone, progesterone alone, and progesterone following a preliminary dosage with oestrone, did not produce mucification. Freud (1937) brought about vaginal mucification in normal mice by giving, during periods of between 10 and 20 days, daily doses of from 100 to 500 r.u. of a gonadotrophic extract prepared from pregnancy urine. This extract is rich in LH and it seems probable that its mucifying effect on the vagina was caused indirectly through an enhanced secretion of progestin by the ovary.

Freud (1937) also induced mucification of the vagina in spayed mice by the daily simultaneous injection of 100 to 500 i.u. of oestrone and 1 to 2 mg. of progesterone, or 2 mg. of pregnanediol. These doses were continued for periods of 10 to 20 days. His results have a special added interest, inasmuch as they seem to show that pregnanediol is not biologically inert.

Shorr (1940) studied the effects of oestrogen and progesterone on the vagina in women after the menopause by means of smears and biopsies. Oestrogen given by itself caused cornification of the vaginal epithelium. When this stage had been reached the same doses of oestrogen were continued and progesterone was added with the result that the cornified cells disappeared from the smears and were replaced by squamous cells with large nuclei, accompanied by mucus and leucocytes.

MacDonald & Robson (1939) showed that the artificially produced oestrogenic compound triphenylethylene when combined in suitable proportions with progesterone or androgen acted like the natural oestrogen in causing vaginal mucification.

In considering the causes of vaginal mucification it is well to remember that the ovary is not the only source of progestin, which may be derived also from the placenta and the adrenal. In the later stages of pregnancy the placenta in some species may perhaps be more important than the ovary as the source of progestin. Klein (1935) has shown that if in the rat the gravid uterus be removed between the 11th and 18th days of gestation, the ovaries being left intact, the mucified vaginal epithelium is rapidly replaced by one which is squamous and cornified.

The fact that androgen and corticosterone, in co-operation with oestrogen, act like progestins in bringing about vaginal mucification has been mentioned elsewhere (pp. 214, 451).

B. *The uterus.* The influence of progesterone on the uterine functions during the oestrous cycle and menstruation has been discussed in connection with its effect on the ovary (p. 408), and the only point that need concern us here, apart from its role in pregnancy, is the action of progesterone on the uterus in the absence of the ovary and of oestrogen. Although to produce its most striking effects on the uterus progesterone requires the co-operation of oestrogen, it seems to have some influence when the latter is absent. Hooker (1940) gave daily injections of 0·25 or 0·5 mg. of progesterone to mice which had been spayed 3 weeks earlier. After a few days (3 to 6) the mice were killed. Examination showed that their uteri were hyperaemic and slightly enlarged, and mitosis was visible in all the uterine tunics.

THE ROLE OF PROGESTIN IN PREGNANCY

(i) *Progestational changes and nidation of ova.* The constant association between the presence of corpora lutea and the uterine changes which accompany the earliest stages of pregnancy suggested that there might be a close connection between the two, and that the induction of progestational changes, enabling the subsequent nidation of ova, might be a special function of the corpus luteum.

Fraenkel (1903) removed the ovaries of rabbits at periods varying from 1 to 6 days after coitus and noted that nidation of ova in these circumstances did not occur. Further he discovered that excision of the corpora lutea, or their destruction by the cautery without removing the ovaries, prevented pregnancy, whereas partial removal of the ovaries with retention of some corpora lutea was followed by the maintenance of pregnancy in about half of all the rabbits so treated.

Ancel & Bouin (1909; Bouin & Ancel, 1909) carried out a series of experiments remarkable in the circumstances for the clarity of thought which guided them and for the finality of the information obtained. They used the rabbit because this animal does not ovulate spontaneously and therefore in the absence of pregnancy or pseudopregnancy its ovary is free from corpora lutea. To avoid the complication of pregnancy with developing ova in the uterus, Ancel & Bouin studied the relationship between the corpora lutea and progestational changes in the uterus during pseudopregnancy following coitus with males whose vasa deferentia had been tied. They observed that during the 5 or 6 days following such coitus the corpora lutea increase rapidly in volume; though still showing glandular activity they then remain stationary in size for 7 or 8 days after which they cease to secrete and undergo involution. The progestational development in the uterus corresponds in time precisely with the period of activity of the corpora lutea; during the first 13 days of pseudopregnancy the uterus continues to increase in size and vascularity, the muscle is hypertrophied, the submucosa becomes much thickened, especially along the mesometrial border, and its vascular sinuses become enlarged; the mucosa undergoes great mitotic activity and becomes folded into the lumen of the uterus so as to produce the 'dentelle' formation which is a striking feature of uterine progestational changes in the rabbit. On or about the 13th day after coitus, when the corpora lutea cease secreting, the progestational activities in the uterus are arrested, involution begins, and the uterus gradually returns to its normal size and condition of repose. In further experiments, Bouin & Ancel (1909), having induced pseudopregnancy in rabbits, destroyed the corpora lutea with a fine thermocautery at various intervals of time after coitus. If the corpora lutea were destroyed within a few hours no progestational changes occurred in the uterus. If the cauterization were done some days after coitus, the progestational development in the uterus ceased at once and involution began. Coitus, they found, was not essential for the development of corpora lutea and pseudopregnancy; if the ripe ovarian follicles of a virgin rabbit were ruptured by a needle or fine scissors, corpora lutea developed and the uterine changes of pseudopregnancy followed. The exact correspondence between the condition of the corpora lutea and the progestational changes in the uterus leave no doubt that the latter are dependent on the former.

According to Corner (1928) ovulation occurs in the rabbit about 10 hours after mating and the ova enter the uterus at the end of another 10 hours or thereabout; nidation occurs on the 8th day. He found that removal of both ovaries between 14 and 18 hours after mating prevented progestational responses in the uterus, though ova were seen in its lumen between $4\frac{1}{2}$ and $7\frac{3}{4}$ days after coitus. Partial resection of the ovaries so as to retain at least one corpus luteum did not prevent nidation of ova and pregnancy. Removal of all the corpora lutea, the ovaries being retained, prevented progestational changes in the uterus and nidation of the ova so that pregnancy did not follow.

A considerable advance of our knowledge in this field was made by the discovery that luteal extracts induce the uterine changes associated with the presence of the corpora lutea themselves. Hisaw, Meyer & Weichert (1928) found that a progestational condition could be induced in the non-pregnant rat's uterus by subcutaneous injections of an extract of corpora lutea obtained from sows. Corner & Allen (1929) allowed adult rabbits to mate and 18 hours later removed their ovaries and a portion of one uterine horn. Thereafter an extract prepared from sows' corpora lutea was injected subcutaneously on each of the 5 ensuing days. The rabbits were killed on the day after the final dose and progestational changes in the uterus were observed in all. In another experiment Corner & Allen gave injections of luteal extract to immature rabbits of about 8 to 12 weeks old and ranging in weight from 600 to 1,500 g. In some the treatment with luteal extract was preceded by injections of oestrin and in every one of the latter cases complete progestational development took place in the uterus.

Courrier (1935) showed that with the help of progesterone nidation can occur even though the ovaries have been removed. He spayed rabbits on the 6th day after coitus and gave them four doses of progesterone on the day of operation and the next day, and killed the animals on the 8th day. The fact of nidation was revealed by examination of the uterine horns. These observations on rabbits have been repeated and confirmed by Pincus & Werthessen (1938), who noted also that the action of progesterone in these circumstances could be prevented by oestrone.

One of the conditions essential for the implantation of ova in the uterus appears to be a secretory condition of the endometrium, and this depends on progesterone. Bloch (1939) studied the nidation of ova in mice. She says that transverse sections of a uterus with a blastocyst ready for implantation show that the endometrial epithelium in the immediate vicinity of the blastocyst is secreting and a dense mass of granules surrounds each nucleus. This intense secretory condition, she states, is seen only in uteri in the neighbourhood of blastocysts ready for implantation, and the secretion seems to migrate into the cells of the ovum. Adjacent to the actively secreting epithelium the cellular stroma of the uterus shows a syncytial cytoplasm with large round nuclei. Courrier (1935) has suggested that the occurrence or non-occurrence of nidation of ova following successful mating in rabbits might be used for the assay of progesterone. The method he suggests is to remove the ovaries from a rabbit on the 6th day after coitus, the presence of corpora lutea being verified. The material to be assayed for its content of progesterone is injected in four doses on the 6th and 7th days after coitus and the uterus is examined on the 8th day, when the presence or absence of nidification can be readily determined without the aid of a microscope.

The favouring influence of progesterone in preparing the uterus to receive and maintain the ova may perhaps be used to increase fertility. D'Amour, D'Amour & Gustavson (1933) gave daily doses of 50 mg. of luteal hormone to rats during the first few days after copulation. The treatment was followed, they say, by a much higher percentage of living embryos than is usual.

Apparently the action of progesterone on the uterus is direct and does not need the mediation of the pituitary, though it may be that in the absence of the pituitary larger amounts of progesterone would be needed to produce a given effect. Rowlands & McPhail (1936), experimenting on cats, found that progestational effects could still be produced by progesterone after hypophysectomy and removal of the ovaries. Reynolds, Firor & Allen (1936) have reported that in five hypophysectomized rabbits 1 rb.u. of progesterone failed to cause progestational proliferation in the uterus, whereas in one hypophysectomized rabbit 4 rb.u. of progesterone caused marked progestational changes.

The experiments just referred to establish the fact that corpora lutea by supplying progesterone are responsible for the progestational changes in the uterus which lead to nidation of the ova and pregnancy. They also seem to show that progesterone can induce these changes in the uterus only if the organ has been subjected previously to the influence of oestrogen. Evidence will be adduced later to prove that this successful co-operation between progesterone and oestrogen depends upon the relative quantities in which they are provided, and that an excess of oestrogen will prevent the normal action of progesterone on the uterus.

Retarded nidation of ova. Although corpora lutea are essential for the nidation of ova they do not necessarily expedite the process; in the presence of corpora lutea a long interval may intervene between the formation of a blastocyst and its attachment to the uterus. Lataste (1891) noticed that in some animals gestation is prolonged if conception occurs during lactation, and yet the newborn are normal in weight. In such instances the duration of pregnancy varies with the number of young which are being suckled at the beginning of gestation. As there is no increase in the size of the newborn young in spite of the prolonged gestation, development must have been arrested or retarded at some stage, and Lataste, whose observations were made on five different rodents, found that the delay followed immediately after the entry of the ova into the uterus. He also discovered that delayed development of the ova could be caused by injuries received by the mother. By wounding the uterus on the 2nd, 3rd or 4th day after coitus gestation could be prolonged by as much as 10 days in some instances, and a still greater extension of the period of gestation was caused by repeated cautery of the mother's thigh.

Lataste's observations have since been confirmed and extended (Daniel, 1910; King, 1913; Kirkham, 1916). Daniel, like Lataste, noticed that in mice the prolongation of gestation by suckling depended to some extent on the number of the young which were being nursed. If more than three were being suckled gestation was prolonged approximately by about 1 day for each additional nursling. Kirkham (1916), who also made his observations on mice, found that ova fertilized during lactation developed normally up to the blastula stage, which was reached on the 6th day. At this stage, in mice which were suckling, development of the embryos was temporarily suspended; the blastulae underwent little if any visible change and they remained free in the uterus, nidation being delayed. Eventually nidation took place and embryological development was resumed in a normal manner. The prolongation of pregnancy corresponded with this delay during the blastula stage.

Enzmann, Saphir & Pincus (1932; see also Enzmann, 1935), using mice, have obtained results like those just quoted. Normally the implantation of ova in the mouse occurs on the 6th day after mating. If the mother is suckling at the time, implantation, they say, may be postponed up to the 16th day after mating; meanwhile the blastocysts remain at rest and show little or no development beyond that attained on the 6th day in a normal pregnancy.

In some animals the fertilized ova, as blastocysts, may remain at rest for long periods in normal circumstances. This phenomenon has been discussed by Hamlett (1935). In an armadillo (*Dasypus novemcinctus*), he says, the average date of ovulation is 15 July. A blastocyst forming a few days later will lie free in the uterus without further development until November, when nidation takes place and development is resumed. During the interval the corpus luteum is the most conspicuous feature of the ovary though it appears to be inactive. When implantation occurs secretory droplets are seen in the luteal cells. Removal of the corpus luteum during the resting stage does not lead to the immediate destruction of the blastocyst, though it prevents its ultimate nidation. Hamlett mentions other animals in which a delayed implantation of the blastocyst is a normal occurrence. The roe deer (*Capreolus capreolus*), he says, mates in August and implantation of the ovum does not occur until December. The European badger (*Meles meles*) mates in July or August and nidation of ova does not take place till January. Similar delays of development in the blastocyst stage normally occur in some other animals (martin, bear, weasel, shrew).

In connection with delayed nidation of the blastocyst reference may be made to a phenomenon known to breeders, namely the bearing of a second litter in the absence of postparturitional mating. The occurrence appears to depend upon copulation during pregnancy, fertilization occurring at the postparturitional ovulation in cases where gestation has been confined to one uterine horn. Among inbred Wistar rats under the author's observation the phenomenon occurred in 2 per cent of 473 pregnancies (Burrows, 1941). The average interval between the successive litters was 24·4 days, the limits being 21 to 28. It is evident either that the survival of spermatozoa in an active state in the special circumstances exceeded the usual limits, or that there was a delayed nidation of the blastocysts, probably due to the presence of lactational corpora lutea.

(ii) *The role of progestin in the formation and maintenance of the placenta.* The nature of the stimuli which lead to placental formation was investigated by Loeb (1907). He discovered that if slight injuries were inflicted on the endometrium of guinea-pigs between the 4th and 6th days after coitus, nodules of decidual tissue appeared at the sites of injury. These nodules could be caused apart from the presence of an ovum in the uterus, as shown by the fact that the numbers produced corresponded with the number of injuries and not with the number of ova present. Between the 3rd and 4th weeks of gestation the artificially induced deciduomata underwent necrosis. Hammond (1917) performed the same kind of experiment on rabbits. He mated females with vasectomized males and on the 7th day after coitus he injured the uterus by snips or slits and noted that placentomata grew at the site of injury if corpora lutea were present; in the absence of corpora lutea no placentomata developed. Long & Evans (1922 *a, b*) found that placentomata were not caused by uterine trauma at any stage of the short oestrous cycle in the rat; in this animal the corpora lutea which develop during normal oestrus do not become functional as they do in pregnancy, pseudo-pregnancy and lactation, or in animals which have more prolonged oestral cycles. It was shown by Corner & Warren (1919) that placentomata can be induced in the presence of the lactational corpora lutea which follow oestrus immediately after parturition in the rat, and Evans (1923) reported that they can be caused when corpora lutea of pseudopregnancy are present.

Placenta formation may be induced in rats (Teel, 1926) apart from pregnancy, pseudopregnancy or lactation, by giving repeated doses of a luteinizing pituitary extract. Under this treatment oestrus is delayed while the corpora lutea become enlarged, functional and persist beyond the natural period. The presence of an ovary is essential in this experiment, which proves that placentomata in the rat owe their existence to corpora lutea and not to a direct influence on the uterus of a pituitary hormone. Parkes (1929) found that placentomata could be caused in the mouse by passing a thread transversely through the uterus during pregnancy, pseudopregnancy or lactation; they could be brought about also during dioestrum after subcutaneous injection of LH. Courrier (1930b) showed that placentomata could be induced in the guinea-pig's uterus in the presence of corpora lutea of lactation, and Brouha (1932) obtained similar results in the pseudopregnant rabbit.

As with progestational changes, the formation of a placenta needs the co-operation of progesterone with oestrogen and the two hormones must be available in the correct relative proportions. If oestrogen is either deficient or in excess, decidual tissue will not develop under the influence of progesterone.

Courrier & Kehl (1930), in a series of experiments, transfixed with thread the sterile horn of a unilaterally pregnant rabbit's uterus at various stages of gesta-tion; if the operation were done on the 5th, 6th or 8th day after mating, a placentoma was induced, but no such result followed if the uterine trauma were inflicted on the 12th, 15th or 20th day. Apparently the presence of a placenta inhibits a decidual response to uterine trauma, and this fact might be explained by the formation in the placenta of oestrogen in large amount relatively to the available progesterone. Comparable observations have been made on rats by Selye & McKeown (1935), who limited pregnancy to a single uterine horn by ligating one of the oviducts, and then placed threads in the sterile horn or in the pregnant horn between the embryos. If this operation were done on the 5th day of gestation placentomata formed at the sites of trauma, but no decidual reaction followed if the uterus were injured between the 9th and 15th days. Selye & McKeown found also that placentomata could be induced in the rat's uterus during lactation if the uterus were injured between the 4th and 13th days of lactation, but not if done during the first 3 days after parturition. In this instance the prevention of a decidual response may be attributed perhaps to an excess of oestrogen accompanying the postparturitional oestrus.

Selye & McKeown (1935) state that, in the rat, blood sinuses form at the site of an artificially produced placentoma through liquefaction of decidual cells; from this observation it seems that the sinuses are not caused by foetal chorionic villi as some have thought, for in these experiments there were no embryos. The placental sinuses appear to be purely maternal structures induced by progesterone.

Another histological change described by Selye & McKeown in the maternal placenta is a metaplasia in the arterial endothelium, the cells of which become cylindrical and project into the lumen of the vessel. This change also must be included, apparently, among the biological reactions induced by progesterone.

(iii) *The effect of progesterone on the glycogen content of the endometrium.* Among the changes caused by progesterone in the endometrial epithelium is an accumulation of glycogen. Lendrum & Hisaw (1936) observed that in monkeys under treatment with oestrone glycogen is absent from the endometrial cells except perhaps in mere traces. On giving progesterone, they say, one of the earliest manifestations in the uterine epithelium is a migration of the nuclei toward the lumina, leaving a clear zone between them and the basement membrane. In this zone glycogen is concentrated. If the treatment with progesterone is continued the glycogen masses become transferred to the luminal ends of the cells and the nuclei return to their former position near the basement membrane.

Hisaw & Greep (1938) found that oestradiol and progesterone when given together caused an accumulation of glycogen in the uterine glands of the spayed monkey. With larger doses of oestrogen the accumulation of glycogen increased, but the glycogen was not released into the lumina of the glands. A function of progesterone on the endometrium appeared to be the accumulation and the release of glycogen, the amount of which in the endometrial cells was greatest during the secretory phase. Zondek & Stein (1940) estimated the glycogen content of the human uterine mucosa at different stages of the oestral cycle and found that the concentration of glycogen was 0·1 per cent of fresh tissue in the follicular phase and between 0·25 and 0·66 per cent in the luteal phase. They state further that uterine glycogen was deficient in 18·4 per cent of the cases of sterility in women which they investigated.

Apropos of these observations attention may be given to some experiments by Brunelli (1935), in which he found that, whereas oestrin causes a mobilization of hepatic glycogen with a consequent hyperglycaemia, progestin favours the accumulation of glycogen in the liver with hypoglycaemia.

(iv) *The influence of progesterone on the maintenance and duration of gestation.*
(a) *Maintenance of pregnancy.* Parkes (1928a) showed that corpora lutea are essential for the maintenance of pregnancy to its normal term in mice. In twenty-five mice when 3 weeks old the right ovary was sterilized by exposure to X-rays. The mice were allowed to become pregnant and between the 11th and 17th days of pregnancy the healthy left ovary was removed from seventeen of the mice, and in the other eight mice only the right sterilized ovary was removed. Depriving the mice of the normal ovary alone led to the termination of pregnancy; removal of the sterilized ovary alone did not affect gestation. Pencharz & Long (1932) performed similar experiments on rats with like result. Johnson & Challans (1930) found that if rats were spayed between the 6th and 21st days of gestation abortion followed within 48 hours; this result, they showed, could be prevented by giving daily injections of an extract of corpora lutea. Harris & Pfiffner (1929) had already performed a similar experiment with the same consequences.

Courrier & Kehl (1938 a, b) found that daily doses of 2·5 to 3 mg. of progesterone prevented abortion in rabbits which were spayed between the 6th and 14th days of pregnancy. With daily doses of 0·75 or 1 mg. of progesterone given in similar conditions gestation was partially maintained, i.e. some of the embryos survived and some died. Courrier & Jost (1939) showed that the influence of progesterone in sustaining pregnancy in the rabbit after removal of the ovaries could be annulled by oestrone if given in a proportion of 1 of oestrone to 325 of progesterone by weight.

(b) *Duration of gestation.* Progesterone not only has an influence in the maintenance of pregnancy, but plays an important part also in the onset of parturition. By a continued supply of progesterone labour may be postponed beyond its normal date. Teel (1926) noted that the daily administration of an alkaline extract of bovine anterior pituitary caused in rats a prolongation of gestation for 2 to 6 days beyond the normal period, the foetuses on delivery being above the normal in size. In these circumstances the ovaries were enlarged, weighing 126 mg. as compared with 75 mg. in the untreated rat at term, and consisted almost entirely of corpora lutea. Evans & Simpson (1929a) also observed that daily implantations of fresh pituitary substance or injections of a pituitary extract toward the end of pregnancy in rats led to the persistence of corpora lutea and the prolongation of gestation. Nelson, Pfiffner & Haterius (1930) gave extracts of sows' corpora lutea to pregnant rats, beginning the injections on the 16th, 17th or 18th day of pregnancy. By this treatment gestation could be maintained for as much as 150 hours beyond its normal period, which for the rat is 21 or 22 days. If parturition were delayed in this way for more than 70 hours still-births occurred, but with less delay the young were viable.

Katzman, Levin & Doisy (1931) and Levin, Katzman & Doisy (1931) reported that oestrin-free chorionic gonadotrophin when given to pregnant rats during the second half of gestation caused parturition to be deferred. The foetuses usually died *in utero* and were often larger than normal. The result was attributed to the maintenance of luteal activity under the influence of chorionic gonadotrophin. The ovaries were larger and showed more luteinization than those of normal pregnant rats. The same extracts given to pregnant rats after removal of the ovaries did not cause gestation to be prolonged.

The conclusion that continued activity of the corpora lutea is the cause of postponed labour in the experiments just quoted has been questioned by Hain (1932). She confirmed that gestation in the rat can be prolonged by the administration of pituitary substance or gonadotrophic extracts, but it seemed difficult to attribute the result entirely to persistent luteal activity, because in three of her rats in which pregnancy had been prolonged by the daily administration of gonadotrophin the ovaries consisted of a mass of very large follicles and contained no corpora lutea; moreover, parturition was delayed for several days by injections of chorionic gonadotrophin even in rats which had been spayed.

Whatever part the ovaries may play in determining the date of parturition it appears that the placenta also has an influence on this. Newton (1935) killed all the foetuses in pregnant mice between the 12th and 15th days of gestation by squeezing them *in situ* under anaesthesia. This operation left the placentae intact, and they were delivered at the normal term, oestrus following within the next 2 days. If all the foetuses and their placentae were removed between the 12th and 15th days of gestation oestrus ensued between 3 and 4 days afterwards. Kirsch (1938) made comparable observations on pregnant rats. Between the 14th and 21st days of gestation the embryos were removed from the uterus, the placentae being left undisturbed. Regardless of the date at which the operation was performed, the placentae remained until full term had been reached and were then delivered. If the placentae and embryos were both removed and replaced by paraffin pellets of the same size and shape these were delivered between 40 and 48 hours later. But if placentae only were left in one pregnant horn, and paraffin pellets placed in the other horn from which the embryos and placentae had been removed, gestation continued until.full term, when the placentae and pellets were delivered.

Snyder (1938) says that if ovulation is induced in a rabbit on the 25th day of gestation by an intravenous dose of chorionic gonadotrophin, parturition fails to occur at the normal time (32 days) because a new set of corpora lutea will have formed. In this event labour will be postponed till the 40th day, the increased length of gestation corresponding with the active life of the new corpora lutea (12–14 days). The newly formed corpora lutea in this experiment can be shown to be functional by microscopical and biological tests.

The experiments on gestation just described can be explained on the grounds that (1) progesterone inhibits labour and (2) that it is formed, in some animals, by the placenta as well as by the corpora lutea. The delay of parturition by progesterone has, been demonstrated by several independent workers. Portman (1934) gave daily injections of 2 or 3 rb.u. of progesterone to rabbits towards the end of gestation, with the result that parturition was delayed for 2 or 3 days beyond its normal time, the young being born alive. With daily doses of 4 rb.u., gestation was prolonged for as much as 9 days and the young were still-born. A similar experiment was carried out by Heckel & Allen (1937), who gave daily doses of progesterone to pregnant rabbits, beginning on the 20th, 25th or 28th day of gestation. Daily doses of 1·5 mg. or more caused a delay of parturition, and in nearly every instance in which delivery occurred later than the 34th day of gestation the young were larger and had a more advanced development of hair than young rabbits born at the end of a normal gestation period of 32 days. The ovaries of the mother in these examples of delayed parturition showed atrophic corpora lutea, which therefore could not have been the cause of the postponement of labour.

Cope (1940b) has recorded a case in which the corpus luteum together with an ovarian cyst was removed from a woman on the 67th day of pregnancy. The operation was followed by abortion and by a rapid fall in the excretion of pregnanediol, the urinary output of which was 7·2 mg. on the 2nd day, 4 mg. on the 3rd day, and 1·2 mg. and 1 mg. on the 5th and 6th days.

When considering these results two possible factors in the termination of gestation have to be remembered, namely (1) the dependence of the placenta on a sufficient supply of progesterone without which it will degenerate and (2) the effect of progesterone in preventing the action of oxytocin and pitressin on the uterine muscle, which will be discussed presently. Whether the hour of labour depends primarily on the placental or the muscular factor may be uncertain, but there seems little doubt that parturition is preceded by a fall in the supply of progesterone relative to that of oestrogen, though other factors may share in starting the process.

(v) *The effect of progesterone on the motility of the uterus.* In 1906 Cushny and Dale independently reported that during pregnancy the uterus of the cat does not respond to adrenalin or to stimulation of the hypogastric nerves like the uterus of a non-pregnant cat. In the latter adrenalin or hypogastric stimulation inhibits the tone and the rhythmical contractions which characterize the normal unim-pregnated uterus. During pregnancy adrenalin or hypogastric stimulation causes the uterus to contract. These striking observations acquired additional interest when Knaus (1926) demonstrated that the response of the uterus to pituitrin is inverted during the earlier stages of pregnancy. While testing the effects of the posterior lobe of the pituitary on pregnant rabbits, Knaus found that the uterine responses varied greatly at different stages of gestation. At full term (32 days) 0·075 mg. of moist posterior lobe tissue caused the birth of the whole litter within 10 minutes. On the 29th day 0·6 mg. was required to produce the same effect. Between the 28th and 29th days a sudden change occurred in the responsiveness of the uterus to pituitrin, so that before the 28th day of gestation no amount of posterior pituitary lobe induced parturition on the same day as the injection, though a dose of 1·3 mg. caused foetal death accompanied by bleeding in the placenta. The earlier the stage of gestation the larger was the dose of posterior pituitary substance required to induce parturition; from the 1st to the 17th day 15 mg. per kilo of the rabbit's bodyweight failed to terminate pregnancy. From these experiments it appears that the uterus of the pregnant rabbit becomes in-creasingly responsive to the secretions of the posterior lobe of the pituitary after the 17th day of gestation, attaining a maximum sensitivity at the time of parturi-tion. Knaus (1929, 1930 a, b) further showed that when a non-pregnant rabbit had been given subcutaneous or intramuscular injections of extracts prepared from the corpora lutea of pigs and cows, the uterus failed to contract in response to posterior pituitary extracts. He suggested that this phenomenon might be used as a criterion for the standardization of luteal extracts. Following Knaus, Robson & Illingworth (1931) made extensive tests of an extract prepared from the corpora lutea of cows on the reactions of the rabbit's uterus. To adult spayed rabbits they gave six injections of the extract at intervals of 12 hours and the animals were killed 12 hours after the last injection. One uterine horn of each rabbit so treated was examined microscopically and the other was placed in Ringer's solution and its reactions to the pressor hormones of the posterior lobe of the pituitary were examined. Even though the progestational changes were small as seen with the microscope, the reactions of the uterus to both oxytocic and pressor fractions of the pituitary were found to be suppressed. Doses of oestrone up to 3,000 m.u. did not affect these results, although oestrone by itself increases the reactivity of the uterus to oxytocin (Robson, 1933 a). Robson & Illingworth (1932) found also that contraction of the uterus in response to pituitrin could be prevented by an extract of placenta.

Reynolds & Allen (1932) observed the movements of the rabbit's uterus by means of a uterine fistula. In an adult rabbit ready for mating the uterus showed continued rhythmical contractions. Shortly after sterile copulation these move-ments ceased. They could also be arrested in most instances by daily subcutaneous injections of an extract of sows' corpora lutea, each injection representing 0·2 of a rb.u. of progestin. Uterine motility reappeared from 2 to 5 days after the dosage had been stopped. During the first 5 days after mating 200 r.u. of oestrone given daily by intravenous injection failed to restore the uterine movements. Klein & Klein (1933), having investigated the influence of the corpora lutea on uterine motility, state that during pregnancy, pseudopregnancy, or treatment with a luteal extract, the uterine muscle of the rabbit is insensitive to pituitrin, in spite of the fact that the spontaneous contractions and relaxations of the uterus continue. On the 12th day of pseudopregnancy, the uterus being refractory to pituitrin, the corpora lutea were destroyed by thermocautery. Three days later the uterus was no longer irresponsive to pituitrin. Klein & Klein made the further observation that the irresponsiveness of the uterus to pituitrin can be prolonged beyond the period normal to pseudopregnancy by injecting extracts of corpora lutea.

Robson (1935a) showed that progesterone—in contrast with oestrone, which has the opposite effect—inhibits the response of the uterine muscle to oxytocin in the rabbit whether spayed or deprived of its pituitary. Some doubt has arisen (Robson, 1936a) whether these results occur in every species alike. However this may be, the effects of progesterone in man appear to be the same as those observed in the rabbit. Falls, Lackner & Krohn (1935, 1936) tested the action of progesterone on the motility of the human uterus by inserting a bag into the cavity of the organ and connecting it by tubing with a recording tambour and drum. They found that as small a dose as 1 rb.u. of progestin prevented the oxytocic action of 1 c.c. of pituitrin given intramuscularly. This inhibitory action occurred whether the progesterone was given before or after the pituitrin. The effect was apparent in a few minutes and lasted for some hours.

Kurzrok, Wiesbader, Mulinos & Watson (1937), by a method similar to that of Falls, Lackner & Krohn, examined the uterine movements in fifty women. They observed that spontaneous contractions and relaxations of the uterus continue throughout the menstrual cycle, and that at every stage of the cycle an injection of pituitrin causes uterine contraction. In other words they could detect no inhibition of the uterine response to pituitrin attributable to the presence of corpora lutea during the menstrual cycle. Wilson & Kurzrok (1938) by the same methods could detect a difference in the activity of the uterine muscle in women at different stages of the oestrous cycle; the follicular phase, they say, is characterized by a high tonus with small, rapid contractions, whereas the luteal phase is accompanied by a low tonus with large, slow contractions.

The reversed action of hypogastric stimulation on the cat's uterus during pregnancy, as reported by Dale & Cushny, can be brought about in the nonpregnant cat by progesterone (Kennard, 1937). Cats were spayed and some days later were given an intraperitoneal injection of progesterone; at varying intervals afterward the abdomen was opened and the effects of stimulating the hypogastric nerves recorded. In this way Kennard found that doses of progesterone varying from 0·4 to 1 mg. brought about contraction instead of relaxation of the uterus in response to hypogastric stimulation. This reversal of response ensued between 3 and 6 hours after the injection of progesterone.

(vi) *The adaptation of the uterus to accommodate the growing embryo.* The presence of a living embryo is not the only cause of uterine hypertrophy during pregnancy. In cases of extrauterine gestation the uterus enlarges, though not to the same extent as in a normal pregnancy; and inert pellets inserted into the cavity of a non-pregnant animal's uterus will induce a local hypertrophy of the organ. We have therefore to consider two factors in the enlargement of the uterus during pregnancy, namely (1) a mechanical stimulus and (2) a hormonal agent. These two factors appear to act in co-operation.

Reynolds (1937) introduced cylindrical paraffin pellets into the uterus of spayed rabbits; the result was a uterine hypertrophy which reached its maximum in 4 days. The amount of hypertrophy induced depended on the size of the pellets. If these were too large or too small hypertrophy of the uterus did not occur or was of small amount; the greatest response was obtained when the diameters of the pellets were between one-half and twice that of the uterus. If the rabbit carrying one of these intrauterine pellets were given oestrone, the uterine hypertrophy was hindered, whereas progesterone encouraged the hypertrophy and enabled the uterus to enlarge in response to pellets which would have been too large if progesterone had not been given. Reynolds suggests that during pregnancy progesterone acts in this way because it reduces the tone of the uterine muscle, and so enables the uterus to become accommodated to the growing foetus. Reynolds & Allen (1937), as a result of this sort of experiment, learned that to cause enlargement of the uterus more distension was needed if progesterone were given than if it were withheld, and if oestrone were given the distension required was less.

Co-operation and Antagonism between Progesterone and other Gonadal Hormones

References have been made on other pages to the co-operative and antagonistic effects of oestrogen and progesterone when acting simultaneously, and to the way in which these effects are determined by the relative proportions in which the hormones are available and the sequence of time during which they are in action (pp. 117, 413). It may be useful here to amplify the subject so far as concerns the uterus.

(i) *Co-operation between progestin and oestrogen in the production of pro-gestational changes and placentomata in the uterus.* Hisaw (1929) reported that progestational changes in the uterus could be brought about by an extract of sow's corpora lutea in spayed rats and guinea-pigs, a result which may be explained to-day by the fact that corpora lutea in their earlier stages supply oestrogen in addition to progesterone. Later Hisaw & Leonard (1930) showed that, in order to evoke progestational changes by an oestrin-free luteal extract the uterus must first be sensitized with oestrin. They spayed rabbits within a few hours of copulation, and during the next few days gave them injections of oestrin-free luteal extract, or oestrin, or both of these hormones together or in succession. They found that the luteal extract induced progestational changes in the uterus only if oestrin had been administered previously. They gave 20 r.u. of oestrin twice daily for 5 days, and then, discontinuing the oestrin, they gave injections of luteal extract 4 times daily for 4½ days. After this treatment the uterus showed a progestational response. Clauberg (1930), using immature rabbits weighing about 600 g., also found that progestational changes could be induced by a luteal extract only if this were given after a preliminary treatment with oestrogen.

The naturally formed oestrogens are not the only substances which are capable of sensitizing the uterus to the action of progesterone; De Fremery & Geerling (1939) found that a similar sensitization could be caused by diethylstilboestrol. To immature rabbits during a period of 8 days they gave daily doses of the oestrogen to be tested, and for the next 5 days two doses, each of o·3 mg. of progesterone, were given daily. In every instance full progestational effects were produced (Table 192).

TABLE 192.   Sensitization of rabbit's uterus to the action of progesterone by oestrogen (De Fremery & Geerling, 1939)

Number of rabbits	Oestrogen	Daily dose of oestrogen during 8 days (γ)	Daily dose of progesterone during 5 days (mg.)	Full progestational changes in uterus in all
5	Diethylstilboestrol	15	o·6	+
2	Oestrone	15	o·6	+
3	Oestrone	15	o·6	+

Experiments have been recorded which seem to show that the co-operation of oestrogen may not be essential for the progestational effects induced in the uterus by progesterone if large enough doses of the latter be given. Selye (1940a) spayed rats and 9 days later began to give them 15 mg. of progesterone daily; at the end of a further 10 days these large doses had caused progestational changes in the uterus. The same doses permitted the development of deciduomata in response to uterine trauma. Howard & Gengradom (1940) gave 1·5 mg. of progesterone intraperitoneally to spayed mice every day for a fortnight, at the end of which time the diameter of the uterus had trebled or quadrupled and advanced progestational changes were present. In these experiments, it will be noted, the doses of progesterone were very large and the results would hardly condemn the belief, which appears well established by previous work, that in natural circumstances the co-operation of oestrogen is required for the production of progestational changes in the uterus by progesterone.

(ii) *Antagonism between oestrogen and progesterone.* (a) *Inhibition of the action of oestrogen by progesterone.* Macht & Stickels (1931) found that in mature non-pregnant mice the ratio of the weight of the uterus to that of the whole body was not more than 1 : 400. After daily injections of 1 to 2 m.u. of oestrin given for a period of 10 days this ratio was increased to 1 : 282. In another group of mice which had received the same doses of oestrin together with an extract of corpora lutea the ratio was 1 : 465. In this experiment the enlargement of the uterus which normally follows the administration of oestrone had been prevented by simultaneous doses of an extract of corpora lutea.

Hisaw & Lendrum (1936) tested the antagonistic actions between progesterone and oestrin in immature or young adult spayed monkeys. Oestrin given daily to these animals caused a squamous metaplasia in the glands of the uterine cervix. This reaction could be quite inhibited by progesterone ('corporin'). At the end of 59 days, during which the monkeys had received 100 r.u. of oestrin and 4 rb.u. of progesterone daily, there was no squamous epithelium to be found in the uterus. For this result 1 rb.u. of progesterone per diem was enough.

Dessau (1937) found that the earlier hyperplastic effects of oestrogen on the uterus could be prevented by progesterone when given in sufficient doses. His observations have been corroborated by Korenchevsky and Hall (1938b, 1940). They spayed 26 rats soon after weaning and to some of them gave daily injections of oestrogen alone and to others oestrogen and progesterone, during a period of 21–27 days. The daily doses given were of oestrone or oestradiol 30γ, and of progesterone 400γ. In the rats which received progesterone hyperplasia of the uterus in response to the oestrogen did not occur.

There is an indirect way by which progesterone can interfere with the action of oestrogen. Herlant (1939) and Burrows (1940) have shown that progesterone may inhibit the output of FRH from the pituitary and by this means check the formation of oestrogen by the ovary.

Furthermore, it has been suggested that progesterone may perhaps accelerate the excretion of oestrogen by the kidney. Smith and Smith (1931) noticed that when a dose of 600 or 700 r.u. of oestrin were injected into normal, non-pregnant rabbits the amount of oestrogen subsequently detected in the urine was much less than when a similar dose had been accompanied by progesterone. They also observed that the injection of LH into a normal female rabbit increased the urinary excretion of oestrogen tenfold, and a comparable increase followed injections of progesterone. LH had no such effect on spayed rabbits. Frank, Goldberger and Spielman (1932) were unable to confirm these observations.

In a later report Smith and Smith (1938) say that progesterone to some extent protects oestrogen from inactivation, so that it is available in increased amount for utilization in the body and for excretion in the urine.

Pincus and Zahl (1937) reported the results of some experiments which have a bearing on this problem. They determined the amounts of oestrone and oestriol respectively in the urine of rabbits under various conditions. Rabbits which were pregnant or pseudopregnant, and therefore had active corpora lutea, excreted with the urine between three and four times as much oestriol as that excreted by female rabbits without active corpora lutea. A similar increase of oestriol in the urine followed the injection of progesterone. Hysterectomy prevented this enhanced output of oestriol and Pincus and Zahl concluded that the uterus under the influence of progesterone converts oestrone into oestriol, which has less oestrogenic potency and is more easily excreted than oestrone.

(b) *Inhibition of the action of progesterone by oestrogen: the placentoma reaction.* Loeb (1907) demonstrated that a slight injury inflicted on the guinea-pig's uterus at a time when functional corpora lutea are present will result in the development of a deciduoma (placentoma) at the site of injury. Courrier (1930b) noticed that in the guinea-pig's ovary, a few hours after parturition follicles rupture and corpora lutea of lactation are formed. These corpora lutea are capable of bringing about the decidual reaction of Loeb in response to uterine trauma. If, however, oestrin be given in sufficiently large doses this decidual formation is entirely prevented.

Brouha (1932) caused sterile mating in rabbits and on the 5th, 6th or 7th day passed a thread through the uterus. In normal rabbits this procedure always caused a decidual growth in the uterus, but this reaction was prevented by daily doses of 100 r.u. of oestrin.

Votquenne (1934) also observed that oestrin prevented the formation of placentomata. To be sure of the presence of active corpora lutea, he used lactating rats and ascertained that in them uterine injuries were constantly followed by the formation of placentomata. If, however, oestrin were given in doses ranging from 2 to 18 r.u. on each of the 5 days following the operation, the decidual action might be entirely prevented. With daily doses of half a rat unit a partial inhibition of the decidual reaction occurred. No such effect followed daily doses of less than 0·4 r.u. of oestrin.

Progestational changes in the uterus and the nidation of ova. In normal conditions the changes in the uterus which are necessary for the nidation and welfare of the ova require a preliminary submission of the uterus to the action of oestrogen followed by that of progesterone. Hisaw & Leonard (1930; see also Leonard, Hisaw and Fevold, 1932), using spayed rabbits, noted that if too much oestrogen be given progestational changes in the uterus will be prevented.

Courrier (1930b) had previously reported that injections of oestrone given 24 hours after coitus entirely prevented the appearance of progestational changes in the rabbit's uterus. Courrier (1933), Courrier & Kehl (1933a, b) and Courrier & Raynaud (1934a, b) showed that in the rabbit progestational changes in the uterus and nidation of ova could both be prevented by suitable doses of oestrone given on each of the 6 days following coitus (Table 193).

TABLE 193.   The effects of oestrone on the uterus and nidation of ova in rabbits (Courrier & Raynaud, 1934a)

Dose of oestrone (i.u.)	Progestational changes in the uterus	Nidation of ova
180	+	+
400	I	—
1,200	—	—

Courrier (1936) made a quantitative estimation of the amount of an oestrogen required to inhibit the effects of progesterone on the uterus. Rabbits were mated and spayed on the next day. Doses of 1·5 mg. of progesterone together with varying amounts of oestradiol were given twice a day for 5 days and the rabbits were killed on the 6th day. The results (Table 194) demonstrate the absence of inhibition with smaller doses of oestradiol and its presence with larger doses.

TABLE 194.   The inhibition by oestradiol of the action of progesterone on the spayed rabbit's uterus (Courrier, 1936)

Doses of progesterone (mg.)	Doses of oestradiol (mg.)	Progestational changes in uterus
1·5	0·08	—
1·5	0·04	—
1·5	0·03	—
1·5	0·02	—
1·5	0·01	+
1·5	0·005	+ +
1	0·0025	+

A curious species-difference in the reactivity of the uterus to sensitizing doses of oestrone has been described by Rowlands & McPhail (1936), who estimated by experiment that 70 times the amount of oestrone is required to sensitize the cat's uterus to progesterone as will be effective for this purpose in the rabbit.

The antagonism between oestrogen and progesterone as shown by the prevention of progestational changes in the uterus is well exemplified by some experiments by Robson (1937a). He spayed rabbits and 12 or 13 days later gave them 0·01 mg. of oestrone daily for a week, at the end of which period some of the rabbits were given progesterone alone and others were given oestrone in addition to progesterone. In the former progestational changes developed, in the latter they did not. Doses of 0·75 mg. of progesterone were nullified by 0·01 mg. of oestrone, and doses of 1·5 mg. of progesterone were nullified by 0·04 mg. of oestrone.

W. M. Allen (1937) states that a large dose of oestrin given to a rabbit during the first 5 days after ovulation, or to a spayed rabbit which has received injections of oestrin and progestin, will quite prevent the occurrence of progestational changes in the uterus. If given after the 6th day, that is to say when the uterus has already attained a progestational state, single large doses of oestrin will cause extensive degeneration of the endometrium. The action of 3 rb.u. of progestin on the rabbit's uterus will be suppressed, Allen says, by 675 r.u. of oestrin.

THE ROLE OF PROGESTERONE IN MAMMARY DEVELOPMENT

(a) *The nipple.* Folley & Bottomley (1941; see also Folley, Scott Watson & Bottomley, 1941) observed that in the kid the teats grow isometrically with the body for a while and then grow more rapidly in relation to the body, perhaps because of gonadal activity. During the breeding season, however, the teats cease to grow. This inhibition was thought perhaps to be an effect of progestin. Experiment showed that the growth of teats in a male could be inhibited by pregneninolone, and that the stimulating effect of oestrogen on growth of the teats could be prevented by progesterone.

(b) *The effect of corpora lutea on the formation of alveoli in the breast.* It is now recognized that the main initial factor in mammary development is a sufficient though not excessive supply of oestrogen. The constant association of corpora lutea with full development and secretory activity of the breast suggests that progesterone also may play a part in these phenomena. Ancel & Bouin (1909; see also Bouin & Ancel, 1909) performed experiments on rabbits which seemed to prove that corpora lutea are essential both for the development of the mamma and for lactation in this animal. They found that the virgin rabbit's mammae are between 2 and 2·5 cm. in diameter. To study the relationship between the formation and activity of corpora lutea and mammary development they induced pseudopregnancy in rabbits by mating them with vasectomized males. In this way they avoided having to consider any possible effect of foetus and placenta. Periodical examinations of the mammae and ovaries of these animals showed a rapid growth of mammary ducts and acini following the formation of corpora lutea. This mammary development continued until the 14th day, that is while the corpora lutea remained active. With cessation of luteal activity, growth and secretion in the mammae came to an end. As the time elements corresponded so closely, and as the corpora lutea were the only new factor in the organism, Ancel & Bouin concluded that mammary growth and secretory activity in the rabbit depend on the corpora lutea. Pursuing the matter they destroyed the corpora lutea by thermocautery in pseudopregnant rabbits at various intervals after mating. If this destruction was done a few hours after the appearance of corpora lutea no mammary development occurred; if done some days later all further growth in the mammae was immediately arrested. It was demonstrated clearly that coitus did not provide an initial stimulus for mammary development because corpora lutea could be induced in the virgin rabbit's ovary merely by opening ripe follicles with a needle or fine scissors, and in these circumstances mammary development again coincided with the formation of luteal tissue.

So far as concerns the rabbit the conclusions of Ancel & Bouin have been confirmed by subsequent inquiries, and have been found applicable to some but not, perhaps, to every other species. O'Donoghue (1911) studied the development of the mammae in the marsupial spotted cat (*Dasyurus viverrinus*) of Australia and observed that the most active growth of the mamma after coitus occurs before the embryo has become attached to the uterus and therefore is not dependent on a placental stimulus. On the other hand, he noticed that mammary development is closely connected with the presence of active corpora lutea. Hammond (1917) submitted rabbits to sterile coitus and later examined their ovaries and mammae. He found that when, as usual, the mating was followed by the development of corpora lutea, mammary development also occurred, whereas in the absence of corpora lutea no mammary changes were present. Asdell & Salisbury (1933) observed that if the ovaries were removed from a rabbit after sterile coitus mammary development was at once arrested. Marshall & Halnan (1917) had already shown that in the dog, as in the rabbit, growth of the mammae coincided with secretory activity of the corpora lutea and that mammary involution closely followed their degeneration.

Gardner (1935b) has pointed out that in untreated female mice, as in the rabbit, mammary acini develop only during pregnancy or pseudopregnancy, that is to say in the presence of active corpora lutea. Mice have a very short oestrous cycle, during which the corpora lutea acquire little or no functional potency; in the absence of pseudopregnancy the virgin rabbit has no corpora lutea because she has no oestrous cycles. The evidence suggests that in these animals the growth of mammary acini, as distinct from ducts, depends largely on the presence of active corpora lutea.

Some caution is required before accepting this explanation as though it were complete. Gardner, Diddle, Allen & Strong (1934) and Burrows (1935e) have shown that acini may sometimes develop in the mammae of male mice which have been treated with oestrogen alone. Gardner, Smith & Strong (1935) gave 500 i.u. of oestrone benzoate once a week to male mice of the C_3H and A strains, beginning before the mice had been weaned. These large doses led to a stunted development of the mammary ducts with an extensive formation of acini. Bonser (1936) and Bonser & Robson (1940) have found that in the males of some inbred strains of mice (R_3, CBA), though not in others (Strong A), oestrogens produce well-developed mammary acini. In guinea-pigs oestrogen alone will cause full mammary development, and in the monkey (*Macacus rhesus*) also the development of acini has been induced by giving oestrogen alone. Gardner & Van Wagenen (1938) gave oestrone in weekly doses varying from 2,000 to 8,000 i.u. to male and immature spayed female monkeys. The first effect of this treatment was to cause an extension of the duct system, but with continued injections an abundant development of acini ensued in both sexes. Hypothetically, two possible solutions of these apparent anomalies may be offered. First, as already stated, oestrogens help to maintain the corpora lutea in a functional state and to that extent assist in the formation of progesterone; secondly, other organs than the corpora lutea may perhaps affect the development of lobules in the breast. In certain species and circumstances the adrenal cortex perhaps may provide sufficient progesterone, or an efficient substitute, e.g. androgen or corticosterone: the biological activities of these hormones have more than one feature in common with that of progesterone. We shall revert later to this possibility.

(c) *The effect of luteal extracts, or of oestrogen and progesterone together, on the formation of mammary acini.* Progesterone by itself appears to have little or no effect on the mamma. Corner (1930) gave progesterone alone to non-pregnant spayed rabbits and found that growth of the mammae was not induced thereby even when the injections were continued for the period of a normal pregnancy. Turner & Schultze (1931) and Turner & Frank (1932) have reported the same results in rats and rabbits. Selye, Brown & Collip (1936) say that 4 mg. of progesterone given daily to rats for a period of 12 days did not cause any recognizable change in the mammae.

The mouse appears to be an exception to this rule, for Gardner & Hill (1936) observed an extensive mammary duct system in castrated and non-castrated male mice which had received injections of progesterone alone.

Many inquirers have found that full mammary development may be induced by appropriate supplies of oestrogen plus progesterone. In the guinea-pig full mammary development follows the administration of oestrogen alone. Nelson (1936a) did not observe any difference between the mammae of guinea-pigs which had received injections of oestrone and progesterone and those of guinea-pigs which had received injections of oestrone only. This result does not invalidate the rule that oestrogen requires the co-operation of progesterone or some other hormone to bring the breasts to full functional maturity. It suggests rather that the effective accessory hormone, whether progesterone, androgen or other compound, may be derived in the guinea-pig from some source apart from the ovary, perhaps from the adrenals, which in the guinea-pig are large compared with those of rats, mice and rabbits.

Observations which have shown the influence of corpora lutea on the growth of the mammary gland may be explained by the fact that corpora lutea in their earlier stages produce oestrogen in addition to progestin, and at the later stages of pregnancy when they form progestin with little or no oestrogen the latter is supplied by the placenta. It may be of interest to consider further experiments which have a bearing on the subject. Nelson & Pfiffner (1930, 1931) caused hypertrophy of the mammae and nipples in gonadectomized immature male and female guinea-pigs by daily subcutaneous injections of an extract of sows' corpora lutea. Such an extract, it is now known, contains both oestrogen and progesterone. Turner & Frank (1931) found that the daily administration of oestrin to gonadectomized male or female rabbits caused extension of the duct system without the formation of acini. However, in a castrated male rabbit which was given

12 r.u. of oestrin daily for 30 days followed by 1 c.c. of a luteal extract plus 12 r.u. of oestrin daily a development of the mammae ensued like that which accompanies pregnancy. A still more pronounced result of the same kind was induced in another castrated male rabbit by giving the same dose of luteal extract with a dose of oestrin larger than that used in the former experiment. Turner & De Moss (1934) induced an extensive development of the mammary duct system in two male cats by the prolonged administration of oestrone (100 r.u. daily for 60 days). After this stage had been reached, daily injections of 1 rb.u. of progestin and 25 r.u. of oestrone were given for 13 days to one and 20 days to the other. At the end of these periods a pronounced lobular system had developed in both animals. Freud & De Jongh (1935) gave 100 i.u. of oestrone daily to young spayed rats for a fortnight, after which they were given 1 rb.u. of progesterone daily, the injections of oestrone being continued. Other spayed rats were treated in the same way except that no progesterone was given. In the animals which had received progesterone and oestrone the mammae resembled those of pregnancy; the milk ducts had narrow lumina, the epithelial cells having rounded nuclei and bulky cytoplasm. In the rats which had received oestrone alone, the mammary ducts showed acini only at their ends, the lumina were wide, the epithelial cells small with flat nuclei and little cytoplasm. When these experiments were repeated on hypophysectomized rats the same results were obtained.

Asdell & Seidenstein (1935) spayed four rabbits and 2 months later removed their pituitaries. After this they were given 25 r.u. of oestradiol benzoate and 4 rb.u. of progesterone daily. Under this treatment mammary enlargement with the development of acini occurred in each instance, and was almost equal in degree to that in rabbits whose pituitaries had not been removed. These changes in the mamma were accompanied, Asdell & Seidenstein say, by progestational changes in the uterus.

Apparently the different species do not all respond alike to comparable doses of progesterone. When considering the experimental results mentioned above particular regard may be paid to the fact that co-operation or antagonism between two different types of gonadal hormone may depend on the relative quantities of each which are available at the same time, and it may be that some of the failures to induce the development of mammary acini by giving progesterone have been due to a deficiency or excess of oestrogen. Lyons & McGinty (1941) gave eighteen doses of oestrone and progesterone during 4 weeks to immature male rabbits. The same doses of oestrone, namely 120 i.u., were given throughout, but in different groups of rabbits the doses of progesterone were 0·25, 1·0, 4·0 and 8·0 i.u. To other rabbits oestrone alone was given. The latter treatment caused growth and dilatation of the ducts without the formation of acini. In the rabbits receiving progesterone in addition to oestrone acini developed. The greatest effect of this kind was induced by doses of 1 i.u. of progesterone combined with 120 i.u. of oestrone, both ducts and acini being well formed. Larger doses of progesterone were followed by poor mammary growth and apparently were inhibitory. Scharf & Lyons (1941) have reported the same results.

Selye (1940a) states that when given to spayed rats in large enough doses progesterone will cause extensive mammary development even if no oestrogen be given. He spayed young adult rats and 2 days later began to give them 15 mg. of progesterone daily. This treatment was continued for 10 days and the rats were killed 24 hours after the last doses. The mammae showed a lobular development like that of late pregnancy. In another series of rats treated in the same way except that the injections of progesterone were not begun until 9 days after the spaying, only slight mammary development was induced.

Deoxycorticosterone has some of the properties of progesterone and may perhaps act as its substitute in assisting development of the mamma. Gardner (1940) gave daily subcutaneous injections of various hormones during a fortnight to hypophysectomized male mice, and found that whereas progesterone and deoxycorticosterone both amplified the action of oestrogen on the mamma, testosterone propionate hindered this action (Table 195). The adrenals perhaps have a larger share of gonadal action in one species than in another.

A compound which is difficult to classify because it has activities which are common to all the three main types of gonadal hormone is pregneninolone. This resembles progesterone in its action on the breast, for in spayed mice it causes development of the mammary lobules, its action in this respect being favoured by a simultaneous supply of oestrone (Mixner & Turner, 1941 b).

TABLE 195. The effect of hormones on mammary development in hypophysectomized mice (Gardner, 1940)

Hormones used	Number of mice	Daily dose	Number of mice showing enhanced mammary development
Progesterone	4	0·125 to 0·25 mg.	1
Deoxycorticosterone	7	0·25 mg.	2
Testosterone propionate	11	0·25 to 1·25 mg.	0
Oestrone	10	0·05 to 1 γ	5
Oestrone plus Progesterone	12	Varied	11
Oestrone plus Deoxycorticosterone	14	Varied	12
Oestrone plus Testosterone propionate	7	Varied	0

Gardner's experiments seem to show that in mice the action of oestrogen and progesterone on the mouse's mamma is direct, and independent of the pituitary. Other workers, however, believe that the action of these ovarian hormones is indirect and is caused by their capacity to stimulate the production of mamma-trophin by the pituitary (p. 335).

The reader may be reminded once more that the term hypophysectomy as used in this kind of experimental work is relative and not absolute. The pars tuberalis is rarely removed and the part remaining may have sufficient potency to confuse the outcome.

The influence of progesterone in lactation. Apart from its co-operation with oestrogen in causing the mammary development requisite for the formation of milk it is possible that progesterone assists lactation in some other way. Such an idea is supported by the fact that active corpora lutea persist during suckling, although when it has been once established lactation may continue in the absence of the ovaries. Indeed there is evidence that when the mammae have been brought to a prelactational state, the secretion of milk will follow removal of the ovaries or the corpora lutea. Hammond (1917) found that in the rabbit lactation follows removal of corpora lutea provided that mammary development is enough advanced. Selye, Collip & Thomson (1933a) caused heavy luteinization in the ovaries of non-pregnant rats by subcutaneous injections of an anterior pituitary extract. If then the luteinized ovaries were excised lactation commenced within 36 hours, though not if the pituitary were, removed at the same time as the ovaries. One may suspect that in this case it was the sudden withdrawal of oestrogen that caused the onset of lactation, for it is known that oestrogen inhibits the flow of milk. On the other hand, it has been shown that progesterone, even when given in large amount, has little if any effect by itself on lactation. Folley & Kon (1937, 1938) gave 1 mg. of progesterone daily for 12 days to lactating rats, beginning on the 5th day after parturition, and were unable to detect any effect on lactation, as judged by the rate of growth of the young. So far as any opinion can be formed on the available evidence it seems that progesterone has a definite role in the maintenance of lactation, namely to inhibit the formation of FRH by the pituitary and the consequent production of oestrogen by the ovary.

Actions common to Progesterone and Adrenal Cortical Hormones

As mentioned earlier (p. 143), there is some community of function between corpora lutea and the adrenal cortices.

(a) *Maintenance of life after adrenalectomy.* Rogoff & Stewart (1927, 1928), while studying the effects of adrenalectomy in dogs, observed that if the operations were done on females during pregnancy or pseudopregnancy the animals did not succumb as they would have done otherwise. In seventeen dogs whose adrenals had been removed during pregnancy life was prolonged in one instance for 58 days and in another for 47 days. One of the dogs lived for 26 days after littering and in this case survival could not be ascribed to the action of the adrenals of the foetuses. No such survival or anything approaching it was seen among seventy-four control dogs after adrenalectomy. These observations were supported by Swingle, Parkins, Taylor & Morrell (1936), who removed the adrenals from five dogs during oestrus. Although cortical hormone was not given these dogs survived for 60, 57, 47, 45 and 40 days respectively; at these dates signs of adrenocortical deficiency supervened. Pseudopregnancy is a regular sequence of unfertile oestrus in dogs, and in one bitch after removal of her adrenals, oestrus and pseudopregnancy were induced by gonadotrophin, after which though no cortical hormone was given the dog remained in good health for 25 days. The results of these experiments suggested that the functional corpora lutea of pregnancy or pseudopregnancy might act as substitutes for the adrenal cortex. In a later paper Swingle, Parkins, Taylor, Hays & Morrell (1937) report that daily doses of 2 mg. of progesterone did not prolong the lives of adrenalectomized dogs and cats in the absence of pregnancy or pseudopregnancy. In view of other work in this field it seems probable that their doses of progesterone were too small. The same kind of experiment was performed on rats by Emery & Schwabe (1936), who excised the adrenals of 200 rats when they were 30 days old, and then gave them implants of pituitary taken from castrated male rats, or extracts of pituitary gland or extracts of pregnancy urine. In females this treatment caused a marked prolongation of life, but only if they possessed intact ovaries. In spayed females or in males the administration of gonadotrophin had no such beneficial effect after removal of the adrenals. Emery & Schwabe also found that 5 r.u. of oestrone given daily failed to prolong life in the absence of the adrenals, and they regarded the corpora lutea which developed under the influence of the artificially supplied gonadotrophin as the sources of the derived benefit. Fischer & Engel (1939) and D'Amour & D'Amour (1939) also obtained an increased survival time in female rats after adrenalectomy by treating them with gonadotrophin.

Gaunt (1937) experimented with the ferret. This animal does not ovulate spontaneously and once in oestrus it remains in that condition for long periods, unless mated. Meanwhile corpora lutea do not develop. If pseudopregnancy is induced in this animal, by means of sterile mating or the injection of gonado-trophin, it lasts for about 6 weeks and is accompanied by the presence of well-developed, active corpora lutea. The effects of adrenalectomy done during oestrus and during pseudopregnancy were compared. Normally ferrets die within 10 days after being deprived of their adrenals, the average survival period being about 6 days. The coexistence of oestrus accelerated death. But when the adrenals were removed during pseudopregnancy, when active corpora lutea were present, the average period of survival was 37·5 days, no other treatment having been given.

In later work Gaunt & Hays (1938) removed the adrenals of five ferrets, one being a castrated male and the others spayed and non-spayed females. All of them were maintained in excellent health by giving 5 mg. of progesterone daily without other treatment. The minimal life-saving daily dose of progesterone was about 2 mg. All the ferrets died when the injections of progesterone were dis-continued, showing that adrenalectomy had been efficiently performed. Gaunt (1937) and Gaunt, Nelson & Loomis (1938) were able to maintain life in rats adrenalectomized when 30 days old, by daily injections of between 1 and 2 mg. of progesterone. Unlike cortical hormone, neither progesterone nor testosterone prevented the fatal effects produced by excessive doses of water given to rats after adrenalectomy. In this respect progesterone appears to differ from cortical hormone. Greene, Wells & Ivy (1939) removed the adrenals of rats weighing between 36 and 49 g. For 15 days after the operation some of the rats were given 4 mg. of progesterone daily, while their littermate controls received no post-operative treatment. The latter died within 7 days, whereas those injected with progesterone lived longer and gained weight (Table 196).

TABLE 196. The effect of progesterone in prolonging life in adrenalectomized rats (Greene, Wells & Ivy, 1939)

Number of rats	Treatment	Average duration of life (days)	Limits (days)	Gained weight after operation
11	Adrenalectomized only (controls)	5·1	3–7	−
10	Adrenalectomized and given progesterone (4 mg. daily for 15 days)	18·0	11–23	+

Schwabe & Emery (1939) and Emery & Greco (1940) found that daily doses of 1 mg. of progesterone were enough to maintain life in rats which had been adrenalectomized in two stages between the 25th and 30th days after birth.

(b) *The effect of progesterone on the excretion of water and electrolytes.* Thorn & Harrop (1937) reported that both progesterone and pregnanediol reduced the urinary excretion of sodium in the normal male dog. Thorn & Engel (1938) kept an adrenalectomized dog of 13·35 kg. on a constant diet and gave him a single injection of 20 mg. of progesterone subcutaneously. The injection was followed by a reduced excretion of water, sodium and chloride, and an increased output of potassium, the maximum effect being reached after 24 hours (Table 197). No comparable effects were obtained by similar doses of progesterone in normal dogs.

TABLE 197.   The effect of progesterone on the excretion of water and electrolytes in an adrenalectomized dog (Thorn & Engel, 1938)

24-hour periods following a dose of 20 mg. of progesterone	Body-weight (kg.)	Volume of urine (c.c.)	Na (mol. equiv.)	Cl (mol. equiv.)	K (mol. equiv.)	N (mol. equiv.)
1st	13·35	530	56·8	56·5	18·7	9·5
2nd	13·40	400	42·8	42·1	19·4	9·7
3rd	13·40	625	56·1	54·8	16·2	9·6

Dow & Zuckerman (1939 b) reported that progesterone, like oestrone and testosterone propionate, caused an increase in the weight of axolotls through retention of water.

In contrast with the effects on animals deprived of their adrenals, progesterone in large doses appears to have a diuretic effect in some animals with intact adrenals. Selye & Bassett (1940) gave 10 mg. of progesterone daily to normal and hypophysectomized rats and noted that under such treatment the daily output of urine was increased, even reaching 50 per cent of the rat's bodyweight. In this connection might be mentioned the deturgescent action of progesterone on the sex-skin of the baboon as reported by Gillman (1940), who noticed that 3 mg. of progesterone given to a baboon (*Papio porcarius*) in the turgid stage of the oestrous cycle caused deturgescence of the sex-skin.

These experiments seem to indicate that the influence of progesterone on the excretion of water is not the same after adrenalectomy as before that operation. The effects of deoxycorticosterone, it seems, also may be affected in the same way by adrenal activity. R. F. Loeb (1941) studied the effects of deoxycorticosterone on patients with adrenal cortical deficiency (Addison's disease). He noticed great variations among these in the degree of response to esters of deoxycorticosterone, though a characteristic result was a gain in weight caused by retention of water. Normal individuals under the same treatment gained little if any weight.

Menstrual oedema. In spite of the different effects of progesterone on animals with and without intact adrenals it may be possible, perhaps, to connect the water-retaining effects of progesterone with the slight general oedema noticed in women just before menstruation. Sweeney (1934) recorded periodically the weights of forty-two normal women 30 per cent of whom showed a gain in weight of 3 pounds or more some time in the menstrual cycle, usually just before menstruation, on the onset of which the weight began to fall. In some instances a pitting oedema accompanied the increase of bodyweight. Selye & Stone (1943) found that progesterone will cause general oedema in chicks. This action is favoured by the simultaneous administration of NaCl. Salt favours also the causation of general oedema by deoxycorticosterone acetate.

Apart from the metabolism of salts and water, there are other biological activities which are common to progesterone and corticosterone, including their effects on the mamma and uterus (p. 143). Some of these will be mentioned in the chapter concerned with adrenal hormones (p. 451).

Miscellaneous Actions of Progesterone

Our knowledge of progesterone is still primitive and much more work is required before any satisfactory and complete account can be given of its biological actions. Meanwhile a few effects which have not been mentioned in the earlier part of this chapter may be added.

(a) *The pituitary.* Reference has been made on previous pages to the influence of progestin on the functions and cytology of the pituitary (pp. 57, 91). One effect of special interest however may be mentioned here; it is concerned with the resemblance between progesterone and deoxycorticosterone as regards both their chemical structure and their biological action. This resemblance apparently extends to the influence of these two types of compound on the pituitary. It is known that repeated doses of adrenal cortical extract inhibit the supply of adrenocorticotrophin from the pituitary and so bring about atrophy of the adrenal cortex. Clausen (1940) has found that progesterone has a similar effect. He gave 4 mg. of progesterone daily for 15 days to male rats and killed them 2 days after the last injection. Comparison with controls showed that under this treatment the fasciculate and reticular zones of the adrenals had undergone some degree of atrophy as shown by the diminished size of the cells.

(b) *The liver.* Engelhart & Riml (1934) found an increased concentration of glycogen in the livers of guinea-pigs which had been given injections of an extract of corpora lutea, and Brunelli (1935) reported that a luteal extract, in contrast with oestrone which causes a diminution of hepatic glycogen, leads to an increase of glycogen in the liver of the spayed rat. Gaunt, Remington & Edelmann (1939) detected little if any action by large doses (35 mg.) of progesterone on the liver glycogen of rats, but in ferrets which had fasted for 24 hours progesterone raised the liver glycogen and increased the blood sugar, though the muscle glycogen showed little if any alteration.

(c) *The adrenal.* When young mice become pregnant a rapid disappearance of the x zone of the adrenal occurs, and at one time it seemed possible that progesterone might be the cause of the degeneration. This, however, has been disproved by Howard & Gengradom (1940), who implanted pellets of progesterone subcutaneously into spayed mice of a month old. At the end of 4 weeks the mice were killed. The average daily absorption of progesterone as estimated by weighing the residual pellets had been 0·137 mg. The adrenals of these mice all showed undiminished x zones, equal in size to those of control mice. Similar results were obtained in non-spayed mice after treatment with progesterone.

During gestation in the woman, whether a male or female embryo is present, an increased excretion of androgen in the urine apparently takes place, and it is perhaps to a gestational androgen that the disappearance of the x zone during pregnancy in mice may be attributed (p. 233).

(d) *The copulatory reflex.* In spayed guinea-pigs the copulatory reflex may be obtained by suitable doses of progesterone if these have been preceded by a sensitizing dose of oestrogen. Hertz, Meyer & Spielman (1937) spayed young adult virgin guinea-pigs and a week later gave them 1γ of oestrone twice a day for 2 days. On the 3rd day 0·2 mg. of progesterone was given. This caused the guinea-pigs to give the copulatory reflex within a period varying from $3\frac{1}{2}$ to $7\frac{1}{4}$ hours. Other compounds related to progesterone, including pregnanedione, pregnanediol, oestriol, androstenedione, dehydroandrosterone and testosterone, did not produce a copulatory reflex in otherwise similar circumstances. Collins, Boling, Dempsey & Young (1938) confirmed this action of progesterone in the spayed female guinea-pig. It appears that a preparatory dose of oestrogen is necessary, a single dose of oestrogen being effective in this respect for the next 5 days. Beach (1942) gave 500 r.u. of oestradiol benzoate to spayed rats and tested them for sexual reciprocity 66–68 hours later. The males were attracted and tried to mount but were not accepted. If given 500 r.u. of oestrogen followed 48 hours later by 0·5 or 1 mg. of progesterone all the female rats showed signs of receptivity. Witschi & Pfeiffer (1935) had already shown that female rats kept in constant oestrus by parabiosis with a castrated partner or by repeated injections of FRH do not ovulate, form corpora lutea or copulate. The injection of LH in these circumstances is followed by ovulation, luteinization and mating.

(e) *The minnow's ovipositor.* Although all the three main types of gonadal hormone bring about hypertrophy of the minnow's ovipositor, progesterone is so much more effective than oestrogen or androgen that it may be used, according to De Wit (1938), for the assay of progesterone. Oestrogens need a latent period of between 10 and 12 hours, whereas androgens and progesterone cause a measurable enlargement of the ovipositor after a latent period of 1 hour. Androgen, however, is so much less influential than progesterone that 50 times the amount is needed to give a response.

PART VI. SEX HORMONES OF THE ADRENAL CORTEX

Chapter XXIII. *Sex Hormones of the Adrenal Cortex*

General considerations. Pituitary-adrenal relationship. Adrenal-gonad relationship. Adrenal virilism and feminism. Inactivation of deoxycorticosterone by the liver.

General Considerations

THE adrenal cortex has such an important share in the physiology of reproduction that it must be regarded as a sexual organ, though it has other than sexual functions to perform. As Kendall (1941) remarked, the adrenal cortex does not elaborate any single substance which can be described as the vital hormone of the gland. It produces numerous steroid compounds with different biological actions, and substitution therapy after adrenalectomy requires a combination of compounds.

Before considering those adrenal activities which bear directly on sexual life a few anatomical details may be recalled.

The adrenal cortex consists of three chief zones: the glomerular which is outermost, the fasciculate which is intermediate and the reticular which is inmost and immediately surrounds the medulla, which consists of nerve tissue. In young animals there is an additional zone lying next the medulla and described as the *x* zone (Elliott & Armour, 1911). This might perhaps be regarded not as a separate zone but as a reticular zone in its earlier form, when it differs greatly in appearance from that seen in the normal adult. Between the glomerular and fasciculate zones is a thin layer of smaller cells with condensed nuclei which has been described as the *demarcation zone*.

In the adult the adrenal cortex is replenished by mitoses of those glomerular cells which lie next the capsule (Zwemer, 1936; Grollman, 1936). As they pass inward from this region the cells become occupied by lipoid granules or droplets and eventually disintegrate, or disappear in some other manner, in the region of the reticular zone. In early life, and perhaps later, cell regeneration probably occurs also in or near the reticular zone.

Changes in the adrenals accompany advancing age. The *x* zone soon disappears —during early infancy in man and about puberty in the mouse—though its disappearance may take place at different ages in the two sexes. With advanced age, especially in females, the zona fasciculata undergoes a characteristic change, the cells as they pass inward toward the medulla become increasingly laden with lipoid-like material, until as they approach the reticular region they may lose their cell walls and coalesce to form rounded masses of fat-like material containing pycnotic nuclei (Burrows, 1936*b*).

At present an attempt to allocate the different functions of the adrenal to particular zones would be hazardous, though there is reason to think that corticosterone and its closest allies are formed in the outer zones and that the sexual hormones are produced nearer to the medulla. But our knowledge is too limited to regard as more than tentative such an ill-defined proposition as this. Bennett (1939) states that the three ketonic steroids, corticosterone, deoxycorticosterone and dehydroxycorticosterone, can be identified only in the 'spongiocytes' of the outer part of the fasciculate zone. In the cat these hormones, he says, cannot be detected in other parts of the cortex.

Sex hormones isolated from the adrenal. In the discussion of the sources of androgen, oestrogen and progestin (pp. 163, 257, 391) the ability of the adrenal to produce all these kinds of hormone was mentioned and little more attention need be given to the subject here except to refer once again to the constitutional resemblance between the special steroids produced by the adrenal cortex and those produced by the gonads (Fig. 1, pp. 98, 99).

Pituitary-adrenal Relationship

(a) *The action of pituitary implants or extracts on the adrenal: adrenocorticotrophin.* As with the gonads, the development and functional activity of the adrenal cortex are directly controlled by the pituitary, through the agency of one or more adrenocorticotrophins. Delille (1909) in a well-documented paper called attention to the fact that hyperplasia of the adrenal cortex follows the injection of pituitary extracts. This observation has been confirmed by additional experiments in which extracts or implants of pituitary were introduced (Evans, 1923; Smith & Engle, 1927; Emery & Atwell, 1933; Collip, Anderson & Thomson, 1933, and others). Such treatment causes not merely enlargement of the adrenal cortex but stimulates also its production of sexual hormones. Davidson & Moon (1936; see also Davidson, 1937) showed that an adrenocorticotrophic extract, free from gonadotrophin or growth hormone, caused enlargement of the adrenal cortex and accessory generative organs in rats whose testes and pituitaries had been removed. The experimenters castrated eighteen rats 30 days after birth and performed hypophysectomy 4 days later. Of these rats seven were kept as controls, and the remaining eleven were injected twice a day for a fortnight with 25 units of the adrenocorticotrophin, and were killed on the day after the last injection. Davidson noted that adrenocorticotrophin did not cause enlargement of the adrenal medulla. Some of his results are given in Table 198.

TABLE 198. The effect of adrenocorticotrophin on the size of the adrenals and their output of androgen (Davidson, 1937)

		Average weights of organs in mg.	
Condition of rats	Number of rats	Adrenals	Prostate and Seminal vesicles
Castrated and hypophysectomized	7	9·5	40
Castrated, hypophysectomized and given adrenocorticotrophin	11	67·0	72
Normal untreated rats of same age	—	25–30	—

Emery & Atwell (1933) observed that the pronounced adrenal hypertrophy induced by adrenocorticotrophin, whether in castrated or non-castrated rats, is caused less by proliferation than by enlargement of the glandular cells whose cytoplasm becomes unusually rich in lipoid. A few mitoses are seen, but in their experiments adrenocorticotrophin seems to have affected function more than cellular multiplication.

Anselmino, Hoffmann & Herold (1934) gave injections of adrenocorticotrophin to infantile mice and noted the following results in the adrenals: (a) an increase in the number and size of the cells forming the fasciculate and glomerular zones, (b) an increase of secretory activity in these cells as shown by their lipoid contents, and (c) an intense hyperaemia of the gland. McQueen-Williams (1934b) implanted finely ground beef pituitary into rats. This treatment caused hypertrophy of the adrenal cortex, the cells of the reticular and fasciculate zones being much increased in size and containing abnormally large amounts of lipoid. Nelson (1941a) gave daily injections of adrenocorticotrophin (10–25 mg. per diem) for periods ranging from 10 to 30 days to immature male and female rats. Among the results of this treatment were noticed atrophy of the thymus, enlargement of the mammae and of the accessory reproductive organs in both sexes. These effects were obtained even in the absence of the pituitary and the gonads, and suggest that the adrenocorticotrophin had stimulated the adrenals to produce hormones with biological potencies like those of hormones formed by the gonads. When considering this kind of experiment the difficulty of preparing adrenocorticotrophin in a pure state, free from other pituitary trophins, has to be borne in mind.

In man there is evidence that the adrenal is governed by the pituitary. Cushing & Davidoff (1927) reported that in every one of four cases of acromegaly associated with acidophylic adenomas of the pituitary the adrenals were hypertrophied, the largest glands being 4 times the normal weight. They say that hypertrophy of the adrenal cortex is one of the most constant expressions of hyperpituitarism. In contrast with this finding they mention that in one of two instances of hypopituitarism in man the adrenals were very small, weighing together only 4 g. Cushing (1932) has said later that all known disorders of the pituitary are apt to cause pronounced secondary changes in the adrenal cortex.

(b) *The effect of hypophysectomy on the adrenal cortex.* The atrophic changes in the adrenal which follow hypophysectomy have been recorded by successive observers since the pioneer experiments of P. E. Smith and his colleagues. Evans (1923–24) noted that removal of the pituitary is followed by a reduction in size of the adrenal cortex without any corresponding shrinkage of the medulla. Collip, Anderson & Thomson (1933) found that after hypophysectomy atrophy of the adrenal cortex ensued and that if in these circumstances one adrenal were excised the other failed to undergo compensatory hypertrophy. The administration of pituitary extracts free from growth hormone caused enlargement of the surviving adrenal.

The histological effects of hypophysectomy in the adrenal of the mouse were studied by Leblond & Nelson (1937 a, b). The first notable changes, they say, appear in the first 48 hours, though the rapidity of response varies with different individuals. These early changes are seen first in the reticular zone, in the cells of which vacuoles appear and increase in size until the cells lose their boundaries and become confluent, forming siderophil masses in which are scattered pycnotic nuclei. The siderophil masses are often yellow; in sections for the microscope they become coloured by fat stains. The changes resemble in the main those which are induced in the adrenal cortex by oestrogen (Burrows, 1936b) and are apt to accompany old age in both sexes, though more markedly in females. General atrophy of the adrenal cortex follows the earlier changes caused by hypophysectomy, and the cortex becomes reduced to a thin layer, the shrinkage of the fasciculate zone advancing more rapidly than that of the glomerular zone. The individual cells of the atrophied cortex are lessened in size and no longer contain secretory granules or vacuoles; nor do they take up fat stains. The medulla remains little if at all changed.

(c) *The effect of adrenocortical hormones on the pituitary.* It will be remembered that the repeated administration of gonadal hormones hinders the output of gonadotrophin from the pituitary and so brings about gonadal atrophy. The same sort of mechanism for regulating glandular activity in accordance with the requirements of the individual is evident in the pituitary-adrenal relationship. Ingle, Higgins & Kendall (1938) gave large doses of cortical hormone ('cortin') daily to rats for periods extending from 7 to 28 days. To one group receiving the largest doses adrenocorticotrophin was also given. In the rats which had received cortin only the adrenal cortex was atrophied, and the atrophy was most pronounced in the animals which had received the largest doses, resembling that seen in hypophysectomized rats. In the rats which had received cortin plus adrenocorticotrophin atrophy of the cortex had not occurred. Thus it appears that an adrenal cortical hormone has an inhibitory influence on the output of adrenocorticotrophin from the pituitary. Selye (1940e) has called attention to this general phenomenon, for which he suggests the term 'compensatory atrophy'. As additional examples he mentions atrophy of the thyroid induced by thyroid feeding, atrophy of the islands of Langerhans under the influence of insulin, of the ovaries and testes caused by gonadal hormones, and of the parathyroids caused by parathyroid extract.

It may be noted that, according to Selye, the atrophic effects of deoxycorticosterone in the rat are present in equal or nearly equal degree in all three zones of the adrenal. He observed also that 3 mg. of deoxycorticosterone given daily by subcutaneous injection for a period of 20 days to gonadectomized mice 1 month old caused complete degeneration of the x zone of the adrenal.

The hormones of the adrenal cortex not only control the output of adrenocorticotrophin from the pituitary but they influence also the production of gonadotrophin; in fact they appear to be essential for the formation of FRH. Corey & Britton (1931) showed that precocious puberty can be induced by an extract of the adrenal cortex. They gave daily intraperitoneal injections of an adrenal cortical extract to Wistar rats, beginning when these were 28 days old. At the end of 14 days of this treatment the ovaries were much increased in size and contained large follicles, and the uterus also was enlarged. The effects on the testes were not so striking and appeared later; nevertheless the male rats too showed precocious development, spermatogenesis being present at the age of 35 days.

(d) *The effects of adrenalectomy on the pituitary.* The chief effect of adrenalectomy on the sexual functions of the pituitary seems to be a hindrance to the supply of FRH. This matter will be discussed more fully in connection with the effects of adrenalectomy on the gonads (p. 448).

(e) *The administration of deoxycorticosterone causes enlargement of the pituitary.* Turner & Meites (1941) treated guinea-pigs with deoxycorticosterone acetate for a period of 15 days, after which the weights of their pituitaries were compared with those of untreated guinea-pigs. In all cases the pituitaries of the treated guinea-pigs were heavier than those of the untreated (Table 199). It will be noticed that atrophy of the adrenals was apparently not caused in this experiment by the administration of deoxycorticosterone.

TABLE 199.   The effect of deoxycorticosterone acetate on the weight of the guinea-pig's pituitary (Turner & Meites, 1941)

Number of guinea-pigs	Sex	Amount of deoxycorticosterone given (mg.)	Duration of treatment (days)	Average body-weight (g.)	Average weight of adrenals (mg.)	Average weight of pituitary (mg.)
5	Female	0	—	405	207·9	11·46
5	Female	15	15	439	206·7	15·46
5	Male	0	—	340	164·9	9·56
5	Male	15	15	344	170·0	12·36

Adrenal-gonad Relationship

The functions of the gonads and adrenals to some extent overlap. For example, the capacities to maintain life after adrenalectomy and to conserve water and electrolytes in the body, which are recognized properties of the adrenal cortex, are shared to some extent and in certain circumstances by the gonads. The production by the adrenals of all the three main types of gonadal hormone, namely androgen, oestrogen and progestin, has also been proved. Further evidence that the adrenals are sexual organs may be given under several headings.

(a) *Sexual dimorphism of the adrenals.   The relative size of the adrenals differs in the two sexes.* This difference was noticed by Hatai (1913 b), whose observations were made on rats. He found that the mean weight of the two adrenals was less in the male than in the female. This difference was not apparent during infancy but became discernible when the rats were about 50 days old. Hatai further observed that gonadectomy caused a reduction in size of the adrenals in the female rat and an increase in their size in the male.

The x *zone.*   In some species, including the mouse, the most striking difference between the male and female adrenals concerns that part of the adrenal known as the *x* zone. This structure was first discovered in the human infant and described by Elliott & Armour (1911) and Thomas (1911), and was investigated also by Lewis & Pappenheimer (1916). In the human infant at birth the adrenal cortex is hypertrophied so that the gland is large as compared with the kidney. At this stage of existence the adrenal consists of (1) a narrow zone of tissue representing the permanent cortex, (2) a thin core of medulla and (3), situated between these layers, a thick zone of foetal cortex. In man this foetal cortex or *x* zone begins to degenerate soon after birth, and by the end of the first postnatal year has almost or quite disappeared according to the observers mentioned above.

Possibly this disappearance may not be complete in females, for in the adrenals of women Goormaghtigh (1922) observed juxtamedullary cortical cells which have characters like those of the x zone.

In certain animals the fate of the x zone is different in the two sexes. Howard-Miller (1927) states that in the mouse at the 4th or 5th week after birth the x zone in the female occupies a larger proportion of the adrenal than it ever occupies in the male. At the age of 4 weeks in the male the x zone ceases to grow and rapidly disappears, so that by the end of the next 10 days little or no trace of it remains. In females it persists for a much longer period. In the absence of pregnancy the x zone of the female mouse gradually degenerates until, after a period varying from 80 to 200 days, the adrenal resembles that of the male. If pregnancy ensues the x zone rapidly disappears and in most cases is not seen after the 12th day of gestation, but this disappearance does not accompany pseudopregnancy. As it exists in the immature animal the x zone consists of cells which are smaller than those of the fasciculate zone and the cytoplasm is more intensely coloured by eosin. At this period, though the cells of the fasciculate zone may be coloured deeply by fat stains (Sudan III) those of the x zone remain unstained. While undergoing degeneration, however, the cells of this zone in the female may or may not (Whitehead, 1932–33) become full of droplets which take up fat stains; the cells so affected may coalesce to form lipoid masses in the reticular region. In the male degeneration is not accompanied by this lipoid change.

Goormaghtigh (1922) has pointed out that the lipoid in the cells of the fasciculate zone is doubly refracting and labile, and differs from that of the juxtamedullary zone.

Castration in the young male leads to a persistence of the x zone, whereas in the female no change is caused by removal of the ovaries. For the most part these observations have been confirmed by Deanesly (1928) and others. Martin (1930) reported that in mice the administration of pituitary gonadotrophin caused enlargement of the x zone in female mice and its persistence in males, whereas androgen caused its disappearance in males and females. Apropos of the effect of gonadotrophin on the adrenal, Deanesly (1938a) made the interesting observation that in pituitary-deficient dwarf mice the x zone is absent.

The different changes which normally take place in the x zone in the male and female respectively are almost certainly attributable to the gonadal hormones supplied by the two sexes. Androgens cause a rapid disappearance of the zone (Whitehead, 1932–33) and oestrogens given in excess bring about a lipoid change within the cells of which it is composed.

The adrenals of adult male and female mice differ in another way. In the male there is a well-defined layer of fibrous tissue separating the cortex from the medulla. In females this fibrous barrier is less well developed, so that cells of the cortex almost touch and in places may seem to intermingle with those of the medulla. In mice this sexual dimorphism is well marked, and sections of adrenals examined under the microscope usually enable the sex of the animal to be recognized by this difference alone. The fibrous layer is thought to represent the condensed framework of connective tissue belonging to the inner part of the cortex from which the glandular cells have been lost, perhaps when the x zone disappeared.

The fate of the x zone and the degree of separation of the cortex from the medulla by fibrous tissue are not the only features in which the male and female adrenals differ. Hill (1933) found that in monkeys the zona fasciculata is larger in the female than in the male and so is the adrenal as a whole. Zalesky (1936) made similar observations on the adrenals of the guinea-pig and noted moreover that this sexual difference is not abolished by gonadectomy and therefore does not depend entirely on gonadal activity. Whitehead (1935 a) states that in the guinea-pig the mean thickness of that part of the cortex in which lipoid is present, as shown by staining with Sudan III, has a proportion of 29 : 63 in males and females respectively. Giroud & Santa (1940) compared the relative proportion between the medulla and cortex of the adrenals of different animals and found in this respect a distinct sexual dimorphism, the cortex forming a larger part in females than in males of the total weight of the whole adrenal (Table 200).

TABLE 200.   Comparison of the weights of medulla and cortex in the adrenals of male and female animals (Giroud & Santa, 1940)

(Mean weight of cortex in g. per 100 g. of total adrenal.)

Animal	Male	Female	Castrate
Pig	86·0	90·2	90·5
Horse	82·1	86·4	91·8
Bull, Cow and Ox	72·0	87	83·2
Sheep	85·3	90·2	88·6

(b) *Seasonal changes in the adrenal cortex.* In both male and female the adrenals, like the gonads, undergo changes in accordance with the sexual activities of the host. Such changes have been noted in reference to the breeding season, oestrus, pregnancy and lactation.

The breeding season. Riddle (1923) noticed that in female pigeons the adrenals increase in weight with each nesting cycle, the maximum weight corresponding with the time of ovulation. In four groups of pigeons of different kinds the mean increase in weight of the adrenals at ovulation was 40 per cent. Moore, Simmons, Wells, Zalesky & Nelson (1934; see also Zalesky, 1934), when studying the reproductive organs of the ground squirrel, noted that enlargement of the adrenal cortex accompanies the breeding season in both sexes. The hypertrophy affects chiefly the outer part of the reticular zone. A similar change, they discovered, can be induced during the non-breeding period by giving pituitary or chorionic gonadotrophin. Comparable observations have been made by Kayser & Aron (1938) on the hamster. In this rodent they find that in March the reticular zone of the adrenal is swollen and consists of large cells arranged in several layers and containing intracellular inclusions of lipoid, whereas in September the reticular zone is narrow and lipoid inclusions are absent. In a man, the administration of 20 mg. daily of deoxycorticosterone acetate is said to have caused intense libido (Lawrence, 1943).

Oestrus. Andersen & Kennedy (1932) studied the changes in the adrenal cortex which take place during the oestral cycle. For this purpose they killed virgin rats (1) 60 hours after oestrus and (2) at oestrus. In dioestrus, they say, there is a wide glomerular zone defined from the fasciculate zone by a row of cells with condensed nuclei. The cells of the outer part of the fasciculate zone are somewhat larger than those of the inner part and contain a few vacuoles which become stained by Scarlet-red. Near the medulla are small groups of cells the cytoplasm of which is stained deeply by eosin. These cells also become stained by Scarlet-red. At oestrus the glomerular zone is narrower than in dioestrus and is less clearly demarcated from the fasciculate zone. The cells of the outer half or two-thirds of the zona fasciculata are much enlarged, vacuolated and pale, as seen in sections stained with haematoxylin and eosin. The eosinophilic cells near the medulla are enlarged, contain vacuoles and resemble the cells of the outer portion of the zona fasciculata. A few cells in the rim of the medulla and some adjacent cortical cells take up the Scarlet-red stain freely, but most of the cortex merely becomes stained a pale red. During oestrus, these workers say, the cortex is significantly increased in weight. Bourne & Zuckerman (1941) have made confirmatory observations.

Pregnancy and lactation. Guieysse (1899) examined the adrenals of guinea-pigs at various stages of pregnancy. He found that during gestation the adrenal cortex is enlarged. The hypertrophy is caused by an increase in size of its component cells and not by their multiplication. After about the 12th day of pregnancy, he says, vacuoles indicating secretory activity begin to form in the fasciculate zone, the cells of which attain their greatest dimensions at the 30th day or perhaps later. Guieysse noticed also that the adrenal cortex is enlarged during lactation.

(c) Mutual reactions between the adrenals and gonads: the effect of gonadectomy on the adrenals. According to Hatai (1913 b, 1915) the effect of gonadectomy on the adrenal in rats differs in the two sexes, the gland becoming enlarged after the operation in the male and diminished in the female. Altenburger (1924) castrated four mice between 3 and 8 weeks old and killed them 3 months later when he found hypertrophy of the adrenal cortex with histological abnormalities in its structure. In the normal adult male mouse, as mentioned earlier, there is a well-defined fibrous boundary between the cortex and medulla; in the castrated mouse this connective tissue barrier is less substantial. The hypertrophy which follows castration involves the cortex only (Table 201). Similar changes in the adrenal after castration have been recorded in the dog and rabbit.

TABLE 201.   Hypertrophy of the adrenal cortex in the young male mouse
3 months after castration (Altenburger, 1924)

Condition of mice	Mean diameters of medulla (mm.)	Mean width of cortex (mm.)
Normal	1·0 × 0·66	0·26
Castrated	0·88 × 0·52	0·44

Masui & Tamura (1927) also recorded an enlargement of the adrenal in the male mouse after castration, though in the female removal of the ovaries was not followed by an increase in the size of the gland. They noted further that if male mice were castrated before puberty the *x* zone was preserved. According to Andersen & Kennedy castration of adult males is followed by little if any change in the adrenals. Spaying of the adult female causes in the next few days an enlargement of the adrenal cortex resembling that seen during oestrus. Six to

8 weeks after spaying the adrenal cortex has shrunk, the lipoid content of the fasciculate zone has decreased and degenerative changes are present in the zona reticularis. In the ground squirrel Moore, and his collaborators (1934) have seen degeneration of the zona reticularis in both sexes after removal of the gonads. Zalesky (1936) studied the effect of gonadectomy on the guinea-pig's adrenal and found an increase of lipoid in the cells of the fasciculate and reticular zones. These changes following gonadectomy were less pronounced in male than in female guinea-pigs; the animals were immature when the operation was done, and the adrenals were examined 18 months later. Hall & Korenchevsky (1938) castrated rats when they were between 21 and 27 days old. The operation was followed by hypertrophy of the adrenals. The histological changes affected the fasciculate and reticular zones, and included an increase in the vacuolation and size of the individual cells. The vacuolation and accumulation of lipoid granules in the outermost cells of the zona fasciculata caused the demarcation layer between the glomerular and fasciculate zones to disappear. The administration of androgens brought about a return of the adrenal cortex to its normal size, or to less than its normal size, by a decrease in the vacuolation and in the number of lipoid granules within the cells. All the androgens tested produced these results, having been administered twice a day for 3 weeks. The compounds proved to be effective in this way were androsterone, androstanediol, *trans*dehydroandrosterone, testosterone, testosterone propionate, Δ^4-androstenedione and Δ^5-androstenediol. Testosterone propionate injected into non-castrated rats did not induce definite changes in the size or weight of the adrenals.

The effects of adrenalectomy on the ovary and testis. Jaffe & Marine (1923) noticed that in most female rabbits which had survived adrenalectomy for 30 or more days ovarian enlargement was present. The increase in size was due to hypertrophy of the interstitial cells, and was regarded as a compensatory reaction. Jaffe and Marine observed no corresponding change in the testes of rabbits after adrenalectomy. Kitagawa (1930) and Wyman (1928) discovered that double adrenalectomy in the rat caused a suppression of oestrus, whereas removal of one adrenal alone had no obvious effect on the ovary. Martin (1932) performed adrenalectomy in two stages, with a 3- or 4-day interval between the operations, on 121 rats. In 74 anoestrus followed, in 32 the oestral cycles were prolonged, and in 15 oestrus continued as usual, possibly because accessory adrenal tissue was present. Others have confirmed these observations and have shown that the anoestrus following removal of the adrenals can be cured by substitution therapy. Pencharz & Olmsted (1931) demonstrated that the effects of adrenalectomy could be overcome by the implantation of an adrenal into an ovary, and others have obtained reversal of the effects of adrenalectomy on the ovary by giving adrenal cortical extracts (Martin, 1932; Corey & Britton, 1934). Additional experiments seem to show that the atrophy of the gonads after adrenalectomy is an indirect effect in which the pituitary is unable to form gonadotrophin, or at least FRH, without assistance from the adrenals.

Corey & Britton (1931) and Nice & Shiffer (1931) found that precocious sexual maturity as shown by opening of the vagina could be induced in rats, provided that they were at least 20 days old, by means of adrenocortical extracts. The effects resembled those caused by pituitary extracts. Martin (1932) tested the gonado-trophic potency of the rat's pituitary after adrenalectomy by means of implanta-tions into immature female rats, and found the potency reduced. The ovaries of the rats in which anoestrus had been caused by adrenalectomy were about one-third or a half the size of the ovaries of control rats and contained solid masses of corpora lutea. When in these circumstances pituitary extracts were administered, maturation of follicles and oestral cycles were resumed. The pituitaries of rats after adrenalectomy, Martin states, are reduced in size and show a reduction of the eosinophile cells. Shumacker & Firor (1934) have recorded that adrenalectomy is followed by a reduction of the gonadotrophic potency of the rat's pituitary and that pituitary implants will induce premature sexual maturity in the absence of the adrenals. Corey & Britton (1934) found that a chorionic gonadotrophin rich in LH ('*Antuitrin S*') failed to restore oestral cycles after adrenalectomy; whereas Swingle, Parkins, Taylor & Morrell (1936) found that FRH obtained from postmenopausal urine caused a resumption of oestrus in the absence of the adrenals.

According to Jaffe & Marine (1923) adrenalectomy affects the testis differently from the ovary, for the testis becomes smaller after the operation and the inter-stitial cells are not enlarged. The reduction in size of the testis may be partly due to the arrest of spermatogenesis.

There is evidence that after adrenalectomy the interstitial glandular cells of the testis will yet react to gonadotrophin. Freed, Brownfield & Evans (1931) record the same results of adrenalectomy in rats as those found by Jaffe & Marine in rabbits, and have further noted that injections of LH, though not restoring spermatogenesis, apparently will stimulate the interstitial cells to secrete, so that the seminal vesicles become doubled in size.

The action of gonadal hormones on the adrenal has been already discussed in the separate chapters devoted to the actions of androgens, oestrogens and progestins.

The action of adrenal cortical hormones on the gonads. Corey &·Britton (1931) gave adrenocortical extracts to young male rats and thereby caused precocious sexual maturity. The seminal tubules became enlarged and showed increased activity with precocious spermatogenesis, though little response was seen until the rats were 35 days old. The results, they say, were like those caused by extracts of the anterior lobe of the pituitary. Besides its own specific hormones the adrenal produces androgen, oestrogen and progestin. Although in chemical constitution the gonadal hormones produced by the adrenal may differ somewhat from the compounds formed by the gonads, their biological actions are sufficiently like those of the latter to require no separate analysis in this chapter. When the combined action of the adrenal secretions is being considered we are faced with the fact that there are many of them; moreover, they have co-operative, antagonistic, or independent effects according to the relative proportions in which they are supplied. Further, we do not yet know what influences determine the output by the adrenal of each individual type of hormone. In view of these difficulties the writer will refrain from any attempt to analyse in detail the tissue activities which are controlled by the sexual hormones of the adrenal.

What knowledge we have of the subject is chiefly derived from (1) the effects of adrenalectomy which have been discussed a few pages earlier and (2) the abnormalities which accompany the conditions known in man as adrenal virilism and feminism.

Overlapping activities of the adrenal and gonadal hormones. The fact that gonadal hormones can perform to some extent the special functions of the adrenal cortex has been mentioned (p. 143) and our present discussion will be confined to the capacity of the adrenal cortex to act as a substitute for gonadal tissue. This capacity is not explained only by the recognized ability of the adrenals to produce androgen, oestrogen and progestin, but is due also to some similarity of action between the corticosterones and the gonadal hormones, especially progestin. Most of what we shall say will be directed to the resemblance between the cortical hormones and progestin.

The action of deoxycorticosterone, like that of several other hormones when given by subcutaneous injection, may be amplified by esterification. This was shown by Miescher, Fischer & Tschopp (1938). They used groups of adrenalectomized rats as test objects, giving to each rat a single injection of 10 mg. of the compound to be tested. The results are indicated by comparing the average periods of survival in each group so treated (Table 202).

TABLE 202. The effect of esterification on the capacity of deoxycorticosterone to maintain the adrenalectomized rat (Miescher, Fischer & Tschopp, 1938)

Compound injected		Days of survival after one injection of 10 mg.
None (control)		6·5
Deoxycorticosterone		9
,,	acetate	10
,,	propionate	11
,,	valerianate	13
,,	benzoate	20
,,	palmitate	25

(a) *Progestational changes in the uterus and effects on muscular responses.* In the same paper these authors reported that deoxycorticosterone acetate in doses of less than 10 mg. will cause some degree of progestational change in the rabbit's uterus. Van Heuverswyn, Collins, Williams & Gardner (1939) have confirmed this observation. Immature female rabbits were given daily doses of 8·3 or 20γ of oestradiol benzoate for 6 days and then a single dose of deoxycorticosterone. In this way it was shown that 3 mg. of deoxycorticosterone had a progestational influence equal to that of 0·5 mg. of progesterone. Wells & Greene (1939) have obtained comparable results in the rabbit by the same method. Leathem & Crafts (1940) spayed six cats and removed the right adrenal from each at the same time. After ten daily injections of 0·033 mg. of oestradiol benzoate the left adrenal was removed, and the cats were then given daily doses of deoxycorticosterone acetate ranging from 10 to 20 mg. After 5–9 days of this treatment all the cats showed a progestational endometrium comparable with that caused by progesterone.

Robson (1939) made more extended observations. He found that deoxycorticosterone acetate in doses of 5 to 10 mg. not only induced progestational pro-

liferation in the rabbit's endometrium but, like progesterone, inhibited *in vitro*
and *in vivo* the response of the uterine muscle to pituitrin. Daily doses of 3 mg.
of deoxycorticosterone acetate maintained pregnancy in mice after removal of
the ovaries on the 14th day of gestation, and daily doses of 7·5 mg. maintained
pregnancy in rabbits after removal of the pituitary or ovaries.

Another feature which deoxycorticosterone shares with progesterone is its
capacity to prevent the uterine bleeding which ordinarily follows the cessation
of a supply of oestrogen given to a mature spayed monkey (Zuckerman, 1940).

(*b*) *Inhibition of oestrus and vaginal cornification.* In the paper cited above
Robson reports that daily doses of 0·1 or 0·2 mg. of deoxycorticosterone acetate
inhibit oestrus in the mouse and cause mucification of the vaginal epithelium.
Deoxycorticosterone also prevents cornification of the vagina under the influence
of oestrone. Salmon (1939) found that deoxycorticosterone in doses ranging
from 0·1 to 0·5 mg. had no cornifying effect on the rat's vagina. On the other
hand 5 or 10 mg. of deoxycorticosterone acetate given 3 times a week to ten postmenopausal women, whose vaginae gave typical oestrogen deficiency smears,
caused in all at the end of 8 days a full oestrogenic effect.

(*c*) *Copulation.* Van Heuverswyn, Collins, Williams & Gardner (1939)
demonstrated that deoxycorticosterone causes another effect which is induced
also by progesterone, namely the copulatory reflex in guinea-pigs. They gave
1 mg. of deoxycorticosterone to each of ten spayed guinea-pigs 24 hours after
they had received 50 i.u. of oestrone. Six of the ten guinea-pigs treated in this
way showed a copulatory response after an interval averaging 7·3 hours. Comparable results have been obtained in guinea-pigs by Torstveit & Mellish (1941),
who used an extract of adrenal cortex 'containing the life-maintaining principle'.

(*d*) *Lactation, mammary development and function.* Lactation ceases after
adrenalectomy, but Brownell, Lockwood & Hartman (1933) say that after this
operation lactation may be maintained in rats by giving an extract of adrenal
cortex. They tested two different extracts and found that one of these ('*Cortin*')
was able to support life but did not enable the rats to lactate. The other extract,
when given in addition to cortin, permitted lactation. Experiments of another
kind have revealed an influence of the adrenal cortex over lactation. Nelson &
Gaunt (1936) found that a purified lactogenic extract of the pituitary, unlike crude
pituitary extracts, will not by itself cause lactation in the hypophysectomized
guinea-pig, but it will do so if adrenal cortical hormones are given in addition.
This observation has been confirmed by Gomez & Turner (1936 *a, b*). Nelson
& Gaunt (1937*b*) carried their earlier observations a stage further. They caused
full development of the mammary glands in male and female guinea-pigs by the
daily administration of 30 r.u. or more of oestrone for not less than 1 month. In
these guinea-pigs they found that (1) a lactogenic pituitary extract would not
start lactation after hypophysectomy unless adrenal cortical hormone were given
in addition, whereas (2) a crude pituitary extract alone induced lactation. They
concluded that the factor missing from the purified lactogenic extract and present
in the crude extract was probably adrenocorticotrophin. Further tests seemed to
prove the correctness of this assumption, because either galactogen and adrenocorticotrophin or galactogen and adrenal cortical extract were shown to be

essential for milk formation. The question remained whether the influence of the adrenal cortex on lactation depended on its control over the metabolism of salt and water. After inducing full development of the mammae in guinea-pigs as before with oestrone, Nelson & Gaunt (1937 b) performed hypophysectomy and then gave the animals galactogen together with copious supplies of sodium chloride and water; all the guinea-pigs treated in this way lactated. If, however, guinea-pigs in addition to the treatment just mentioned were deprived of their adrenals lactation, though detectable, was never normal in amount.

Adrenal hormones do not appear by themselves to act as galactogens, for Hurst, Meites & Turner (1942) tested several adrenal cortical extracts on the pigeon's crop-gland and failed to obtain evidence of prolactational activity.

It might be permissible here to revert to a perpetual difficulty in the interpretation of results obtained with hormones, a difficulty exemplified by the various experiments just described. (1) The isolation of individual pituitary trophins in a pure state has not yet been achieved; only approximations to purity have been attained. Therefore when we talk of the injection of galactogen, adrenocorticotrophin or gonadotrophin, we cannot ignore the possibility that other trophins may be present if only in traces. (2) The overlapping of the effects of different hormones has to be considered; in the experiments now being considered the extent to which corpora lutea can act as substitutes for the adrenal cortex has to be borne in mind. (3) The mutual co-operations and antagonisms of the various hormones may be controlling factors in any result obtained by their use; for example, the consequences of giving progestin may depend upon whether the administration is or is not preceded or accompanied by oestrone in some particular relative proportion. Other complexities might be quoted, but it would be wearisome. Reference has been made to the matter here only because the type of experimental work under discussion affords a good example of a difficulty by which the investigator of sex hormones is perpetually opposed.

After this digression we may return to our original train of thought, and consider some effects, apart from lactation, which adrenal hormones exert upon the mamma. It has been found by Van Heuverswyn, Folley & Gardner (1939) that deoxycorticosterone brings about *development of the mamma* in both normal and castrated male mice. The method used was to administer by subcutaneous injection 0·5 mg. of deoxycorticosterone acetate dissolved in oil every other day for 16 days. Chamorro (1940) has confirmed this observation, but finds that deoxycorticosterone has no effect on the mamma in the absence of the pituitary. These results are of interest in connection with gynaecomastia as an occasional accompaniment of adrenal tumours, and of the therapeutic use of corticosterone in man.

Apparently both adrenosterone and corticosterone will cause enlargement of the nipple in the male guinea-pig (Jadassohn, Uehlinger & Margot, 1938)— a biological ability which they share with several androgens and oestrogens.

Androgenic effects. The capacity of the adrenal cortex to cause androgenic effects is recognized, and specific androgens have been isolated from adrenal extracts (p. 163). Whether corticosterone or its congeners share in producing these effects is doubtful. Hooker & Collins (1940) tested a synthetic deoxycorticosterone acetate on capons and castrated rats and mice and found that doses of

2 mg. given on 5 successive days caused comb growth in the capon, and daily doses of 1 mg. maintained to some degree the weights of the prostate and seminal vesicles of the castrated rat. Similar effects were produced in mice by daily doses of 0·25 and 0·5 mg. Other experimenters have obtained contrary results. Greene & Burrill (1940a) did not find any signs of prostatic stimulation in castrated rats which had been given daily doses of 2·5 mg. of deoxycorticosterone for 8 days; Chamorro (1940) and Paschkis (1941) also failed to observe androgenic effects after giving deoxycorticosterone to castrated rats, mice or to capons. Probably we shall not err in ascribing the masculine features of adrenal virilism to the excessive production of specific androgens rather than to deoxycorticosterone.

Riddle (1937) has made an observation which suggests that the adrenals may supply enough androgen to maintain masculine behaviour in the absence of testes. He examined sixteen adult pigeons or doves in which no gonads could be found although the birds displayed male behaviour equal to that of those with functional testes.

Adrenal Virilism and Feminism

The term 'virilism' is now so generally applied to the androgenic effects of adrenal overactivity in women that it is perhaps too late to expect a better one. The term is bad because in the adult excessive adrenal activity does not necessarily convey psychological implications. The woman who grows a beard may have quite as gentle and feminine a mind as her normal sister, although in children the condition may be accompanied by an unusual mentality.

Objection can be made also against the term 'feminism' as applied to the manifestations of adrenal overactivity in men; impotence, though a common symptom of the condition, is hardly to be regarded as a feministic phenomenon.

These preliminary remarks have been made to free the writer, and possibly the reader, from confused thinking, and to justify the use of the condemned words in the following pages on the ground that they have become established by custom.

Etiology.   The symptoms of adrenal virilism and feminism are caused by an excessive production of androgen or oestrogen by the adrenal. Such an excess may accompany (1) an alteration of adrenal function without naked-eye changes in the gland, (2) simple hyperplasia, or (3) neoplasia of the adrenal. Beyond this point the etiology has not yet been traced; the pituitary may be suspected of taking some part in the causation.

Though virilism may be present without any very obvious change in the adrenals, Broster & Vines (1933) have described a staining reaction founded on a special affinity of the adrenal cortex in cases of virilism for fuchsin. In cases of virilism, whether hyperplasia is or is not present, the fuchsinophilia is most pronounced in the reticular zone, less marked in the fasciculate zone and least in the zona glomerulosa. If symptoms of virilism have been present, a cortex of normal size, they say, will give a positive response to their method of staining with fuchsin. Vines (1938) examined thirty-four adrenals removed by operation on account of virilism and all these glands showed fuchsinophilia. The weights of the glands varied from 2·3 to 25 g., the mean weight being 8·1 g., and 52 per cent fell within normal limits, namely between 4 and 7 g. Broster & Vines (see also Vines, 1938) have observed this staining reaction in the adrenals of the normal human foetus, where it is present for a shorter period and in a less pronounced degree in the female than in the male (Table 203). Their findings in this respect possibly may be correlated with the increased output of androgen which accompanies pregnancy.

TABLE 203.   Fuchsinophile staining of the adrenal cortex in the
human male and female foetus (Broster & Vines, 1933)

Age of foetus in weeks	Number		Number showing fuchsinophile staining of adrenal cortex	
	Male	Female	Male	Female
7–10	3	2	1	0
10–14	4	3	4	3
14–20	4	6	4	2
Full term	6	3	0	0

Carnes (1940) assayed for their androgen and oestrogen content the adrenals of twenty still-born and newborn babies, the gestation periods having ranged from 18 to 40 weeks. The tests were made by applying extracts of the glands directly to the combs of day-old chicks. In no instance was any androgen revealed.

Goormaghtigh (1940) has investigated the relationship between cytology and function in cases of adrenal tumour, and he concludes that virilizing tumours contain siderophile and fuchsinophile granules, whereas these are absent from feminizing tumours and tumours which do not produce endocrine effects. The cytology of a virilizing adrenal tumour, he says, may closely resemble that of an active corpus luteum, and it appears to him that the juxtamedullary cells of the adrenal cortex and the luteal cells have features in common, including siderophilia and fuchsinophilia.

Sudds (1940), having demonstrated fuchsinophile granules in three adrenal tumours which had been associated with virilism, made an inquiry into the frequency with which fuchsinophilia might be detected in the adrenal apart from symptoms of endocrine disturbance, and he found that the characteristic granules were present in the adrenals of 24 per cent of the adult males and 28 per cent of the adult females. In females the granules showed a steady increase with advancing age; and the same was true to a less extent in males. They were not found in either sex before the age of 24 years (Table 204).

TABLE 204.   The presence of fuchsinophile granules in the adrenal cortex
of men and women at various ages (Sudds, 1940)

Age in years	Percentage of cases in which fuchsinophile granules were present	
	Male	Female
0–20	0	0
20–30	12·5	22
30–40	33	23
40–50	38	28
50–60	30	42
60–70	34	45
70–80	50	50

The hormonal disturbances caused by adrenal abnormalities.   The adrenals produce numerous hormones, and it is not surprising that disorders of these glands may bring about various effects in the host, nor that several different hormonal compounds have been identified in the urine of patients with overactive, hyperplastic or neoplastic adrenals.

When a patient displays signs of virilism a biological excess of androgen may be expected in the urine. On the other hand, adrenal tumours may cause an excessive output of oestrogen, in which event the features of the disease may correspond with the effects known to be caused by this type of hormone. In the urine of two women with adrenal carcinoma Frank (1934) found a large excess of oestrogen, and Burrows, Cook & Warren (1936; see also Burrows, Cook, Roe & Warren, 1937) made the same observation in a case of adrenal carcinoma in a man. In the latter case an assay of the urine yielded 3,000 i.u. of oestrogen, probably oestrone, per litre. In addition an androgen was isolated and identified as $\Delta^{3:5}$-androstadiene-17-one. Other cases are on record in which the presence of an adrenal tumour has been accompanied by an increased output of both androgen and oestrogen in the urine; in this event the symptoms probably will depend on the relative amounts and biological activities of the compounds which are present. Examples of a simultaneous increase in the output of androgen and oestrogen by patients with adrenal tumours are given in Table 205, which has been abstracted from a paper by Dingemanse & Laqueur (1938); from the figures given it will be seen that oestrogen preponderated in the male and one of the females while in the other female androgen was in excess.

TABLE 205. Excretion of sex hormones by patients with adrenal tumour
(Dingemanse & Laqueur, 1938)

Case	Sex	Hormones excreted given as international units per litre		
		Androgen	Oestrogen	Gonadotrophin (LH)
1	Female	1,000	13,000	400
2	Female	2,200	100	Less than 400
3	Male	600	1,000	Less than 80

The number of sterones and allied substances which have been isolated and identified from patients with adrenal disease is already large. This fact suggests the need of a greater provision than we possess at present for chemical investigation of the urine in cases showing evidence of hormonal disturbance of any kind. Hormones are very potent agents and small amounts may have great effects. To obtain from one patient enough material for analysis and identification, a large amount of urine must be extracted, and if this extraction is to be done rapidly and effectively special chemical facilities and machinery must be available.

Variations of symptoms in adrenal disease. As mentioned above, the adrenals produce many different hormones, and when they are overactive in a pathological sense, the symptoms which arise in each case will depend on the particular kind of hormone which predominates. Clinically it is recognized that in women and in children of either sex overaction of the adrenal cortex usually, though not invariably, brings about what is described as virilism. The chief symptoms in such cases are attributable to an excessive supply of androgen, and in the adult woman include a thickening of the skin with hypertrophy of the cutaneous structures.

Acne is apt to occur and the distribution of hair becomes like that in the male, outstanding features being its coarseness and copious growth on the face. The distribution of fat over the body also assumes that of the male. Amenorrhoea may be an additional accompaniment. When the disease occurs in early life the effects may be still more striking, especially in girls. The voice becomes gruff, and the face bearded, there may be an absence of the lower end of the vagina, the clitoris is enlarged, and the general muscular system becomes precociously developed. Additional symptoms appear and are such as may be ascribed to an abnormally abundant supply of androgen.

Although the cause of virilism of the kind now under consideration is an overactivity of the adrenal in producing androgen, this excessive action is not necessarily associated with an increase in the size of the gland, though some enlargement is present in most cases.

In men a pathological degree of virilism is not often seen. Perhaps it occurs but is overlooked. With adrenal tumours in men the usual hormonal disturbance seems to be an excessive production of oestrogen, and the symptoms include an absence of libido, impotence with atrophy of the testicles, and a diminution in the size of the penis. There may be some tendency to obesity. Gynaecomastia, sometimes with pigmentation of the areolae, has been present in many cases (Holl, 1930; Lisser, 1936).

Inactivation of Deoxycorticosterone by the Liver

Much has yet to be learned about the metabolism of the adrenal steroids. Probably their fate resembles that of the gonadal hormones. The liver certainly seems to take part in their inactivation, for Burrill & Greene (1942a) found that pellets of deoxycorticosterone, when implanted into the mesentery so as to be submitted almost at once on their solution to hepatic action, lose much of their power, as compared with similar pellets implanted subcutaneously, to maintain life in animals after adrenalectomy.

APPENDIX

ABBREVIATIONS

A-Z test	Aschheim-Zondek test for pregnancy	i.u.	International unit
FRH	Follicle ripening hormone (synonymous with FSH)	m.u.	Mouse unit
		r.u.	Rat unit
LH	Luteinizing hormone	rb.u.	Rabbit unit
ICSH	Interstitial cell stimulating hormone	c.u.	Capon unit

PROPRIETARY SEX HORMONE PREPARATIONS AND SYNONYMS

A. TROPHINS

ADRENOCORTICOTROPHIN

	RAW MATERIAL	FIRM
Corticotrophin	Ox pituitaries	Organon Laboratories

GONADOTROPHIN

	BIOLOGICAL SOURCE	RAW MATERIAL	FIRM
Ambinon	Pituitary	Pig pituitaries	Organon Laboratories
Antex	Pituitary	Pregnant mare serum*	Løvens kemische Fabrik
Antostab	Pituitary	Pregnant mare serum*	Boots
Antoxylin S	Placenta	Pregnancy urine	Oxo
Antuitrin S	Placenta	Human pregnancy urine	Parke Davis & Co.
Atregone	Placenta	Human pregnancy urine	Abbott Laboratories
Follutein	Placenta	Human pregnancy urine	—
Gestyl	Pituitary	Pregnant mare serum*	Organon Laboratories
Glanduantin	Pituitary	—	Richter
Gonadotraphon	Pituitary	Ox or sheep pituitaries	Paines & Byrne
Gonadotraphon S	Placenta	Human pregnancy urine	Paines & Byrne
Gonadyl	Pituitary	Pregnant mare serum*	Roussel Laboratories
Gonan	Placenta	Human pregnancy urine	British Drug Houses
Hebin	Pituitary	Sheep pituitaries	
Luteo-antin	Pituitary	Pregnant mare serum*	Richter
Phyostab	Placenta	—	Boots
Pregnyl	Placenta	Human pregnancy urine	Organon Laboratories
Prolan	Placenta	—	Bayer Products
Serogan	Pituitary	Pregnant mare serum*	British Drug Houses

* The supposed origin of the gonadotrophin found in the serum of pregnant mares has recently been questioned.

GALACTOGEN, GALACTIN, LACTOGEN, PROLACTIN

	BIOLOGICAL SOURCE	RAW MATERIAL	FIRM
Physolactin	Pituitary	—	Glaxo Laboratories
Prolactin	Pituitary	—	Organon Laboratories
Prolactin	Pituitary	—	Allen & Hanburys

B. ADRENAL AND GONADAL HORMONES

ADRENAL CORTICAL HORMONES

	IDENTITY	FIRM
Cortigen	Extract of adrenal cortex	Richter
Cortin organ	Extract of adrenal cortex	Organon Laboratories
Cortiron	Deoxycorticosterone	Schering
D.O.C.A.	Deoxycorticosterone acetate	Organon Laboratories
Eschatin	Extract of adrenal cortex	Parke Davis & Co.
Eucortone	Extract of adrenal cortex	Allen & Hanburys
Percorten	Deoxycorticosterone acetate	Ciba Co.
Sapracort	Extract of adrenal cortex	Paines & Byrne
Syncortyl	Deoxycorticosterone acetate	Roussel Laboratories

ANDROGENS

	IDENTITY	FIRM
Androfort	Androsterone	Richter
Neo-hombreol	Testosterone propionate	Organon Laboratories
Neo-hombreol (M)	Methyltestosterone	Organon Laboratories
Oreton	Testosterone propionate	Schering Corporation
Perandren	Testosterone propionate	Ciba Co.
Proviron	Androsterone benzoate	Schering Kahlbaum
Sterandryl	Testosterone propionate	Roussel Laboratories
Testoviron	Testosterone propionate	Schering
Virormone	Testosterone	Paines & Byrne

OESTROGENS

	IDENTITY	FIRM
Amniotin	—	Squibb
Benzo-Gynoestryl	Oestradiol benzoate	Roussel Laboratories
Clinestrol	Stilboestrol	Glaxo Laboratories
Dimenformon benzoate	Oestradiol benzoate	Organon Laboratories
Dimenformon dipropionate	Oestradiol dipropionate	Organon Laboratories
Folliculin	Oestrone	—
Glandubolin	Oestrone	Richter
Menformon	Oestrone	Organon Laboratories
Ovendosyn	Stilboestrol	Menley & James
Ovocyclin	Oestradiol	Ciba Co.

	IDENTITY	FIRM
Ovocyclin P	Oestradiol dipropionate	Ciba Co.
Pabestrol	Stilboestrol	Paines & Byrne
Pabestrol D	Stilboestrol dipropionate	Paines & Byrne
Progynon	Oestrone	British Schering Co.
Progynon B	Oestradiol benzoate	British Schering Co.
Stilboestroform	Diethylstilboestrol	British Drug Houses
Syntestrin	Stilboestrol	Richter
Theelin	Oestrone	—
Theelol	Oestriol	—

PROGESTERONE

	IDENTITY	FIRM
Gestone	Progesterone	Paines & Byrne
Glanducorpin	Progesterone	Richter
Lipo-lutin	Progesterone	Parke Davis & Co.
Luteostab	Progesterone	Boots
Lutocyclin	Progesterone	Ciba Co.
Lutogyl	Progesterone	Roussel Laboratories
Lutren	Progesterone	Bayer Co.
Progestin B.D.H.	Progesterone	British Drug Houses
Progestin Organon	Progesterone	Organon Laboratories
Progestone	Progesterone	Carnrick
Proluton	Progesterone	British Schering Co.

ETHISTERONE

(PREGNENINOLONE, ETHINYLTESTOSTERONE)

	IDENTITY	FIRM
Lutocyclin Oral	Pregneninolone	Ciba Co.
Lutogyl tablet	Pregneninolone	Roussel Laboratories
Progestoral	Pregneninolone	Organon Laboratories
Proluton C	Pregneninolone	Schering

REFERENCES

REFERENCE ABBREVIATIONS

ABN	Acta brevia Néerl.
AE	Arch. f. Entwicklung. d. Organ.
AEPP	Arch. exper. Path. u. Pharm.
AG	Arch. f. Gynäk.
AJA	Amer. J. Anat.
AJC	Amer. J. Cancer
AJOG	Amer. J. Obst. Gyn.
AJP	Amer. J. Physiol.
AR	Anat. Rec.
ARP	Arch. Pathol.
AS	Arch. Surg.
AZE	Arch. de Zool. Expér.
BC	Bull. du Cancer
BJ	Biochem. J.
BJEB	Brit. J. exp. Biol.
BJH	Bull. Johns Hopkins Hosp.
BJS	Brit. J. Surg.
BMJ	Brit. Med. J.
CI	Chem. & Indus.
CMA	Canad. med. Ass. J.
CR	Cancer Research
CRA	Compt. rend. Acad. Sci.
CRS	Compt. rend. Soc. Biol.
CSH	Cold Spring Harbor Symposia
DMW	Deutsch. med. Woch.
E	Endocrinology
GO	Gynéc. et Obstét.
JA	J. Anat.
JAMA	J. Amer. med. Ass.
JBC	J. biolog. Chem.
JCI	J. clin. Invest.

JCR	J. Cancer Res.
JE	J. Endocrinol.
JEB	J. exp. Biol.
JEM	J. exp. Med.
JEZ	J. exp. Zool.
JM	J. Morph.
JN	J. Nutrition
JNCI	J. Nat. Cancer Inst.
JOG	J. Obst. & Gynaec.
JP	J. Physiol.
JPB	J. Path. & Bact.
JPET	J. Pharm. & exp. Therap.
JPL	J. Pharmacol.
JU	J. Urology
KW	Klin. Woch.
L	Lancet
N	Nature
NW	Naturwiss.
PB	Proc. Soc. exp. Biol. & Med.
PHR	Public Health Reps. U.S.A.
PR	Proc. Roy. Soc. B
PRSM	Proc. Roy. Soc. Med.
PT	Phil. Trans. Roy. Soc. B
QJ	Quart. J. exp. Physiol.
QJMS	Quart. J. Micros. Sci.
S	Science
SGO	Surg. Gyn. & Obst.
VA	Virchow's Archiv f. path. Anat.
YJ	Yale J. Biol. & Med.
ZG	Zbl. f. Gynäk.
ZGG	Zeits. Geburt. u. Gynäk.

REFERENCES

Abarbanel, A.R., *PB*, 1940, **44**, 452

Abramowitz, A.A., & Hisaw, F.L., *E*, 1939, **25**, 633

Ackart, R.J., & Leavy, S., *PB*, 1939, **42**, 720

Adair, F.E., & Bagg, H.J., *Internat. Clinics*, 1925, **4**, 19

Addison, W.H.F., *J. Comp. Neurol.*, 1917, **28**, 441

Adler, A.A., de Fremery, P., & Tausk, M., *N*, 1934, **133**, 293

Adler, L., *AG*, 1912, **95**, 349

Alexiu, M., & Herrnberger, K., *ZG*, 1938, **62**, 9

Allan, H., Dickens, F., Dodds, E.C., & Howitt, F.O., *BJ*, 1928, **22**, 1526

Allan, H., & Wiles, P., *JP*, 1932, **75**, 23

Allanson, M., *PR*, 1937, **124**, 196

Allen, B. M., *AJA*, 1904, **3**, 89

Allen, E., *AJA*, 1922, **30**, 297

Allen, E., *AJA*, 1923, **31**, 439

Allen, E., *Contr. Embryol. Carneg. Instn.*, 1927, **19**, 1

Allen, E., (a) *AJA*, 1928, **42**, 467

Allen, E., (b) *AJP*, 1928, **85**, 471

Allen, E., (c) *J. Morph. & Physiol.*, 1928, **46**, 479

Allen, E., 'Sex and Internal Secretions', 1932, 434

Allen, E., *CSH*, 1937, **5**, 104

Allen, E., *JAMA*, 1940, **114**, 2107

Allen, E., *JAMA*, 1941, **116**, 405

Allen, E., & Diddle, A.W., *AJOG*, 1935, **29**, 83

Allen, E., Diddle, A.W., Burford, T.H., & Elder, J.H., *E*, 1936, **20**, 546

Allen, E., Diddle, A.W., & Elder, J.H., *AJP*, 1935, **110**, 593

Allen, E., & Doisy, E.A., *JAMA*, 1923, **81**, 819

Allen, E., & Doisy, E.A., *AJP*, 1924, **69**, 577
Allen, E., & Doisy, E.A., *Physiol. Rev.*, 1927, **7**, 600
Allen, E., Francis, B.F., Robertson, L.L., Colgate, C.E., & Johnston, C.G., *AJA*, 1924, **34**, 133
Allen, E., & Gardner, W.U., *CR*, 1941, **1**, 359
Allen, E., Gardner, W.U., & Diddle, A.W., *E*, 1935, **19**, 305
Allen, E., Pratt, J.P., & Doisy, E.A., *JAMA*, 1925, **85**, 399
Allen, E., Pratt, J.P., Newell, Q.U., & Bland, L.J., *AJP*, 1930, **92**, 127
Allen, E., Smith, G.M., & Gardner, W.U., *AJA*, 1937, **61**, 321
Allen, E., Smith, G.M., & Reynolds, S.R.M., *PB*, 1937, **37**, 257
Allen, W.M., *AJP*, 1930, **92**, 612
Allen, W.M., *CSH*, 1937, **5**, 66, 70
Allen, W.M., Butenandt, A., Corner, G.W., & Slotta, K.H., *N*, 1935, **136**, 303
Allen, W.M., & Heckel, G.P., *S*, 1936, **84**, 161
Allen, W.M., & Meyer, R.K., *AJP*, 1933, **106**, 55
Allen, W.M., & Meyer, R.K., *AR*, 1935, **61**, 427
Allen, W.M., & Wintersteiner, O., *S*, 1934, **80**, 190
Altenburger, H., *Pflügers Archiv*, 1924, **202**, 668
Ancel, P., & Bouin, P., (a) *CRA*, 1903, **137**, 1288
Ancel, P., & Bouin, P., (b) *CRS*, 1903, **55**, 1680
Ancel, P., & Bouin, P., (a) *J. Physiol. et Path.*, 1904, **6**, 1039
Ancel, P., & Bouin, P., (b) *CRA*, 1904, **138**, 110, 138, 168
Ancel, P., & Bouin, P., *CRA*, 1906, **142**, 232, 298
Ancel, P., & Bouin, P., *CRS*, 1909, **67**, 454, 605
Andersen, D.H., *JP*, 1935, **83**, 15
Andersen, D.H., & Kennedy, H.S., *JP*, 1932, **76**, 247
Andersen, D.H., & Kennedy, H.S., *JP*, 1933, **79**, 1
Anderson, A.F., Hain, A.M., & Patterson, J., *JPB*, 1943, **55**, 341
Andervont, H.B., *JNCI*, 1940, **1**, 135, 147
Andervont, H.B., *JNCI*, 1941, **2**, 307
Andervont, H.B., & McEleney, W.J., *PHR*, 1939, **54**, 1597
Andervont, H.B., & McEleney, W.J., *JNCI*, 1941, **2**, 7 and 13
Andrews, C.T., *BMJ*, 1942 (ii), 185
Andrews, F.N., *PB*, 1940, **45**, 867
Anselmino, K.J., & Hoffmann, F., *ZG*, 1936, **60**, 501
Anselmino, K.J., Hoffmann, F., & Herold, L., *KW*, 1934, **13**, 209
Arai, H., *AJA*, 1920, **27**, 405 and **28**, 59
Aron, E., *CRS*, 1921, **85**, 482
Aron, E., *Arch. de Biol.*, 1925, **34**, 1
Aron, E., *CRS*, 1933, **113**, 1069
Arthus, A., & Malan, J., *CRS*, 1936, **121**, 196

Aschheim, S., *Med. Klin.*, 1926, **22**, 2023
Aschheim, S., *ZGG*, 1933, **104**, 527
Aschheim, S., & Varangot, J., *CRS*, 1939, **130**, 827, 830
Aschheim, S., & Zondek, B., *KW*, 1927, **6**, 1322
Aschheim, S., & Zondek, B., *KW*, 1928, **7**, 8
Aschoff, L., *VA*, 1894, **138**, 119
Asdell, S.A., *BJEB*, 1924, **1**, 473
Asdell, S.A., & Crowell, M.F., *JN*, 1935, **10**, 13
Asdell, S.A., & Hammond, J., *AJP*, 1933, **103**, 600
Asdell, S.A., & Salisbury, G.W., *AJP*, 1933, **103**, 595
Asdell, S.A., & Seidenstein, H.R., *PB*, 1935, **32**, 931
Askew, F.A., & Parkes, A.S., *BJ*, 1933, **27**, 1497
Asmundson, V.S., & Wolfe, M.J., *PB*, 1935, **32**, 1107
Astwood, E.B., *E*, 1938, **23**, 25
Astwood, E.B., (a) *AJP*, 1939, **126**, 162
Astwood, E.B., (b) *JE*, 1939, **1**, 49
Astwood, E.B., Geschickter, C.F., & Rausch, E.O., *AJA*, 1937, **61**, 373
Astwood, E.B., & Greep, R.O., *PB*, 1938, **38**, 713
Astwood, E.B., & Weil, P.G., see Marker & Hartman, *JBC*, 1940, **133**, 529
Aub, J.C., Karnofsky, D., & Towne, L.E., *CR*, 1941, **1**, 737
Aykroyd, O.E., & Zuckerman, S., *JP*, 1938, **94**, 13

Bachman, C., *PB*, 1936, **33**, 551
Bachman, C., Collip, J.B., & Selye, H., *PR*, 1934, **32**, 544
Bachman, C., Collip, J.B., & Selye, H., *PR*, 1935, **117**, 16
Bachman, C., Collip, J.B., & Selye, H., *PB*, 1936, **33**, 549
Bacon, A.R., *AJOG*, 1930, **19**, 352, 426
Bacsich, P., & Folley, S.J., *JA*, 1939, **73**, 432
Bagg, H.J., *Amer. Naturalist*, 1926, **60**, 234
Bagg, H.J., *AJC*, 1936, **27**, 542
Bagg, H.J., & Hagopian, F., *AJC*, 1939, **35** 175
Baker, B.L., & Johnson, G.E., *E*, 1936, **20**, 219
Baker, J.R., & Ranson, R.M., *PR*, 1932, **110** 313
Baker, J.R., & Ranson, R.M., *Proc. Zool. Soc Lond.*, 1938, **108**, 101
Baldwin, F.M., Goldin, H.S., & Metfessel, M. *PB*, 1940, **44**, 373
Ball, J., *PB*, 1941, **46**, 669
Barks, O.L., & Overholser, M.D., *AR*, 1938 **70**, 401
Barnes, D.O., Regan, J.F., & Nelson, W.O. *JAMA*, 1933, **101**, 926
Barrie, M.M.O., *L*, 1937 (ii), 251
Bates, R.W., Riddle, O., & Lahr, E.L., *PB* 1934, **31**, 1223
Bates, R.W., Riddle, O., & Lahr, E.L., *AJP* 1937, **119**, 610
Battaceano, G., & Vasiliu, C., *CRS*, 1936, **121** 1541

Beach, F.A., *PB*, 1942, **51**, 369
Beadle, O.A., *Med. Res. Council, Spec. Reps.*, No. 161, 1931, p. 21
Beall, D., *N*, 1939, **144**, 76
Beall, D., (a) *JE*, 1940, **2**, 81
Beall, D., (b) *BJ*, 1940, **34**, 1293
Beall, D., & Reichstein, T., *N*, 1938, **142**, 479
Beatson, G.T., *L*, 1896 (ii), 162
Beilly, J.S., *E*, 1939, **25**, 128, 275
Beilly, J.S., *E*, 1940, **26**, 959
Beilly, J.S., & Solomon, S., *E*, 1940, **26**, 236
Bellerby, C.W., *BJ*, 1933, **27**, 615
Bennett, H.S., *PB*, 1939, **42**, 786
Benoit, J., (a) *CRS*, 1935, **118**, 672
Benoit, J., (b) *CRS*, 1935, **120**, 133, 136, 905, 1323
Benoit, J., *PB*, 1937, **36**, 782
Benoit, J., (a) *CRS*, 1938, **127**, 909
Benoit, J., (b) *CRS*, 1938, **129**, 231, 234
Benoit, J., & Aron, E., *CRS*, 1934, **116**, 221
Benoit, J., & Ott, L., *CRS*, 1938, **127**, 906
Berdnikoff, A., & Champy, C., *CRS*, 1934, **116**, 515
Berger, M., *KW*, 1935, **14**, 1601
Bergman, A.J., & Turner, C.W., *E*, 1942, **30**, 11
Berthold, A.A., *Arch. f. Anat. u. Physiol.*, 1849, p. 42
Bialet-Laprida, Z., *CRS*, 1933, **114**, 377
Bialet-Laprida, Z., *CRS*, 1934, **117**, 456
Biddulph, C., *AR*, 1939, **73**, 447
Biddulph, C., & Meyer, R.K., *AJP*, 1941, **132**, 259
Biddulph, C., Meyer, R.K., & Gumbreck, L.G., *E*, 1940, **26**, 280
Biggart, J.H., *BJH*, 1934, **54**, 157
Bischoff, F., *E*, 1942, **30**, 525
Bishop, P.M.F., *BMJ*, 1938 (i), 939
Bishop, P.M.F., & McKeown, T., *JE*, 1941, **2**, 339
Biskind, G.R., *PB*, 1940, **43**, 259
Biskind, G.R., (a) *PB*, 1941, **46**, 452
Biskind, G.R., (b) *PB*, 1941, **47**, 266
Biskind, M.S., & Biskind, G.R., *E*, 1942, **31**, 109
Bissonnette, T.H., (a) *AJA*, 1930, **45**, 289, 307
Bissonnette, T.H., (b) *AJA*, 1930, **46**, 477
Bissonnette, T.H., *JEZ*, 1931, **58**, 281
Bissonnette, T.H., *PR*, 1932, **110**, 322
Bittner, J.J., *AJC*, 1935, **25**, 791
Bittner, J.J., *AJC*, 1937, **30**, 530
Bittner, J.J., (a) *AJC*, 1939, **36**, 44
Bittner, J.J., (b) *PHR*, 1939, **54**, 1827
Bittner, J.J., (a) *JNCI*, 1940, **1**, 155
Bittner, J.J., (b) *PB*, 1940, **45**, 805
Bittner, J.J., *CR*, 1941, **1**, 290
Bittner, J.J., & Little, C.C., *J. Heredity*, 1937, **28**, 117
Bloch, P.W., *E*, 1936, **20**, 307
Bloch, S., *JE*, 1939, **1**, 399
Boling, J.L., & Hamilton, J.B., *AR*, 1939, **73**, 1
Bond, C.J., *BMJ*, 1906 (ii), 121
Bonser, G.M., (a) *12th Ann. Rep., Brit. Emp. Cancer Camp.*, 1935, p. 95
Bonser, G.M., (b) *JPB*, 1935, **41**, 33

Bonser, G.M., *JPB*, 1936, **42**, 169
Bonser, G.M., *Proc. 7th Internat. Genetic Cong., Edin.*, 1939, p. 80
Bonser, G.M., *JEB*, 1942, **54**, 149
Bonser, G.M., & Connal, K.I., *JPB*, 1939, **48**, 263
Bonser, G.M., & Robson, J.M., *JPB*, 1940, **51**, 9
Bonser, G.M., Stickland, L.H., & Connal, K.I., *JPB*, 1937, **45**, 709
Borchardt, H., Dingemanse, E., & Laqueur, E., *NW*, 1934, **22**, 190
Bottomley, A.C., & Folley, S.J., (a) *JP*, 1938, **92**, 33 P.
Bottomley, A.C., & Folley, S.J., (b) *JP*, 1938, **94**, 26
Bottomley, A.C., & Folley, S.J., (c) *PR*, 1938, **126**, 224
Bouin, P., & Ancel, P., (a) *CRS*, 1903, **55**, 1682, 1688
Bouin, P., & Ancel, P., (b) *AZE*, 1903, **1**, 437
Bouin, P., & Ancel, P., *CRA*, 1904, **138**, 231
Bouin, P., & Ancel, P., *AZE*, 1905, **3**, 391
Bouin, P., & Ancel, P., *CRA*, 1906, **142**, 232, 298
Bouin, P., & Ancel, P., *CRS*, 1909, **67**, 505, 689
Bourg, R., (a) *CRS*, 1930, **103**, 916
Bourg, R., (b) *CRS*, 1930, **104**, 106, 1048
Bourne, A. W., & Burn, J. H., *L*, 1928 (ii), 1020
Bourne, G., & Zuckerman, S., *JE*, 1941, **2**, 283
Bowman, K.M., & Bender, L., *Amer. J. Psychiat.*, 1932, **11**, 867
Boycott, M., & Rowlands, I.W., *BMJ*, 1938 (i), 1097
Boycott, M., & Smiles, J.M., *L*, 1939 (i), 1419
Bradbury, J.T., *PB*, 1932, **30**, 212
Bradbury, J.T., & Gaensbauer, F., *PB*, 1939, **41**, 128
Brambell, F.W.R., Fielding, U., & Parkes, A.S., *PR*, 1928, **102**, 385
Brambell, F.W.R., & Hall, K., *JA*, 1935, **70**, 339
Brambell, F.W.R., & Parkes, A.S., *PR*, 1927, **101**, 316
Breneman, W.R., *AR*, 1936, **64**, 211
Breneman, W.R., *E*, 1939, **24**, 55
Brindeau, A., Hinglais, H., & Hinglais, M., *CRS*, 1934, **115**, 1509
Brooks, C.McC., *AJP*, 1938, **121**, 157
Brooks, C.McC., *Res. Publ. Ass. nerv. ment. Dis.*, 1940, **20**, 525
Brooksby, J.B., *PB*, 1938, **38**, 235 and 832
Broster, L.R., *BMJ*, 1941 (i), 117
Broster, L.R., & Vines, H.W.C., 'The Adrenal Cortex', London, 1933
Brouha, L., *CRS*, 1932, **110**, 1023
Brouha, L., *CRS*, 1933, **113**, 406
Brouha, L., & Simonnet, H., *CRS*, 1926, **95**, 674
Brouha, L., & Simonnet, H., *CRS*, 1927, **96**, 96
Brouha, L., & Simonnet, H., *CRS*, 1928, **99**, 41
Browman, L.G., *PB*, 1937, **36**, 205
Browne, J.S.L., Henry, J.S., & Venning, E.H., *AJOG*, 1939, **38**, 927

Browne, J.S.L., & Venning, E.M., *L*, 1936 (ii), 1507

Browne, J.S.L., & Venning, E.H., *AJP*, 1938, **123**, 209

Brownell, K.A., Lockwood, J.E., & Hartman, F.A., *PB*, 1933, **30**, 783

Brown-Séquard, C.E., *CRS*, 1889, Sér. 9, i, 415, 420, 430 and 454

Bruehl, R., *KW*, 1929, **8**, 1766

Brunelli, B., *Arch. Internat. de Pharmacodynamie*, 1935, **49**, 214, 243, 262, and 295

Brunschwig, A., & Bissel, A.D., *AS*, 1936, **33**, 515

Bryan, W.R., Kahler, H., Shimkin, M.B., & Andervont, H.B., *JNCI*, 1942, **2**, 451

Buchheim, V., *CRS*, 1932, **109**, 1290

Buckner, G.D., Insko, W.M., & Martin, J.H., *Poultry Science*, 1934, **71**, 365

Bugbee, E.P., & Simond, A.E., *E*, 1926, **10**, 360

Bulliard, H., & Ravina, A., *CRS*, 1937, **125**, 965

Bullough, W.S., (a) *JE*, 1942, **3**, 141, 150, 211

Bullough, W.S., (b) *N*, 1942, **149**, 271

Bullough, W.S., *N*, 1943, **151**, 531

Bullough, W.S., & Carrick, R., *N*, 1939, **144**, 33

Bullough, W.S., & Gibbs, H.F., *N*, 1941, **148**, 439

Bunde, C.A., & Greep, R.O., *PB*, 1936, **35**, 235

Burdick, H.O., *E*, 1940, **27**, 825

Burns, E.L., Moskop, M., Suntzeff, V., & Loeb, L., *AJC*, 1936, **26**, 56 and 761

Burns, E.L., & Schenken, J.R., *PB*, 1939, **40**, 497

Burns, E.L., & Schenken, J.R., *PB*, 1940, **43**, 608

Burns, E.L., Suntzeff, V., & Loeb, L., *AJC*, 1938, **32**, 534

Burns, R.K., *JEZ*, 1925, **42**, 31

Burns, R.K., *AR*, 1928, **39**, 177

Burns, R.K., *JEZ*, 1930, **55**, 123

Burns, R.K., *JEZ*, 1931, **60**, 339

Burns, R.K., (a) *PB*, 1939, **41**, 60 and 270

Burns, R.K., (b) *AR*, 1939, **73**, 73

Burns, R.K., (c) *JM*, 1939, **65**, 79

Burns, R.K., & Buyse, A., *JEZ*, 1934, **67**, 115

Burr, H.S., Hill, R.T., & Allen, E., *PB*, 1935, **33**, 109

Burr, H.S., Musselman, L.K., Barton, D., & Kelly, N.B., *YJ*, 1937, **10**, 155

Burrill, M.W., & Greene, R.R., *PB*, 1939, **40**, 327

Burrill, M.W., & Greene, R.R., (a) *PB*, 1940, **44**, 273

Burrill, M.W., & Greene, R.R., (b) *E*, 1940, **26**, 645

Burrill, M.W., & Greene, R.R., (a) *E*, 1942, **30**, 142

Burrill, M.W., & Greene, R.R., (b) *E*, 1942, **31**, 73

Burrows, H., 'Some Factors in the Localization of Disease in the Body', London, 1932

Burrows, H., (a) *N*, 1934, **134**, 570

Burrows, H., (b) *BJS*, 1934, **21**, 507

Burrows, H., (a) *PR*, 1935, **118**, 485

Burrows, H., (b) *JPB*, 1935, **41**, 43, 218 and 423

Burrows, H., (c) *AJC*, 1935, **23**, 490

Burrows, H., (d) *AJC*, 1935, **24**, 613

Burrows, H., (e) *BJS*, 1935, **23**, 191

Burrows, H., (f) *JP*, 1935, **85**, 159

Burrows, H., (a) *JPB*, 1936, **42**, 161

Burrows, H., (b) *JPB*, 1936, **43**, 121

Burrows, H., (c) *BJS*, 1936, **23**, 658

Burrows, H., (d) *AJC*, 1936, **28**, 741

Burrows, H., (e) *PRSM*, 1936, **29** (Comp. Med.), 22

Burrows, H., (f) *N*, 1936, **138**, 164

Burrows, H., *JPB*, 1937, **44**, 481 and 699

Burrows, H., (a) *JE*, 1939, **1**, 417

Burrows, H., (b) *JPB*, 1939, **48**, 242

Burrows, H., (c) *N*, 1939, **143**, 858

Burrows, H., *JPB*, 1940, **51**, 385

Burrows, H., *Veterinary J.*, 1941, **97**, 225

Burrows, H., Cave, A.J.E., & Parbury, K., *Brit. J. Radiol.*, 1943, **16**, 87

Burrows, H., Cook, J.W., Roe, E.M.F., & Warren, F.L., *BJ*, 1937, **31**, 950

Burrows, H., Cook, J.W., & Warren, F.L., *CI*, 1936, **55**, 1031

Burrows, H., & Kennaway, N.M., *AJC*, 1934, **20**, 48

Butenandt, A., *DMW*, 1929, **55**, 2171

Butenandt, A., *Ber. dtsch. chem. Ges.*, 1930, **63**, 659

Butenandt, A., *Z. angew., Chem.*, 1931, **44**, 905

Butenandt, A., *NW*, 1936, **24**, 15 and 545

Butenandt, A., & Dannenbaum, H., *Ztschr. f. physiol. Chem.*, 1934, **229**, 192

Butenandt, A., & Hanisch, G., (a) *Hoppe-Seyl. Zeit.*, 1935, **237**, 89

Butenandt, A., & Hanisch, G., (b) *Ber. dtsch. chem. Ges.*, 1935, **68**, 1859

Butenandt, A., & Kudszus, H., *Hoppe-Seyl. Zeit.*, 1935, **237**, 75

Butenandt, A., Westphal, U., & Hohlweg, W., *Z. physiol. Chem.*, 1934, **227**, 84

Butler, G.C., & Marrian, G.F., *JBC*, 1937, **119**, 565

Butler, G.C., & Marrian, G.F., *JBC*, 1938, **124**, 237

Buxton, C.L., & Westphal, U., *PB*, 1939, **41**, 284

Byerly, T.C., & Burrows, W.H., *E*, 1938, **22**, 366

Calatroni, C.J., *CRS*, 1934, **117**, 452

Callow, N.H., *BJ*, 1939, **33**, 559

Callow, N.H., & Callow, R.K., *BJ*, 1938, **32**, 1759

Callow, N.H., & Callow, R.K., *BJ*, 1939, **33**, 931

Callow, N.H., & Callow, R.K., *BJ*, 1940, **34**, 276

Callow, N.H., Callow, R.K., & Emmens, C.W., *JE*, 1939, **1**, 99

Callow, N.H., Callow, R.K., & Emmens, C.W., *JE*, 1940, **2**, 88

Callow, N.H., Callow, R.K., Emmens, C.W., & Stroud, S.W., *JE*, 1939, **1**, 76

Callow, R.K., (a) *JP*, 1936, **86**, 49 P.
Callow, R.K., (b) *L*, 1936 (ii), 565
Callow, R.K., (c) *CI*, 1936, **55**, 1030
Callow, R.K., *PRSM*, 1938, **31**, 841
Callow, R.K., & Deanesly, R., (a) *L*, 1935 (ii), 77
Callow, R.K., & Deanesly, R., (b) *BJ*, 1935, **29**, 1424
Callow, R.K., & Parkes, A.S., *JP*, 1936, **87**, 28 P.
Campbell, N. R., Dodds, E. C., & Lawson, W., *N*, 1938, **142**, 1121
Camus, L., & Gley, E., *CRS*, 1896, **3**, 787; *CRS*, 1897, **4**, 787
Cantarow, A., Rakoff, A.E., Paschkis, K.E., & Hansen, L.P., *PB*, 1942, **49**, 707
Cardoso, D.M., *CRS*, 1934, **115**, 1347
Carmichael, E.S., & Marshall, F.H.A., *JP*, 1908, **36**, 431
Carnes, W.H., *PB*, 1940, **45**, 502
Carroll, W.R., *PB*, 1942, **49**, 50
Cartland, G.F., Meyer, R.K., Miller, L.C., & Rutz, M.H., *JBC*, 1935, **109**, 213
Casida, L.E., *E*, 1934, **18**, 715
Casida, L.E., *PB*, 1936, **33**, 570
Catchpole, H.R., Cole, H.H., & Pearson, P.B., *AJP*, 1935, **112**, 21
Chadwick, C.S., *AJA*, 1936, **40**, 129
Chamorro, A., *CRS*, 1940, **133**, 546
Champy, C., *Arch. Morph. gén. et exp.*, 1922, **4**, 1
Champy, C., *CRS*, 1937, **125**, 634
Champy, C., & Coujard, R., *CRS*, 1937, **125**, 632
Champy, C., & Coujard, R., *CRS*, 1939, **130**, 250
Champy, C., & Kritch, N., *CRS*, 1925, **92**, 683
Champy, C., & Lavedan, J.P., *CRA*, 1938, **207**, 99
Chanton, L.R., *CRS*, 1938, **129**, 1047
Chapin, C.L., *JEZ*, 1917, **23**, 453
Charipper, H.A., *PB*, 1934, **32**, 402
Cheatle, Sir G. Lenthal, *BJS*, 1921, **8**, 149 and 285
Ch'en, G., & Van Dyke, H.B., *PB*, 1939, **40**, 172
Chiodi, H., *CRS*, 1938, **129**, 866
Chiodi, H., *CRS*, 1939, **130**, 457
Chiodi, H., *E*, 1940, **26**, 107
Chow, B.F., Greep, R.O., & Van Dyke, H.B., *JE*, 1939, **1**, 439
Chu, J.P., *JE*, 1940, **2**, 21
Clark, H.M., *AR*, 1935, **41**, 175 and 193
Clauberg, C., *ZG*, 1930, **54**, 2757
Clauberg, C., *ZG*, 1932, **56**, 964
Clausen, H.J., *AR*, 1940, **76**, Suppl. 2, 14
Cleveland, R., & Wolfe, J.M., *AJA*, 1933, **53**, 191
Cloudman, A.M., & Little, C.C., *J. Genetics*, 1936, **32**, 487
Cohen, S.L., Marrian, G.F., & Watson, M., *L*, 1935 (i), 674
Cole, H.H., *PR*, 1933, **114**, 136
Cole, H.H., *AJA*, 1936, **59**, 299
Cole, H.H., *AJP*, 1937, **119**, 704

Cole, H.H., Hart, G.H., Lyons, W.R., & Catchpole, H.R., *AR*, 1933, **56**, 275
Cole, R.K., & Furth, J., *CR*, 1941, **1**, 957
Collins, V.J., Boling, J.L., Dempsey, E.W., & Young, W.C., *E*, 1938, **23**, 188
Collip, J.B., *CMA*, 1930, **22**, 761
Collip, J.B., *JAMA*, 1935, **104**, 556
Collip, J.B., *CSH*, 1937, **5**, 210
Collip, J.B., & Anderson, E.M., *JAMA*, 1935, **104**, 965
Collip, J.B., Anderson, E.M., & Thomson, D.L., *L*, 1933 (ii), 347
Collip, J.B., McEuen, C.S., & Selye, H., *AJP*, 1936, **116**, 29
Collip, J.B., Selye, H., & Anderson, E.M., *JAMA*, 1933, **101**, 1553
Collip, J.B., Selye, H., & Thomson, D.L., (a) *N*, 1933, **131**, 56
Collip, J.B., Selye, H., & Thomson, D.L., (b) *PB*, 1933, **30**, 589, 913
Collip, J.B., Selye, H., & Thomson, D.L., *PB*, 1934, **31**, 682
Collip, J.B., Selye, H., Thomson, D.L., & Williamson, J.E., *PB*, 1933, **30**, 665
Collip, J.B., Selye, H., & Williamson, J.E., *E*, 1938, **23**, 279
Collip, J.B., Thomson, D.L., McPhail, M.K., & Williamson, J.E., *CMA*, 1931, **24**, 201
Collip, J.B., & Williamson, J.E., *CMA*, 1936, **34**, 458
Commins, W.D., *JEZ*, 1932, **63**, 573
Compte, L., *Beitr. path. Anat. allg. Path.*, 1898, **23**, 90
Cook, J.W., & Dodds, E.C., *N*, 1933, **131**, 205
Cook, J.W., Dodds, E.C., Hewett, C.L., & Lawson, W., *PR*, 1934, **114**, 272
Cooke, J.V., *JAMA*, 1942, **119**, 547
Cope, C.L., (a) *BMJ*, 1940 (ii), 545
Cope, C.L., (b) *Clinical Science*, 1940, **4**, 217
Corey, E.L., *AR*, 1928, **41**, 40
Corey, E.L., & Britton, S.W., *AJP*, 1931, **99**, 33
Corey, E.L., & Britton, S.W., *AJP*, 1934, **107**, 207
Cori, C.F., *JEM*, 1927, **45**, 983
Corner, G.W., *AJP*, 1928, **86**, 74
Corner, G.W., *AJP*, 1930, **95**, 43
Corner, G.W., *AJP*, 1935, **113**, 238
Corner, G.W., & Allen, W.M., *AJP*, 1929, **88**, 326
Corner, G.W., & Warren, S.L., *AR*, 1919, **16**, 168
Cornil, L., Paillas, J.E., & Rosanoff, H., *CRS*, 1938, **129**, 981
Cotte, G., & Noël, R., *Gynéc. et Obstét.*, 1936, **34**, 294
Courrier, R., *CRS*, 1921, **85**, 486
Courrier, R., (a) *Proc. 2nd Internat. Cong. Sex Research*, London, 1930, p. 352
Courrier, R., (b) *CRS*, 1930, **104**, 280 and 1178
Courrier, R., *CRS*, 1933, **115**, 299
Courrier, R., *CRS*, 1934, **117**, 1117
Courrier, R., *CRS*, 1935, **120**, 1263
Courrier, R., *CRS*, 1936, **122**, 661

Courrier, R., & Cohen-Solal, G., *CRS*, 1936, **121**, 903

Courrier, R., & Cohen-Solal, G., *CRS*, 1937, **124**, 925 and 961

Courrier, R., & Gros, G., *CRS*, 1934, **115**, 1097

Courrier, R., & Gros, G., *CRS*, 1935, **118**, 683 and 686

Courrier, R., & Gros, G., *CRS*, 1938, **128**, 194

Courrier, R., & Jost, A., *CRS*, 1939, **130**, 726 and 1162

Courrier, R., & Kehl, R., *CRS*, 1930, **104**, 1180

Courrier, R., & Kehl, R., *CRS*, 1931, **107**, 1547

Courrier, R., & Kehl, R., *CRS*, 1932, **109**, 877

Courrier, R., & Kehl, R., (a) *CRS*, 1933, **112**, 675

Courrier, R., & Kehl, R., (b) *CRS*, 1933, **113**, 607

Courrier, R., & Kehl, R., (a) *CRS*, 1938, **127**, 529

Courrier, R., & Kehl, R., (b) *CRS*, 1938, **128**, 188

Courrier, R., Kehl, R., & Raynaud, R., *CRS*, 1929, **100**, 1103

Courrier, R., & Potvin, R., *CRS*, 1926, **94**, 878

Courrier, R., & Raynaud, R., *CRS*, 1932, **109**, 881

Courrier, R., & Raynaud, R., (a) *CRS*, 1934, **115**, 299

Courrier, R., & Raynaud, R., (b) *CRS*, 1934, **116**, 1073

Crabtree, C., *S*, 1940, **111**, 299

Craig-Bennett, A., *PT*, 1931, **219**, 197

Cramer, W., & Horning, E.S., *L*, 1936 (i), 247

Cramer, W., & Horning, E.S., (a) *JPB*, 1937, **44**, 633

Cramer, W., & Horning, E.S., (b) *L*, 1937 (i), 1330

Cramer, W., & Horning, E.S., (a) *Proc. 7th Internat. Genetic Cong., Edin.*, 1939, p. 91

Cramer, W., & Horning, E.S., (b) *AJC*, 1939, **37**, 343

Cramer, W., & Horning, E.S., (c) *L*, 1939 (i), 192

Craver, L.F., & Stewart, F.W., *JAMA*, 1936, **106**, 1802

Crew, F.A.E., *J. Genetics*, 1921, **11**, 141

Crew, F.A.E., *JA*, 1922, **56**, 98

Crew, F.A.E., (a) *BMJ*, 1936 (i), 363

Crew, F.A.E., (b) *BMJ*, 1936 (ii), 1092

Crew, F.A.E., & Mirskaia, L., *QJ*, 1930, **20**, 105

Crooke, A.C., & Callow, R.K., *Quart. J. Med.*, 1939, **32**, 233

Cushing, H., *BJH*, 1932, **50**, 137

Cushing, H., & Davidoff, L.M., *Monogr. Rockefeller Inst. med. Res.*, 1927, No. 22, 109

Cushing, H., & Goetsch, E., *PB*, 1913, **11**, 25

Cushing, H., & Goetsch, E., *JEM*, 1915, **22**, 25

Cushny, A.R., *JP*, 1906, **35**, 1

Cutuly, E., *PB*, 1941, **47**, 290

Cutuly, E., & Cutuly, E.C., *E*, 1938, **22**, 568

Cutuly, E., McCullagh, D.R., & Cutuly, E.C., (a) *AJP*, 1937, **119**, 121

Cutuly, E., McCullagh, D.R., & Cutuly, E.C., (b) *E*, 1937, **21**, 241

Daels, J., & Heymans, C., *CRS*, 1938, **127**, 1109

Dale, H.H., *JP*, 1906, **34**, 163

D'Amour, F.E., *E*, 1940, **26**, 88

D'Amour, F.E., D'Amour, M.C., & Gustavson, R.G., *JPET*, 1933, **49**, 146

D'Amour, M.C., & D'Amour, F.E., *PB*, 1939, **40**, 417

Danby, M., *ABN*, 1938, **8**, 90 and 186

Danby, M., *E*, 1940, **27**, 236

Danforth, C.H., (a) *Proc. 2nd Internat. Cong. Sex Research, London*, 1930

Danforth, C.H., (b) *Biolog. generalis*, 1930, **6**, 100

Danforth, C.H., & Foster, F., *JEZ*, 1929, **52**, 443

Daniel, J.F., *JEZ*, 1910, **9**, 865

Dantchakoff, V., *CRS*, 1936, **122**, 168 and 1307

Dantchakoff, V., (a) *CRA*, 1937, **204**, 195

Dantchakoff, V., (b) *CRS*, 1937, **124**, 195, 516 and 519

Dantchakoff, V., (a) *CRS*, 1938, **127**, 674 and 1255

Dantchakoff, V., (b) *CRS*, 1938, **129**, 946

Darwin, C., 'The Descent of Man', London, 1871

David, J.C., *JPL*, 1931, **43**, 1

David, K., *ABN*, 1938, **8**, 133

David, K., Dingemanse, E., Freud, J., & Laqueur, E., *Zeits. physiol. Chem.*, 1935, **233**, 281

David, K., Freud, J., & De Jongh, S.E., *BJ*, 1934, **28**, 1360

Davidson, C.S., *PB*, 1937, **36**, 703

Davidson, C.S., & Moon, H.D., *PB*, 1936, **35**, 281

Davis, M.E., & Koff, A.K., *AJOG*, 1938, **36**, 183

Deanesly, R., *PR*, 1928, **103**, 523

Deanesly, R., (a) *N*, 1938, **141**, 79

Deanesly, R., (b) *PR*, 1938, **126**, 122

Deanesly, R., *JE*, 1939, **1**, 36 and 307

Deanesly, R., Fee, A.R., & Parkes, A.S., *JP*, 1930, **70**, 38

Deanesly, R., & Newton, W.H., *JE*, 1941, **2**, 317

Deanesly, R., & Parkes, A.S., (a) *BMJ*, 1936 (i), 257

Deanesly, R., & Parkes, A.S., (b) *L*, 1936 (i), 837

Deanesly, R., & Parkes, A.S., (c) *BJ*, 1936, **30**, 291

Deanesly, R., & Parkes, A.S., (a) *PR*, 1937, **124**, 279

Deanesly, R., & Parkes, A.S., (b) *CI*, 1937, **56**, 447

Deanesly, R., & Parkes, A.S. (c) *QJ*, 1937, **26**, 393

De Fremery, P., *JP*, 1936, **87**, 50 P.

De Fremery, P., *ABN*, 1937, **7**, 9

De Fremery, P., & Geerling, M.C., *ABN*, 1939, **9**, 17

De Fremery, P., Lucks, A., & Tausk, M., *Pflügers Archiv*, 1932, **231**, 341

De Fremery, P., & Scheygrond, B., *N*, 1937, **139**, 1015
De Jongh, S.E., *ABN*, 1933, **3**, 52, 88 and 112
De Jongh, S.E., (*a*) *ABN*, 1934, **4**, 69 and 98
De Jongh, S.E., (*b*) *Nederl. Tijd. v. Geneesk.*, 1934, **78**, 489
De Jongh, S.E., (*a*) *ABN*, 1935, **5**, 28
De Jongh, S.E., (*b*) *Arch. Intern. Pharmak. Therap.*, 1935, **50**, 348
De Jongh, S.E., Freud, J., & Laqueur, E., *KW*, 1930, **9**, 2344
De Jongh, S.E., & Kok, D.J., *ABN*, 1935, **5**, 177
De Jongh, S.E., & Korteweg, R., *ABN*, 1935, **5**, 126
De Jongh, S.E., & Van der Woerd, L.A., *ABN*, 1939, **9**, 153
del Castillo, E.B., & Calatroni, C.J., *CRS*, 1930, **104**, 1024
del Castillo, E.B., & Pinto, A., *CRS*, 1938, **129**, 868
del Castillo, E.B., & Pinto, A., *CRS*, 1939, **131**, 297
del Castillo, E.B., & Sammartino, R., *CRS*, 1938, **129**, 870
Delille, A., 'L'Hypophyse et la médication hypophysaire', Paris, 1909
Dempsey, E.W., *AJP*, 1937, **120**, 126
Dempsey, E.W., *AJP*, 1939, **126**, 758
Dempsey, E.W., Hertz, R., & Young, W.C., *AJP*, 1936, **116**, 201
Desclin, L., *CRS*, 1934, **115**, 439
Desclin, L., *CRS*, 1939, **131**, 837
Desclin, L., & Brouha, L., *CRS*, 1931, **107**, 895
Desclin, L., & Dessiennes, R., *CRS*, 1940, **133**, 682
Dessau, F., *ABN*, 1937, **7**, 126
Dessau, F., & Freud, J., *ABN*, 1936, **6**, 9
Dessau, F., & Freud, J., *ABN*, 1938, **8**, 53
Deulofeu, V., *CRS*, 1939, **130**, 458
Deulofeu, V., & Ferrari, J., *N*, 1934, **133**, 835
Dickens, F., *BJ*, 1941, **35**, 1011
Dingemanse, E., *ABN*, 1938, **8**, 55
Dingemanse, E., Borchardt, H., & Laqueur, E., *BJ*, 1937, **31**, 500
Dingemanse, E., & Freud, J., *ABN*, 1939, **9**, 157
Dingemanse, E., Freud, J., & Laqueur, E., *N*, 1935, **135**, 184
Dingemanse, E., & Laqueur, E., *Nederl. Tijd. v. Geneesk.*, 1938, **82**, 4166
Dingemanse, E., & Laqueur, E., *JU*, 1940, **44**, 530
Dobrovolskaya-Zavadskaya, N., & Pezzini, Z.M., *CRS*, 1939, **131**, 240
Dobrovolskaya-Zavadskaya, N., & Zéphiroff, P., *CRS*, 1938, **128**, 971
Dodds, E.C., *Helv. chim. Acta*, 1936, **19**, 49
Dodds, E.C., *BMJ*, 1937 (i), 398
Dodds, E. C., Goldberg, L., Lawson, W., & Robinson, R., *N*, 1938, **141**, 247
Dodds, E.C., & Lawson, W., *N*, 1936, **137**, 996
Dodds, E.C., & Lawson, W., *N*, 1937, **139**, 627 & 1068
Dodds, E.C., & Noble, R.L., *BMJ*, 1936 (ii), 824

Dohrn, M., Faure, W., Poll, H., & Blotevogel, W., *Med. Klin.*, 1926, **22**, 1417
Doisy, E.A., Curtis, J., & Collier, W.D., *PB*, 1931, **28**, 885
Doisy, E.A., Veler, C.D., & Thayer, S., *JBC*, 1930, **86**, 499
Domm, L.V., *JEZ*, 1927, **48**, 31
Domm, L.V., *AE*, 1929, **119**, 171
Domm, L.V., *PB*, 1931, **29**, 310
Domm, L.V., *CSH*, 1937, **5**, 241
Domm, L.V., *PB*, 1939, **42**, 310
Domm, L.V., & Van Dyke, H.B., *PB*, 1932, **30**, 349 and 351
Donahue, J.K., *E*, 1940, **27**, 149
Donaldson, H.H., & Hatai, S., *J. Comp. Neurol.*, 1911, **21**, 155
Dorfman, R.I., *PB*, 1940, **45**, 739
Dorfman, R.I., *PB*, 1941, **46**, 351
Dorfman, R.I., & Greulich, W.W., *YJ*, 1937, **10**, 79
Dorfman, R.I., Greulich, W.W., & Solomon, C.I., *E*, 1937, **21**, 741
Dow, D., & Zuckerman, S., (*a*) *E*, 1939, **25**, 525
Dow, D., & Zuckerman, S., (*b*) *JE*, 1939, **1**, 396
Drummond, J.C., Noble, R.L., & Wright, M.D., *JE*, 1939, **1**, 275
Dublin, L.I., & Lotka, A.J., 'A study of the Mortality Experience among the Industrial Policyholders of the Metropolitan Life Insurance Company, 1911 to 1935', New York
Du Shane, G.P., Levine, W.T., Pfeiffer, C.A., & Witschi, E., *PB*, 1935, **33**, 339
Duyvené De Wit, J.J., *KW*, 1938, **17**, 376, 660, 792
Duyvené De Wit, J.J., *JE*, 1940, **2**, 141

Edwards, E.A., Hamilton, J.B., & Duntley, S.Q., *New Engl. J. Med.*, 1939, **220**, 865
Edwards, J., *PR*, 1940, **128**, 407
Eidelsberg, J., & Ornstein, E.A., *E*, 1940, **26**, 46
Elden, C.A., *E*, 1936, **20**, 47
Elder, J.H., *PB*, 1941, **46**, 57
Elliott, T.R., & Armour, R.G., *JPB*, 1911, **15**, 481
Ellison, E.T., & Burch, J.C., *E*, 1936, **20**, 746
Ellison, E.T., Campbell, M., & Wolfe, J.M., *AR*, 1932, **52**, Suppl., 54
Emery, F.E., *PB*, 1937, **36**, 731
Emery, F.E., & Atwell, W.J., *AR*, 1933, **58**, 17
Emery, F.E., & Greco, P.A., *E*, 1940, **27**, 473
Emery, F.E., & Schwabe, E.L., *E*, 1936, **20**, 550
Emmens, C.W., (*a*) *Med. Res. Counc. Spec. Reps.*, 1939, No. 234
Emmens, C.W., (*b*) *JE*, 1939, **1**, 142 and 373
Emmens, C.W., (*c*) *N*, 1939, **143**, 476
Emmens, C.W., (*d*) *JP*, 1939, **95**, 379
Emmens, C.W., *JE*, 1940, **2**, 63
Emmens, C.W., *JE*, 1941, **2**, 444
Emmens, C.W., *JE*, 1942, **3**, 64, 168, 174
Emmens, C.W., & Bradshaw, T.E.T., *JE*, 1939, **1**, 378
Emmens, C.W., & Parkes, A.S., (*a*) *JE*, 1939, **1**, 321 and 323
Emmens, C.W., & Parkes, A.S., (*b*) *N*, 1939, **143**, 1064

Engelhart, E., *KW*, 1930, **9**, 2114
Engelhart, E., & Riml, O., *KW*, 1934, **13**, 101
Engle, E.T., *PB*, 1927, **25**, 84
Engle, E.T., (*a*) *JAMA*, 1929, **93**, 276
Engle, E.T., (*b*) *AJP*, 1929, **88**, 101
Engle, E.T., (*a*) *PB*, 1932, **29**, 1224
Engle, E.T., (*b*) *E*, 1932, **16**, 506 and 513
Engle, E.T., *AJP*, 1933, **106**, 145
Engle, E.T., *CSH*, 1937, **5**, 111
Engle, E.T., & Crafts, R.C., *PB*, 1939, **42**, 293
Engle, E.T., & Smith, P.E., *AR*, 1935, **61**, 471
Engle, E.T., Smith, P.E., & Shelesnyak, M.C., *AJOG*, 1935, **29**, 787
Entenman, C., Lorenz, F.W., & Chaikoff, I.L., *JBC*, 1938, **126**, 133
Entwisle, R.M., & Hepp, J.A., *JAMA*, 1935, **104**, 395
Enzmann, E.V., *AR*, 1935, **62**, 31
Enzmann, E.V., Saphir, N.R., & Pincus, G., *AR*, 1932, **54**, 325
Erdheim, J., & Stumme, E., *Beitr. path. Anat. allg. Path.*, 1909, **46**, 1
'Espinasse, P.G., *N*, 1934, **134**, 738
'Espinasse, P.G., (*a*) *N*, 1939, **144**, 1013
'Espinasse, P.G., (*b*) *Proc. Zool. Soc. Lond.*, 1939, **109**, 247
Evans, H.M., *The Harvey Lectures, Philadelphia*, 1923–24, Series xix, 212
Evans, H.M., *JBC*, 1928, **77**, 651
Evans, H.M., *AJP*, 1931, **99**, 477
Evans, H.M., & Bishop, K.S., *J. Metab. Research*, 1922, **1**, 335
Evans, H.M., & Burr, G.O., *AJP*, 1926, **77**, 518
Evans, H.M., & Burr, G.O., *Mem. Univ. California*, 1927, viii
Evans, H.M., & Gorbman, A., *PB*, 1942, **49**, 674
Evans, H.M., Kohls, C.L., & Wonder, D.H., *JAMA*, 1937, **108**, 287
Evans, H.M., & Long, J.A., (*a*) *AR*, 1922, **23**, 19
Evans, H.M., & Long, J.A., (*b*) *Proc. Nat. Acad. Sci.*, 1922, **8**, 38
Evans, H.M., Meyer, K., & Simpson, M.E., *PB*, 1931, **28**, 845
Evans, H.M., Meyer, K., & Simpson, M.E., *AJP*, 1932, **100**, 141
Evans, H.M., & Simpson, M.E., (*a*) *PB*, 1929, **26**, 595, 598
Evans, H.M., & Simpson, M.E., (*b*) *AJP*, 1929, **89**, 371, 375, 379 and 381
Evans, H.M., & Simpson, M.E., *AR*, 1930, **45**, Suppl., 215, 216
Evans, H.M., & Simpson, M.E., *AJP*, 1931, **98**, 511 and 536
Evans, H.M., & Simpson, M.E., *AR*, 1934, **60**, 405 and 423
Evans, H.M., & Simpson, M.E., *AR*, 1937, **35**, 36
Evans, H.M., & Simpson, M.E., *E*, 1940, **27**, 305
Evans, H.M., Simpson, M.E., & Austin, P.R., (*a*) *JEM*, 1933, **57**, 897
Evans, H.M., Simpson, M.E., & Austin, P.R., (*b*) *JEM*, 1933, **58**, 545 and 561

Evans, H.M., Simpson, M.E., Austin, P.R., & Ferguson, R.S., *PB*, 1933, **31**, 21
Evans, H.M., Simpson, M.E., & Lyons, W.R., *PB*, 1941, **46**, 586
Evans, H.M., Simpson, M.E., & Pencharz, R.I., *PB*, 1935, **32**, 1048
Evans, H.M., Simpson, M.E., & Pencharz, R.I., *CSH*, 1937, **5**, 229
Evans, H.M., Simpson, M.E., Tolksdorf, S., & Jensen, H., *E*, 1939, **25**, 529
Ewald, F.K., *ZG*, 1936, **60**, 559
Fagin, J., & Reynolds, S.R.M., *AJP*, 1936, **117**, 86
Falls, F.H., Lackner, J.E., & Krohn, L., *PB*, 1935, **32**, 1451
Falls, F.H., Lackner, J.E., & Krohn, L., *JAMA*, 1936, **106**, 271
Fee, A.R., Marrian, G.F., & Parkes, A.S., *JP*, 1929, **67**, 377
Fee, A.R., & Parkes, A.S., *JP*, 1929, **67**, 383
Fekete, E., & Green, C.V., *AJC*, 1936, **27**, 513
Fekete, E., & Little, C.C., *CR*, 1942, **2**, 525
Fekete, E., Woolley, G., & Little, C.C., *JEM*, 1941, **74**, 1
Fell, H.B., *QJ*, 1922, **13**, 145
Fels, E., *AG*, 1930, **141**, 3
Fels, E., Slotta, K.H., & Ruschig, H., *KW*, 1934, **13**, 1207
Ferguson, R.S., (*a*) *AJC*, 1933, **18**, 269
Ferguson, R.S., (*b*) *JAMA*, 1933, **101**, 1933
Ferguson, R.S., Downes, H.R., Ellis, E., & Nicholson, M.E., *AJC*, 1931, **15**, 835
Fevold, H.L., *CSH*, 1937, **5**, 93
Fevold, H.L., & Fiske, V.M., *E*, 1939, **24**, 823
Fevold, H.L., Hisaw, F.L., & Greep, R.O., *AR*, 1934, **60**, Suppl., 51
Fevold, H.L., Hisaw, F.L., & Greep, R.O., (*a*) *AJP*, 1936, **114**, 508
Fevold, H.L., Hisaw, F.L., & Greep, R.O., (*b*) *AJP*, 1936, **117**, 68
Fevold, H.L., Hisaw, F.L., & Greep, R.O., *E*, 1937, **21**, 343
Fevold, H.L., Hisaw, F.L., Hellbaum, A., & Hertz, R., *AJP*, 1933, **104**, 710
Fevold, H.L., Hisaw, F.L., & Leonard, S.L., *AJP*, 1931, **97**, 291
Fichera, G., (*a*) *Arch. Ital. de Biol.*, 1905, **43**, 405
Fichera, G., (*b*) *Il Policlinico, Sez. Chir.*, 1905, **12**, 250
Fiessinger, N., & Moricard, R., *CRS*, 1934, **115**, 1602
Finlay, G.F., *BJEB*, 1925, **2**, 439
Fischer, A., & Engel, M., *L*, 1939 (i), 354
Fish, W.R., & Dorfman, R.I., *JBC*, 1941, **140**, 83
Flaks, J., Himmel, I., & Zotnik, A., *Presse Méd.*, 1938, **2**, 1506
Fleischmann, W., & Kann, S., *S*, 1938, **87**, 305
Fluhmann, C.F., *E*, 1931, **15**, 177
Fluhmann, C.F., *AJP*, 1933, **106**, 238
Fluhmann, C.F., *AJOG*, 1935, **30**, 584
Fluhmann, C.F., *PB*, 1941, **47**, 378

Fluhmann, C.F., & Laqueur, G.L., *E*, 1942, **31**, 375

Folley, S.J., *BJ*, 1936, **30**, 2262

Folley, S.J., *E*, 1939, **24**, 814

Folley, S.J., *L*, 1941 (i), 40

Folley, S.J., & Bottomley, A.C., *JP*, 1941, **99**, Proc.

Folley, S.J., Guthkelch, A.N., & Zuckerman, S., *PR*, 1939, **126**, 469

Folley, S.J., & Kon, S.K., *N*, 1937, **139**, 1107

Folley, S.J., & Kon, S.K., *PR*, 1938, **124**, 476

Folley, S.J., Scott Watson, H.M., & Bottomley, A.C., *JP*, 1940, **98**, 15 P.

Folley, S.J., Scott Watson, H.M., & Bottomley, A.C., (a) *J. Dairy Res.*, 1941, **12**, 241

Folley, S.J., Scott Watson, H.M., & Bottomley, A.C., (b) *JP*, 1941, **100**, Proc.

Folley, S.J., & White, P., *N*, 1937, **140**, 505

Folley, S.J., & Young, F.G., *PR*, 1938, **126**, 45

Folley, S.J., & Young, F.G., (a) *N*, 1941, **148**, 563

Folley, S.J., & Young, F.G., (b) *L*, 1941 (i), 380

Foote, C.L., *PB*, 1940, **43**, 519

Forbes, T.R., *S*, 1938, **87**, 282

Fortner, H.C., & Owen, S.E., *AJC*, 1935, **25**, 89

Foss, G.L., *L*, 1939 (i), 502

Foss, G.L., & Phillips, P., *BMJ*, 1938 (ii), 887

Foster, M.A., Foster, R.C., & Hisaw, F.L., *E*, 1937, **21**, 249

Fraenkel, L., *AG*, 1903, **68**, 438

Fraenkel, L., & Martins, T., *CRS*, 1938, **127**, 466

Fraenkel, L., Martins, T., & Mello, R.F., *E*, 1940, **27**, 836

Fraenkel-Conrat, H.L., Herring, V.V., Simpson, M.E., & Evans, H.M., *PB*, 1941, **48**, 333

Fraenkel-Conrat, H.L., Simpson, M.E., & Evans, H.M., *PB*, 1940, **45**, 627

Frank, R.T., *PB*, 1934, **31**, 1204

Frank, R.T., *JAMA*, 1940, **114**, 1504

Frank, R.T., Frank, M.L., Gustavson, R.G., & Weyerts, W.W., *JAMA*, 1925, **85**, 510

Frank, R.T., & Goldberger, M.A., *JAMA*, 1926, **86**, 1686

Frank, R.T., Goldberger, M.A., & Salmon, U.J., *PB*, 1936, **33**, 615

Frank, R.T., Goldberger, M.A., & Spielman, F., *PB*, 1932, **29**, 1229

Frank, R.T., Goldberger, M.A., & Spielman, F., *JAMA*, 1934, **103**, 393

Frank, R.T., & Klempner, E., *PB*, 1937, **36**, 763

Frank, R.T., & Salmon, U.J., (a) *PB*, 1935, **32**, 1237

Frank, R.T., & Salmon, U.J., (b) *PB*, 1935, **33**, 311

Frank, R.T., Salmon, U.J., & Friedman, R., *PB*, 1935, **32**, 1666

Frazer, J.E., 'Anatomy of the Human Skeleton', London, 1920

Frazier, C.N., & Mu, J.W., *PB*, 1935, **32**, 997

Freed, S.C., Brownfield, B., & Evans, H.M., *PB*, 1931, **29**, 1

Freed, S.C., & Coppock, A., *E*, 1936, **20**, 81

Freed, S.C., Greenhill, J.P., & Soskin, S., *PB*, 1938, **39**, 440

Freud, J., (a) *ABN*, 1933, **3**, 123

Freud, J., (b) *BJ*, 1933, **27**, 1438

Freud, J., *ABN*, 1935, **5**, 97

Freud, J., *ABN*, 1937, **7**, 72 and 115

Freud, J., *ABN*, 1938, **8**, 127, 131 and 176

Freud, J., *ABN*, 1939, **9**, 11, 161 and 202

Freud, J., & de Jongh, S.E., *ABN*, 1935, **5**, 47

Freud, J., Dingemanse, E., & Polak, J., *ABN*, 1935, **5**, 179

Freud, J., Dingemanse, E., & Polak, J., *Arch. Int. Pharmacodyn.*, 1937, **57**, 369

Freud, J., & Laqueur, E., *ABN*, 1934, **4**, 100

Freud, J., Levie, L.H., & Kroon, D.B., *JE*, 1939, **1**, 56

Friedman, M.H., (a) *AJP*, 1929, **89**, 438

Friedman, M.H., (b) *AJP*, 1929, **90**, 617

Friedman, M.H., *AJP*, 1932, **99**, 332

Friedman, M.H., *PB*, 1938, **37**, 645

Friedman, M.H., & Weinstein, G.L., *E*, 1937, **21**, 489

Furth, J., & Butterworth, J.S., *AJC*, 1936, **28**, 66

Fussgänger, *Mediz. u. Chemie (I.G. Farbenind.)*, *Leverhusen*, 1934, 194

Gaarenstroom, J.H., *ABN*, 1937, **7**, 156

Gaarenstroom, J.H., *ABN*, 1939, **9**, 13

Gaarenstroom, J.H., *JE*, 1940, **2**, 47

Gaarenstroom, J.H., & De Jongh, S.E., *ABN*, 1939, **9**, 178

Gaarenstroom, J.H., & Freud, J., *ABN*, 1938, **8**, 178

Gaarenstroom, J.H., & Levie, L.H., *JE*, 1939, **1**, 420

Gallagher, T.F., Kenyon, A.T., Peterson, D.H., Dorfman, R.I., & Koch, F.C., *JAMA*, 1937, **108**, 586

Gallagher, T.F., & Koch, F.C., *JPET*, 1930, **40**, 327

Gallagher, T.F., Peterson, D.H., Dorfman, R.I., Kenyon, A.T., & Koch, F.C., *JCI*, 1937, **16**, 695

Gardner, W.U., (a) *PB*, 1935, **33**, 104

Gardner, W.U., (b) *E*, 1935, **19**, 656

Gardner, W.U., *AJA*, 1936, **59**, 459

Gardner, W.U., *AR*, 1937, **68**, 339

Gardner, W.U., *PB*, 1940, **45**, 230 and 835

Gardner, W.U., *E*, 1941, **28**, 53

Gardner, W.U., & Allen, E., *E*, 1937, **21**, 727

Gardner, W.U., Allen, E., & Smith, G.M., *PB*, 1941, **46**, 511

Gardner, W.U., Allen, E., Smith, G.M., & Strong, L.C., *JAMA*, 1938, **110**, 1182

Gardner, W.U., Allen, E., & Strong, L.C., *AR*, 1936, **64**, Suppl., 17

Gardner, W.U., & De Vita, J., *YJ*, 1940, **13**, 213

Gardner, W.U., Diddle, A.W., Allen, E., & Strong, L.C., *AR*, 1934, **60**, 265 and 457

Gardner, W.U., Gomez, E.T., & Turner, C.W., *AJP*, 1935, **112**, 673

Gardner, W.U., Kirschbaum, A., & Strong, L.C., *ARP*, 1940, **29**, 1

Gardner, W.U., & Pfeiffer, C.A., (a) *PB*, 1938, **37**, 678

Gardner, W.U., & Pfeiffer, C.A., (b) *PB*, 1938, **38**, 599

Gardner, W.U., Smith, G.M., Allen, E., & Strong, L.C., *ARP*, 1936, **21**, 265

Gardner, W.U., Smith, G.M., & Strong, L.C., *PB*, 1935, **33**, 148

Gardner, W.U., Smith, G.M., Strong, L.C., & Allen, E., (a) *ARP*, 1936, **21**, 504

Gardner, W.U., Smith, G.M., Strong, L.C., & Allen, E., (b) *JAMA*, 1936, **107**, 656

Gardner, W.U., & Strong, L.C., *AJC*, 1935, **25**, 282

Gardner, W.U., & Strong, L.C., *YJ*, 1940, **12**, 543

Gardner, W.U., Strong, L.C., & Smith, G.M., *AJC*, 1936, **26**, 541

Gardner, W.U., & Turner, C.W., *Missouri Agric. Exper. Station Res. Bull.*, 1933, No. 196

Gardner, W.U., & Van Wagenen, G., *E*, 1938, **22**, 164

Gaunt, R., *CSH*, 1937, **5**, 396

Gaunt, R., & Hays, H.W., *S*, 1938, **88**, 576

Gaunt, R., Nelson, W.O., & Loomis, E., *PB*, 1938, **39**, 319

Gaunt, R., Remington, J.W., & Edelmann, A., *PB*, 1939, **41**, 429

Geist, S.H., Gaines, J.A., & Salmon, U.J., *PB*, 1940, **44**, 319

Geist, S.H., Mintz, M., & Salmon, U.J., *PB*, 1939, **41**, 556

Geist, S.H., Salmon, U.J., Gaines, J.A., & Walter, R.I., *JAMA*, 1940, **114**, 1539

Geist, S.H., & Spielman, F., *PB*, 1934, **31**, 662

Gemelli, A., *Archiv. sci. med.*, Turin, 1906, **30**, 341

Genther, I.T., *AJA*, 1931, **48**, 99

Geschickter, C.F., *S*, 1939, **89**, 35

Geschickter, C.F., Lewis, D., & Hartman, C.G., *AJC*, 1934, **21**, 828

Gibson, L.M., *JCR*, 1930, **14**, 570

Gilder, H., & Phillips, R.A., *AJP*, 1939, Proc., 86

Girard, A., Sandulesco, G., & Fridenson, A., *CRS*, 1933, **112**, 964

Giroud, A., & Santa, N., *CRS*, 1940, **133**, 420

Glass, S.J., Edmondson, H.A., & Soll, S.N., *E*, 1940, **27**, 749

Gley, P., *CRS*, 1928, **98**, 504 and 656

Gley, P., & Delor, J., *CRS*, 1937, **125**, 52 and 813

Godard, E., *CRS*, 1859, **1**, 313

Goetsch, E., *BJH*, 1916, **27**, 29

Goetsch, E., & Cushing, H., *PB*, 1913, **11**, 26

Goetsch, E., Cushing, H., & Jacobson, C., *BJH*, 1911, **22**, 165

Goldblatt, H., & Benischek, M., *JEM*, 1927, **46**, 699

Golden, J.B., & Severinghaus, E.L., *PB*, 1938, **39**, 361

Golding, G.T., & Ramirez, F.T., *E*, 1928, **12**, 804

Goldmann, E.E., *Beitr. z. klin. Chir.*, 1909, **64**, 192

Gomez, E.T., & Turner, C.W., (a) *PB*, 1936, **34**, 320 and 404

Gomez, E.T., & Turner, C.W., (b) *PB*, 1936, **35**, 365

Gomez, E.T., & Turner, C.W., (a) *PB*, 1937, **36**, 59 and 78

Gomez, E.T., & Turner, C.W., (b) *Rep. Internat. Cancer Foundation*, 1937, p. 121

Gomez, E.T., & Turner, C.W., *PB*, 1938, **37**, 607

Gomez, E.T., Turner, C.W., Gardner, W.U., & Hill, R.T., *PB*, 1937, **36**, 287

Gomez, E.T., Turner, C.W., & Reece, R.P., *PB*, 1937, **36**, 286

Goodale, H.D., *S*, 1914, **40**, 594

Goodall, J.R., *PRSM*, 1920, **13**, 63

Goodall, J.R., *J. Internat. Coll. Surg.*, 1939, **2**, 196

Goodman, Le R., *AR*, 1934, **59**, 223

Goormaghtigh, N., 'Le cortex surrénal humain dans les plaies de l'abdomen et aux périodes intéressants de la vie sexuelle', Liège, 1922

Goormaghtigh, N., *AJC*, 1940, **38**, 32

Goormaghtigh, N., & Amerlinck, A., *BC*, 1930, **19**, 527

Gorer, P.A., *JPB*, 1940, **50**, 17

Grafflin, A.L., *E*, 1942, **30**, 571

Graham, W.R., *BJ*, 1934, **28**, 1368

Granel, F., *CRS*, 1939, **131**, 1255

Gray, L.A., & Lawson, H., *PB*, 1939, **41**, 108

Greene, H.S.N., *JEM*, 1941, **73**, 273

Greene, H.S.N., & Saxton, J.A., *JEM*, 1938, **67**, 691

Greene, R., *L*, 1938 (ii), 79

Greene, R.R., & Burrill, M.W., *PB*, 1939, **42**, 585

Greene, R.R., & Burrill, M.W., *PB*, 1940, **43**, 382

Greene, R.R., Burrill, M.W., & Ivy, A.C., (a) *S*, 1938, **87**, 396

Greene, R.R., Burrill, M.W., & Ivy, A.C., (b) *PB*, 1938, **38**, 1

Greene, R.R., Burrill, M.W., & Ivy, A.C., (c) *S*, 1938, **88**, 130

Greene, R.R., Burrill, M.W., & Ivy, A.C., (a) *PB*, 1939, **41**, 169

Greene, R.R., Burrill, M.W., & Ivy, A.C., (b) *E*, 1939, **24**, 351

Greene, R.R., Burrill, M.W., & Ivy, A.C., *PB*, 1940, **43**, 32

Greene, R.R., & Harris, S.C., *PB*, 1940, **45**, 34

Greene, R.R., & Ivy, A.C., *S*, 1937, **86**, 200

Greene, R.R., Wells, J.A., & Ivy, A.C., *PB*, 1939, **40**, 83

Greenwood, A.W., *BJEB*, 1925, **2**, 165

Greenwood, A.W., & Blyth, J.S.S., (a) *QJ*, 1935, **25**, 267

Greenwood, A.W., & Blyth, J.S.S., (b) *PR*, 1935, **118**, 97

Greep, R.O., *CSH*, 1937, **5**, 136
Greep, R.O., *AR*, 1939, **73**, Suppl., 23
Greep, R.O., *AR*, 1940, **76**, Suppl., 25
Greep, R.O., & Fevold, H.L., *E*, 1937, **21**, 611
Greep, R.O., Fevold, H.L., & Hisaw, F.L., *AR*, 1936, **65**, 261
Greep, R.O., & Hisaw, F.L., *PB*, 1938, **39**, 359
Greep, R.O., Van Dyke, H.B., & Chow, B.F., *PB*, 1941, **46**, 644
Greulich, W.W., & Burford, T.H., *AJC*, 1936, **28**, 496
Griffiths, M., Marks, H.P., & Young, F.G., *N*, 1941, **147**, 359
Griffiths, M., & Young, F.G., *N*, 1940, **146**, 266
Griffiths, M., & Young, F.G., *JE*, 1942, **3**, 96
Groher, H., *AG*, 1938, **165**, 347
Grollman, A., 'The Adrenals', London, 1936
Grüter, F., *Proc. 2nd Internat. Cong. Sex Res.*, 1930, p. 443
Guieysse, M.A., *CRS*, 1899, **51**, 898
Guirdham, A., *Bristol Med. & Chir. J.*, 1941, **58**, 19
Gurin, S., *PB*, 1942, **49**, 48
Gustavson, R.G., & Green, D.F., *JBC*, 1934, **105**, Proc., 34
Gustavson, R.G., Wood, T., & Hays, E., *JBC*, 1936, **114**, Proc., 46
Gustus, E.L., Meyer, R.K., & Dingle, J.H., *PB*, 1935, **33**, 257
Gutman, A.B., *JAMA*, 1942, **120**, 1112
Gutman, A.B., & Gutman, E.B., (a) *PB*, 1938, **39**, 529
Gutman, A.B., & Gutman, E.B., (b) *JCI*, 1938, **17**, 473
Gutman, E.B., Sproul, E.E., & Gutman, A.B., *AJC*, 1936, **28**, 485
Guyénot, É., Ponse, K., & Wietrzykowska, J., *CRA*, 1932, **194**, 1051
Guyon, L., *CRS*, 1939, **131**, 1167

Haddow, A., *L*, 1942, **2**, 230
Hain, A.M., *QJ*, 1932, **22**, 249
Hain, A.M., (a) *QJ*, 1935, **25**, 131 and 303
Hain, A.M., (b) *Edin. Med. J.*, 1935, **42**, 101
Hain, A.M., *QJ*, 1936, **26**, 29 and 37
Hain, A.M., *QJ*, 1937, **27**, 293
Hain, A.M., *QJ*, 1938, **28**, 353
Hain, A.M., *JE*, 1940, **2**, 104
Hain, A.M., *JE*, 1942, **3**, 10
Hain, A.M., & Robertson, E.M., (a) *L*, 1939 (i), 1324
Hain, A.M., & Robertson, E.M., (b) *BMJ* 1939 (i), 1226
Halban, J., & Köhler, R., *AG*, 1914, **103**, 575
Hall, B.V., & Lewis, R.M., *E*, 1936, **20**, 210
Hall, K., *JPB*, 1938, **47**, 19
Hall, K., & Korenchevsky, V., *N*, 1937, **140**, 318
Hall, K., & Korenchevsky, V., *JP*, 1938, **91**, 365
Halpern, S.R., & D'Amour, F.E., *PB*, 1934, **32**, 108

Hamblen, E.C., Ashley, C., & Baptist, M., *E*, 1939, **24**, 1
Hamblen, E.C., Cuyler, W.K., & Baptist, M., *E*, 1940, **27**, 16
Hamblen, E.C., Cuyler, W.K., & Hirst, D.V., *E*, 1940, **27**, 33 and 35
Hamblen, E.C., Pattee, C.J., & Cuyler, W.K., *E*, 1940, **27**, 734
Hamblen, E.C., Ross, R.A., Cuyler, W.K., Baptist, M., & Ashley, C., *E*, 1939, **25**, 491
Hamburger, C., *Acta path. et microbiol. Scand.*, 1933, Suppl. xvii
Hamburger, C., *CRS*, 1933, **112**, 99
Hamburger, C., *PB*, 1936, **34**, 519
Hamilton, H.L., *PB*, 1940, **45**, 571
Hamilton, J.B., *PB*, 1936, **35**, 386
Hamilton, J.B., *E*, 1937, **21**, 649 and 744
Hamilton, J.B., (a) *E*, 1938, **23**, 53
Hamilton, J.B., (b) *AR*, 1938, **70**, 533
Hamilton, J.B., & Gardner, W.U., *PB*, 1937, **37**, 570
Hamilton, J.B., & Gilbert, J.B., *JCR*, 1942, **2**, 125
Hamilton, J.B., & Hubert, G., *S*, 1938, **88**, 481
Hamilton, J.B., & Leonard, S.L., *AR*, 1938, **71**, 105
Hamilton, J.B., & Wolfe, J.M., *E*, 1938, **22**, 360
Hamlett, G.W.D., *Quart. Rev. Biol.*, 1935, **10**, 432
Hammond, J., *PR*, 1917, **89**, 534
Hammond, J., & Marshall, F.H.A., 'Reproduction in the Rabbit', London, 1925, pp. 123 and 128
Handovsky, H., & Daels, J., *CRS*, 1937, **126**, 428
Handovsky, H., & Daels, J., *CRS*, 1938, **127**, 1112
Hanes, F.M., & Hooker, C.W., *PB*, 1937, **35**, 549
Hansen, L.P., McCahey, J.F., & Soloway, D., *Amer. J. Med. Sci.*, 1936, **192**, 149
Harde, E., *CRS*, 1934, **116**, 999
Harper, E.H., *AJA*, 1904, **3**, 349
Harrenstein, R.J., *Centralbl. f. Chir.*, 1928, **55**, 1734
Harris, R.G., & Pfiffner, J.J., *AR*, 1929, **44**, 205
Harsh, R., Overholser, M.D., & Wells, L.J., *JE*, 1939, **1**, 261
Hart, G.H., & Cole, H.H., *AJP*, 1934, **109**, 320
Hartman, C., Dupre, C., & Allen, E., *E*, 1926, **10**, 291
Hartman, C.G., *AJA*, 1925, **35**, 1
Hartman, C.G., *PB*, 1937, **37**, 87
Hartman, C.G., *E*, 1940, **26**, 449
Hartmann, M., & Wettstein, A., *Helv. Chim. Acta*, 1934, **17**, 878, 1365
Harvey, William, 'The Works of William Harvey', Sydenham Soc., 1847
Haskins, A.L., *PB*, 1939, **42**, 624
Hatai, S., (a) *JEZ*, 1913, **15**, 297
Hatai, S., (b) *AJA*, 1913, **15**, 87
Hatai, S., *AJA*, 1914, **16**, 251
Hatai, S., *JEZ*, 1915, **18**, 1
Haterius, H.O., (a) *PB*, 1932, **29**, 962
Haterius, H.O., (b) *AR*, 1932, **54**, 343

Haterius, H.O., *AJP*, 1933, **103**, 97
Haterius, H.O., *PB*, 1935, **33**, 101
Haterius, H.O., *AJP*, 1936, **114**, 399
Haterius, H.O., & Charipper, H.A., *AR*, 1931, **51**, 85
Haterius, H.O., & Nelson, W.O., *JEZ*, 1932, **61**, 175
Heape, W., *QJMS*, 1901, **44**, 1
Heape, W., *PR*, 1905, **76**, 260
Heape, W., *JP*, 1906, **34**, Proc. i
Heape, W., 'Emigration, Migration and Nomadism', Cambridge, 1931
Hechter, O., Fraenkel, M., Lev, M., & Soskin, S., *E*, 1940, **26**, 680
Hechter, O., Krohn, L., & Harris, J., *E*, 1942, **30**, 598
Hechter, O., Lev, M., & Soskin, S., *E*, 1940, **26**, 73
Heckel, G.P., & Allen, W.M., *AJP*, 1937, **119**, 330
Heckel, G.P., & Allen, W.M., *E*, 1939, **24**, 137
Heidrich, L., Fels, E., & Mathias, E., *Bruns Beitr. klin. Chir.*, 1930, **150**, 349
Heim, C., *KW*, 1931, **10**, 1598
Hellbaum, A.A., *PB*, 1933, **30**, 641
Hellbaum, A.A., *AR*, 1935, **63**, 147
Hellbaum, A.A., *PB*, 1936, **33**, 568
Hellbaum, A.A., & Greep, R.O., *E*, 1943, **32**, 33
Heller, C.G., *E*, 1940, **26**, 619
Heller, R.E., *PB*, 1930, **27**, 751
Heller, R.E., *AJA*, 1932, **50**, 73
Henderson, W.R., & Rowlands, I.W., *BMJ*, 1938 (i), 1094
Heringa, G.C., & De Jongh, S.E., *Ztschr. Zellforsch. u. micros. Anat.*, 1934, **21**, 629
Herlant, M., *CRS*, 1939, **131**, 1315 and 1318
Herrmann, E., *Monat. f. Geburts. u. Gynäk.*, 1915, **41**, 1
Hertz, R., & Hisaw, F.L., *AJP*, 1934, **108**, 1
Hertz, R., & Meyer, R.K., *E*, 1937, **21**, 756
Hertz, R., Meyer, R.K., & Spielman, M.A., *E*, 1937, **21**, 533
Hesser, F.H., Langworthy, O.R., & Vest, S.A., *E*, 1940, **26**, 241
Heston, W.E., & Andervont, H.B., *JNCI*, 1944, **4**, 403
Heyl, H.L., *S*, 1939, **89**, 540
Hill, M., & Parkes, A.S., *PR*, 1930, **107**, 30 and 39
Hill, M., & Parkes, A.S., *PR*, 1932, **110**, 180
Hill, M., & Parkes, A.S., *PR*, 1933, **113**, 530, 537 and 541
Hill, R.T., *JEZ*, 1932, **63**, 203
Hill, R.T., *E*, 1933, **17**, 414
Hill, R.T., *E*, 1937, **21**, 495 and 633
Hill, R.T., *E*, 1941, **28**, 426
Hill, R.T., & Gardner, W.U., *AR*, 1936, **64**, Suppl., 21
Hill, R.T., & Strong, M.T., *E*, 1940, **27**, 79
Hill, W.C.O., *JA*, 1933, **68**, 19
Hirschmann, H., *JBC*, 1939, **130**, 421
Hisaw, F.L., *JEZ*, 1925, **42**, 411
Hisaw, F.L., *Physiol. Zool.*, 1929, **2**, 59
Hisaw, F.L., *AJOG*, 1935, **29**, 638

Hisaw, F.L., Fevold, H.L., Foster, M.A., & Hellbaum, A.A., *AR*, 1934, **60**, Suppl., 52
Hisaw, F.L., & Greep, R.O., *E*, 1938, **23**, 1
Hisaw, F.L., Greep, R.O., & Fevold, H.L., *AR*, 1936, **67**, Suppl., 50
Hisaw, F.L., Hertz, R., & Fevold, H.L., *E*, 1936, **20**, 40
Hisaw, F.L., & Lendrum, F.C., *E*, 1936, **20**, 228
Hisaw, F.L., & Leonard, S.L., *AJP*, 1930, **92**, 574
Hisaw, F.L., Meyer, R.K., & Fevold, H.L., *PB*, 1930, **27**, 400
Hisaw, F.L., Meyer, R.K., & Weichert, C.K., *PB*, 1928, **25**, 754
Hogben, L., *Trans. Roy. Soc. of S. Africa*, 1934, **22**, Proc., 17
Hohlweg, W., *KW*, 1934, **13**, 92
Hohlweg, W., & Chamorro, A., *KW*, 1937, **16**, 196
Holl, G., *Deutsch. Ztschr. f. Chir.*, 1930, **226**, 276
Holst, G., & Turner, C.W., *PB*, 1939, **41**, 198
Hooker, C.W., *PB*, 1940, **45**, 270
Hooker, C.W., *E*, 1942, **30**, 77
Hooker, C.W., & Collins, V.J., *E*, 1940, **26**, 269
Hooker, C.W., Gardner, W.U., & Pfeiffer, C.A., *JAMA*, 1940, **115**, 443
Hooker, C.W., & Pfeiffer, C.A., *CR*, 1942, **2**, 759
Hooker, C.W., & Pfeiffer, C.A., *E*, 1943, **32**, 69
Hoskins, W., & Webster, B., *PB*, 1940, **43**, 604
Houssay, B.A., *CRS*, 1935, **120**, 496 and 502
Houssay, B.A., & Biasotti, A., *E*, 1931, **15**, 511
Howard, E., *AR*, 1940, **77**, 181
Howard, E., & Gengradom, S., *E*, 1940, **26**, 1048
Howard-Miller, E., *AJA*, 1927, **40**, 251
Hu, C.K., & Frazier, C.N., *PB*, 1935, **33**, 326
Huffman, M.N., MacCorquodale, D.W., Thayer, S.A., Doisy, E.A., Smith, G.v.S., & Smith, O.W., *JBC*, 1940, **134**, 591
Huggins, C., & Clark, P.J., *JEM*, 1940, **72**, 747
Huggins, C., & Hodges, C.V., *CR*, 1941, **1**, 293
Huggins, C., Masina, M.H., Eichelberger, L., & Wharton, J.D., *JEM*, 1939, **70**, 543
Huggins, C., Scott, W.W., & Hodges, C.V., *JU*, 1941, **46**, 997
Huggins, C., Stevens, R.E., & Hodges, C.V., *AS*, 1941, **43**, 209
Humphrey, R.R., *AR*, 1928, **40**, 67
Hundley, J.M., Diehl, W.K., & Diggs, E.S., *AJOG*, 1942, **44**, 858
Hunter, J., (*a*) 'Treatise on the Blood, Inflammation, and Gun-shot Wounds', 1794, pp. 208, 224
Hunter, J., (*b*) 'The Works of John Hunter', ed. by J. F. Palmer, 1837, **4**, 49
Hunter, R.H., *JA*, 1935, **70**, 68
Hurst, V., Meites, J., & Turner, C.W., *PB*, 1942, **49**, 592

Ihrke, I.A., & D'Amour, F.E., *AJP*, 1931, **96**, 289

Ingle, D.J., Higgins, G.M., & Kendall, E.C., *AR*, 1938, **71**, 363

Ingleby, H., *ARP*, 1942, **33**, 573

Inhoffen, H.H., & Hohlweg, W., *NW*, 1938, **26**, 96

Israel, S.L., Meranze, D.R., & Johnston, C.G., *Amer. J. Med. Sci.*, 1937, **194**, 835

Itho, M., & Kon, T., *CRS*, 1935, **120**, 678

Ivanova, S., *AEPP*, 1935, **179**, 349

Jadassohn, W., Uehlinger, E., & Margot, A., *J. Investig. Dermatol.*, 1938, **1**, 31

Jaffe, H.L., & Marine, D., *JEM*, 1923, **38**, 93 and 107

Janes, R.G., & Nelson, W.O., *PB*, 1940, **43**, 340

Jedlicky, A., Lipschütz, A., & Vargas, L., *CRS*, 1939, **130**, 1466

Jeffcoate, T.N.A., *JOG*, 1932, **39**, 67

Jensen, H., Hauschildt, J.D., & Evans, J.S., *PB*, 1942, **50**, 356

Jesse, E., 'Gleanings in Natural History', London, 1835

Johnson, E., & Challans, J.S., *AR*, 1930, **47**, 300

Johnston, R.L., *E*, 1934, **18**, 123

Jones, G.E.S., Gey, G.O., & Gey, M.K., *BJH*, 1943, **72**, 26

Jones, M.S., & MacGregor, T.N., *L*, 1936 (ii), 974

Kallas, H., (a) *CRS*, 1929, **100**, 979

Kallas, H., (b) *CRS*, 1929, **102**, 280

Karp, L., *CRS*, 1933, **114**, 357

Katzman, P.A., *PB*, 1932, **29**, 700

Katzman, P.A., & Doisy, E.A., *JBC*, 1934, **106**, 125

Katzman, P.A., Levin, L., & Doisy, E.A., *PB*, 1931, **28**, 873

Katzman, P.A., Wade, N.J., & Doisy, E.A., *E*, 1937, **21**, 1

Kayser, C., & Aron, M., *CRS*, 1938, **129**, 225

Kellar, R.J., & Sutherland, J.K., *JOG*, 1939, **46**, 1

Kelly, G.L., *SGO*, 1931, **52**, 713

Kemp, T., *CRS*, 1925, **92**, 1318

Kemp, T., & Pedersen-Bjergaard, K., *L*, 1937 (ii), 842

Kendall, E.C., *JAMA*, 1941, **116**, 2394

Kennard, J.H., *AJP*, 1937, **118**, 190

Kennedy, W.P., *QJ*, 1925, **15**, 103

Kenyon, A.T., *E*, 1938, **23**, 121

Kenyon, A.T., Gallagher, T.F., Peterson, D.H., Dorfman, R.I., & Koch, F.C., *JCI*, 1937, **16**, 705

Kenyon, A.T., Knowlton, K., Sandiford, I., Koch, F.C., & Lotwin, G., *E*, 1940, **26**, 26

Kenyon, A.T., Sandiford, I., Bryan, A.H., Knowlton, K., & Koch, F.C., *E*, 1938, **23**, 135

Kerly, M., *BJ*, 1940, **34**, 814

Khayyal, M.A., & Scott, C.M., *JP*, 1931, **72**, 13 P.

Khayyal, M.A., & Scott, C.M., *QJ*, 1934, **24**, 249

Khayyal, M.A., & Scott, C.M., *QJ*, 1935, **25**, 77

Kidd, F., 'Common Infections of the Kidneys', London, 1920

King, H.D., *Biol. Bull.*, 1913, **24**, 377

Kinosita, R., *Trans. Soc. Pathol. Japon*, 1937, **27**, 665

Kippen, A.A., & Loeb, L., *E*, 1936, **20**, 201

Kirkham, W.B., *AR*, 1916, **11**, 31

Kirsch, R.E., *AJP*, 1938, **122**, 86

Kirschbaum, A., *AR*, 1933, **57**, Suppl., 62

Kirschbaum, A., & Pfeiffer, C.A., *PB*, 1941, **46**, 649

Kirschbaum, A., & Ringoen, A.R., *AR*, 1936, **64**, 453

Kitagawa, W., *Biol. Abstr.*, 1930, **4**, 4253

Klein, M., (a) *CRS*, 1933, **112**, 819

Klein, M., (b) *CRS*, 1933, **113**, 441

Klein, M., *CRS*, 1935, **119**, 577, 579

Klein, M., *PR*, 1937, **124**, 23

Klein, M., (a) *PR*, 1938, **125**, 348

Klein, M., (b) *CRS*, 1938, **127**, 915, 918

Klein, M., & Klein, L., *CRS*, 1933, **112**, 821

Klein, M., & Parkes, A.S., (a) *CI*, 1936, **55**, 236

Klein, M., & Parkes, A.S., (b) *PR*, 1936, **121**, 574

Klein, M., & Parkes, A.S., *PR*, 1937, **121**, 574

Kleiner, I.S., Weisman, A.I., & Mishkind, D.I., *JAMA*, 1936, **106**, 1643

Klempner, E., Frank, R.T., & Hollander, F., *PB*, 1940, **44**, 633

Knauer, E., *AG*, 1900, **60**, 322

Knaus, H., *JP*, 1926, **61**, 383

Knaus, H., *AG*, 1929, **138**, 201

Knaus, H., (a) *AEPP*, 1930, **151**, 371

Knaus, H., (b) *KW*, 1930, **9**, 961

Koch, F.C., *JAMA*, 1931, **96**, 937

Koch, F.C., *JAMA*, 1937, **109**, 1312

Koch, F.C., *Bull. New York Acad. Med.*, 1938, **14**, 655

Kochakian, C.D., *E*, 1936, **21**, 60

Kochakian, C.D., *E*, 1939, **24**, 331

Kochakian, C.D., *E*, 1940, **26**, 54

Kok, D.J., *Vlaamsch Diergeneesk. Tijds.*, 1936, 5

Koref, O., Lipschütz, A., & Vargas, L., *CRS*, 1939, **130**, 303

Korenchevsky, V., *JPB*, 1930, **33**, 607

Korenchevsky, V., *N*, 1935, **136**, 185

Korenchevsky, V., (a) *JP*, 1937, **90**, 371

Korenchevsky, V., (b) *BMJ*, 1937 (ii), 896

Korenchevsky, V., *JPB*, 1941, **52**, 341

Korenchevsky, V., & Dennison, M., *BJ*, 1934, **28**, 1474 and 1486

Korenchevsky, V., & Dennison, M., *JPB*, 1935, **41**, 323

Korenchevsky, V., & Dennison, M., (a) *JPB*, 1936, **42**, 91

Korenchevsky, V., & Dennison, M., (b) *JPB*, 1936, **43**, 345

Korenchevsky, V., Dennison, M., & Brovsin, I., *BJ*, 1936, **30**, 558

Korenchevsky, V., Dennison, M., & Eldridge, M., *BJ*, 1937, **31**, 467 and 475

Korenchevsky, V., Dennison, M., & Hall, K., *BJ*, 1937, **31**, 780

Korenchevsky, V., Dennison, M., & Kohn-Speyer, A., *BJ*, 1932, **26**, 2097

Korenchevsky, V., Dennison, M., & Simpson, S.L., *BJ*, 1935, **29**, 2131, 2522 and 2534

Korenchevsky, V., & Hall, K., *JPB*, 1937, **45**, 681

Korenchevsky, V., & Hall, K., (a) *JOG*, 1938, **45**, 22

Korenchevsky, V., & Hall, K., (b) *N*, 1938, **142**, 998

Korenchevsky, V., & Hall, K., *BMJ*, 1939 (i), 4

Korenchevsky, V., & Hall, K., *JPB*, 1940, **50**, 295

Korenchevsky, V., Hall, K., Burbank, R.C., & Cohen, J., *BMJ*, 1941 (i), 396

Korenchevsky, V., Hall, K., Burbank, R.C., & Ross, M.A., *BJ*, 1939, **33**, 36

Korenchevsky, V., Hall, K., & Ross, M.A., *BJ*, 1939, **33**, 213

Korenchevsky, V., & Ross, M.A., *BMJ*, 1940 (i), 645

Korteweg, R., *Genetica*, 1936, **18**, 350

Kozelka, A.W., & Gallagher, T.F., *PB*, 1934, **31**, 1143

Kraus, E.J., *KW*, 1930, **9**, 1943

Kreyberg, L., *AJC*, 1935, **24**, 554

Kreyberg, L., *Proc. 7th Internat. Genetical Cong., Edin.*, 1939, p. 178

Kreyberg, L., & Eker, R., *Avhandl. utgitt av det Norske Videnskaps-Akademi i Oslo*, I. Mat.-Naturv. Klasse, 1939, No. 3

Kroszczynski, S., & Bychowska, M., *CRS*, 1939, **130**, 570

Kun, H., & Peczenik, O., *Wien. klin. Wschr.*, 1937, **50**, 439

Kunde, M.M., D'Amour, F.E., Gustavson, R.G., & Carlson, A.J., *PB*, 1930, **28**, 122

Kuramitsu, C., & Loeb, L., *AJP*, 1921, **56**, 40

Kurzrok, R., Kirkman, I.J., & Creelman, M., *AJOG*, 1934, **28**, 319

Kurzrok, R., & O'Connell, C.P., *E*, 1938, **23**, 476

Kurzrok, R., Wiesbader, H., Mulinos, M.G., & Watson, B.P., *E*, 1937, **21**, 335

Kuschinsky, G., *AEPP*, 1931, **162**, 183

Kuschinsky, G., & Tang-sü, *AEPP*, 1935, **179**, 717, 722

Kutscher, W., & Wolbergs, H., *Ztschr. f. physiol. Chem.*, 1935, **236**, 237

Kuzell, W.C., & Cutting, W.C., *E*, 1940, **26**, 537

Lacassagne, A., *CRA*, 1932, **195**, 630

Lacassagne, A., (a) *CRS*, 1933, **113**, 590

Lacassagne, A., (b) *CRS*, 1933, **114**, 427, 870

Lacassagne, A., (a) *CRS*, 1934, **115**, 579, 937

Lacassagne, A., (b) *CRS*, 1934, **116**, 95

Lacassagne, A., (a) *CRS*, 1935, **120**, 585, 1156

Lacassagne, A., (b) *Paris Méd.*, 1935, **1**, 233

Lacassagne, A., (a) *CRS*, 1936, **121**, 607

Lacassagne, A., (b) *CRS*, 1936, **122**, 183, 1060

Lacassagne, A., (c) *AJC*, 1936, **27**, 217

Lacassagne, A., (d) *AJC*, 1936, **28**, 735

Lacassagne, A., (e) 'Certain Biological Problems Relating to Cancer, Hormones and Radiation', 1936

Lacassagne, A., *CRS*, 1937, **126**, 190 and 385

Lacassagne, A., *CRS*, 1938, **129**, 641

Lacassagne, A., (a) *CRS*, 1939, **132**, 222, 365, 395 and 431

Lacassagne, A., (b) *BC*, 1939, **28**, 951

Lacassagne, A., *CRS*, 1940, **133**, 180, 227 and 539

Lacassagne, A., & Chamorro, A., *CRS*, 1939, **131**, 1077

Lacassagne, A., & Danysz, S., *CRS*, 1939, **132**, 395

Lacassagne, A., Fehr, A., & Nyka, W., *CRS*, 1936, **121**, 385

Lacassagne, A., & Nyka, W., *CRS*, 1934, **116**, 844

Lacassagne, A., & Nyka, W., *CRS*, 1937, **126**, 1112

Lacassagne, A., & Raynaud, A., *CRS*, 1937, **126**, 579

Lamar, J.K., *AR*, 1937, **70**, Suppl., 45

Lane, C.E., & Hisaw, F.L., *AR*, 1934, **60**, Suppl., 52

Lane-Claypon, J.E., 'Cancer of the Breast', *Ministry of Health Reps.*, 1926, No. 32

Laqueur, E., *KW*, 1927, **6**, 390

Laqueur, E., *Verh. Intern. Kongr. Physiol.*, 1932

Laqueur, E., *ABN*, 1936, **6**, 147

Laqueur, E., Borchardt, E., Dingemanse, E., & De Jongh, S.E., *DMW*, 1928, **54**, 465

Laqueur, E., David, K., Dingemanse, E., & Freud, J., *ABN*, 1935, **5**, 84

Laqueur, E., & De Jongh, S.E., *JAMA*, 1928, **91**, 1169

Laqueur, E., & De Jongh, S.E., *JPET*, 1929, **36**, 1

Laqueur, E., Dingemanse, E., Hart, P.C., & De Jongh, S.W., *KW*, 1927, **6**, 1859

Laqueur, E., Hart, P.C., & De Jongh, S.E., *DMW*, 1926, **53**, 1247

Laqueur, G.L., *E*, 1943, **32**, 81

Lataste, F., *CRS*, 1891, **43**, 879 (Mém., p. 21)

Lathrop, A.E.C., & Loeb, L., *PB*, 1913, **11**, 38

Lathrop, A.E.C., & Loeb, L., *JCR*, 1916, **1**, 1

Lathrop, A.E.C., & Loeb, L., *JEM*, 1918, **28**, 475

Launois, P.E., & Roy, P., *CRS*, 1903, **55**, 22

Lauson, H.D., Golden, J.B., & Severinghaus, E.L., *E*, 1939, **25**, 47

Lawless, J.J., *AR*, 1936, **66**, 455

Lawrence, C.H., & Werthessen, N.T., *E*, 1940, **27**, 755

Lawrence, R.D., *BMJ*, 1943 (i), 12

Leathem, J.H., & Crafts, R.C., *AR*, 1940, **76**, Suppl. 2, 90

Leblond, C.P., (a) *PB*, 1938, **38**, 66

Leblond, C.P., (b) *CRS*, 1938, **127**, 1248

Leblond, C.P., & Nelson, W.O., (a) *CRS*, 1937, **124**, 9 and 1064

Leblond, C.P., & Nelson, W.O., (b) *AJP*, 1937, **120**, 167

Ledingham, J.C.G., *JPB*, 1940, **50**, 201

Le Gros Clark, W.E., McKeown, T., & Zuckerman, S., *PR*, 1939, **126**, 449

Lehmann, H., *Arch. f. Physiol.*, 1927, **216**, 729

Lehmann, J., *VA*, 1928, **268**, 346
Leiby, G.M., *PB*, 1933, **31**, 15
Lendrum, F.C., & Hisaw, F.L., *PB*, 1936, **34**, 364
Leonard, S.L., (*a*) *PB*, 1933, **30**, 403 and 1251
Leonard, S.L., (*b*) *AR*, 1933, **55**, 26
Leonard, S.L., (*c*) *AR*, 1933, **57**, 45
Leonard, S.L., *PB*, 1937, **37**, 566
Leonard, S.L., *PB*, 1939, **41**, 229
Leonard, S.L., Hisaw, F.L., & Fevold, H.L., *AJP*, 1932, **100**, 111
Leonard, S.L., Hisaw, F.L., & Fevold, H.L., *PB*, 1935, **33**, 319
Leonard, S.L., Meyer, R.K., & Hisaw, F.L., *E*, 1931, **15**, 17
Leonard, S.L., & Reece, R.P., *E*, 1942, **30**, 32
Leonard, S.L., & Smith, P.E., *PB*, 1933, **30**, 1246, 1248
Lespinasse, V.D., *Endocrinology and Metabolism*, 1924, **2**, 491
Levie, L.H., *ABN*, 1938, **8**, 53
Levin, L., Katzman, P.A., & Doisy, E.A., *E*, 1931, **15**, 207
Levine, W.T., & Witschi, E., *PB*, 1933, **30**, 1152
Lewis, A.A., & Turner, C.W., *PB*, 1941, **48**, 439
Lewis, A.A., & Turner, C.W., *E*, 1942, **30**, 585
Lewis, A.A., Turner, C.W., & Gomez, E.T., *E*, 1939, **24**, 157
Lewis, D., & Geschickter, C.F., *JAMA*, 1934, **103**, 1212
Lewis, R.M., *AJOG*, 1933, **26**, 593
Lewis, R.M., & Weinstein, L., *SGO*, 1936, **63**, 640
Lewis, R.W., & Pappenheimer, A.M., *J. Med. Res.*, 1916, **34**, 81
Liebhart, S., *ZG*, 1934, **58**, 1896
Li, R.C., *PB*, 1940, **43**, 598
Lillie, F.R., *JEZ*, 1917, **23**, 371
Lillie, F.R., 'Sex & Internal Secretions', ed. by E. Allen, 1932, pp. 5 and 133
Lillie, F.R., & Bascom, K.F., *S*, 1922, **55**, 624
Lipschütz, A., *JP*, 1917, **51**, 283
Lipschütz, A., *PR*, 1922, **94**, 83
Lipschütz, A., (*a*) *BJEB*, 1925, **2**, 331
Lipschütz, A., (*b*) *CRS*, 1925, **93**, 1066
Lipschütz, A., (*a*) *Biol. Rev.*, 1927, **2**, 263
Lipschütz, A., (*b*) *BJEB*, 1927, **4**, 227
Lipschütz, A., *VA*, 1932, **285**, 35
Lipschütz, A., *CRS*, 1935, **118**, 331
Lipschütz, A., *CRA*, 1936, **203**, 1025
Lipschütz, A., (*a*) *N*, 1937, **140**, 892
Lipschütz, A., (*b*) *GO*, 1937, **36**, 407
Lipschütz, A., (*c*) *Bull. et Mém. de la Section d'Endocrinologie*, 1937, **3**, 1
Lipschütz, A., *GO*, 1938, **37**, 17
Lipschütz, A., & Del-Pino, M., *CRS*, 1936, **121**, 208
Lipschütz, A., & Iglesias, R., *CRS*, 1938, **129**, 519
Lipschütz, A., Iglesias, R., & Vargas, L., *CRS*, 1939, **130**, 1536
Lipschütz, A., Iglesias, R., & Vargas, L., *PB*, 1940, **45**, 788

Lipschütz, A., Murillo, R., & Vargas, L., *L*, 1940 (ii), 420
Lipschütz, A., Palacios, L., & Akel, J., *CRS*, 1936, **121**, 205
Lipschütz, A., Rodriguez, F., & Vargas, L., *CRS*, 1939, **130**, 939
Lipschütz, A., Rodriguez, F., & Vargas, L., *E*, 1941, **28**, 664
Lipschütz, A., & Tütso, M., *CRS*, 1925, **92**, 143
Lipschütz, A., & Vargas, L., *CRS*, 1939, **130**, 596
Lipschütz, A., & Vargas, L., *L*, 1940 (i), 541
Lipschütz, A., Vargas, L., & Palma, J., *CR*, 1941, **1**, 575
Lipschütz, A., Vargas, L., & Ruz, O., *L*, 1939 (ii), 867
Lipschütz, A., & Vivaldi, T., *CRS*, 1934, **116**, 87
Lipschütz, A., Wagner, C., & Tamm, R., *CRS*, 1922, **86**, 240
Lisser, H., *E*, 1936, **20**, 567
Little, C.C., *S*, 1933, **78**, 465
Little, C.C., *JAMA*, 1936, **106**, 2234
Little, C.C., & Pearsons, J., *AJC*, 1940, **38**, 224
Liu, S.H., & Noble, R.L., *JE*, 1939, **1**, 7 and 15
Loeb, H.G., *PB*, 1942, **49**, 340
Loeb, L., *PB*, 1907, **4**, 93
Loeb, L., *Biol. Bull.*, 1914, **27**, 1
Loeb, L., *J. Med. Res.*, 1919, **40**, 477
Loeb, L., *JAMA*, 1921, **77**, 1646
Loeb, L., *AJP*, 1927, **83**, 202
Loeb, L., *E*, 1928, **12**, 161
Loeb, L., *Australian J. exp. Biol. and Med. Sci.*, 1932, **9**, 141
Loeb, L., Burns, E.L., Suntzeff, V., & Moskop, M., *AJC*, 1937, **30**, 47
Loeb, L., & Genther, I.T., *PB*, 1928, **25**, 809
Loeb, L., & Haven, F.L., *AR*, 1929, **43**, 1
Loeb, L., & Kirtz, M.M., *AJC*, 1939, **36**, 56
Loeb, L., & Kountz, W.B., *AJP*, 1928, **84**, 283
Loeb, L., Saxton, J., & Hayward, S.J., *E*, 1936, **20**, 511
Loeb, L., Suntzeff, V., & Burns, E.L., *AJC*, 1939, **35**, 159
Loeb, R.F., *JAMA*, 1941, **116**, 2495
Loeser, A.A., *L*, 1938 (i), 373
Loeser, A.A., *L*, 1943 (i), 518
Loesser, A., *AEPP*, 1934, **176**, 697
Loewe, S., *KW*, 1925, **4**, 1407
Loewe, S., & Lange, F., *KW*, 1926, **5**, 1038
Loewe, S., & Voss, H.E., *KW*, 1926, **5**, 1083
Loewe, S., & Voss, H.E., *KW*, 1928, **7**, 1376
Loewy, A., *Ergebn. der Physiol.*, 1903, **2**, 130
Logie, J.W., *CR*, 1942, **2**, 394
Long, J.A., & Evans, H.M., *Mem. Univ. California*, 1922, **6**, 1
Lorenz, F.W., Chaikoff, I.L., & Entenman, C., *JBC*, 1938, **126**, 763
Lynch, C.J., *JEM*, 1925, **42**, 829
Lyon, R., *PB*, 1942, **51**, 156
Lyons, W.R., (*a*) *CSH*, 1937, **5**, 198 and 208
Lyons, W.R., (*b*) *PB*, 1937, **37**, 207
Lyons, W.R., & McGinty, D.A., *PB*, 1941, **48**, 83

Lyons, W.R., & Page, E., *PB*, 1935, **32**, 1049
Lyons, W.R., & Pencharz, R.I., *PB*, 1936, **33**, 589
Lyons, W.R., & Sako, Y., *PB*, 1940, **44**, 398
Lyons, W.R., Simpson, M.E., & Evans, H.M., *PB*, 1943, **52**, 134
Lyons, W.R., & Templeton, H.J., *PB*, 1936, **33**, 587

MacBryde, C.M., *JAMA*, 1939, **112**, 1045
MacBryde, C.M., Castrodale, D., Helwig, E.B., & Bierbaum, O., *JAMA*, 1942, **118**, 1003 and 1278
McCahey, J.F., Soloway, D., & Hansen, L.P., *Pennsylvania Med. J.*, 1936, **39**, 228
McCullagh, D.R., & Walsh, E.L., *PB*, 1934, **31**, 678
McCullagh, D.R., & Walsh, E.L., *E*, 1935, **19**, 466
McCullagh, E.P., *JAMA*, 1939, **112**, 1037
McCullagh, E.P., & Lilga, H.V., *E*, 1940, **26**, 753
MacDonald, A.M., & Robson, J.M., *JPB*, 1939, **48**, 95
McEuen, C.S., Selye, H., & Collip, J.B., (a) *L*, 1936 (i), 775
McEuen, C.S., Selye, H., & Collip, J.B., (b) *PB*, 1936, **35**, 56
McEuen, C.S., Selye, H., & Collip, J.B., *PB*, 1937, **36**, 213, 390
McGee, L.C., *Proc. Inst. Med. Chicago*, 1927, **6**, 242
McGinty, D.A., Anderson, L.P., & McCullough, N.B., *E*, 1939, **24**, 829
McGinty, D.A., McCullough, N.B., & Wolter, J.C., *PB*, 1936, **34**, 176
McGrath, E.J., *JAMA*, 1935, **105**, 854
Macht, D.I., & Stickels, A.E., *PB*, 1931, **28**, 801
McKeown, T., & Zuckerman, S., *PR*, 1937, **124**, 362 and 464
Macpherson, A.I.S., & Haultain, W.F.T., *BMJ*, 1942 (i), 770
McQueen-Williams, M., (a) *AR*, 1934, **58**, Suppl., 77
McQueen-Williams, M., (b) *PB*, 1934, **32**, 296
McQueen-Williams, M., *PB*, 1935, **33**, 406
McShan, W.H., & Meyer, R.K., *JBC*, 1938, **126**, 361
McShan, W.H., & Meyer, R.K., *PB*, 1939, **40**, 699
Makepeace, A.W., Weinstein, G.L., & Friedman, M.H., *PB*, 1936, **35**, 269
Makepeace, A.W., Weinstein, G.L., & Friedman, M.H., *AJP*, 1937, **119**, 512
Malley, K., & Bradshaw, T.G., *BMJ*, 1941 (i), 454
Mandel, J., & Grisewood, E.N., *PB*, 1934, **32**, 155
Marine, D., & Rosen, S.H., *AJC*, 1940, **39**, 315
Markee, J.E., *AR*, 1933, **55**, Abst., 66
Marker, R.E., *J. Amer. Chem. Soc.*, 1938, **60**, 2442
Marker, R.E., & Hartman, C.G., *JBC*, 1940, **133**, 529

Marker, R.E., Kamm, O., & McGrew, R.V., *J. Amer. Chem. Soc.*, 1937, **59**, 616
Marker, R.E., Wittle, E.L., & Lawson, E.J., *J. Amer. Chem. Soc.*, 1938, **60**, 2931
Marrian, G.F., *BJ*, 1929, **23**, 1090
Marrian, G.F., Cohen, S.L., & Watson, M., *JBC*, 1935, **109**, Proc., 59
Marrian, G.F., & Parkes, A.S., *J. Roy. Micros. Soc.*, 1928, **48**, 257
Marrian, G.F., & Parkes, A.S., (a) *PR*, 1929, **105**, 248
Marrian, G.F., & Parkes, A.S., (b) *JP*, 1929, **77**, 389
Marsh, M.C., *JCR*, 1929, **13**, 313
Marshall, F.H.A., 'The Physiology of Reproduction', London, 1910
Marshall, F.H.A., *PR*, 1912, **85**, 27
Marshall, F.H.A., *PT*, 1936, **226**, 423
Marshall, F.H.A., *JEB*, 1940, **17**, 139
Marshall, F.H.A., *Biol. Rev.*, 1942, **17**, 68
Marshall, F.H.A., & Bowden, F.P., *JEB*, 1934, **11**, 409
Marshall, F.H.A., & Bowden, F.P., *JEB*, 1936, **13**, 383
Marshall, F.H.A., & Halnan, E.T., *PR*, 1917, **89**, 546
Marshall, F.H.A., & Jolly, W.A., *PT*, 1905, **198**, 99
Marshall, F.H.A., & Verney, E.B., *JP*, 1936, **86**, 327
Marshall, F.H.A., Verney, E.B., & Vogt, M., *JP*, 1939, **97**, 128
Marsman, W.S., *Acta Néerl. Morph. Norm. Path.*, 1937, **1**, 115
Martin, S.J., *PB*, 1930, **28**, 41
Martin, S.J., *AJP*, 1932, **100**, 180
Martins, T., *CRS*, 1930, **105**, 789
Martins, T., *CRS*, 1931, **106**, 508
Martins, T., *CRS*, 1935, **119**, 753
Martins, T., *CRS*, 1936, **123**, 702
Martins, T., *CRS*, 1939, **131**, 299
Martins, T., & Cardoso, D.M., *CRS*, 1931, **106**, 509
Martins, T., & Rocha, A., *CRS*, 1930, **105**, 793 and 795
Martins, T., & Rocha, A., *CRS*, 1931, **106**, 510
Martins, T., Rocha, A., & Silva, A., *E*, 1931, **15**, 421
Martins, T., & Valle, J.R., *CRS*, 1938, **129**, 1122
Marx, L., *Zeits. f. Zellforsch. u. mikros. Anat.*, 1932, **16**, 48
Mason, K.E., *AJA*, 1933, **52**, 153
Mason, K.E., & Wolfe, J.M., *AR*, 1930, **45**, 232
Masui, K., & Tamura, Y., *Biol. Abst.*, 1927, **1**, 9266
Matthews, L.H., *PR*, 1939, **126**, 557
Matthews, L.H., *PT*, 1941, **230**, 1
Maxwell, L.C., *AJP*, 1934, **110**, 458
Mazer, M., & Mazer, C., *E*, 1939, **24**, 175
Mazer, M., & Mazer, C., *E*, 1940, **26**, 662
Meites, J., & Turner, C.W., *PB*, 1942, **49**, 190
Mercier, L., *CRS*, 1938, **127**, 92
Merckel, C., & Nelson, W.O., *AR*, 1940, **76**, 391

Meyer, R.K., & Allen, W.M., *AR*, 1933, **56**, 321

Meyer, R.K., & Gustus, E.L., *S*, 1935, **81**, 208

Meyer, R.K., Leonard, S.L., & Hisaw, F.L., *PB*, 1929, **27**, 340

Meyer, R.K., Leonard, S.L., Hisaw, F.L., & Martin, S.J., *PB*, 1930, **27**, 702

Meyer, R.K., Leonard, S.L., Hisaw, F.L., & Martin, S.J., *E*, 1932, **16**, 655

Miescher, K., Fischer, W.H., & Tschopp, E., *N*, 1937, **140**, 726

Miescher, K., Fischer, W.H., & Tschopp, E., *N*, 1938, **142**, 435

Miescher, K., & Tschopp, E., *Schweiz. med. Wschr.*, 1938, **68**, 1258

Miescher, K., Wettstein, A., & Tschopp, E., *CI*, 1936, **55**, 238

Miller, E.W., Orr, J.W., & Pybus, F.C., *JPB*, 1943, **55**, 137

Miller, E.W., & Pybus, F.C., Personal communication, 1941

Miller, E.W., & Pybus, F.C., *JPB*, 1942, **54**, 155

Mills, C.A., *S*, 1939, **89**, 11

Minoura, T., *JEZ*, 1921, **33**, 1

Mixner, J.P., & Turner, C.W., (*a*) *PB*, 1941, **46**, 436

Mixner, J.P., & Turner, C.W., (*b*) *PB*, 1941, **47**, 453

Mixner, J.P., & Turner, C.W., *E*, 1942, **30**, 591

Miyazaki, H., Quoted by Rowan, W., *Biol. Rev.*, 1938, **13**, 374

Mohs, F.E., *AJC*, 1937, **29**, 356

Moir, J., *BMJ*, 1942 (i), 681

Montpellier, J., & Herlant, M., *CRS*, 1933, **114**, 277

Moore, C.R., *JEZ*, 1919, **28**, 137 and 459

Moore, C.R., *JEZ*, 1921, **33**, 365

Moore, C.R., *Biol. Bull.*, 1922, **43**, 285

Moore, C.R., *AJA*, 1924, **34**, 269

Moore, C.R., (*a*) *JEZ*, 1928, **50**, 455

Moore, C.R., (*b*) *Biol. Bull.*, 1928, **55**, 339

Moore, C.R., *Proc. 2nd Internat. Cong. Sex Res.*, London, 1930, p. 293

Moore, C.R., 'Sex & Internal Secretions', ed. by E. Allen, 1932, p. 346

Moore, C.R., (*a*) *JAMA*, 1935, **104**, 1405

Moore, C.R., (*b*) *New Engl. J. Med.*, 1935, **212**, 422

Moore, C.R., *AJA*, 1936, **59**, 63

Moore, C.R., *CSH*, 1937, **5**, 115

Moore, C.R., & Gallagher, T.F., (*a*) *JPET*, 1930, **40**, 341

Moore, C.R., & Gallagher, T.F., (*b*) *AJA*, 1930, **45**, 39

Moore, C.R., Gallagher, T.F., & Koch, F.C., *E*, 1929, **13**, 367

Moore, C.R., Hughes, W., & Gallagher, T.F., *AJA*, 1930, **45**, 109

Moore, C.R., Lamar, J.K., & Beck, N., *JAMA*, 1938, **111**, 11

Moore, C.R., & Price, D., *PB*, 1930, **28**, 38

Moore, C.R., & Price, D., *AJA*, 1932, **50**, 13

Moore, C.R., & Price, D., *E*, 1937, **21**, 313

Moore, C.R., & Price, D., *AR*, 1938, **71**, 59

Moore, C.R., Price, D., & Gallagher, T.F., *AJA*, 1930, **45**, 71

Moore, C.R., & Samuels, L.T., *AJP*, 1931, **96**, 278

Moore, C.R., Simmons, G.F., Wells, L.J., Zalesky, M., & Nelson, W.O., *AR*, 1934, **60**, 279

Moore, R.A., & McLellan, A.M., *JU*, 1938, **40**, 641

Moricard, R., *CRS*, 1933, **112**, 1045

Moricard, R., 'Proliférine sexuelle Femelle', Paris, 1934

Moricard, R., & Cauchoix, J., *CRS*, 1938, **129**, 556

Morpurgo, B., *Münch. med. Woch.*, 1908, **55**, 2447

Morris, H.P., & Lippincott, S.W., *JNCI*, 1941, **2**, 47

Morris, H.P., & Voegtlin, C., *JBC*, 1940, **133**, Proc., 69

Morton, D.G., *AJOG*, 1942, **44**, 799

Moszkowska, A., *CRS*, 1935, **118**, 516

Mühlbock, O., *ABN*, 1938, **8**, 50 and 142

Mühlbock, O., *ABN*, 1939, **9**, 7

Mulligan, R.M., *PB*, 1943, **54**, 21

Murray, W.S., *JCR*, 1928, **12**, 18

Murray, W.S., *AJC*, 1934, **20**, 573

Murray, W.S., *AJC*, 1937, **30**, 517

Murray, W.S., *CR*, 1941, **1**, 738 and 790

Murray, W.S., & Little, C.C., *AJC*, 1936, **27**, 516

Murray, W.S., & Little, C.C., *AJC*, 1939, **37**, 536

Nathanson, I.T., & Andervont, H.B., *PB*, 1939, **40**, 421

Nathanson, I.T., Brues, A.M., & Rawson, R.W., *PB*, 1940, **43**, 737

Nathanson, I.T., Franseen, C.C., & Sweeney, A., *PB*, 1938, **39**, 385

Nathanson, I.T., Shaw, D.T., & Franseen, C.C., *PB*, 1939, **42**, 652

Nathanson, I.T., Towne, L.E., & Aub, J.C., *E*, 1939, **24**, 335

Nelson, W.O., *S*, 1929, **70**, 453

Nelson, W.O., *AR*, 1933, **56**, 241

Nelson, W.O., (*a*) *AR*, 1934, **58**, Abst., 30

Nelson, W.O., (*b*) *PB*, 1934, **31**, 1192

Nelson, W.O., (*c*) *PB*, 1934, **32**, 452

Nelson, W.O., (*d*) *E*, 1934, **18**, 33

Nelson, W.O., (*a*) *E*, 1935, **19**, 187

Nelson, W.O., (*b*) *AR*, 1935, **64**, Abst., 52

Nelson, W.O., (*c*) *PB*, 1935, **32**, 1605

Nelson, W.O., (*a*) *AJA*, 1936, **50**, 341

Nelson, W.O., (*b*) *AR*, 1936, **64**, Suppl., 34

Nelson, W.O., (*c*) *Physiol. Rev.*, 1936, **16**, 488

Nelson, W.O., (*a*) *CSH*, 1937, **5**, 123 and 130

Nelson, W.O., (*b*) *AR*, 1937, **68**, 99

Nelson, W.O., *E*, 1939, **24**, 50

Nelson, W.O., (*a*) *AR*, 1941, **81**, 97

Nelson, W.O., (*b*) *AJP*, 1941, **133**, Proc., 398

Nelson, W.O., & Gallagher, T.F., *AR*, 1935, **64**, 129

Nelson, W.O., & Gallagher, T.F., *S*, 1936, **84**, 230

Nelson, W.O., & Gaunt, R., *PB*, 1936, **34**, 671

Nelson, W.O., & Gaunt, R., (a) *PB*, 1937, **36**, 136

Nelson, W.O., & Gaunt, R., (b) *CSH*, 1937, **5**, 398

Nelson, W.O., & Merckel, C.G., *PB*, 1937, **36**, 823 and 825

Nelson, W.O., & Merckel, C.G., *PB*, 1938, **38**, 737

Nelson, W.O., & Overholser, M.D., *PB*, 1934, **32**, 150

Nelson, W.O., & Pfiffner, J.J., *PB*, 1930, **28**, 1

Nelson, W.O., & Pfiffner, J.J., *AR*, 1931, **51**, 51

Nelson, W.O., Pfiffner, J.J., & Haterius, H.O., *AJP*, 1930, **91**, 690

Neumann, H., & Peter, F., *Z. Kinderheilk.*, 1931, **52**, 24

Newton, W.H., *JP*, 1934, **81**, 277

Newton, W.H., *JP*, 1935, **84**, 196

Newton, W.H., & Richardson, K.C., *JE*, 1941, **2**, 323

Nice, L.B., & Shiffer, A.L., *E*, 1931, **15**, 205

Nicholson, G.W., *BJS*, 1921, **8**, 527

Noble, G.K., & Wurm, M., *E*, 1940, **26**, 837

Noble, R.L., (a) *JP*, 1938, **94**, 177

Noble, R.L., (b) *L*, 1938 (ii), 192

Noble, R.L., *JE*, 1939, **1**, 128, 184 and 216

Noble, R.L., & Greenberg, B., *PB*, 1940, **44**, 460

Noble, R.L., Rowlands, I.W., Warwick, M.H., & Williams, P.C., *JE*, 1939, **1**, 22

Novak, E., *Amer. J. Surg.*, 1934, **24**, 595

Novak, E., *JAMA*, 1941, **116**, 947

Novak, E., *AJOG*, 1944, **47**, 20

Novak, E., & Gray, L.A., *AJOG*, 1936, **31**, 213

Novak, E., & Richardson, E.H., *AJOG*, 1941, **42**, 564

Oberling, C., Guérin, M., & Guérin, P., *CRS*, 1936, **123**, 1152

O'Donoghue, C.H., *QJMS*, 1911, **57**, 187

Ogston, A.G., Philpot, J. St L., & Zuckerman, S., *JE*, 1939, **1**, 231

Olch, I.Y., *AJC*, 1937, **30**, 563

Osborne, T.B., & Mendel, L.B., *JBC*, 1914, **17**, 325; *ibid.* 1916, **26**, 1

Österreicher, W., *KW*, 1933, **12**, 896

Österreicher, W., *KW*, 1934, **13**, 1019

Overholser, M.D., & Allen, E., *PB*, 1933, **30**, 1322

Overholser, M.D., & Allen, E., *AR*, 1936, **64**, Suppl., 35

Overholser, M.D., & Warren, O.N., *AR*, 1935, **62**, 247

Owen, S.E., & Cutler, M., *AJC*, 1936, **27**, 308

Palmer, A., *PB*, 1937, **37**, 273

Palmer, A., *JE*, 1940, **2**, 70

Papanicolaou, G.N., *AR*, 1920, **18**, 251

Papanicolaou, G.N., *JAMA*, 1926, **86**, 1422

Papanicolaou, G.N., *AJA*, 1933, **52**, 519

Papanicolaou, G.N., & Falk, E.A., *PB*, 1934, **31**, 750

Papanicolaou, G.N., & Falk, E.A., *S*, 1938, **87**, 238

Papanicolaou, G.N., Ripley, H.S., & Shorr, E., *PB*, 1938, **37**, 689

Papanicolaou, G.N., & Shorr, E., *PB*, 1935, **32**, 585

Papanicolaou, G.N., & Stockard, C.R., *PB*, 1920, **17**, 143

Parhon, C.I., & Coban, B., *CRS*, 1935, **119**, 219

Parker, F., & Tenney, B., *E*, 1940, **26**, 527

Parkes, A.S., *PR*, 1926, **100**, 172

Parkes, A.S., (a) *PR*, 1927, **101**, 71 and 421

Parkes, A.S., (b) *PR*, 1927, **102**, 51

Parkes, A.S., (a) *JP*, 1928, **65**, 341

Parkes, A.S., (b) *QJ*, 1928, **18**, 397

Parkes, A.S., *PR*, 1929, **104**, 183 and 189

Parkes, A.S., *PR*, 1930, **107**, 188

Parkes, A.S., *CI*, 1935, **54**, 928

Parkes, A.S., *N*, 1937, **139**, 965

Parkes, A.S., *JE*, 1942, **3**, 220

Parkes, A.S., *JE*, 1943, **3**, 268

Parkes, A.S., & Bellerby, C.W., *JP*, 1927, **62**, 145, 301 and 385

Parkes, A.S., & Bellerby, C.W., *JP*, 1928, **64**, 233

Parkes, A.S., Dodds, E.C., & Noble, R.L., *BMJ*, 1938 (ii), 557

Parkes, A.S., & Drummond, J.C., *BJEB*, 1926, **3**, 251

Parkes, A.S., & Hammond, J., *PRSM*, 1940, **33**, 483

Parkes, A.S., & Rowlands, I.W., *JP*, 1936, **88**, 305

Parkes, A.S., & White, W.E., *JP*, 1933, **79**, 226

Parkes, A.S., & Zuckerman, S., *L*, 1935 (i), 925

Parkes, A.S., & Zuckerman, S., *JP*, 1938, **93**, Proc., 16

Paschkis, K.E., *PB*, 1941, **46**, 336

Passey, R.D., *B.E. Cancer Campaign*, 19th Ann. Rep., 1942, p. 35

Pattee, C.J., Venning, E.H., & Browne, J.S.L., *E*, 1940, **27**, 721

Patterson, J., *BMJ*, 1937 (ii), 522

Pearl, R., *JEZ*, 1912, **13**, 153

Pearl, R., & Schoppe, W.F., *JEZ*, 1921, **34**, 101

Pearson, P.B., *AJP*, 1937, **118**, 786

Pedersen-Bjergaard, K., *ZG*, 1936, **60**, 372

Pencharz, R.I., *S*, 1940, **91**, 554

Pencharz, R.I., & Long, J.A., *AJA*, 1932, **50**, 1

Pencharz, R.I., & Long, J.A., *AJA*, 1933, **53**, 89, 117 and 191

Pencharz, R.I., & Lyons, W.R., *PB*, 1934, **31**, 1131

Pencharz, R.I., & Olmsted, J.M.D., *PB*, 1931, **28**, 600

Perloff, W.H., & Kurzrok, R., *PB*, 1941, **46**, 262

Perry, I.H., *PB*, 1936, **35**, 325

Perry, I.H., & Ginzton, L.L., *AJC*, 1937, **29**, 680

Pézard, A., *CRA*, 1911, **153**, 1027

Pézard, A., *E*, 1920, **4**, 527

Pézard, A., *CRA*, 1922, **174**, 1573

Pfeiffer, C.A., *AJA*, 1936, **58**, 195

Pfeiffer, C.A., *E*, 1937, **21**, 260

Pfeiffer, C.A., Emmel, V.M., & Gardner, W.U., *YJ*, 1940, **12**, 493

Pfeiffer, C.A., & Gardner, W.U., *E*, 1938, **23**, 485

Pfiffner, J.J., & North, H.B., *JBC*, 1940, **132**, 459

Phelps, D., Burch, J.C., & Ellison, E.T., *E*, 1938, **23**, 458

Philipp, E., *ZG*, 1929, **53**, 2386

Philipp, E., *ZG*, 1930, **54**, 450 and 1858

Philipp, E., *ZG*, 1931, **55**, 491

Phillips, W.A., *AJP*, 1937, **119**, 623

Pincus, G., *CSH*, 1937, **5**, 44

Pincus, G., *AR*, 1940, **77**, 1

Pincus, G., & Graubard, M., *E*, 1940, **26**, 427

Pincus, G., & Martin, D.W., *E*, 1940, **27**, 838

Pincus, G., & Werthessen, N.T., *AJP*, 1933, **103**, 631

Pincus, G., & Werthessen, N.T., *JEZ*, 1938, **78**, 1

Pincus, G., & Zahl, P.A., *J. Gen. Physiol.*, 1937, **20**, 879

Plum, C.M., *Acta med. Scand.*, 1942, **112**, 151

Pomerat, G.R., & Coe, R.C., *E*, 1941, **29**, 1015

Poncet, *L' Ass. Franç. p. l'Avanc. d. Sc.*, 1877, **6**, 893

Portes, L., Dalsace, J., & Wallich, R., *CRS*, 1939, **130**, 1100

Portman, K., *CRS*, 1934, **115**, 89

Pratt, J.P., *E*, 1927, **11**, 195

Price, D., *AJA*, 1936, **60**, 79

Price, D., *PB*, 1939, **41**, 580

Pybus, F.C., & Miller, E.W., *Newcastle Med. J.*, 1936, **16**, 49

Pybus, F.C., & Miller, E.W., (a) *N*, 1938, **142**, 872

Pybus, F.C., & Miller, E.W., (b) *AJC*, 1938, **34**, 248 and 252

Pybus, F.C., & Miller, E.W., *AJC*, 1940, **40**, 47

Pybus, F.C., & Miller, E.W., Personal communication, 1941

Rasmussen, A.T., *E*, 1921, **5**, 33

Rasmussen, A.T., *PB*, 1931, **28**, 716

Rathery, F., & Turiaf, J., *CRS*, 1938, **128**, 155

Rauther, M., *Jenaische Z. Naturwiss.*, 1904, p. 377

Rawson, R.W., Sterne, G.D., & Aub, J.C., *E*, 1942, **30**, 240

Raynaud, A., (a) *CRS*, 1937, **126**, 866

Raynaud, A., (b) *CRA*, 1937, **205**, 1453

Raynaud, A., (a) *Bull. Biol.*, 1938, **72**, 297

Raynaud, A., (b) *CRS*, 1938, **127**, 993

Raynaud, A., (c) *CRS*, 1938, **129**, 289, 528, 632, 637 and 1033

Raynaud, A., (a) *CRS*, 1939, **130**, 872, 1012, and 1061

Raynaud, A., (b) *CRS*, 1939, **131**, 218

Raynaud, A., & Lacassagne, A., *CRS*, 1937, **126**, 868

Raynaud, A., Marill, F.G., & Xicluna, R., *CRS*, 1939, **130**, 655

Raynaud, R., *CRS*, 1930, **104**, 284

Reece, R.P., *PB*, 1938, **39**, 77

Reece, R.P., *PB*, 1941, **46**, 265

Reece, R.P., Bartlett, J.W., Hathaway, I.L., & Davis, H.P., *PB*, 1940, **43**, 183

Reece, R.P., & Bivins, J.A., *PB*, 1942, **49**, 582

Reece, R.P., & Leonard, S.L., *PB*, 1939, **42**, 200

Reece, R.P., & Leonard, S.L., *PB*, 1942, **49**, 660

Reece, R.P., & Mixner, J.P., *PB*, 1939, **40**, 66

Reece, R.P., & Turner, C.W., *PB*, 1936, **34**, 402

Reece, R.P., & Turner, C.W., *PB*, 1937, **36**, 283

Reece, R.P., Turner, C.W., & Hill, R.T., *PB*, 1936, **34**, 204

Reese, J.D., *AR*, 1932, **52**, 74

Reichert, F.L., Pencharz, R.I., Simpson, M.E., Meyer, K., & Evans, H.M., *PB*, 1931, **28**, 843

Reichert, F.L., Pencharz, R.I., Simpson, M.E., Meyer, K., & Evans, H.M., *AJP*, 1932, **100**, 157

Reichstein, T., *Helv. chim. Acta*, 1936, **19**, 29 and 223

Reinhardt, W.O., & Wainman, P., *PB*, 1942, **49**, 257

Reisel, J.H., *ABN*, 1936, **6**, 24

Retterer, E., & Lelièvre, A., *CRS*, 1912, **73**, 184

Retterer, E., & Lelièvre, A., *CRS*, 1913, **74**, 1403

Retterer, E., & Neuville, H., *CRS*, 1913, **75**, 165

Reynolds, S.R.M., *AJP*, 1931, **97**, 706

Reynolds, S.R.M., *CSH*, 1937, **5**, 84

Reynolds, S.R.M., *S*, 1938, **87**, 537

Reynolds, S.R.M., *JP*, 1939, **95**, 258

Reynolds, S.R.M., *J. Investig. Dermat.*, 1941, **4**, 7

Reynolds, S.R.M., & Allen, W.M., *AJP*, 1932, **102**, 39

Reynolds, S.R.M., & Allen, W.M., *PB*, 1937, **36**, 455

Reynolds, S.R.M., Firor, W.M., & Allen, W.M., *E*, 1936, **20**, 681

Riboulleau, J., *CRS*, 1938, **129**, 914

Richon, L., & Jeandelize, P., *CRS*, 1903, **55**, 1684

Richon, L., & Jeandelize, P., *CRS*, 1905, **57**, 1086 and 1087

Riddle, O., *AJP*, 1923, **66**, 322

Riddle, O., *E*, 1931, **15**, 307

Riddle, O., *CSH*, 1937, **5**, 218 and 321

Riddle, O., Bates, R.W., Lahr, E.L., & Moran, C.S., *AJP*, 1936, **116**, 128

Riddle, O., & Braucher, P.F., *AJP*, 1931, **97**, 617

Riddle, O., & Dotti, L.B., *S*, 1936, **84**, 557

Riddle, O., & Dunham, H.H., *E*, 1942, **30**, 959

Riddle, O., & Honeywell, H.E., *AJP*, 1923, **66**, 340

Riddle, O., Lahr, E.L., & Bates, R.W., *PB*, 1935, **32**, 730

Riddle, O., & Reinhart, W.H., *AJP*, 1926, **76**, 660

Riddle, O., & Tange, M., *AJP*, 1928, **87**, 97

Riley, G.M., & Fraps, R.M., *E*, 1942, **30**, 537
Robertson, D.C., Maddux, W.P., & Allen, E., *E*, 1930, **14**, 77
Robson, J.M., *JP*, 1930, **71**, Suppl., 3
Robson, J.M., (*a*) *JP*, 1933, **79**, 139
Robson, J.M., (*b*) *QJ*, 1933, **22**, 209
Robson, J.M., (*a*) *JP*, 1935, **85**, 145
Robson, J.M., (*b*) *QJ*, 1935, **24**, 337
Robson, J.M., (*a*) *BMJ*, 1936 (i), 1033
Robson, J.M., (*b*) *JP*, 1936, **86**, 415
Robson, J.M., (*c*) *PB*, 1936, **35**, 49
Robson, J.M., (*a*) *JP*, 1937, **87**, 100
Robson, J.M., (*b*) *JP*, 1937, **90**, 435 and 15 *P*.
Robson, J.M., (*c*) *PB*, 1937, **36**, 153
Robson, J.M., (*d*) *QJ*, 1937, **26**, 355
Robson, J.M., *QJ*, 1938, **28**, 49 and 195
Robson, J.M., *JP*, 1939, **96**, 21 *P*.
Robson, J.M., & Bonser, G.M., *N*, 1938, **142**, 836
Robson, J.M., & Illingworth, R.E., *QJ*, 1931, **21**, 93
Robson, J.M., MacGregor, T.N., Illingworth, R.E., & Steere, N., *BMJ*, 1934 (i), 888
Robson, J.M., & Schönberg, A., *N*, 1937, **140**, 196
Robson, J.M., & Wiesner, B.P., *QJ*, 1931, **21**, 225
Rock, J., Reboul, J., & Snodgrass, J.M., *AJOG*, 1938, **36**, 733
Rogers, P.V., *PB*, 1936, **34**, 644
Rogers, P.V., *PB*, 1937, **37**, 212
Rogers, P.V., (*a*) *E*, 1938, **22**, 35
Rogers, P.V., (*b*) *AJP*, 1938, **121**, 565
Rogers, P.V., & Allen, E., *E*, 1937, **21**, 629
Rogoff, J.M., & Stewart, G.N., *AJP*, 1927, **79**, 508
Rogoff, J.M., & Stewart, G.N., *AJP*, 1928, **86**, 20
Romeis, R., *Anat. Anz.*, 1923, **57**, Suppl., 263
Ross, M., & Dorfman, R.I., *CR*, 1941, **1**, 52
Ross, M.A., & Korenchevsky, V., *JPB*, 1941, **52**, 349
Rost, F., *KW*, 1933, **31**, 1213
Rous, P., *JEM*, 1914, **20**, 433
Rouzet, F.G.L., 'Recherches et Observations sur Le Cancer', Paris, 1818, p. 317
Rowan, W., *N*, 1925, **115**, 494
Rowan, W., *Proc. Boston Soc. Nat. Hist.*, 1929, **39**, 151
Rowan, W., *N*, 1937, **139**, 668
Rowan, W., (*a*) *Proc. Zool. Soc. Lond.*, 1938, Series A, **108**, 51
Rowan, W., (*b*) *Biol. Rev.*, 1938, **13**, 374
Rowlands, I.W., *PT*, 1936, **226**, 99
Rowlands, I.W., *PR*, 1937, **121**, 517
Rowlands, I.W., *PR*, 1938, **126**, 76
Rowlands, I.W., & McPhail, M.K., *QJ*, 1936, **26**, 109
Rowlands, I.W., & Parkes, A.S., *L*, 1937 (i), 924
Rowlands, I.W., & Sharpey-Schafer, E.P., *BMJ*, 1940 (i), 205
Rowlands, I.W., & Singer, E., *JP*, 1936, **86**, 323
Rowlands, I.W., & Spence, A.W., *BMJ*, 1939 (ii), 947

Rowlands, I.W., & Williams, P.C., *JE*, 1943, **3**, 310
Rubinstein, H.S., Abarbanel, A.R., & Kurland, A.A., *E*, 1939, **25**, 397
Rubinstein, H.S., Abarbanel, A.R., & Nader, D.N., *PB*, 1938, **39**, 20
Rubinstein, H.S., Kurland, A.A., & Goodwin, M., *E*, 1939, **25**, 724
Rubinstein, H.S., & Solomon, M.L., *PB*, 1940, **45**, 745
Rugh, R., *S*, 1937, **85**, 588
Rusch, H.P., *E*, 1937, **21**, 511
Ruth, E.B., *AR*, 1934, **60**, 279
Ruzicka, L., Goldberg, M.W., Meyer, J., Brüngger, H., & Eichenberger, E., *Helv. chim. Acta*, 1934, **17**, 1395
Ruzicka, L., & Wettstein, A., *Helv. chim. Acta*, 1935, **18**, 986, 1264

Sager, V.J., & Leonard, S.L., *PB*, 1936, **35**, 242
Salmon, U.J., (*a*) *PB*, 1938, **37**, 488
Salmon, U.J., (*b*) *PB*, 1938, **38**, 352
Salmon, U.J., *PB*, 1939, **41**, 515
Salmon, U.J., & Frank, R.T., *PB*, 1936, **34**, 463
Salmon, U.J., Geist, S.H., & Salmon, A.A., *PB*, 1941, **47**, 279
Salmon, U.J., & Salmon, A.A., *PB*, 1940, **43**, 709
Sammartino, R., & Arenas, N., *CRS*, 1940, **133**, 321
Samuels, L.T., Reinecke, R.M., & Petersen, W.E., *PB*, 1941, **46**, 379
Sand, K., (*a*) *Arch. f. Physiol.*, 1919, **173**, 1
Sand, K., (*b*) *JP*, 1919, **53**, 257
Sandes, F.P., *Proc. Linnean Soc. New S. Wales*, 1903, **28**, 364
Sardi, J.L., *CRS*, 1935, **120**, 503
Saunders, F.J., & Cole, H.H., *PB*, 1936, **33**, 504
Saunders, F.J., & Cole, H.H., *E*, 1938, **23**, 302
Schacher, J., Browne, J.S.L., & Selye, H., *PB*, 1937, **36**, 488
Schachter, B., & Marrian, G.F., *PB*, 1936, **35**, 222
Scharf, G., & Lyons, W.R., *PB*, 1941, **48**, 86
Schenken, J.R., Burns, E.L., & McCord, W.M., *E*, 1942, **30**, 344
Schilling, W., & Laqueur, G.L., *E*, 1942, **30**, 753
Schlacta, J., *Arch. f. mikr. Anat.*, 1904, **64**, 405
Schmidt, I.G., *AR*, 1936, **64**, 255
Schmidt, I.G., *E*, 1937, **21**, 461
Schmidt, P.P.H., *L*, 1942 (ii), 84
Schockaert, J.A., *CRS*, 1933, **112**, 733
Schockaert, J.A., & Delrue, G., *CRS*, 1936, **123**, 306
Schoeller, W., Schwenk, E., & Hildebrandt, F., *NW*, 1933, **21**, 286
Schwabe, E.L., & Emery, F.E., *PB*, 1939, **40**, 383
Schweizer, M., Charipper, H.A., & Haterius, H.O., *E*, 1937, **21**, 30
Schweizer, M., Charipper, H.A., & Kleinberg, W., *E*, 1940, **26**, 979
Scipiades, E., *PB*, 1937, **37**, 242

Segaloff, A., & Nelson, W.O., *PB*, 1941, **48**, 33

Selye, H., *PB*, 1934, **31**, 488

Selye, H., (a) *JE*, 1939, **1**, 208

Selye, H., (b) *JU*, 1939, **42**, 637

Selye, H., (a) *PB*, 1940, **43**, 343 and 404

Selye, H., (b) *CMA*, 1940, **42**, 113

Selye, H., (c) *AR*, 1940, **76**, 145

Selye, H., (d) *AJP*, 1940, **130**, 258

Selye, H., (e) *JAMA*, 1940, **115**, 2246

Selye, H., (a) *JE*, 1941, **2**, 352

Selye, H., (b) *PB*, 1941, **46**, 116

Selye, H., *E*, 1942, **30**, 437

Selye, H., & Albert, S., *PB*, 1942, **50**, 159

Selye, H., Bachman, C., Thomson, D.L., & Collip, J.B., *PB*, 1934, **31**, 1113

Selye, H., & Bassett, L., *PB*, 1940, **44**, 502

Selye, H., Browne, J.S.L., & Collip, J.B., *PB*, 1936, **34**, 198 and 472

Selye, H., & Collip, J.B., *PB*, 1933, **30**, 647

Selye, H., & Collip, J.B., *E*, 1936, **20**, 667

Selye, H., Collip, J.B., & Thomson, D.L., (a) *PB*, 1933, **30**, 588

Selye, H., Collip, J.B., & Thomson, D.L., (b) *E*, 1933, **17**, 494

Selye, H., Collip, J.B., & Thomson, D.L., (a) *E*, 1934, **18**, 237

Selye, H., Collip, J.B., & Thomson, D.L., (b) *PB*, 1934, **31**, 487 and 566

Selye, H., Collip, J.B., & Thomson, D.L., *PB*, 1935, **32**, 1377

Selye, H., & Friedman, S., *AJC*, 1940, **38**, 558

Selye, H., Harlow, C., & McKeown, T., *PB*, 1935, **32**, 1253

Selye, H., McEuen, C.S., & Collip, J.B., *PB*, 1936, **34**, 201

Selye, H., & McKeown, T., (a) *PB*, 1934, **31**, 683

Selye, H., & McKeown, T., (b) *AR*, 1934, **60**, 323

Selye, H., & McKeown, T., (c) *SGO*, 1934, **59**, 886

Selye, H., & McKeown, T., *PR*, 1935, **119**, 1

Selye, H., & Stone, H., *PB*, 1943, **52**, 190

Severinghaus, A.E., *AJP*, 1932, **101**, 309

Severinghaus, A.E., (a) *AR*, 1934, **60**, 43

Severinghaus, A.E., (b) *PB*, 1934, **31**, 593 and 1178

Severinghaus, A.E., *CSH*, 1937, **5**, 144

Severinghaus, A.E., 'The Pituitary Gland', Baltimore, 1938

Shapiro, H.A., *CI*, 1936, **55**, 1031

Shapiro, H.A., *N*, 1937, **139**, 588

Sharpey-Schafer, E.P., & Zuckerman, S., *JE*, 1941, **2**, 431

Shelesnyak, M.C., *AR*, 1931, **49**, 179

Shelesnyak, M.C., *AR*, 1933, **56**, 211

Sherwood, T.C., Brend, M.A., & Roper, E.A., *J. Nutr.*, 1936, **11**, 593

Shimkin, M.B., Grady, H.G., & Andervont, H.B., *JNCI*, 1941, **2**, 65 and 611

Shoemaker, H.H., *PB*, 1939, **41**, 299

Shorr, E., *PB*, 1940, **43**, 501

Shorr, E., Almy, T.P., Sloan, M.H., Taussky, H., & Toscani, V., *S*, 1942, **96**, 587

Shorr, E., Bernheim, A.R., & Taussky, H., *S*, 1942, **95**, 606

Shumacker, H.B. & Firor, W. M., *E*, 1934, **18**, 676

Siebke, H., *ZG*, 1930, **54**, 1601

Siebke, H., *ZG*, 1934, **58**, 264

Simonnet, H., & Michel, E., *CRS*, 1939, **129**, 918

Simonnet, H., & Michel, E., *CRS*, 1940, **133**, 358

Simpson, B.T., *25th Ann. Rep. State Inst. for the Study of Malignant Disease*, New York, 1936

Simpson, H.L., De Fremery, P., & Macbeth, A., *E*, 1936, **20**, 363

Simpson, M.E., Li, C.H., & Evans, H.M., *E*, 1942, **30**, 969

Skowron, S., *CRS*, 1935, **119**, 431

Skowron, S., & Skarzynski, B., *CRS*, 1933, **112**, 1604

Slotta, K.H., Ruschig, H., & Fels, E., *Ber. Deutsch. chem. Ges.*, 1934, **67**, B, 1270

Slye, M., *JCR*, 1916, **1**, 109

Slye, M., Holmes, H.F., & Wells, H.G., *AJC*, 1931, **15**, 1387

Smelser, G.K., *AR*, 1933, **57**, Suppl. 28

Smith, G. van S., & Smith, O.W., *AJP*, 1931, **98**, 578

Smith, G. van S., & Smith, O.W., *AJP*, 1933, **103**, 356

Smith, G. van S., & Smith, O.W., *AJP*, 1935, **112**, 340

Smith, G. van S., & Smith, O.W., *AJOG*, 1938, **36**, 769

Smith, G. van S., Smith, O.W., & Pincus, G., *AJP*, 1938, **121**, 98

Smith, P.E., *PB*, 1926, **24**, 131 and 337

Smith, P.E., (a) *AJP*, 1927, **80**, 114

Smith, P.E., (b) *JAMA*, 1927, **88**, 158

Smith, P.E., *AJP*, 1932, **99**, 345 and 349

Smith, P.E., *JAMA*, 1935, **104**, 548

Smith, P.E., & Dortzbach, C., *AR*, 1929, **43**, 277

Smith, P.E., & Engle, E.T., *AJA*, 1927, **40**, 159

Smith, P.E., & Engle, E.T., *AR*, 1929, **42**, 38

Smith, P.E., & Engle, E.T., *PB*, 1932, **29**, 1225

Smith, P.E., Engle, E.T., & Tyndale, H.H., *PB*, 1934, **31**, 745

Smith, P.E., & Leonard, S.L., *PB*, 1933, **30**, 1246 and 1250

Smith, P.E., & Leonard, S.L., *AR*, 1934, **58**, 145

Smith, P.E., Severinghaus, A.E., & Leonard, S.L., *AR*, 1933, **57**, 177

Smith, P.E., & White, W.E., *JAMA*, 1931, **97**, 1861

Snodgrass, J.M., Rock, J., & Menkin, M.F., *AJP*, 1943, **140**, 394

Snyder, F.F., *Physiol. Rev.*, 1938, **18**, 578

Soeken, G., *Z. Kinderheilk.*, 1932, **53**, 339

Solcard, P., Le Chuiton, F., Pervès, J., Berge, C., & Pennanéac'h, J., *BC*, 1936, **25**, 801

Speert, H., *S*, 1940, **92**, 461

Spence, A.W., *Quart. J. Med.*, 1940, **9**, 309

Spence, A.W., Scowen, E.F., & Rowlands, I.W., *BMJ*, 1938 (i), 66

Spencer, J., D'Amour, F.E., & Gustavson, R.G., (a) *E*, 1932, **16**, 647

Spencer, J., D'Amour, F.E., & Gustavson, R.G., (b) *AJA*, 1932, **50**, 129

Spencer, J., Gustavson, R.G., & D'Amour, F.E., *PB*, 1931, **28**, 500

Spielman, M.A., & Meyer, R.K., *PB*, 1937, **37**, 623

Spurrell, W.R., & Ucko, H., *Guy's Hosp. Rep.*, 1938, **88**, 230

Starkey, W.F., Grauer, R.C., & Saier, E., *PB*, 1943, **52**, 227

Starkey, W.F., & Leathem, J.H., *PB*, 1938, **39**, 218

Starkey, W.F., & Schmidt, E.C.H., *E*, 1938, **23**, 339

Steggerda, M., *JEZ*, 1928, **51**, 403

Stein, S.I., *AR*, 1933, **56**, 15

Steinach, E., *AE*, 1916, **42**, 307

Steinach, E., Dohrn, M., Schoeller, W., Hohlweg, W., & Faure, W., *Pflügers Arch.*, 1928, **219**, 306

Steinach, E., & Holzknecht, G., *AE*, 1916, **42**, 307 and 490

Steinach, E., & Kun, H., *Pflügers Arch.*, 1931, **227**, 266

Steinach, E., & Kun, H., *L*, 1937 (ii), 845

Steinach, E., Kun, H., & Peczenik, O., (a) *N*, 1936, **138**, 49

Steinach, E., Kun, H., & Peczenik, O., (b) *Wien. klin. Wschr.*, 1936, **49**, 899 and 903

Steinglass, P., Gordon, A.S., & Charipper, H.A., *PB*, 1941, **48**, 169

Stewart, C.A., Bell, E.T., & Ruehlke, A.B., *AJC*, 1936, **26**, 144

Stockard, C.R., & Papanicolaou, G.N., *AJA*, 1917, **22**, 225

Stone, C.P., *E*, 1940, **26**, 511

Stotsenburg, J.M., *AR*, 1909, **3**, 233

Stover, R.F., & Pratt, J.P., *E*, 1939, **24**, 29

Stricker, P., & Grüter, F., *CRS*, 1928, **99**, 1979

Stricker, P., & Grüter, F., (a) *Presse Méd.*, 1929, **37**, 1268

Stricker, P., & Grüter, F., (b) *CRS*, 1929, **102**, 1076

Strong, L.C., *AJC*, 1935, **25**, 599

Strong, L.C., Gardner, W.U., & Hill, R.T., *E*, 1937, **21**, 268

Sudds, M.V.N., *E*, 1940, **26**, 895

Sulman, F., *JEM*, 1937, **65**, 1

Suntzeff, V., Burns, E.L., Moskop, M., & Loeb, L., *AJC*, 1936, **27**, 229

Sutro, C.J., *PB*, 1940, **44**, 151

Sutton, T.S., & Brief, B.J., *E*, 1938, **23**, 211

Sweeney, J.S., *JAMA*, 1934, **103**, 234

Sweet, J.E., & Hoskins, W.H., *PB*, 1941, **45**, 60

Swezy, O., & Evans, H.M., *S*, 1930, **71**, 46

Swingle, W.W., Parkins, W.M., Taylor, A.R., Hays, H.W., & Morrell, J.A., *AJP*, 1937, **119**, 675

Swingle, W.W., Parkins, W.M., Taylor, A.R., & Morrell, J.A., *PB*, 1936, **34**, 94

Talbot, N.B., *E*, 1939, **25**, 601

Tannenbaum, A., (a) *AJC*, 1940, **38**, 335

Tannenbaum, A., (b) *ARP*, 1940, **30**, 509

Tannenbaum, A., *CR*, 1942, **2**, 460 and 468

Taubenhaus, M., & Soskin, S., *E*, 1941, **29**, 958

Tausk, M., *Brux. Méd.*, 1933, p. 485

Tausk, M., & De Fremery, P., *ABN*, 1935, **5**, 19

Teel, H.M., *AJP*, 1926, **79**, 170 and 184

Thomas, E., *Zieglers Beitr.*, 1911, **50**, 283

Thomas, R.M., *YJ*, 1940, **12**, 415

Thompson, K.W., *PB*, 1937, **35**, 634

Thompson, K.W., & Cushing, H., *PR*, 1937, **121**, 501

Thomsen, E., *N*, 1940, **145**, 28

Thorn, G.W., & Engel, L.L., *JEM*, 1938, **68**, 299

Thorn, G.W., & Harrop, G.A., *S*, 1937, **86**, 40

Tietze, K., *AG*, 1934, **155**, 525

Tolenaar, J., *ABN*, 1939, **9**, 54

Torstveit, O., & Mellish, C.H., *PB*, 1941, **46**, 239

Trentin, J.J., Mixner, J.P., Lewis, A.A., & Turner, C.W., *PB*, 1941, **46**, 440

Trentin, J.J., & Turner, C.W., *E*, 1941, **29**, 984

Tschopp, E., *N*, 1935, **136**, 258

Tschopp, E., *Arch. int. Pharm. et Thérap.*, 1936, **52**, 381

Tuchmann, H., *CRS*, 1937, **125**, 635

Turner, C.D., *AJA*, 1938, **63**, 101

Turner, C.D., & Burkhardt, W.L., *PB*, 1939, **42**, 267

Turner, C.D., Haffen, R., & Struett, H., *PB*, 1939, **42**, 107

Turner, C.W., *Internat. Cancer Res. Foundation, 9th Rept., Phila.*, 1942, p. 141

Turner, C.W., & Allen, E., *AR*, 1933, **55**, Suppl., 80

Turner, C.W., & De Moss, W.R., *Missouri Agr. Exp. Sta. Res. Bull.*, 1934, No. 207

Turner, C.W., & Frank, A.H., *S*, 1931, **73**, 295

Turner, C.W., & Frank, A.H., *Missouri Agric. Exper. Sta. Res. Bull.*, 1932, No. 174

Turner, C.W., Frank, A.H., Gardner, W.U., Schultze, A.B., & Gomez, E.T., *AR*, 1932, **53**, 227

Turner, C.W., & Gomez, E.T., *Missouri Agr. Exp. Sta. Res. Bull.*, 1934, Nos. 206 and 207

Turner, C.W., & Meites, J., *PB*, 1941, **47**, 232

Turner, C.W., & Reineke, E.P., *Missouri Agr. Exp. Sta. Res. Bull.*, 1936, No. 235

Turner, C.W., & Schultze, A.B., *Missouri Agric. Exper. Sta. Res. Bull.*, 1931, No. 157

Turpault, M., *Comp. rend. Soc. Franç. de Gynéc.*, 1937, **7**, 181

Twombly, G.H., *E*, 1936, **20**, 311

Twombly, G.H., *PB*, 1939, **40**, 430

Twombly, G.H., *PB*, 1940, **44**, 617

Twombly, G.H., & Ferguson, R.S., *PB*, 1934, **32**, 69

Tyslowitz, R., *ABN*, 1939, **9**, 15

Tyslowitz, R., & Dingemanşe, E., *E*, 1941, **29**, 817

Uehlinger, E., Jadassohn, W., & Fiertz, H.E., *J. Investig. Dermat.*, 1941, **4**, 331

Uotila, U.U., (a) *PB*, 1939, **41**, 106

Uotila, U.U., (b) E, 1939, **25**, 605
Uotila, U.U., E, 1940, **26**, 123

Van der Woerd, L.A., *ABN*, 1938, **8**, 110
Van Dyke, H.B., 'The Physiology and Pharmacology of the Pituitary Body', Chicago, 1936
Van Dyke, H.B., & Ch'en, G., *AJA*, 1936, **58**, 473
Van Gulik, P.J., & Korteweg, R., *AJC*, 1940, **38**, 506
Van Heuverswyn, J., Collins, V.J., Williams, W.L., & Gardner, W.U., *PB*, 1939, **41**, 552
Van Heuverswyn, J., Folley, S.J., & Gardner, W.U., *PB*, 1939, **41**, 389
Van Horn, W.M., E, 1933, **17**, 152
Van Oort, G.J., & Junge, G.C.A., *ABN*, 1933, **3**, 15
Van Os, P.M., *ABN*, 1936, **6**, 151
Van Wagenen, G., *AR*, 1925, **29**, 398
Van Wagenen, G., *AJP*, 1928, **84**, 461
Van Wagenen, G., (a) S, 1935, **81**, 366
Van Wagenen, G., (b) *AR*, 1935, **63**, 387
Van Wagenen, G., & Aberle, S.B.D., *AJP*, 1931, **99**, 271
Van Wagenen, G., & Folley, S.J., (a) *JE*, 1939, **1**, 367
Van Wagenen, G., & Folley, S.J., (b) *AR*, 1939, **73**, Suppl. 2, 54
Van Wagenen, G., & Morse, A.H., E, 1942, **30**, 459
Vargas, L., & Lipschütz, A., *CRS*, 1938, **129**, 810
Variot, M.G., *CRS*, 1889, **1**, 451
Vasquez-Lopez, E., N, 1940, **146**, 589
Vatna, S., *Biol. Bull.*, 1930, **58**, 322
Venning, E.M., & Browne, J.S.L., (a) *AJP*, 1937, **119**, 417
Venning, E.H., & Browne, J.S.L., (b) E, 1937, **21**, 711
Venning, E.H., & Browne, J.S.L., E, 1940, **27**, 707
Venning, E.M., Henry, J.S., & Browne, J.S.L., *CMA*, 1937, **36**, 83
Venning, E.H., Weil, P.G., & Browne, J.S.L., *JBC*, 1939, **128**, Sci. Proc. cvii
Vest, S.A., & Howard, J.E., *JU*, 1938, **40**, 154
Villaret, M., Justin-Besançon, L., & Rubens-Duval, A., *CRS*, 1938, **127**, 599
Vines, H.W.C., 'The Adrenal Cortex and Intersexuality', London, 1938, p. 137
Vintemberger, P., *Arch. de Biol.*, 1925, **35**, 125
Visscher, M.B., Green, R.G., & Bittner, J.J., *PB*, 1942, **49**, 94
Voegtlin, C., Johnson, J.M., & Thompson, J.W., *PHR*, 1936, **51**, 1689
Voegtlin, C., & Maver, M.E., *PHR*, 1936, **51**, 1436
Voegtlin, C., & Thompson, J.W., *PHR*, 1936, **51**, 1429
Vollmer, E.P., Gordon, A.S., Levenstein, I., & Charipper, H.A., *PB*, 1941, **46**, 409
von Haam, E., & Rosenfeld, I., (a) *J. Immunol.*, 1942, **43**, 109

von Haam, E., & Rosenfeld, I., (b) *PB*, 1942, **49**, 710
von Haam, E., & Rosenfeld, I., (c) *JAMA*, 1942, **118**, 1002
von Haam, E., & Rothermich, O., *PB*, 1940, **44**, 369
von Hansemann, D., *AE*, 1913, **35**, 223
von Wattenwyl, H., *Schweiz. med. Woch.*, 1941, **20**, 1321
Voss, H.E., *Ztschr. f. Zellforsch.*, 1931, **14**, 200
Votquenne, M., *CRS*, 1934, **117**, 1121

Waaler, G.H.M., *CR*, 1932, **7**, Abst., 464
Wade, N.J., & Doisy, E.A., *PB*, 1931, **28**, 714
Wade, N.J., & Doisy, E.A., (a) E, 1935, **19**, 77
Wade, N.J., & Doisy, E.A., (b) *PB*, 1935, **32**, 707
Wainman, P., & Shipounoff, G.C., E, 1941, **29**, 975
Walker, C.E., (a) L, 1908 (i), 934
Walker, C.E., (b) *PRSM*, 1908, Path. **1**, 153
Walker, G., *BJH*, 1910, **21**, 182 and 185
Walker, S.M., & Stanley, A.J., *PB*, 1941, **48**, 50
Wallace, C., *Trans. Path. Soc. Lond.*, 1904, **55**, 255
Wallart, J., *AG*, 1929, **135**, 485
Wallen-Lawrence, Z., *JPET*, 1934, **5**, 263
Walsh, E.L., Cuyler, W.K., & McCullagh, D.R., *PB*, 1933, **30**, 848
Walsh, E.L., Cuyler, W.K., & McCullagh, D.R., *AJP*, 1934, **107**, 508
Wang, G.H., & Guttmacher, A.F., *AJP*, 1927, **82**, 335
Ware, H.H., Main, R.J., & Taliaferro, I., *JAMA*, 1938, **111**, 524
Waring, H., *JE*, 1942, **3**, 123
Warren, F.L., N, 1935, **135**, 234
Wassink, W.F., *Genetica*, 1935, **17**, 103
Wassink, W.F., & Wassink van Raamsdonk, C.P., *Néoplasmes*, 1923, **2**, 145
Webster, B., & Hoskins, W., *PB*, 1940, **45**, 72
Weed, I.G., *PB*, 1936, **34**, 883
Weichert, C.K., *PB*, 1928, **25**, 490
Weichert, C.K., & Boyd, R.W., *AR*, 1933, **58**, 55
Weinstein, L., Gardner, W.U., & Allen, E., *PB*, 1937, **37**, 391
Weinstein, L., Wawro, N.W., Worthington, R.V., & Allen, E., *YJ*, 1938, **11**, 141
Weisman, A.I., *JAMA*, 1941, **117**, 2248
Wells, J.A., & Greene, R.R., *AJP*, 1939, **126**, Proc. 652
Wells, L.J., *AR*, 1935, **62**, 409
Wells, L.J., (a) *AR*, 1936, **64**, 475
Wells, L.J., (b) *PB*, 1936, **34**, 525
Wells, L.J., (a) *PB*, 1937, **36**, 625
Wells, L.J., (b) *AR*, 1937, **67**, 233
Wells, L.J., & Moore, C.R., *AR*, 1936, **66**, 181
Werner, A.A., & Collier, W.D., *JAMA*, 1933, **100**, 633 and **101**, 1466
Werner, S.C., E, 1938, **22**, 291
Werner, S.C., *PB*, 1939, **41**, 101
Westerfeld, W.W., *BJ*, 1940, **34**, 51
Westerfeld, W.W., Thayer, S.A., Mac-Corquodale, D.W., & Doisy, E.A., *JBC*, 1938, **126**, 181, 195

Westman, A., *E*, 1940, **26**, 774
Westphal, U., & Buxton, C.L., *PB*, 1939, **42**, 749
Whitaker, W.L., *PB*, 1936, **34**, 329
Whitehead, R., *JA*, 1932–33, **67**, 387
Whitehead, R., (*a*) *JA*, 1935, **70**, 123
Whitehead, R., (*b*) *JPB*, 1935, **41**, 305
Whitehead, R.H., *AJA*, 1904, **3**, 167
Whitehouse, B., *JOG*, 1926, **33**, 380
Wiesner, B.P., *JOG*, 1935, **41**, 867 and **42**, 8
Wigglesworth, V.B., *N*, 1935, **136**, 338
Williams, P.C., *N*, 1940, **145**, 388
Willier, B.H., *JEZ*, 1921, **33**, 63
Willier, B.H., 'Sex and Internal Secretions', ed. by E. Allen, Baltimore, 1932
Willier, B.H., Gallagher, T.F., & Koch, F.C., *AR*, 1935, **61**, Suppl. 50
Willier, B.H., Rawles, M.E., & Koch, F.C., *Proc. Nat. Acad. Sci.*, 1938, **24**, 176
Willier, B.H., & Yuh, E.C., *JEZ*, 1929, **52**, 65
Wilson, L., *E*, 1940, **27**, 29
Wilson, L., & Kurzrok, R., *E*, 1938, **23**, 79
Wilson, W.K., & Morris, S., *J. Agric. Sci.*, 1932, **22**, 453
Wislocki, G.B., *AR*, 1933, **57**, 133
Witherspoon, J.T., *E*, 1933, **17**, 621
Witherspoon, J.T., (*a*) *AJC*, 1935, **24**, 402
Witherspoon, J.T., (*b*) *SGO*, 1935, **61**, 743
Witschi, E., 'Sex and Internal Secretions', ed. by E. Allen, Baltimore, 1932, p. 222
Witschi, E., *E*, 1940, **27**, 437
Witschi, E., *CSH*, 1942, **10**, 145
Witschi, E., & Fugo, N.W., *PB*, 1940, **45**, 10
Witschi, E., & Keck, W.N., *PB*, 1935, **32**, 598
Witschi, E., & Levine, W.T., *PB*, 1934, **32**, 101
Witschi, E., Levine, W.T., & Hill, R.T., *PB*, 1932, **29**, 1024
Witschi, E., & McCurdy, H.M., *PB*, 1929, **26**, 655
Witschi, E., & Pfeiffer, C.A., *AR*, 1935, **64**, 85
Wolbach, S.B., & Howe, P.R., *JEM*, 1925, **42**, 753
Wolfe, J.K., Fieser, L.F., & Friedgood, H.B., *J. Amer. Chem. Soc.*, 1941, **53**, 582
Wolfe, J.M., *PB*, 1930, **28**, 318
Wolfe, J.M., *PB*, 1933, **30**, 1092
Wolfe, J.M., *AJP*, 1934, **110**, 159
Wolfe, J.M., (*a*) *AR*, 1935, **63**, 3
Wolfe, J.M., (*b*) *PB*, 1935, **32**, 757 and 1192
Wolfe, J.M., Bryan, W.R., & Wright, A.W., *PB*, 1938, **38**, 80
Wolfe, J.M., Burack, E., & Wright, A.W., *AJC*, 1940, **38**, 383
Wolfe, J.M., Burch, J.C., & Campbell, M., *AR*, 1932, **52**, 80
Wolfe, J.M., Campbell, M., & Burch, J.C., *PB*, 1932, **29**, 1263
Wolfe, J.M., & Chadwick, C.S., *E*, 1936, **20**, 503
Wolfe, J.M., & Cleveland, R., *AR*, 1931, **51**, 213
Wolfe, J.M., & Cleveland, R., (*a*) *AR*, 1933, **55**, 233
Wolfe, J.M., & Cleveland, R., (*b*) *AR*, 1933, **56**, 33

Wolfe, J.M., Cleveland, R., & Campbell, M., *AR*, 1932, **52**, Abst., 44
Wolfe, J.M., Ellison, E.T., & Rosenfeld, L., *AR*, 1934, **60**, 357
Wolfe, J.M., & Hamilton, J.B., (*a*) *E*, 1937, **21**, 603
Wolfe, J.M., & Hamilton, J.B., (*b*) *PB*, 1937, **37**, 189
Wolfe, J.M., & Hamilton, J.B., *E*, 1939, **25**, 572
Wolfe, J.M., Phelps, D., & Cleveland, R., *PB*, 1933, **30**, 1092
Wolff, E., *CRS*, 1935, **120**, 1312 and 1314
Wolff, E., (*a*) *CRS*, 1936, **121**, 1474 and 1476
Wolff, E., (*b*) *CRS*, 1936, **123**, 237
Womack, E.B., & Koch, F.C., *Proc. 2nd Internat. Cong. Sex Research*, 1930, p. 329
Womack, E.B., & Koch, F.C., *E*, 1932, **16**, 267 and 273
Womack, E.B., Koch, F.C., Domm, L.V., & Juhn, M., *JPL*, 1931, **41**, 173
Woolley, G.W., Fekete, E., & Little, C.C., *Proc. Nat. Acad. Sci.*, 1939, **25**, 277
Woolley, G.W., Fekete, E., & Little, C.C., *PB*, 1940, **45**, 796
Woolley, G.W., Fekete, E., & Little, C.C., *E*, 1941, **28**, 341
Woolley, G.W., Law, L.W., & Little, C.C., *CR*, 1941, **1**, 955
Wyman, L.C., *AJP*, 1928, **85**, 414

Yarrell, W., *PT*, 1827, **127**, 268
Yerby, L.D., *PB*, 1937, **36**, 496

Zalesky, M., *AR*, 1934, **60**, 291
Zalesky, M., *AR*, 1936, **65**, 467
Zéphiroff, P., & Dobrovolskaya-Zavadskaya N., *CRS*, 1940, **133**, 405
Zéphiroff, P., Drosdovsky, C., & Dobrovolskaya-Zavadskaya, N., *CRS*, 1940, **133**, 236
Zondek, B., *KW*, 1930, **9**, 679, and *Chirurg.*, 1930, **2**, 1072
Zondek, B., *AJOG*, 1932, **24**, 836
Zondek, B., (*a*) *N*, 1934, **133**, 209 and 494
Zondek, B., (*b*) *JP*, 1934, **81**, 472
Zondek, B., (*c*) *L*, 1934 (ii), 356
Zondek, B., 'Hormone d. Ovariums u. d. Hypophysenvorderlappens', Vienna, 2nd ed. 1935, pp. 226 and 272
Zondek, B., (*a*) *L*, 1936 (i), 10 and 776
Zondek, B., (*b*) *L*, 1936 (ii), 842
Zondek, B., *Folia Clin. Orient.*, 1937, **1**, 1
Zondek, B., *L*, 1938 (i), 1107
Zondek, B., (*a*) *JAMA*, 1940, **114**, 1850
Zondek, B., (*b*) *PB*, 1940, **43**, 570
Zondek, B., (*c*) *JE*, 1940, **2**, 12
Zondek, B., & Aschheim, S., *AG*, 1925, **127**, 250
Zondek, B., & Aschheim, S., (*a*) *AG*, 1927, **130**, 1
Zondek, B., & Aschheim, S., (*b*) *KW*, 1927, **6**, 248
Zondek, B., & Bromberg, Y.M., *L*, 1942 (i), 381

Zondek, B., & Marx, L., *N*, 1939, **143**, 378

Zondek, B., & Sklow, J., *PB*, 1941, **46**, 276

Zondek, B., & Sklow, J., *PB*, 1942, **49**, 629

Zondek, B., & Stein, L., *E*, 1940, **27**, 395

Zondek, B., & Sulman, F., *PB*, 1937, **36**, 708

Zondek, B., & Sulman, F., *PB*, 1940, **43**, 86 and 712

Zuckerman, S., *L*, 1936 (ii), 9

Zuckerman, S., (*a*) *L*, 1937 (ii), 676

Zuckerman, S., (*b*) *N*, 1937, **139**, 628

Zuckerman, S., (*c*) *PR*, 1937, **123**, 457

Zuckerman, S., *JA*, 1938, **72**, 264

Zuckerman, S., *JE*, 1939, **1**, 147

Zuckerman, S., *JE*, 1940, **2**, 311

Zuckerman, S., Palmer, A., & Bourne, G., *N*, 1939, **143**, 521

Zuckerman, S., & Parkes, A.S., *JA*, 1935, **69**, 484

Zuckerman, S., & Parkes, A.S., (*a*) *L*, 1936 (i), 242

Zuckerman, S., & Parkes, A.S., (*b*) *JA*, 1936, **70**, 323

Zuckerman, S., & Parkes, A.S., *JA*, 1938, **72**, 277

Zuckerman, S., & Parkes, A.S., *JE*, 1939, **1**, 430

Zuckerman, S., & Sandys, O.C., *JA*, 1939, **73**, 597

Zuckerman, S., & van Wagenen, G., *JA*, 1935, **69**, 497

Zuckerman, W., van Wagenen, G., & Gardiner, R.H., *Proc. Zool. Soc.*, 1938, **108**, 385

Zunz, E., & La Barre, J., *Arch. Int. de Physiol.* (Liège), 1939, **48**, 287

Zwarenstein, H., *N*, 1937, **139**, 112

Zwemer, R.L., *Amer. J. Pathol.*, 1936, **12**, 107

GLOSSARY

Adrenocorticotrophin. A pituitary trophin which governs the development and functional activity of the adrenals.

Bird Unit. A unit of prolactin as tested on the pigeon's crop gland (*v.* International Units).

Chorionepithelioma. A malignant growth of the foetal part of the placenta. A similar form of cancer occurs rarely in the testicle.

Female Hormone. A crude synonym for oestrogen.

Galactogen. A pituitary hormone which induces lactation.

Gonadotrophin. A hormone which governs the development and functional activity of gonads:

(*a*) Pituitary gonadotrophin is formed by the anterior lobe of the pituitary.

(*b*) Placental or chorionic gonadotrophin is derived from placental tissue.

International Units of Sex Hormonal Activity (as adopted by the Health Organization of the League of Nations). Standard preparations of each hormone were selected after extended inquiry, and the international units are based on them. The standard preparations of gonadal hormones are all chemically purified, crystalline compounds. By the use of recognized units the hormonal activity of a tissue extract or other unpurified preparation can be stated in terms of measured effect.

Androgen. One international unit of androgenic activity is that exerted by $100\,\gamma$ of the standard purified crystalline preparation of androsterone. This unit represents the minimum daily dose required to give an easily measurable response in the capon's comb at the end of 5 days.

Gonadotrophin:

(*a*) *Chorionic gonadotrophin.* One international unit of this gonadotrophin prepared from human pregnancy urine is $100\,\gamma$ of the standard preparation. This quantity exerts the minimum activity required to cause cornification of the vaginal epithelium of the normal immature rat.

(*b*) *Gonadotrophin from serum of pregnant mare.* One international unit of this gonadotrophin is $250\,\gamma$ of the standard preparation as shown by its capacity to induce morphological changes in the gonads or in the accessory reproductive organs in animals not deprived of their gonads.

Oestrogen:

(*a*) *Oestrone.* One international unit of oestrone is the specific activity exerted by $0\cdot1\,\gamma$ of the standard preparation, as shown by its capacity to induce cornification of the vaginal epithelium in the spayed rat or mouse. (This quantity represents approximately one-third of the original Allen & Doisy rat unit.)

(b) *Oestradiol benzoate.* One international unit of oestradiol monobenzoate is the specific activity exerted by 0·1 γ of the standard preparation as shown by its capacity to induce cornification of the vaginal epithelium in the spayed rat or mouse.

(It is to be noted that although the weights of the international units of oestrone and oestradiol monobenzoate are identical this does not mean that these two units have equivalent biological activities.)

Progesterone. One international unit of progesterone is the specific progestational activity of 1 mgm. of the standard preparation as shown by its capacity to induce a full progestational reaction in the endometrium of a normal immature rabbit of 600 g. after preparatory treatment with oestrogen.

Prolactin. One international unit of prolactin is the specific activity contained in 100 γ of the standard preparation, as shown by its capacity to induce functional activity of the pigeon's crop gland. This unit is nearly the same as one Riddle unit.

Male Hormone. A crude synonym for androgen.

Pitocin, Pitressin, Pituitrin. Pressor hormones derived from the posterior lobe of the pituitary.

Pregnancy tests:

(a) *Mouse test.* Female infantile mice of 6—9 grams are the test objects, and 0·5 c.c. of the urine to be examined is given to them subcutaneously twice a day for 3 successive days. The mice are killed 96 hours after the first injection. A positive response is shown by the presence of blood-spots or corpora lutea in the ovaries. The urine to be tested should be kept in the refrigerator, the doses being warmed just before injection.

(b) *Rabbit test.* 5 c.c. of the urine to be examined are injected intravenously into mature virgin rabbits which are killed 24 hours later. A positive response is shown by the presence of one or more fresh corpora lutea which have ruptured.

Prolan. The gonadotrophin of pregnancy urine. The FRH factor is Prolan A and the LH factor is Prolan B.

Trophin. A hormone which controls specifically the development and function of a particular structure in the body. The anterior lobe of the pituitary and the placenta are the main known sources.

INDEX

(Attention is drawn to the special indexing of new growths which have been grouped together under the two headings of 'Cancer' and 'Tumours'.)

Printed in the United States
By Bookmasters